HEINEMANN BIOLOGY 1

6TH EDITION

Zoë Armstrong
Christina Bliss
Reuben Bolt
Erin Bruns
Elaine Georges
Caryn Hellberg
Carina Jansen
Bernd Kalinna
Katherine McMahon
Heather Maginn
Troy Potter
Sue Siwinski
Natasha Ward

VCE UNITS 1 AND 2 • 2022–2026

Pearson Australia
(a division of Pearson Australia Group Pty Ltd)
707 Collins Street, Melbourne, Victoria 3008
PO Box 23360, Melbourne, Victoria 8012
www.pearson.com.au

First published 2021 by Pearson Australia
2024 2023 2022 2021
10 9 8 7 6 5 4 3 2 1

Project Leads: Misal Belvedere, Malcolm Parsons
Content Developer: Rebecca Wood
Project Managers: Shaun Birtles, Michelle Thomas
Production Editors: Elizabeth Gosman, Natalie Lincoln, Virginia O'Brien
Lead Development Editors: Fiona Cooke, Zoe Hamilton
Editor: Sam Trafford
Designer, Cover Designer: Anne Donald
Rights and Permissions Editor: Madeleine Roberts
Production Services Design Analysts: JitPin Chong, Jennifer Johnston
Illustrators: Bruce Rankin, Pearson India, Guy Holt, Dimitiros Prokopis, Pat Kermode, Andrew Plant
Proofreader: Jane Fitzpatrick
Indexer: Ann Philpott

Printed in Malaysia by Vivar

National Library of Australia Cataloguing-in-Publication entry

A catalogue record for this book is available from the National Library of Australia

ISBN: 978 0 6557 0005 0

Pearson Australia Group Pty Ltd ABN 40 004 245 943

Disclaimers
The selection of internet addresses (URLs) provided for this book was valid at the time of publication and was chosen as being appropriate for use as a secondary education research tool. However, due to the dynamic nature of the internet, some addresses may have changed, may have ceased to exist since publication, or may inadvertently link to sites with content that could be considered offensive or inappropriate. While the authors and publisher regret any inconvenience this may cause readers, no responsibility for any such changes or unforeseeable errors can be accepted by either the authors or the publisher.

Indigenous Australians
Some of the images used in *Heinemann Biology 1 6e* might have associations with deceased Indigenous Australians. Please be aware that these images might cause sadness or distress in Aboriginal or Torres Strait Islander communities.

HEINEMANN BIOLOGY 1

6TH EDITION

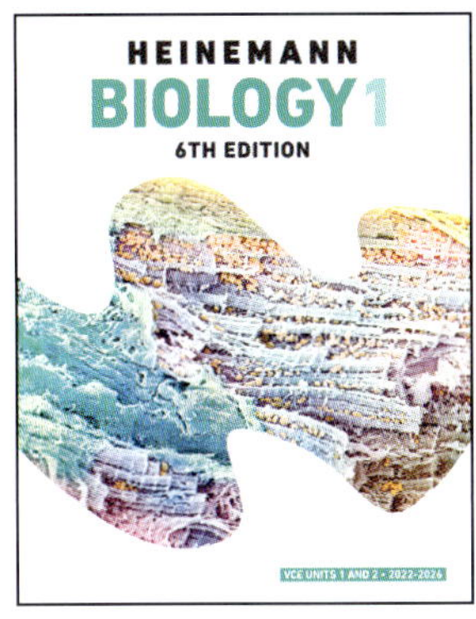

Writing and development team

We are grateful to the following people for their time and expertise in contributing to the **Heinemann Biology 1** project.

Zoë Armstrong
Science communicator and conservation biologist
Author

Doug Bail
Education consultant
SPARKlab developer

Christina Bliss
Teacher
Author

Reuben Bolt
Pro Vice Chancellor Indigenous Leadership, Charles Darwin University
Author, reviewer

Erin Bruns
Teacher, Head of Department
Author, reviewer and digital question content lead

Rima Cilmi
Teacher
Teacher support author

Elaine Georges
Teacher
Author

Rachael Gillespie
Teacher
Digital question author

Martin Hall
Teacher
Series reviewer and teacher support author

Caryn Hellberg
Science communicator, former teacher and genetic counsellor
Author, reviewer and digital question author

Neil van Herk
Teacher
Digital question author

Carina Jansen
Teacher
Author, reviewer and digital question content lead and author

Bernd Kalinna
Teacher, former Associate Professor of Biomedical Science
Author

Jacqueline Kerr
Teacher
Reviewer

Katherine McMahon
Teacher
Author and reviewer

Heather Maginn
Educational travel program development manager
Author, reviewer and digital question author

Jonathan Meddings
Science and medical writer
Reviewer

Troy Potter
Lecturer, The University of Melbourne
Author

Yvonne Sanders
Teacher, former Head of Department
Skills and Assessment author

Sue Siwinski
Teacher, former Head of Department, education consultant
Author, reviewer, digital question author and teacher support author

Natasha Ward
Education officer
Author

Laurence Wooding
Laboratory Technician
Reviewer and teacher support author

The Publisher wishes to thank and acknowledge Pauline Ladiges and Barbara Evans for their contribution in creating the original works of the series and longstanding dedicated work with Pearson and Heinemann.

Contents

AREA OF STUDY 3

Heinemann Biology 1 6th edition includes a comprehensive set of resources to support Area of Study 3 via your Pearson Places bookshelf.

Unit 2 How does inheritance impact on diversity?

AREA OF STUDY 3

Heinemann Biology 1 6th edition includes a comprehensive set of resources to support Area of Study 3 via your Pearson Places bookshelf.

How to use this book

Heinemann Biology 1 6th edition

Heinemann Biology 1 6th edition has been written to the new VCE Biology Study Design 2022–2026. The book covers Units 1 and 2. Explore how to use this book below.

Case study

Case studies place biology in an applied situation or relevant context. They refer to the nature and practice of biology, applications of biology and associated issues, and the historical development of biological concepts and ideas.

CASE STUDY

Salmon: osmoregulation in fresh and salt water

Some animals, like salmon, are able to survive in a relatively wide range of salt concentrations, making them different to other animals, like humans, who can only function efficiently in a narrow range of salt concentrations. The gills of salmon are adapted to transport salt ions through specialised transport mechanisms. This unique adaptation allows salmon to spend part of their lives in fresh water and part in salt water (Figure 5.2.17).

Salmon in salt water

When salmon are in salt water there will be more water molecules inside the body cells than outside the cells, so water will move out of the cells via osmosis. This leads to a high concentration of salt in the blood, which is recognised by receptors in the hypothalamus. Those receptors stimulate the pituitary gland to release into the blood a hormone that initiates the transport of salt ions (Na^+ and Cl^-) out of the body via the gills. Salmon also rely on their kidneys to remove excess salts. Since living in salt water leads to dehydration, salmon have to drink many litres of water a day. In addition, another hormone released by the pituitary gland, antidiuretic hormone (ADH), signals to the kidneys to reabsorb more water before excretion. This leads the kidneys to remove excess salts through the production of very concentrated urine.

Salmon in fresh water

When salmon are living in fresh water there will be more water molecules outside the body cells than inside the cells, so water molecules will move into the body cells via osmosis. This increased movement of water into the body cells leads to salt loss and high water levels. The hypothalamus detects the high water level and stimulates the pituitary gland to release into the blood a hormone that stimulates the gills to actively absorb salts from the water. Due to their high water levels while living in fresh water, salmon do not actively drink water and they excrete large volumes of dilute urine from their kidneys.

FIGURE 5.2.17 Red salmon (*Oncorhynchus nerka*) migrate from a freshwater river to a saltwater ocean as juveniles and return to the same river as adults to spawn (reproduce).

Case Study: Analysis

These case studies include real-world data that can be analysed and evaluated.

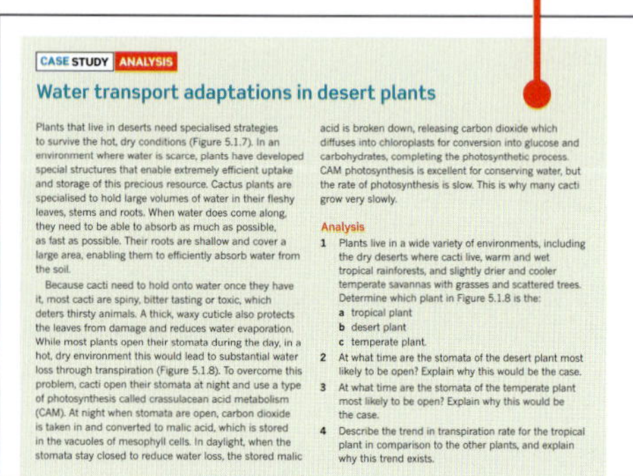

CASE STUDY ANALYSIS

Water transport adaptations in desert plants

Plants that live in deserts need specialised strategies to survive the hot, dry conditions (Figure 5.1.7). In an environment where water is scarce, plants have developed special structures that enable extremely efficient uptake and storage of this precious resource. Cactus plants are specialised to hold large volumes of water in their fleshy leaves, stems and roots. When water does come along, they need to be able to absorb as much as possible, as fast as possible. Their roots are shallow and cover a large area, enabling them to efficiently absorb water from the soil.

Because cacti need to hold onto water once they have it, most cacti are spiny, bitter tasting or toxic, which deters thirsty animals. A thick, waxy cuticle also protects the leaves from damage and reduces water evaporation. While most plants open their stomata during the day, in a hot, dry environment this would lead to substantial water loss through transpiration (Figure 5.1.8). To overcome this problem, cacti open their stomata at night and use a type of photosynthesis called crassulacean acid metabolism (CAM). At night when stomata are open, carbon dioxide is taken in and converted to malic acid, which is stored in the vacuoles of mesophyll cells. In daylight, when the stomata stay closed to reduce water loss, the stored malic acid is broken down, releasing carbon dioxide which diffuses into chloroplasts for conversion into glucose and carbohydrates, completing the photosynthetic process. CAM photosynthesis is excellent for conserving water, but the rate of photosynthesis is slow. This is why many cacti grow very slowly.

Analysis

1 Plants live in a wide variety of environments, including the dry deserts where cacti live, warm and wet tropical rainforests, and slightly drier and cooler temperate savannas with grasses and scattered trees. Determine which plant in Figure 5.1.8 is the:
   a tropical plant
   b desert plant
   c temperate plant.
2 At what time are the stomata of the desert plant most likely to be open? Explain why this would be the case.
3 At what time are the stomata of the temperate plant most likely to be open? Explain why this would be the case.
4 Describe the trend in transpiration rate for the tropical plant in comparison to the other plants, and explain why this trend exists.

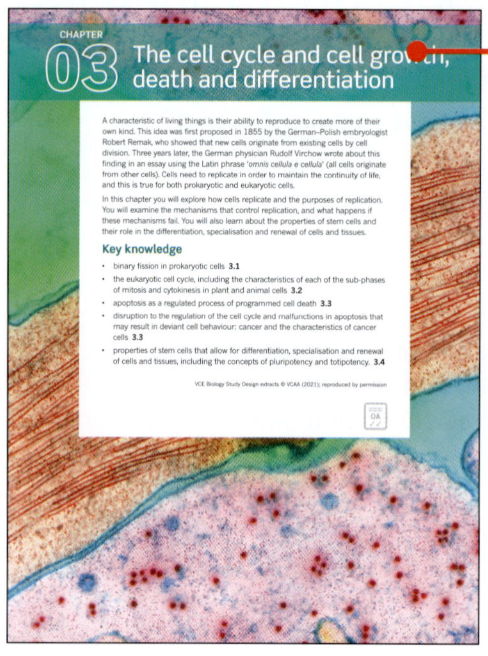

CHAPTER 03 The cell cycle and cell gro death and differentiation

A characteristic of living things is their ability to reproduce to create more of their own kind. This idea was first proposed in 1855 by the German–Polish embryologist Robert Remak, who showed that new cells originate from existing cells by cell division. Three years later, the German physician Rudolf Virchow wrote about this finding in an essay using the Latin phrase '*omnis cellula e cellula*' (all cells originate from other cells). Cells need to replicate in order to maintain the continuity of life, and this is true for both prokaryotic and eukaryotic cells.

In this chapter you will explore how cells replicate and the purposes of replication. You will examine the mechanisms that control replication, and what happens if these mechanisms fail. You will also learn about the properties of stem cells and their role in the differentiation, specialisation and renewal of cells and tissues.

Key knowledge

- binary fission in prokaryotic cells **3.1**
- the eukaryotic cell cycle, including the characteristics of each of the sub-phases of mitosis and cytokinesis in plant and animal cells **3.2**
- apoptosis as a regulated process of programmed cell death **3.3**
- disruption to the regulation of the cell cycle and malfunctions in apoptosis that may result in deviant cell behaviour: cancer and the characteristics of cancer cells **3.3**
- properties of stem cells that allow for differentiation, specialisation and renewal of cells and tissues, including the concepts of pluripotency and totipotency. **3.4**

VCE Biology Study Design extracts © VCAA (2021); reproduced by permission

Chapter opener

Chapter opening pages link the study design to the chapter content. Key knowledge addressed in the chapter is clearly listed. To help you find where each outcome is covered in the chapter, the relevant section numbers are written in bold.

Highlight

Highlight boxes focus on important information such as key definitions and summary points.

BioFile

BioFiles include interesting information and real-world examples.

Plant cells tightly filled with fluid are described as turgid. When they lose water they become flaccid.

BIOFILE

Osmosis in salty environments

There is no biological mechanism for actively transporting water molecules across plasma membranes. Net movement of water across membranes occurs only by passive osmosis. Organisms that are adapted to extremely salty environments survive by retaining much higher ion concentrations within their cells. They also produce small, osmotically active but otherwise inert molecules to reduce the osmotic gradient. This prevents the loss of water to their salty surroundings. Their proteins are specialised to function normally despite a high concentration of salts in the cytosol. Organisms such as these are called halophiles.

The salt lakes of Murray-Sunset National Park in Victoria owe their colour to the presence of red algae. These algae, along with the solid salt bed of the lakes, create the distinctive pinkish red colour of the lake.

If a plant cell absorbs water it swells to some extent, but the cell wall prevents the cell from bursting (Figure 2.4.12f on page 103). Water will continue to enter the cell along an osmotic gradient until the internal fluid pressure equals the osmotic pressure drawing water in, at which point no more water will enter. Plant cells with high internal fluid pressures have a high turgor because they are full of fluid tightly pressing against the cell wall. Turgor in cells provides structural support for plants. Dehydrated plants droop because their cells are flaccid.

In osmosis, we are always comparing solute concentration between two solutions. The terms isotonic, hypertonic and hypotonic solution are often used to describe the differences:

- isotonic solutions—the solutions being compared have equal concentrations of solutes
- hypertonic solution—the solution with a higher concentration of solute (hence lower concentration of free water molecules)
- hypotonic solution—the solution with a lower concentration of solute (hence higher concentration of free water molecules).

Active transport

Diffusion, facilitated diffusion and osmosis are examples of passive transport, because they do not require energy to move particles across the plasma membrane. **Active transport** involves the use of energy by the cell to transport particles across membranes (see Figure 2.4.13).

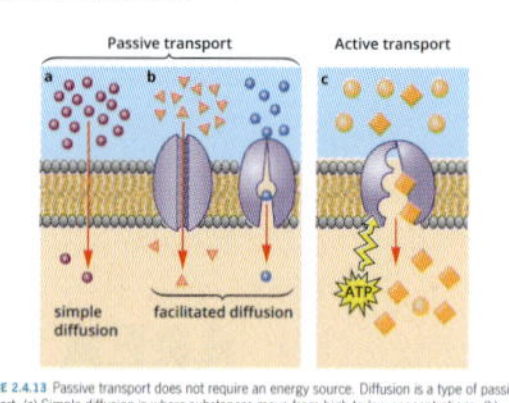

FIGURE 2.4.13 Passive transport does not require an energy source. Diffusion is a type of passive transport. (a) Simple diffusion is where substances move from high to low concentrations. (b) Facilitated diffusion is where substances move from high to low concentrations with help from a carrier protein. (c) Active transport requires an energy source. As a result, active transport usually moves substances from low to high concentrations. This shows a protein pump that assists the movement of substances.

Icons

This icon is used to alert you to engage with auto-corrected questions through Pearson Places.

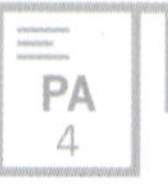

These icons indicate when it is the best time to engage with a worksheet (WS), a practical activity (PA) or exam-style questions (EQ) in *Heinemann Biology 1 Skills and Assessment* book.

Chapter review

Each chapter concludes with a list of key terms and questions that test your understanding of the key knowledge covered in the chapter.

Section summary

Each section includes a summary to help you consolidate key points and concepts.

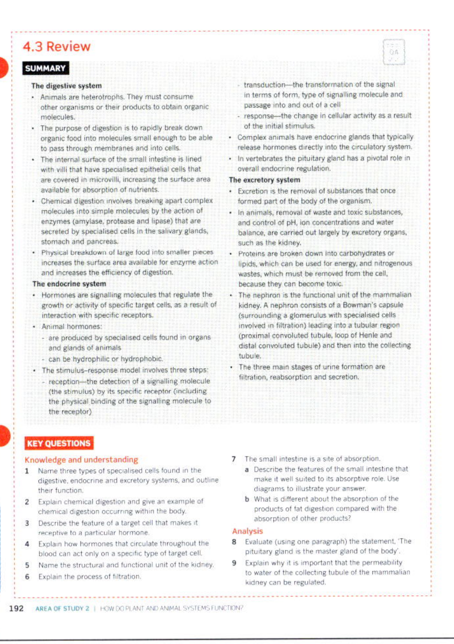

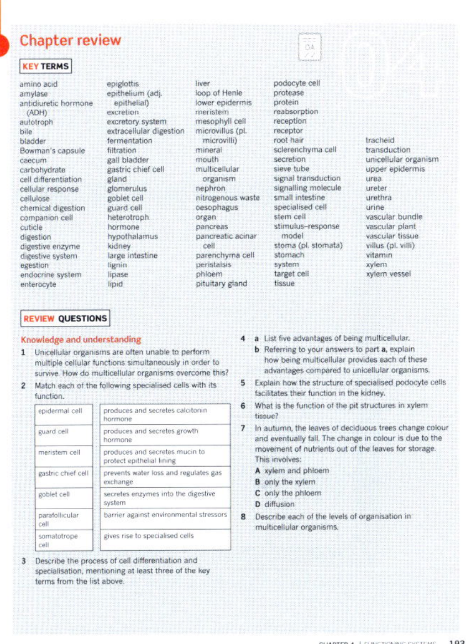

Section review

Each section concludes with questions that test your ability to recall, explain and apply key concepts.

Area of Study review

Each area of study finishes with a comprehensive set of exam-style questions, including multiple choice and short answer, supporting you in your exam preparation.

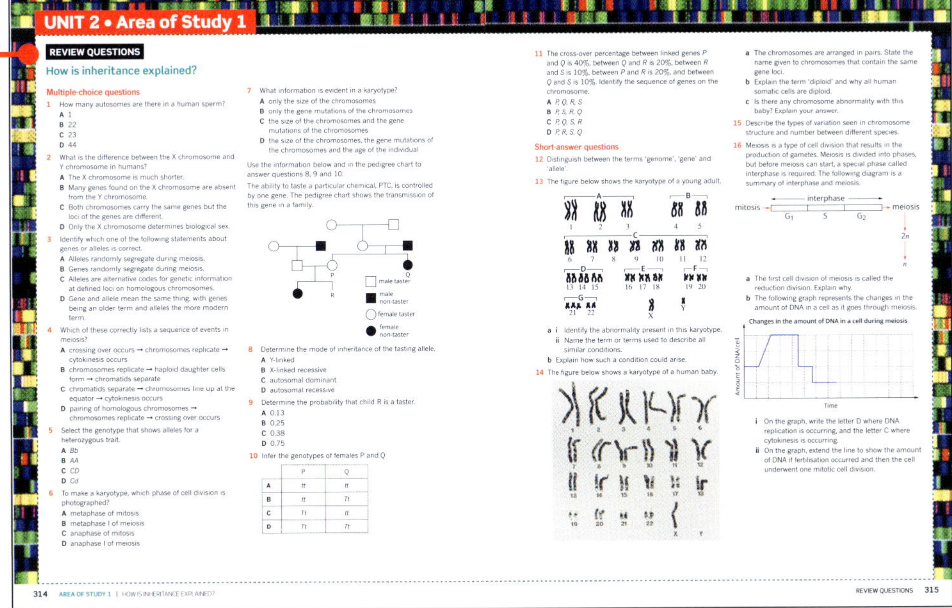

Answers

Answers and fully worked solutions for all section review questions, chapter review questions and Area of Study review questions are provided via *Heinemann Biology 1 eBook* 6th edition.

Glossary

Key terms are shown in **bold** throughout, and are listed at the end of each chapter. A comprehensive glossary at the end of the book defines all key terms.

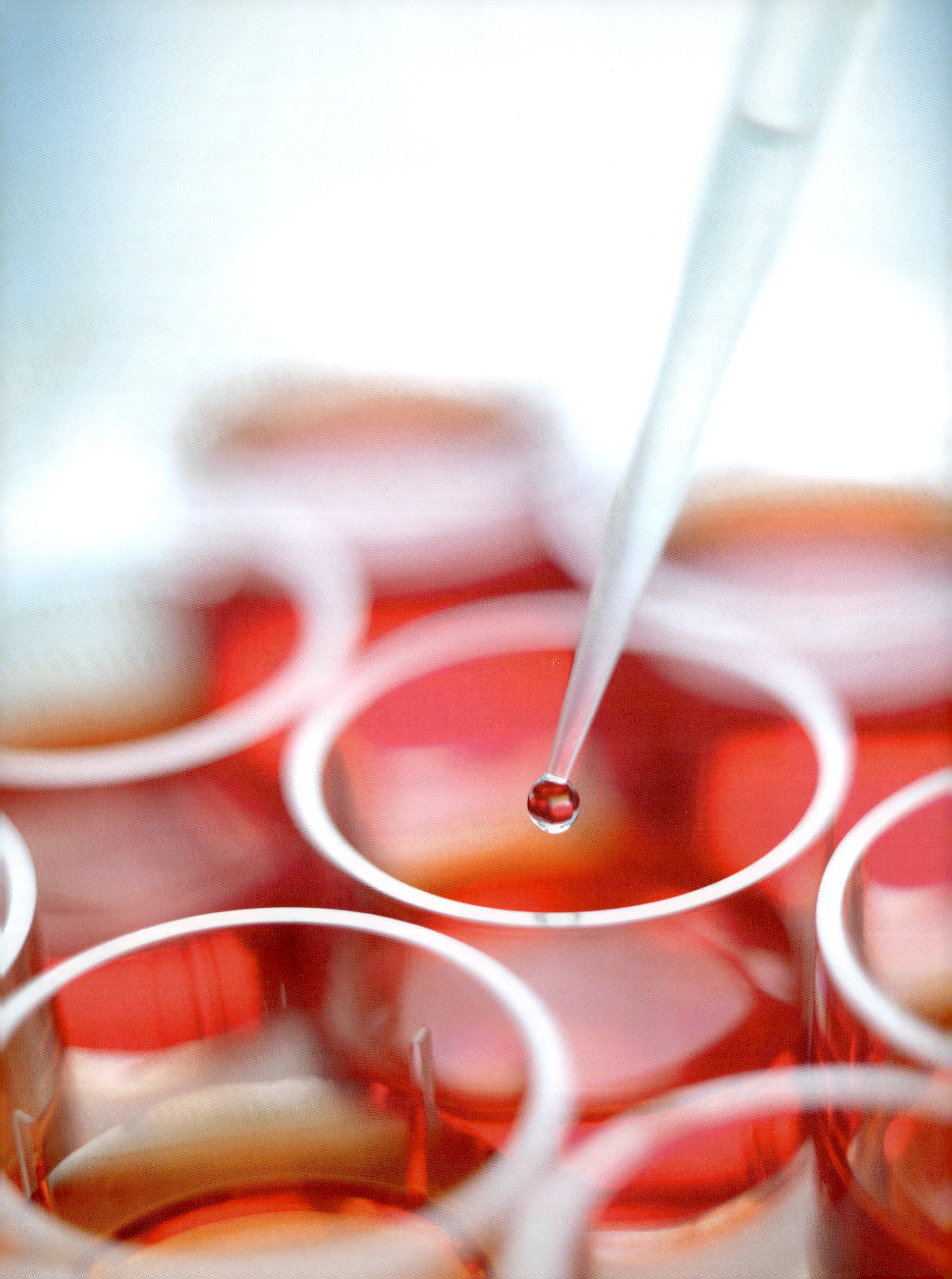

CHAPTER 01

Scientific investigation

Learning outcomes

The development of a set of key science skills is a core component of the study of VCE Biology and applies across Units 1 to 4 in all areas of study. Chapter 1 scaffolds the development of these skills. It is important to develop, use and demonstrate these skills in a variety of contexts before undertaking investigations and when evaluating the research of others.

Although this chapter can be read as a whole, it is best to refer to it and use it when the need arises as you work through other chapters. For example, you may need a refresher on the process of the scientific method. This chapter also contains useful checklists to assist when drawing scientific diagrams, graphing results and completing aspects of your report. Similarly, when performing a practical investigation, refer to this chapter to make sure you collect data properly and that your data is of high quality.

Key science skills

Develop aims and questions, formulate hypotheses and make predictions

- identify, research and construct aims and questions for investigation **1.1, 1.2**
- identify independent, dependent and controlled variables in controlled experiments **1.1, 1.2**
- formulate hypotheses to focus investigation **1.1, 1.2**
- predict possible outcomes **1.1, 1.2**

Plan and conduct investigations

- determine appropriate investigation methodology; case study; classification and identification; controlled experiment; correlational study; fieldwork; literature review; modelling; product, process or system development; simulation **1.1, 1.2**
- design and conduct investigations; select and use methods appropriate to the investigation, including consideration of sampling technique and size, equipment and procedures, taking into account potential sources of error and uncertainty; determine the type and amount of qualitative and/or quantitative data to be generated or collated **1.1, 1.3**
- work independently and collaboratively as appropriate and within identified research constraints, adapting or extending processes as required and recording such modifications **1.1, 1.2**

Comply with safety and ethical guidelines

- demonstrate safe laboratory practices when planning and conducting investigations by using risk assessments that are informed by safety data sheets (SDS), and accounting for risks **1.2**
- apply relevant occupational health and safety guidelines while undertaking practical investigations **1.2**
- demonstrate ethical conduct when undertaking and reporting investigations **1.2**

Generate, collate and record data

- systematically generate and record primary data, and collate secondary data, appropriate to the investigation, including use of databases and reputable online data sources **1.3, 1.4**
- record and summarise both qualitative and quantitative data, including use of a logbook as an authentication of generated or collated data **1.4**
- organise and present data in useful and meaningful ways, including schematic diagrams, flow charts, tables, bar charts and line graphs **1.5, 1.6**
- plot graphs involving two variables that show linear and non-linear relationships **1.5, 1.6**

Analyse and evaluate data and investigation methods

- process quantitative data using appropriate mathematical relationships and units, including calculations of ratios, percentages, percentage change and mean **1.5**
- identify and analyse experimental data qualitatively, handling where appropriate concepts of: accuracy, precision, repeatability, reproducibility and validity of measurements; errors (random and systematic); and certainty in data, including effects of sample size in obtaining reliable data **1.4, 1.5**
- identify outliers, and contradictory or provisional data **1.4, 1.5**
- repeat experiments to ensure findings are robust **1.4**
- evaluate investigation methods and possible sources of personal errors/mistakes or bias, and suggest improvements to increase accuracy and precision, and to reduce the likelihood of errors **1.4, 1.6**

Construct evidence-based arguments and draw conclusions

- distinguish between opinion, anecdote and evidence, and scientific and non-scientific ideas **1.2**
- evaluate data to determine the degree to which the evidence supports the aim of the investigation, and make recommendations, as appropriate, for modifying or extending the investigation **1.4, 1.6**
- evaluate data to determine the degree to which the evidence supports or refutes the initial prediction or hypothesis **1.4, 1.6**
- use reasoning to construct scientific arguments, and to draw and justify conclusions consistent with the evidence and relevant to the question under investigation **1.6**
- identify, describe and explain the limitations of conclusions, including identification of further evidence required **1.6**
- discuss the implications of research findings and proposals **1.6**

Analyse, evaluate and communicate scientific ideas

- use appropriate biological terminology, representations and conventions, including standard abbreviations, graphing conventions and units of measurement **1.4, 1.5, 1.6**
- discuss relevant biological information, ideas, concepts, theories and models and the connections between them **1.1, 1.2, 1.6**
- analyse and explain how models and theories are used to organise and understand observed phenomena and concepts related to biology, identifying limitations of selected models/theories **1.1, 1.6**
- critically evaluate and interpret a range of scientific and media texts (including journal articles, mass media communications and opinions in the public domain), processes, claims and conclusions related to biology by considering the quality of available evidence **1.2, 1.4**
- analyse and evaluate bioethical issues using relevant approaches to bioethics and ethical concepts, including the influence of social, economic, legal and political factors relevant to the selected issue **1.2**
- use clear, coherent and concise expression to communicate to specific audiences for specific purposes in appropriate scientific genres, including scientific reports and posters **1.6**
- acknowledge sources of information and assistance, and use standard scientific referencing conventions **1.6**

OA

1.1 The scientific method

Biology is the study of living organisms. As scientists, biologists extend their understanding using the scientific method, which involves investigations that are carefully designed, conducted and reported (Figure 1.1.1). Well-designed research is based on a sound knowledge of what is already understood about a subject, as well as careful preparation and observation.

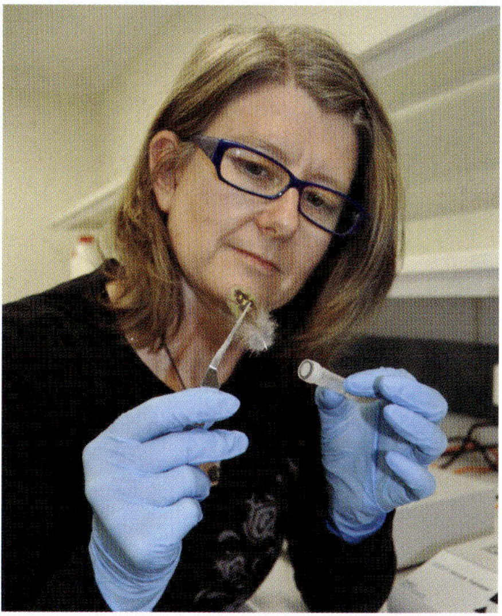

FIGURE 1.1.1 Biological research uses a variety of methodologies and methods. Analysis of DNA extracted from feathers by scientists at the Museum of Western Australia has confirmed that the night parrot (*Pezoporus occidentalis*) is not extinct, as previously thought.

OBSERVATION

Observation includes using all your senses and a wide variety of instruments and laboratory techniques to allow closer observation. Through careful inquiry and observation you can learn a lot about organisms, the ways they function, and their interactions with each other and the environment. For example, animals clearly function very differently from plants. Animals usually move around, take in nutrients and water, and often interact with each other in groups. Plants, however, are stationary, turn their leaves towards the light and grow. Many other distinguishing macroscopic structures and behaviours can be discerned from simple observation. Microscopic observation of cells reveals similarities and differences in the structure of plant and animal cells, as well as the specialisations in the cells of a particular organism.

Practical investigations

The idea for a practical investigation of a complex problem arises from prior learning and observations that raise further questions. For example, indoor plants do not grow well in the long term without artificial lighting, which suggests light is required for photosynthesis in plants (Figure 1.1.2). This aspect of photosynthesis can be researched and the new knowledge applied to other applications, such as methods for growing plants in the laboratory for genetic selection and modification for crop improvement.

FIGURE 1.1.2 Laboratory methods such as plant tissue culture rely on careful observations and data collection about the requirements for growth of plants in natural conditions. Laboratory investigations then provide new information that can be applied to plants growing in the field.

Interpreting observations

How observations are interpreted depends on past experiences and knowledge, but to enquiring minds they will usually provoke further questions such as:

- How do organisms gain and expend energy?
- How do multicellular organisms develop specialised tissues?
- What are the molecular building blocks of cells?
- How do species change and evolve over time?
- How do cells communicate with each other?

Many of these questions cannot be answered by observation alone, but they can be answered through scientific investigations. Good scientists have acute powers of observation and enquiring minds, and they make the most of these chance opportunities, like Alexander Fleming did when he discovered penicillin.

CASE STUDY

Australian discovery—peptic ulcers

In 2005, Australian scientists Professor Barry Marshall and J. Robin Warren (Figure 1.1.3) were awarded the Nobel Prize in Physiology or Medicine with their discovery of the role of the bacteria species *Helicobacter pylori* in gastritis and peptic ulcer disease.

Through careful observations and questioning, the two scientists showed that more than 90% of peptic ulcer cases that led to gastritis were not caused by stress, as previously thought, but were caused by infection with *H. pylori*. *H. pylori* is able to withstand the extreme acidic conditions of the stomach, pass through the protective mucous layer, and attach to and colonise the wall of the stomach. The bacteria release several compounds that lead to damage of the stomach lining (Figure 1.1.4).

The research of Marshall and Warren led to effective treatment of peptic ulcers with antibiotics. An unexpected outcome of their research was the discovery that gastritis is a precursor to stomach cancer. Marshall and Warren's research not only led to effective treatment for peptic ulcers but has also drastically reduced the rate of stomach cancer.

FIGURE 1.1.3 Australian scientists Professor Barry Marshall and Dr Robin Warren

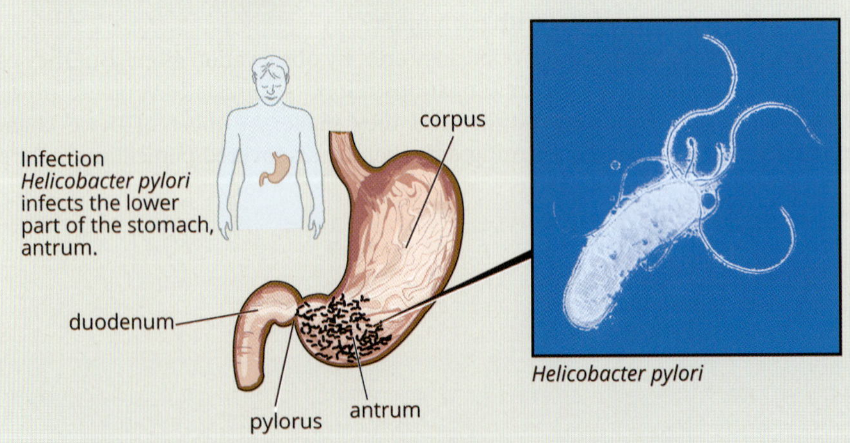

FIGURE 1.1.4 The majority of peptic ulcers are caused by the bacteria *Helicobacter pylori*.

● *You will now be able to answer key question 1.*

THE SCIENTIFIC PROCESS

Scientists observe, study what is already known, and then ask questions. Using their knowledge and experience, scientists suggest possible explanations for the things they observe. A **hypothesis** is a prediction based on scientific reasoning that can be tested experimentally. This is the basis of the **scientific method** (Figure 1.1.5).

A hypothesis is a prediction based on scientific reasoning about what an investigator might expect to see in the results of their experiment.

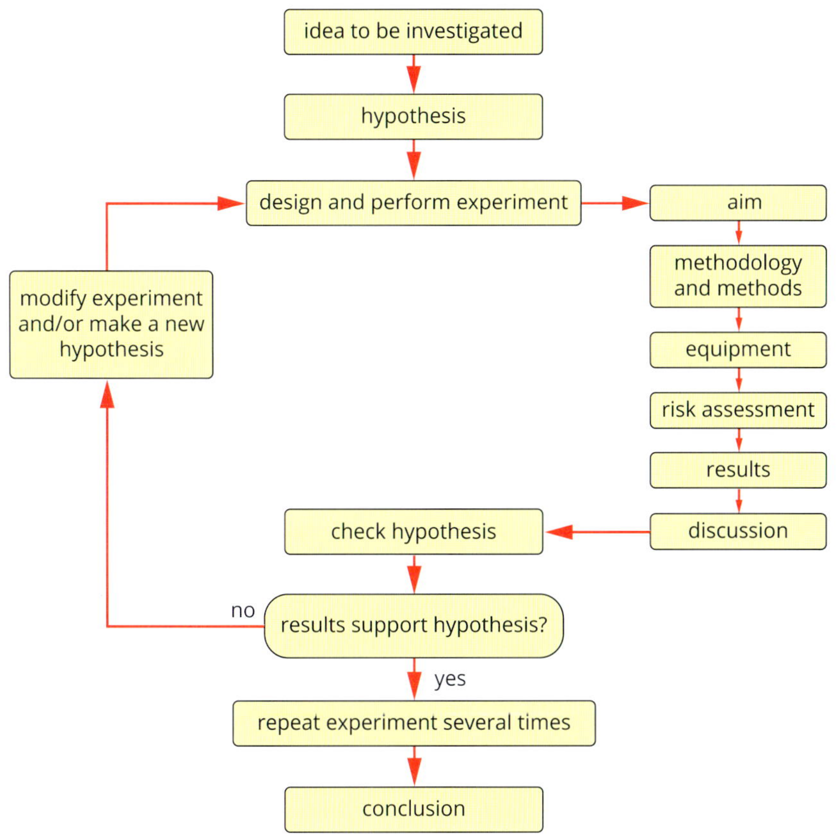

FIGURE 1.1.5 The scientific method

Carefully designed experiments are conducted to determine whether the predictions are accurate or not. If the results of an experiment do not fall within an acceptable range, the hypothesis is rejected. If the predictions are found to be accurate, the hypothesis is supported. If, after many different experiments, one hypothesis is supported by all the results obtained so far, then this explanation can be given the status of a **theory** or **principle**.

There is nothing mysterious about the scientific method. You might use the same process to find out how an unfamiliar machine works if you had no instructions. Careful observation is usually the first step.

You will now be able to answer key question 2.

Research questions

In science, there is little value in asking questions that cannot be answered. A hypothesis must be testable, but your inability to test a particular hypothesis does not mean that the hypothesis cannot be supported.

Your ability to test a hypothesis may be limited by the resources and equipment you have available. If you ask a research question, form and test your hypothesis, and find your hypothesis is supported, that does not mean it is true in all circumstances. Likewise, if your hypothesis is not supported, that does not mean it is never true.

For example, you might hypothesise 'If hydrogen peroxide is a toxic by-product of cellular respiration that is broken down by catalase, then all eukaryotes will contain catalase'. However, there may be a eukaryote that lacks catalase, but testing every eukaryotic organism would be impossible, and just because a eukaryote without catalase hasn't been identified does not mean none exist.

● *You will now be able to answer key question 5.*

Methodology and methods

The **methodology** is a brief description of the general approach taken to investigate the research question or hypothesis and the reasons why this approach is taken. The methodology can be described as the rationale behind your investigative methods. Examples of scientific investigation methodologies are controlled experiments, fieldwork, literature reviews, modelling and simulation. The **methods** (also known as procedures) are the specific steps that are taken to collect data during the investigation. The type of scientific investigation methodology and the methods selected will depend on the aim of the investigation and the research question.

> Scientific investigations must be able to be repeated by other scientists to be considered reliable.

The methodology and methods must be described clearly and in sufficient detail to allow other scientists to repeat the investigation. If other scientists cannot obtain similar results using the same methods and conditions, then the results from the original investigation are considered unreliable. It is also important to avoid personal bias that might affect the collection of data or the analysis of results. A good scientist works hard to be objective (free of personal bias) rather than subjective (influenced by personal views). The results of an investigation must be clearly stated and must be separate from any discussion of the conclusions that are drawn from the results.

Conducting an investigation once or using a small number of samples is not sufficient. You can have little confidence in a single result because you cannot be sure that the result was not due to some unusual circumstance that occurred at the time. The same experiment is usually repeated a number of times over a period of time and the combined results are then analysed statistically. If the statistics show that there is a low probability (usually less than 5%) that the results could have occurred as a result of chance, then the result is accepted as being significant.

● *You will now be able to answer key question 3.*

Experimental controls

> The experimental conditions of the control group are identical to those of the experimental group, except that the independent variable is also kept constant.

It is difficult—sometimes impossible—to eliminate all **variables** that might affect the outcome of an investigation. In biology, time of day, temperature, amount of light, humidity, and unidentified infections in organisms are examples of such variables. A way to eliminate the possibility that random factors affect results is to set up a second group within the experiment (called a **control group**) that is identical in every way to the first group (the **experimental group**) except for the single experimental variable that is being tested. This is a controlled experiment, because it allows you to examine one variable at a time. Controlled experiments are an important way of testing your hypothesis.

The variable that the experimenter is manipulating is the **independent variable**.

The **dependent variable** is what is measured when the independent variable changes. All of the other factors that could vary but must be kept the same in all experimental groups are called **controlled variables**.

When investigating antibacterial activity of compounds extracted from fungi or other sources, the variables to consider include the source, purity and concentration of the extract, the composition and consistency of the agar plates, the type of bacteria tested, the amount of substance on the test disc, the thickness of the discs and the incubation temperature. The independent variable would be the extract being tested. The dependent variable would be the presence and size of the zone of inhibition around the disc. The other variables listed above all need to be controlled. In Section 1.4 you will learn about setting up an investigation with controls.

The independent variable is the only variable that the experimenter changes in a controlled experiment. The dependent variable is measured to determine the effect of changing the independent variable.

In a controlled experiment, controlled (fixed) variables are kept constant.

Forming conclusions

Conclusions are evidence-based statements that are developed from the analysis of results. When drawing conclusions from the results of an investigation, the quality of the data needs to be considered—the data should be accurate, reliable and valid. A conclusion is valid if it provides a response to the research question that the investigation set out to answer. Conclusions should summarise and explain the results of the investigation, and identify the extent to which the investigation addressed the research question or hypothesis.

Speculation involves going beyond the results to make suggestions about what might be occurring. Conclusions are necessary, but speculation is interesting and thought-provoking. Both concluding and speculating are worthwhile, but you must be careful to keep them separate. It is also the usual practice of scientists to accept the simplest hypothesis that accounts for all the evidence available.

The conclusion made by Marshall and Warren, that *H. pylori* can cause gastritis and peptic ulcer disease (see the Case study on page 4), was evidence-based. Their study has been repeated many times, and has led to effective treatment for peptic ulcers.

LIMITATIONS OF THE SCIENTIFIC METHOD

The scientific method is not perfect; however, it remains the best way to understand your surroundings, and to constantly improve on that understanding. Even when the scientific method is strictly adhered to, there is still an element of chance in scientific discovery.

The scientific method can be applied only to hypotheses that can be tested, or to questions that can be answered. A hypothesis that is not testable can be neither supported nor disproved by the scientific method. Such hypotheses therefore remain as possible explanations. For example, Fleming's observation led to the hypothesis that certain fungi can produce chemicals that inhibit the growth of certain bacteria. This was testable for *Penicillium* and other fungi that can be grown on agar plates in the laboratory. If the hypothesis was broadened to 'All fungi produce antibiotics', this might not be testable, as testing it would depend on being able to grow all fungi and all potential bacterial targets in the laboratory.

A hypothesis can never be proven by a scientific study; it can only be supported under the conditions that have been tested.

It is also important to understand that although a hypothesis may be supported by experimental data, the same hypothesis may not be supported in all circumstances—it has only been found to be true under the conditions that have been tested.

The scientific method cannot be used to test morality or ethics. These judgements belong to the fields of philosophy, history, politics and law. Science can, however, provide valuable information that people can take into account when making these judgements. For example, science can be used to predict the environmental consequences of pollution and the medical consequences of chemical weapons, but it cannot itself make value or moral judgements about either.

DETERMINING APPROPRIATE INVESTIGATION METHODOLOGY

When it comes to beginning your scientific investigation, you will need to think about the best way to address your research question. For some investigations, setting up a controlled experiment may require using equipment that is not readily available to you in the school setting. This may mean you need to look at a computer simulation to model the outcomes of the investigation. Other approaches could include a literature review of other studies focussed on a similar research question. The different approaches that you could use are outlined in Table 1.1.1.

TABLE 1.1.1 Scientific investigation methodologies

Type of methodology	Explanation	Example
case study	investigation of a real or hypothetical situation, such as an activity, event, problem or behaviour, often involving analysis of data within a real-world context	looking at the impact of an oil spill in one part of the world and using this analysis to prepare, hypothesise and plan for the impact of an oil spill of similar magnitude in another part of the world
classification and identification	arranging objects, events or organisms into manageable groups by identifying shared or similar features	using morphology (physical features) to group organisms into taxonomic groups based on shared characteristics
controlled experiment	experimental investigation that involves formulating a hypothesis and testing the effect of an independent variable on the dependent variable while controlling all other variables in the experiment	investigating the impact of a change in temperature on the activity of an enzyme
correlational study	making observations and recording events and behaviours to investigate the relationship or association between variables	investigating the correlation between body mass index and the incidence of coronary heart disease
fieldwork	observing and interacting with particular environments to determine if a relationship exists between organisms or environmental factors and organisms; often involves observations and sampling of organisms and environments	chi-square test to investigate whether a relationship exists between two different species of marine molluscs in an intertidal zone
literature review	critical analysis of what has already been investigated and published, using secondary data from other people's investigations to explain events or propose new ideas or relationships	analysis of data looking at the impact of smoking on lung cancer in a variety of research papers to support, refute or develop new hypotheses
modelling	using models as representations of objects, systems or processes to aid understanding or make predictions	model of the connections between neurons in the human brain constructed from brain-scanning technology
product, process or system development	using scientific understanding and advances in technology to design a new tool, method or process to meet the demands or needs of society	developing a new biodegradable packaging material
simulation	using mathematical models or simulations to test hypotheses, conduct virtual experiments or model the complexity of whole cells, systems, organs or organisms	computer simulation of immune cells attacking other cells

EXPERIMENTATION

Once you have a testable hypothesis, you are ready to conduct an experiment to test it. Every experiment has to be designed and planned carefully. You need to be sure that someone else can repeat your experiment exactly the way you did it and get similar results. In Section 1.2 you will learn how to formulate your hypothesis and design an experiment to test it.

● *You will now be able to answer key questions 4 and 6–7.*

MODELS

Scientific models are used to create and test theories and explain concepts. They may also be developed as prototypes for functional devices such as replacement organs. The introduction of computer technology, including two- and three-dimensional animations, has helped to create more detailed and realistic representations of biological processes. Different types of models can be used, but each model has limitations in the type of information it can provide.

Modelling concepts

Models are created to answer specific questions or demonstrate specific processes. How a model is designed will depend on its purpose. The two most familiar types of models are visual models and physical models, but mathematical models and computational models are also common and increasingly important in the biosciences. Models help to make sense of ideas by visualising:

- objects that are difficult to see because of their size (too big or too small) or position, such as ecosystems, organs such as the heart and pancreas, cells, molecules and atoms
- processes that cannot easily be seen directly, such as digestion, feedback loops, biochemical reactions, gene expression and protein folding
- abstract ideas, such as energy transfer and the particulate nature of matter
- complex processes, such as networks of biochemical reactions, genome organisation and regulation, evolution, and brain connectivity and function.

For example, models of all the connections between neurons in the human brain have been constructed from brain scanning technology. The models are used to predict and test signalling and communication between neurons (Figure 1.1.6).

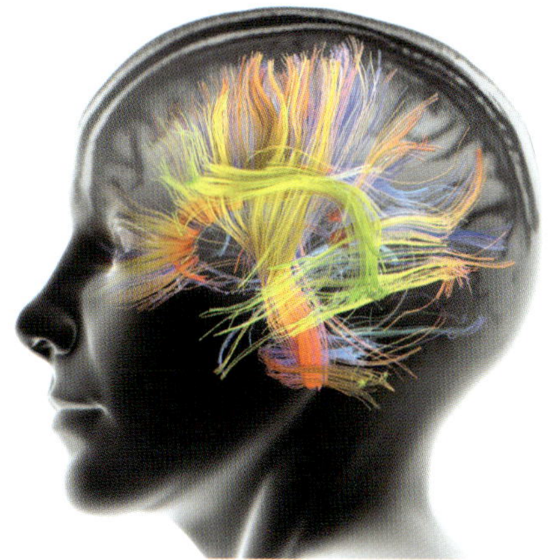

FIGURE 1.1.6 A model of the brain's wiring pattern explored in the Human Connectome Project

A deeper understanding of concepts can be developed through models. However, you need to identify the benefits and limitations of using a particular model to represent a concept. Furthermore, the quality and validity of a model is limited by the depth and accuracy of the information used to construct the model.

Model organisms

Biologists use live bacteria, animals and plants as model organisms for the investigation of cells and systems in situ and in vivo. It is possible to test in animals hypotheses that cannot be tested in humans for ethical reasons. Most of the advances in understanding animal and plant biology, genetics, pathology and medicine result from the use of model organisms. These organisms include the bacterium *Escherichia coli*, the nematode *Caenorhabditis elegans* (Figure 1.1.7), rats and mice, the plant *Arabidopsis thaliana* and the fruit fly *Drosophila melanogaster*.

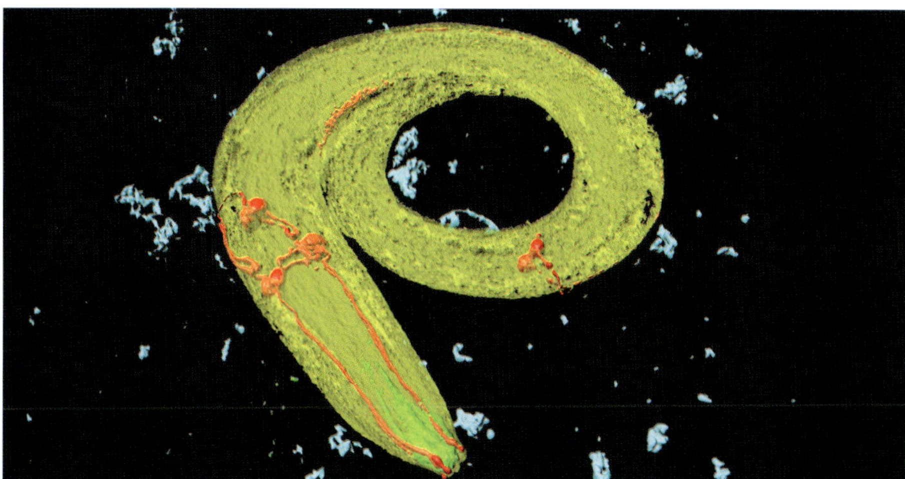

FIGURE 1.1.7 Model organism *Caenorhabditis elegans*, a nematode (roundworm). Confocal laser scanning micrograph of *C. elegans* with neurons stained green and the digestive tract stained red. *C. elegans* is a soil-dwelling nematode worm about 1 mm long and one of the most studied animals in biological and genetic research.

Studies that are in situ are 'in position' or 'in place', such as when studying cells functioning within an intact organ, or molecules in their normal cellular location.

Studies that are in vivo are 'within the living', such as when cells are studied in a living organism.

Studies that are in vitro are 'in glass' or in a dish or test tube, such as when cells are removed from the organism and studied in a culture dish (it doesn't have to be glass).

Efforts are being made to reduce the number of animals used in research, and strict ethical guidelines must be followed in their use. Studies performed in vitro, and advances in computer simulation and 'virtual' cells and organisms that have made *in silico* studies possible, allow for a reduced reliance on live animals. But keep in mind that the value and validity of a virtual model or simulation is only as good as the data and information used to construct the model. This ultimately comes from living cells and organisms.

> Studies that are *in silico* are 'in silicon', which refers to the silicon chips used in computers for computer simulations.

You will now be able to answer key questions 8–9.

1.1 Review

OA ✓✓

SUMMARY

- Well-designed experiments are based on a sound knowledge of what is already understood or known and careful observation.
- The scientific method is an accepted procedure for conducting investigations.
- A hypothesis is a possible explanation for a set of observations that can be used to make predictions, which can then be tested experimentally.
- Controlled experiments allow us to examine one factor at a time; they are a commonly used methodology for testing hypotheses.

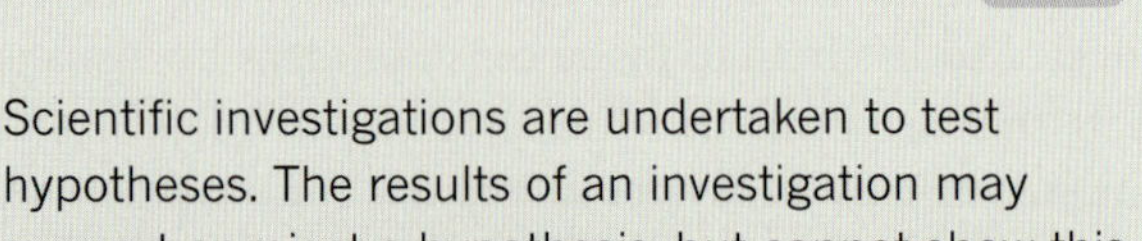

- Scientific investigations are undertaken to test hypotheses. The results of an investigation may support or reject a hypothesis, but cannot show this to be true in all circumstances.
- Science cannot be used to evaluate hypotheses that are not testable, nor can it make value or moral judgements.
- Models are useful tools that can be created and used to assist in a deeper understanding of concepts.

KEY QUESTIONS

Knowledge and understanding

1 What is the scientific method based on?
- **A** observation
- **B** subjective decisions
- **C** manipulation of results
- **D** generalisations

2 Name the key components of the scientific method.

3 **a** What do 'objective' and 'subjective' mean?
b Why must experiments be conducted objectively?

4 Which of the following is an important part of conducting an experiment?
- **A** disregarding results that do not fit the hypothesis
- **B** making sure the experiment can be repeated by others
- **C** producing results that are identical to each other
- **D** changing the results to match the hypothesis

Analysis

5 Write a hypothesis to test whether:
- **a** carrot seeds or tomato seeds germinate quicker
- **b** sourdough, multigrain or white bread goes mouldy the quickest
- **c** Trigg the dog likes dry food or fresh food better.

6 A scientist conducts a set of experiments, analyses the results and publishes them in a scientific journal. Other scientists in different laboratories repeat the experiment, but do not get the same results as the original scientist. Suggest several reasons that could explain this.

7 Design an experiment to test whether temperature is an important factor in the distribution of a mollusc species on a rocky coast. Clearly state the hypothesis that your experiment will test. Explain the methods that you would use. Do not forget to include experimental controls.

8 Explain the benefits of using a torso model to learn about the parts and relative positions of organs in the human body.

9 Explain two limitations of using models. Include an example.

1.2 Planning investigations

FIGURE 1.2.1 Scientists collecting grape vine samples for genetic research on the geographical origins of vines in the Mediterranean Basin

Practical investigations are those for which you gather the raw data yourself. These often take the form of experiments, activities, field trips or surveys (Figure 1.2.1). There are many elements to this type of practical investigation. A step-by-step approach will help you through the process and assist you in completing a solid and worthwhile investigation.

Taking the time to carefully plan and design an investigation before you begin will help you maintain a clear and concise focus throughout. Preparation is essential. In this section you will learn about some of the key steps to take when planning investigations:

- choosing a topic
- defining key terms
- sourcing information
- obtaining ethics approval
- ensuring occupational health and safety
- writing a protocol and schedule.

CHOOSING A TOPIC

Throughout this course you will conduct practical work (laboratory or fieldwork) on a range of topics. For Unit 1 Area of Study 3 you are required to adapt or design and then conduct a scientific investigation related to the function and/or regulation of cells or systems.

When you choose a topic, consider the following:

- Choose a research question you find interesting.
- Start with a topic about which you already have some background information, or some clues about how to perform the experiments.
- Check that your school laboratory has the resources for you to perform the experiments or investigate the topic.
- Choose a topic that can provide clear, measurable data.

A number of topics that may be addressed in the course are suggested in Table 1.2.1 on page 12. You will learn more about useful research techniques for topics like these in Section 1.3.

Before you start

The topics in Table 1.2.1 on page 12 are only suggestions. Select your topic based on what resources are available to you. Before commencing your investigation, check that you have:

- the materials required to grow or culture an organism (e.g. plants, bacteria, yeast, protists or invertebrates)
- equipment such as microscopes, pH meters, spectrophotometers, centrifuges, and data loggers
- the materials needed to perform the experiments, such as biochemical test strips (for glucose, protein), enzymes and substrates, acids and bases.

Also ensure that you:

- can order any materials needed that are not on hand
- have a solid understanding of the theory behind your investigation
- are trained to use the required equipment
- have a detailed plan for the practical components of your investigation
- are able to access the school laboratory when you need to.

TABLE 1.2.1 Potential areas for investigation in Units 1 and 2

Laboratory experiments may be used to investigate cellular structure and function.	Possible topics for laboratory investigation include: • surface area to volume ratio and its role in cellular functions • phagocytosis or endocytosis in living cells • variations in the cell cycle and mitosis in different cell types • the regulation of water balance in plants • modelling the effect of sweating on heat loss.
Fieldwork may be used to investigate cellular processes or the interactions, adaptations and distribution of species.	Possible topics for fieldwork investigation include: • variations in photosynthetic pigments in leaves from different environments • the influence of the environment on phenotypes • variations in stomatal density in leaves along a transect • the biological advantages and disadvantages of different reproductive strategies • the ecological role of keystone species.
The use of data from online databases may facilitate, or be central to, your investigation.	Possible topics for investigation using online databases include: • regulation of the eukaryotic cell cycle • the inheritance patterns of genes or traits • genetic diversity in different populations, using DNA sequence analysis • changes in the distribution and density of species over time.

DEFINING KEY TERMS

When you begin a scientific investigation, you first have to develop and evaluate a research question, determine the associated variables, formulate a hypothesis and define the aims. It is important to understand that each of these can be refined as the planning of your investigation continues.

- The **research question** defines what is being investigated. For example: Is the rate of photosynthesis in plants dependent on temperature?
- The variables are the factors that change during your experiment. For example: Temperature is a variable for the photosynthesis example given earlier.
- The hypothesis is a statement that can be tested and is based on previous knowledge, evidence or observations, and that attempts to answer the research question. For example: If the temperature increases from 20°C to 40°C, then the rate of photosynthesis will increase.
- The **aim** is a statement describing in detail what will be investigated. For example: To investigate the effect of temperature on the rate of photosynthesis in plants at 20°C, 30°C and 40°C.

Determining your research question

Before conducting an experimental investigation you need a research question to address. Once you have come up with a topic or idea of interest, the first thing you need to do is conduct a search of the relevant literature; that is, you must read scientific reports and other articles on the topic to find out what is already known, and what is not known or not yet agreed upon. The literature also gives you important information for the introduction to your report and ideas for experimental methods. Use this information to generate questions.

When you have defined the research question, you are able to formulate a hypothesis, identify the measurable variables, proceed with designing your investigation and suggest a possible outcome of the experiment.

Stop to evaluate the question before you progress; it may need further refinement or even further investigation before it is suitable as a basis for an achievable and worthwhile investigation. Consider the following checklist:

- ☐ relevance—Make sure your question is related to your chosen topic. For your investigation decide whether your question will relate to cellular structure or organisation, or to structural, physiological or behavioural adaptations of an organism to an environment.
- ☐ clarity and measurability—Make sure your question can be framed as a clear hypothesis. If the question cannot be stated as a specific hypothesis, then it is going to be very difficult to complete your research.
- ☐ time frame—Make sure your question can be answered within a reasonable period of time. Ensure your question isn't too broad.
- ☐ knowledge and skills—Make sure you have a level of knowledge and a level of laboratory skills that will allow you to explore the question. Keep the question simple and achievable.
- ☐ safety and ethics—Consider the safety and ethical issues associated with the question you will be investigating. If there are issues, determine if these need to be addressed.
- ☐ advice—Seek advice from your teacher on your question. Their input may prove very useful. Their experience may lead them to consider aspects of the question that you have not thought about.

When writing a research question, it is advisable to include the independent and dependent variables. For example, what is the effect of [the independent variable] on [the dependent variable]?

Defining your variables

The factors that can change during your experiment or investigation are called the variables. An experiment or investigation determines the relationship between variables. There are three categories of variables:

- independent—a variable that is controlled by the researcher (the one that is selected and manipulated)
- dependent—a variable that may change in response to a change in the independent variable, and is measured or observed
- controlled variables—the variables that are kept constant during the investigation.

You should have only one independent variable. Otherwise you could not be sure which independent variable was responsible for changes in the dependent variable.

Making predictions and constructing a hypothesis

The hypothesis is a prediction of what you think will happen during a scientific investigation. It is a statement that can be tested (based on evidence and prior knowledge) to answer your research question. It defines a proposed relationship between two variables. To do this, you will need to identify the dependent and independent variables.

A good hypothesis is written in terms of the dependent and independent variables, like this:

If x happens, then y will happen. The 'if' part of the hypothesis refers to the independent variable—the variable you alter in the experiment. The 'then' part relates to the dependent variable—the variable you measure or observe.

For example:

If yeast is grown in acidic conditions, ***then*** *the rate of cellular respiration will decrease.*

A hypothesis does not need to include 'if' and 'then' in its wording. For example, the previous hypothesis could also be stated the following way:

The rate of cellular respiration in yeast will decrease when yeast cells are grown in acidic conditions.

A good hypothesis can be tested to determine whether it is supported (verified), or not supported (falsified) by the investigation. To be testable, your hypothesis should include variables that are measurable.

When you evaluate your research question, consider the variables, and think about different potential hypotheses; it helps to create a table that outlines them. For example, Table 1.2.2 outlines a research question, the variables, and a potential hypothesis that relates to the effect of glucose on the rate of cellular respiration in yeast.

TABLE 1.2.2 Example of research question, variables and potential hypothesis

Research question	Will the rate of cellular respiration in yeast cells be faster if the cells are exposed to higher amounts of glucose?
Independent variable	glucose concentration
Dependent variable	rate of cellular respiration measured as change in CO_2 released over time
Controlled variables	yeast culture volume, temperature, light conditions
Potential hypothesis	The rate of cellular respiration in yeast will increase as glucose concentration increases.

Determining your aim

The aim is the key step required to test your hypothesis. The aim should directly relate to the variables in the hypothesis, describing how each will be studied or measured. The aim does not need to include the details of the method.

For example:

- Hypothesis: If algae are exposed to low light levels, then the rate of photosynthesis will decrease.
- Aim: To compare the rates of photosynthesis in algae at different distances from a light source.
- Variables: distance from light source, i.e. light intensity (independent) and rate of photosynthesis (dependent).

● *You will now be able to answer key questions 1–3 and 6–7.*

SOURCING INFORMATION

When you are sourcing information during your search of the literature, researching experimental methods and investigating a broader issue, consider whether that information is from primary or secondary sources. You should also consider the advantages and disadvantages of using resources such as books or the internet.

Primary and secondary sources

Primary sources of information are created by a person directly involved in an investigation. Examples of primary sources are results from research and peer-reviewed scientific articles. **Secondary sources** of information are a synthesis, review or interpretation of primary sources. Examples of secondary sources are textbooks, newspaper articles and websites. For example, a scientist's journal article on a clinical trial of treatments for teenage obesity is a primary source, while a general magazine article about teenage obesity written by a journalist and referring to the scientific study is a secondary source. Table 1.2.3 compares primary and secondary sources.

Secondary sources of information may have a bias, so you need to determine if they are reliable sources of information. You will learn about assessing the accuracy, reliability and validity of data in Section 1.4.

TABLE 1.2.3 Summary of primary and secondary sources

	Primary sources	Secondary sources
Characteristics	• first-hand records of events or experiences • written at the time the event happened • original documents	• interpretations of primary sources • written by people who did not see or experience the event • use information from original documents but rework it
Examples	• results of experiments • scientific journal/magazine articles • reports of scientific discoveries • photographs, specimens, maps and artefacts • interviews with experts • websites (if they meet the criteria above)	• textbooks • biographies • newspaper articles • magazine articles • radio and television documentaries • websites that interpret the scientific work of others • podcasts

● *You will now be able to answer key question 8.*

Using books and the internet

Peer-reviewed scientific journals are the best sources of information, but you are unlikely to have access to many of them, and much of the information is difficult to interpret if you are not an expert in the field.

As books, magazines and internet searches will be your most commonly used resources for information, you should be aware of their limitations (Table 1.2.4 on page 16). Reputable science magazines you might find in your school library include *New Scientist*, *Cosmos*, *Scientific American* and *Double Helix* (Figure 1.2.2).

FIGURE 1.2.2 A reputable science magazine you might find in your school library

TABLE 1.2.4 Advantages and disadvantages of book and internet resources

	Book resources	Internet resources
Advantages	• written by experts • authoritative information • reviewed to ensure information is accurate • logical, organised layout • content is relevant to the topic • contain a table of contents and index to help find relevant information	• quick and easy to access • allow access to hard-to-find information • access to the whole world; millions of websites • up-to-date information • may be interactive and use animations to enhance understanding
Disadvantages	• may not have been published recently • usable by only one person at a time	• time-consuming looking for relevant information • a lot of 'junk' sites and biased material • search engines may not display the most useful sites • cannot always tell how up-to-date information is • difficult to tell if information is accurate • hard to tell who has responsibility for authorship • information may not be well ordered • less than 10% of sites are educational

Evaluating books and journals

Your textbook should be your first source of reliable information. Other information should be consistent with it. Articles published in journals and magazines often present findings of new research, which may or may not be confirmed later, so be careful not to treat such sources of information as established fact. Scientific journals are **peer-reviewed** (critically reviewed by other specialist scientists), which gives them more credibility than other sources.

Evaluating websites

Remember that anyone can publish anything on the internet, so it is important to evaluate the credibility, currency and content of online information. To evaluate online information, follow the checklist below.

- ☐ Credibility—consider who the author is, their qualifications and expertise; check for their contact information and for a trusted abbreviation in the web address, such as .gov or .edu; websites using .com may have a bias towards selling a product (but this product could be a reputable science magazine or journal), and .org sites might have a bias towards one point of view (although these sites can be a good starting point for general information).
- ☐ Currency—check the date the information you are using was last revised.
- ☐ Content—consider whether the information presented is fact or opinion; check for properly referenced sources; compare information to other reputable sources, including books and science journals.

ETHICS

Ethics is a set of moral principles by which your actions can be judged as right or wrong. Every society or group of people has its own principles or rules of conduct. Scientists have to obtain approval from an ethics committee and follow ethical guidelines when conducting research that involves animals, including, and especially, humans.

Applying ethical principles means to:

- consider the implications of investigations of organisms and the environment—you should aim to maximise benefit while minimising harm and risk
- recognise the intrinsic value of life and respect the welfare, autonomy, beliefs, perceptions and customs of others
- use integrity when recording and reporting the outcomes of your investigation, and also when using other people's data (such as in a literature review)
- form a conclusion about science-related ethical issues using scientific knowledge and skills, whilst also considering the needs of all parties involved
- recognise the importance of social, economic and political values in forming conclusions using scientific understanding.

Ethics approval

If you work with animals as part of your studies, you may need to obtain a licence. Check with your school, teacher or laboratory technician. All animal use should follow the Victorian Government's guidelines for the care and use of animals in schools. These guidelines recommend that schools consider the '3Rs rule':

- Replace the use of animals with other methods where possible.
- Reduce the number of animals used.
- Refine techniques to reduce the impact on animals.

You should treat animals with respect and care. The welfare of the animal must be the most important factor to consider when determining the use of animals in experiments. If at any time the animal being used in your experiment is distressed or injured, the experiment must stop.

If human volunteers are needed, then the participants need to be fully briefed on the aim of, and methods involved in, the study, and they must give informed consent. They should also be given the opportunity to see the results of the study and their potential impact on the science community.

You will now be able to answer key question 4.

OCCUPATIONAL HEALTH AND SAFETY

While planning for an investigation in the laboratory or outside in the field, it is important for your safety and the safety of others that you consider the potential risks.

Everything we do has some risk involved. **Risk assessments** are performed to identify, assess and control hazards. A risk assessment should be performed for any situation, whether in the laboratory or out in the field, that could cause harm to people or animals. Always identify the risks and control them to keep everyone safe.

To identify risks, think about:

- the activity that you will be conducting
- where in the environment you will be working (e.g. in a laboratory, the school grounds, or a natural environment)
- how you will use equipment, chemicals, organisms or parts of organisms that you will be handling
- what clothing you should wear.

The hierarchy of risk control (Figure 1.2.3) is organised from the most effective risk management measures at the top of the pyramid to the least effective at the bottom of the pyramid.

Take the following steps to manage risks when planning and conducting an investigation.

- Elimination—eliminate dangerous equipment, methods or substances.
- Substitution—find different equipment, methods or substances to use that will achieve the same result, but have less risk associated.

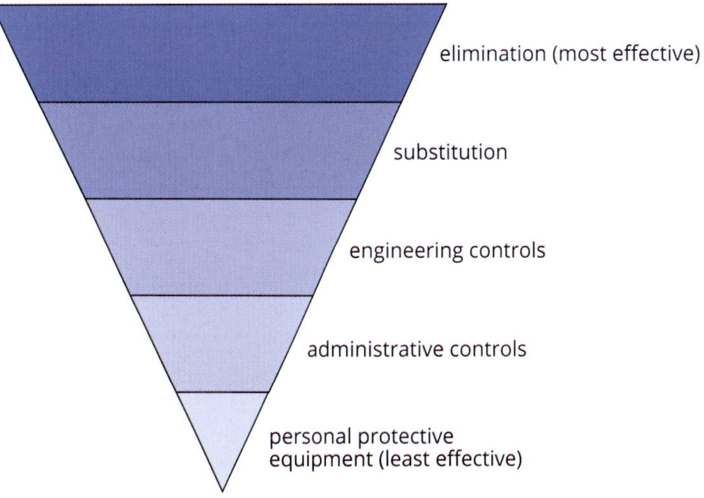

FIGURE 1.2.3 The hierarchy of risk control in this pyramid is shown from top to bottom in order of decreasing effectiveness.

- Engineering controls—modify equipment to reduce risks. Ensure there is a barrier between the person and the hazard. Examples include physical barriers, such as guards in machines, or fume hoods when working with volatile substances.
- Administrative controls—provide guidelines, special procedures, warning signs and safe behaviours for any participants.
- Personal protective equipment (PPE)—wear safety glasses, lab coats, gloves, respirators and any other necessary safety equipment where appropriate, and provide these to other participants. As PPE can be damaged, it is considered the least effective control measure, but it remains an essential safety feature after other control measures are in place (Figure 1.2.4).

FIGURE 1.2.4 A lab coat, gloves and safety glasses are essential items of personal protective equipment in the laboratory.

Science outdoors

Your investigation may involve outdoor fieldwork (Figure 1.2.5). All the potential risks, and ways to minimise them, must be considered when planning fieldwork. Ways to reduce risk include use of suitable protective clothing, knowledge of the terrain, having up-to-date maps, and checking predicted weather and fire risk.

FIGURE 1.2.5 Researchers excavating human fossils at a cave in Atapuerca, Spain. Hard hats, ropes, harnesses, strong clothing and footwear are essential during fossil research in the field.

Chemical safety

Some chemicals used in laboratories are harmful. When you are working with chemicals in the laboratory or at home, it is important to keep them away from your body. Laboratory chemicals can enter the body in three ways:

- ingestion—chemicals that have been ingested (eaten) may be absorbed across cells lining the mouth or enter the stomach, and may then be absorbed into the bloodstream
- inhalation—chemicals that are breathed in (inhaled) can cross the thin cell layer of the alveoli in the lungs and enter the bloodstream
- absorption—some chemicals are able to pass through the skin in a process called absorption.

When working with any type of chemical you should:

- identify the chemical codes and be aware of the dangers they are warning about
- become familiar with the relevant safety data sheets (SDS), formerly known as material safety data sheets (MSDS)
- use personal protective equipment
- wipe up any spills
- wash your hands thoroughly after use.

Chemical codes

The chemicals in laboratories, supermarkets, pharmacies and hardware shops have warning symbols on their labels. These are a chemical code indicating the nature of the contents (Table 1.2.5). From 1 January 2017, the Globally Harmonised System of Classification and Labelling of Chemicals (GHS) pictograms were introduced into Australia. This system is used for labelling containers and in safety data sheets. Some of the pictograms that you may see denote chemicals that are corrosive, pose a health hazard or are flammable. These chemical codes will need to be analysed and addressed when you are planning and conducting scientific investigations. You will perform a risk assessment in which these chemical codes will be provided, then after analysing them, you may need to modify your experimental plan so that safety is improved.

TABLE 1.2.5 GHS pictograms used as warning symbols on chemical labels

GHS pictogram	Use	GHS pictogram	Use	GHS pictogram	Use
	flammable liquids, solids and gases; including self-heating and self-igniting substances		oxidising liquids, solids and gases, may cause or intensify fire		explosion, blast or projection hazard
	corrosive chemicals; may cause severe skin and eye damage and may be corrosive to metals		gases under pressure		fatal or toxic if swallowed, inhaled or in contact with skin
	low level toxicity; this includes respiratory, skin and eye irritation, skin sensitisers and chemicals harmful if swallowed, inhaled or in contact with skin		hazardous to aquatic life and the environment		chronic health hazards; this includes aspiratory and respiratory hazards, carcinogenicity, mutagenicity and reproductive toxicity

Safety data sheets

Every chemical substance used in a laboratory has a **safety data sheet (SDS)**. This contains important information about the possible hazards in using the substance and how it should be handled and stored. An SDS states:

- the name of the hazardous substance
- the chemical and generic names of certain ingredients
- the chemical and physical properties of the hazardous substance
- health hazard information
- how to store the chemical safely
- precautions for safe use and handling
- how to dispose of the chemical safely
- the name of the manufacturer or importer, including an Australian address and telephone number.

An SDS contains important safety and first aid information for teachers and technicians about each chemical you commonly use in the laboratory.

The SDS provides employers, workers and emergency crews with the necessary information to safely manage the risk of hazardous substance exposure.

● *You will now be able to answer key questions 5 and 8.*

1.2 Review

OA ✓✓

SUMMARY

- A research question is a statement that broadly defines what is being investigated.
- The three types of variables are:
  - independent—a variable that is controlled by the researcher (the one that is selected and manipulated)
  - dependent—a variable that may change in response to a change in the independent variable, and is measured or observed
  - controlled variables—variables that are kept constant during the investigation.
- A hypothesis:
  - is a statement that can be tested and is based on previous knowledge and evidence or observations, and addresses the research question
  - often takes the form of a proposed relationship between two or more variables in a cause and effect relationship
  - must be testable; that is, able to be supported (verified) or not supported (falsified) by investigation.
- An aim is a statement describing in detail what will be investigated.
- Primary sources of information are created by a person directly involved in an investigation. Secondary sources of information are a synthesis, review or interpretation of primary sources.
- Ethical and safety considerations must be of the highest priority at all times during a scientific investigation.

KEY QUESTIONS

Knowledge and understanding

1 Define the following terms.
  a research question
  b hypothesis
  c aim

2 Which of the following is a research question?
  A What is the sodium content of baby foods?
  B The sodium content of baby food is less than 100 mg per 100 g.
  C Babies need to have a low sodium diet.
  D Sodium can be measured using gravimetric analysis and atomic absorption spectroscopy.

3 Select the best hypothesis and explain why the other options are not good hypotheses.
  A If light and temperature increase, the rate of photosynthesis increases.
  B Transpiration is affected by temperature.
  C Light is related to the rate of photosynthesis.
  D If bread is exposed to high light intensities, then it will grow mould faster than bread that is exposed to lower light intensities.

4 Explain the reasons for having a safety data sheet (SDS) for every chemical used in the laboratory.

Analysis

5 The following inferences may explain why grass growing near the edge of the concrete path remains green in summer.

Write each of the inferences below as an 'if ... then ...' hypothesis that could be tested in an experiment.
  a This grass receives the rain runoff from the path when it rains.
  b The concrete path insulates the grass roots from the heat and cold.
  c People do not walk on this part of the grass.
  d The soil under the path remains moist while the other soil dries out.
  e More earthworms live under the path than under the open grass.

6 You are conducting an experiment to determine the effect of the pH of water on mussel shell mass. Identify:
  a the independent variable
  b the dependent variable
  c at least one controlled variable.

7 Distinguish between ingestion, inhalation and absorption.

8 Decide whether each of the following is a primary or a secondary source.
  a a newspaper article about global warming
  b an experiment to investigate chemical changes when mixing combinations of chemicals
  c an interview with a forensic scientist about using science in tracking criminals
  d a website with information about genetic engineering

1.3 Techniques used in scientific investigations

In this section you will learn about designing and selecting methods to use in scientific investigations. You will be introduced to different techniques, and understand how selecting appropriate equipment and methods will allow you to obtain accurate and precise measurements. The choice of techniques, sample size and data collection will also be discussed.

MICROSCOPY

Your practical investigation may involve the study of live cells or prepared slides using microscopic methods. You may need to include cell size, number and cellular behaviour as part of your experimental evidence. You probably have access to light microscopes with magnifications up to 400× and possibly 1000× (oil immersion).

Field of view and size of specimens

Biological drawings should include a scale. Calculating the field of view under the microscope is required for estimating the size of specimens viewed. To calculate the field of view you use a minigrid. This is a 1 mm × 1 mm grid with a smaller microgrid of 100 µm × 100 µm in the centre (used with the 40× objective) (Figure 1.3.1).

Typical magnifications and fields of view in a school light microscope are listed. Microscopes usually have 10× eyepieces. The total magnification is the product of the eyepiece (ocular lens) and objective lens.

Objective lens	Total magnification	Field of view
4×	40×	4.5 mm
10×	100×	1.5 mm
40×	400×	450 µm
100×	1000×	150 µm

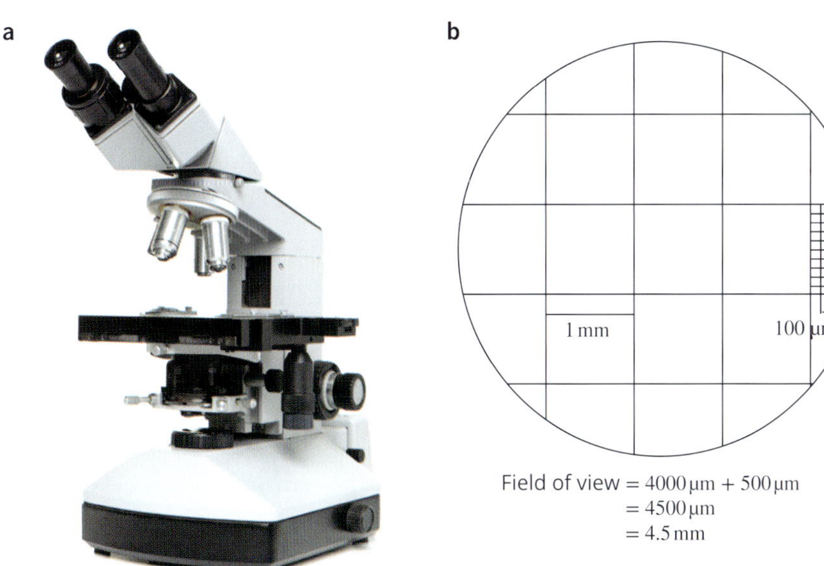

FIGURE 1.3.1 (a) Light microscopes are used extensively in biology. (b) Using a minigrid allows you to measure the field of view and calculate cell size. This is a view of a minigrid at 40× magnification. Each large square is 1 mm^2, so the field of view is 4.5 mm (or 4500 µm).

Once you have calculated your field of view for each lens, you can estimate the size of the cells. For example, you may be studying the processes of phagocytosis and lysosome action in *Amoeba* or *Paramecium* (Figure 1.3.2). If at 400× magnification you estimate that one cell occupies half the diameter of the field of view (or two cells span the field of view) and the field of view is 450 µm, then the estimated size of each cell is

$$\frac{450\,\mu\text{m}}{2\text{ cells}} = 225\,\mu\text{m}.$$

If using microscopy, it is important that a large sample size is used. For example, in an investigation into the effect of an acidic environment on bacterial cell growth, a sample of a culture could be taken and placed on a slide to be stained and viewed using oil immersion microscopy.

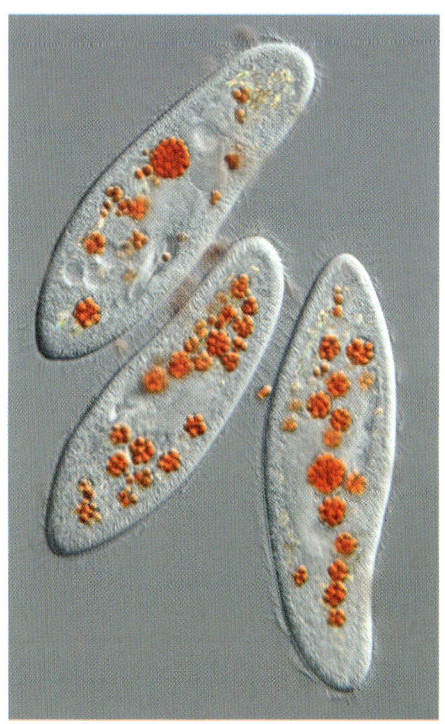

FIGURE 1.3.2 *Paramecium caudatum* viewed through a light microscope. Yeast cells (stained red) that have been engulfed by the *Paramecium* can be seen in the vacuoles.

If only one slide was used, it would be recommended to look at multiple fields of view. The number of bacteria within each field of view could then be calculated and an average number of cells on the slide determined. As the sample size increases, the results will more accurately represent the overall population and the effect of errors and uncertainty in the method will be reduced. In this example, more slides could be prepared from the same bacterial cultures, and multiple fields of view from each slide calculated to obtain an even larger sample size.

● *You will now be able to answer key questions 2 and 5.*

CELL CULTURE

Cell culture is a core technique in biological sciences; unfortunately, animal cell culture is restricted to laboratories that have the specialised equipment and training, as well as the ethics and safety approvals. However, in your school lab you are able to grow cultures of eukaryotic cells, including unicellular algae (e.g. *Chlorella*), protists (e.g. *Paramecium caudatum* and *Amoeba proteus*) and yeast (e.g. *Saccharomyces cerevisiae*, baker's yeast) (Table 1.3.1). Keep in mind that cells take time to grow, so plan early. You can also grow cultures of bacteria (low risk category 1) such as *Escherichia coli*, *Staphylococcus epidermidis* and *Bacillus subtilis* on agar plates or in broth cultures. Live cell cultures can be used to investigate factors affecting cellular processes that may be reflected in cell growth rates, cell responses and other cellular processes.

TABLE 1.3.1 Growing cells for biology investigations

	Bacteria and yeast are cultured in appropriate liquid nutrient broth or nutrient agar plates.
	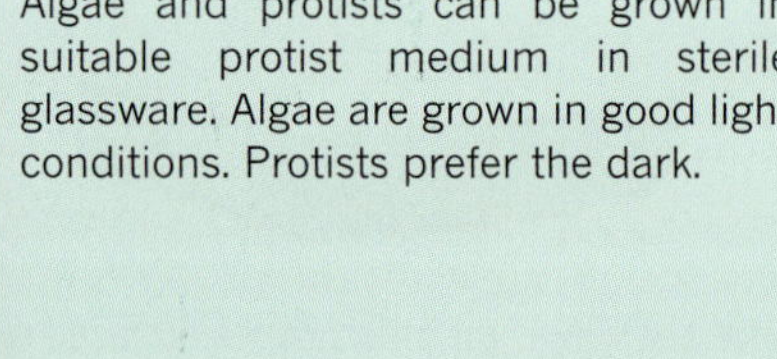 Algae and protists can be grown in suitable protist medium in sterile glassware. Algae are grown in good light conditions. Protists prefer the dark.
	Plant tissue culture. Small segments of stem or leaf are surface sterilised to remove contaminants. Explants (any sample taken from the organism, like a cutting) are cultured on plant nutrient agar over days or weeks.

TOOLS TO SUPPORT YOUR PRACTICAL INVESTIGATIONS

A variety of tools can be used when conducting scientific investigations. The choice of equipment is important to ensure that your measurements are accurate (to minimise error), and that your results are reproducible and reliable (to minimise uncertainty). Equipment that might be of use when conducting investigations is outlined in Table 1.3.2.

TABLE 1.3.2 Tools that may be available for practical investigations

Simple indicator of pH	Measuring pH or temperature	Measuring solutes
Tool: A dipstick test for the full pH range. A strip with pH-sensitive coloured pads is dipped into a solution then read against a reference colour chart after a defined time. **Purpose:** To measure the pH of a solution.	**Tool:** Electronic meters and probes. **Purpose:** To measure pH or temperature.	**Tool:** Strip tests for measuring glucose, protein and other solutes: • Multistix tests for several substances • Uriscan strips test glucose and protein • Glucostix tests glucose only **Purpose:** Usually designed for urine testing. Coloured pads on the strip are dipped into urine or other solutions; colour develops and is read against a reference chart. Detection is often based on an enzyme reaction within the pad.
Data loggers for a range of measurements	**Biochemical/chemical tests to detect molecules**	**Measuring absorbance, optical density or turbidity**
Tool: Common types of probes and capabilities in data loggers include: • pH • temperature • oxygen concentration • CO_2 concentration • absorption colorimeter • concentration of various compounds. **Purpose:** Data loggers enable data collection over significant time periods.	**Tools include:** **a** biuret reagent* for detecting protein (colour change from blue to purple) **b** Benedict's reagent* for detecting reducing sugars such as glucose, maltose, fructose; not sucrose (colour change from blue to orange/red) **c** iodine–potassium iodide (IKI)* reagent for detecting starch (colour change from yellow/orange to deep blue). **Purpose:** To detect different biochemical reactions.	**Tool:** Colorimeter or spectrophotometer. **Purpose:** Used for quantitation of colour reactions, or turbidity for monitoring cell growth.

* Some tests are qualitative; quantitative or 'semi-quantitative' results may be achieved if combined with standards and absorbance readings.

● *You will now be able to answer key questions 1 and 6.*

MEASURING ABSORBANCE FOR QUANTIFYING REACTIONS

If your school has a colorimeter or spectrophotometer, then you may be able to get quantitative results when conducting experiments that use colour-based reactions, such as the detection of protein or starch. You can also use this instrument to measure the turbidity and optical density of bacterial cultures or yeast broths, providing a quantitative measure of their growth rates and an ability to account for sources of error and uncertainty.

A sample is placed in a special tube called a cuvette, which is placed in the instrument. Light of a particular wavelength is shone through the sample, which absorbs some of the light (Figure 1.3.3). The appropriate wavelength of light to select is the one that is maximally absorbed by the sample, and this differs for each substance measured. For example, blue solutions absorb light around 600 nm, and red solutions absorb light around 490 nm (wavelength is measured in nanometres, nm, and represented by the symbol λ). The meter reads the amount of light absorbed by the sample. A sample with a high concentration of the substance will absorb more light and therefore give a higher absorbance reading.

a

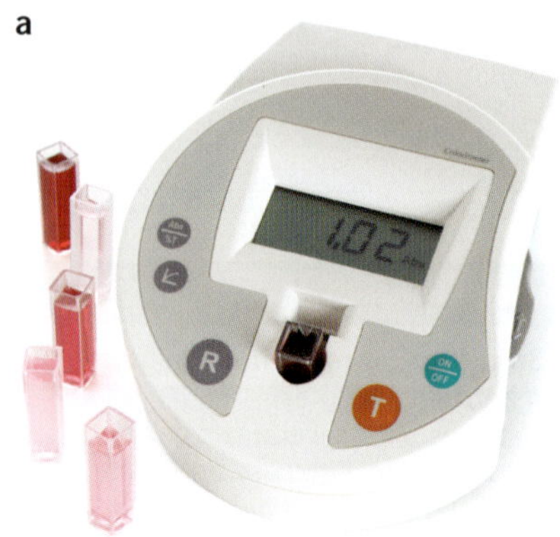

b

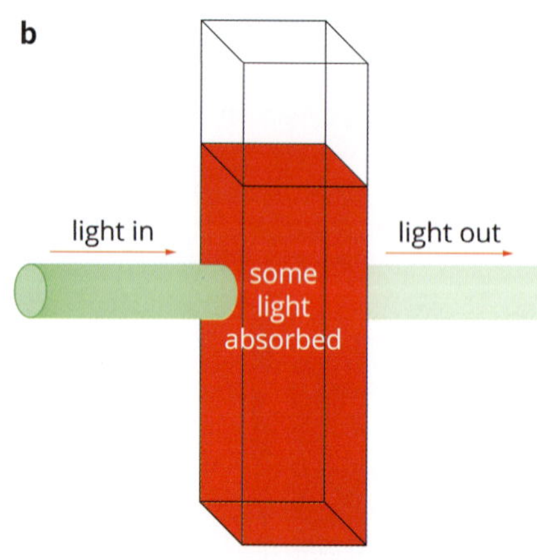

FIGURE 1.3.3 (a) A colorimeter or spectrophotometer reads absorbance of light. A sample is placed in a cuvette and placed in the instrument. (b) Light of a particular wavelength is shone through the sample. The meter reads the amount of light absorbed by the sample. A sample with higher concentration gives a higher absorbance reading.

CHROMATOGRAPHY

Chromatography methods available in your school laboratory may include paper or thin layer chromatography (Figure 1.3.4). Photosynthetic pigments vary in different organisms, such as different plants, algae and cyanobacteria. Different photosynthetic pigments have different properties of light absorbance, so are relevant to the rates of photosynthesis in different conditions.

Amino acids, the building block of proteins, can also be investigated by chromatography with the detection agent ninhydrin, which must be used safely in a fume hood.

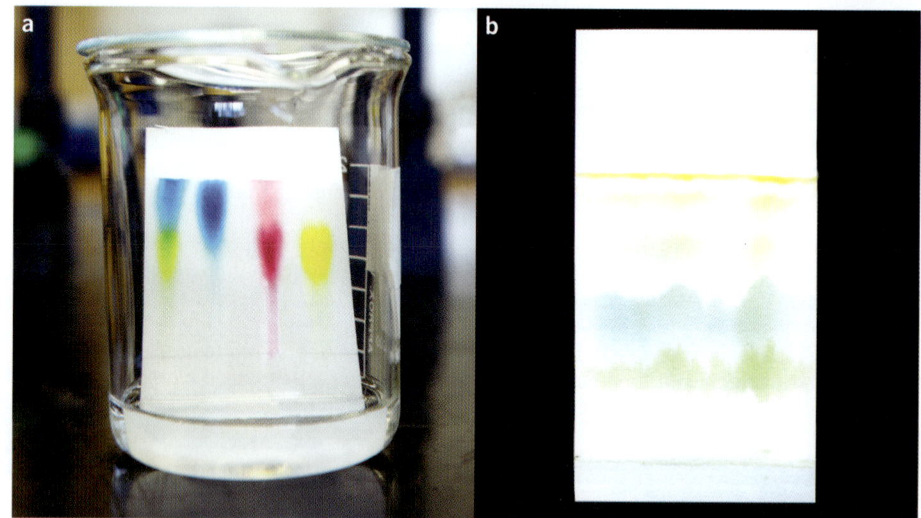

FIGURE 1.3.4 (a) Thin-layer chromatography (TLC) plate in a beaker, showing separated components (colours). TLC is performed on a sheet of glass, plastic or foil coated in a thin layer of adsorbent material. (b) An example of plant pigment molecules separated by paper chromatography. The sample is applied to the plate or paper and a solvent is drawn up the plate or paper via capillary action. Different components move up at different rates, causing them to separate.

FIELDWORK AND OBSERVATIONAL STUDIES

Biological investigations often include fieldwork and observational studies. **Fieldwork** takes place outside the classroom or laboratory and involves investigating organisms in the natural environment. Observational studies of organisms may be conducted in the natural environment as part of fieldwork or may take place in a laboratory or artificial environment, such as a zoo. As with any investigation, it is necessary to consider an appropriate sample size and method for observation of these organisms.

Fieldwork

When studying ecology, it may be necessary to determine the type and number of living organisms in an area. For example, your investigation may look at the population of a particular species in two different areas. There are many different ways to do this, including using quadrats and transects. Whatever method you use, it is important to always leave the environment the way you found it.

In natural environments it is usually impossible to count all the individuals of a species. Even just counting the living things in your school would take a very long time. Sampling gives us a good idea of the organisms in an ecosystem without needing to count each one. Table 1.3.3 on page 27 outlines some sampling techniques and when they are best used. When sampling in the field you should always consider the time and equipment available, the organisms involved and the impact the sampling may have on the environment.

Point sampling

Point sampling involves counting organisms only at selected points (Figure 1.3.5). These points might be selected randomly or regularly, depending on the type of sampling being done. This technique can be used to determine the range of organisms that live in an area and how common they are. Point sampling is quick, but you might miss rare organisms.

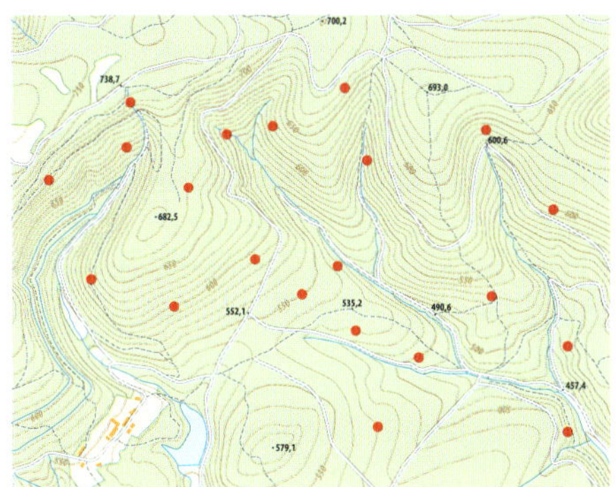

FIGURE 1.3.5 Randomly selected points for sampling marked on a topographical map. Sampling sites are indicated by the red dots.

Quadrats

A **quadrat** is a sampling method that allows you to estimate the number and variety of organisms in a large area by counting their number in a small area (Figure 1.3.6). A quadrat is usually square, rectangular or circular. Keep the following points in mind when using quadrats:

- Quadrats are most useful for sampling immobile organisms such as plants or corals.
- Determine the size of the quadrat based on the size and abundance of the organism that you are sampling.
- The more quadrats you use, the more reliable your results will be.
- Very abundant organisms can be measured as a percentage of the area covered, rather than as a number of organisms.
- Photographing each quadrat can be a useful way of record keeping.

BIOFILE

Biotechnology from plant pigments

Researchers investigate the structure of the different photosynthetic pigments in plants and algae with instruments such as high performance liquid chromatography (see figure below). Better understanding of the structure and function of photosynthetic pigments may lead to biotechnology applications such as silicon-based artificial photosynthesis systems for CO_2 capture, and genetic enhancement of photosynthetic organisms used as food sources and for pharmaceutical production.

Leaf pigment chromatography. Plant physiology researcher extracts a sample of pigments from leaf tissue (green liquid, front right) for analysis in the high-performance liquid chromatography machine in the background. Photographed at the ARS (Agricultural Research Service) Natural Products Utilization Research Unit in Oxford, Mississippi, USA, which conducts research for the US Department of Agriculture.

FIGURE 1.3.6 A botanist using a quadrat frame to monitor a threatened species of aquatic moss. This quadrat frame is a 50 cm square divided into 10 cm squares. The presence or absence of the moss is recorded for each 10 cm square, over a large area marked out by string lines.

Transects

A **transect** is a straight line along which vegetation is sampled. Transects are useful for investigating the distribution of a population of plants, animals or insects across different zones (with different abiotic, or non-living, factors) or in a linear pattern. For example, Figure 1.3.7 shows how a transect could be used to sample and describe the change from eucalypt forest to heathland. A transect running from the sea to the land can also be used to describe the change from seagrass (closest to the sea) to mangrove to saltmarsh communities (inland) associated with the buildup of sediment along a coastline. Physical aspects of the environment, such as soil type and pH, salinity, amount of light, slope angle and height can also be measured along a transect to see if they correlate with changes in biological communities.

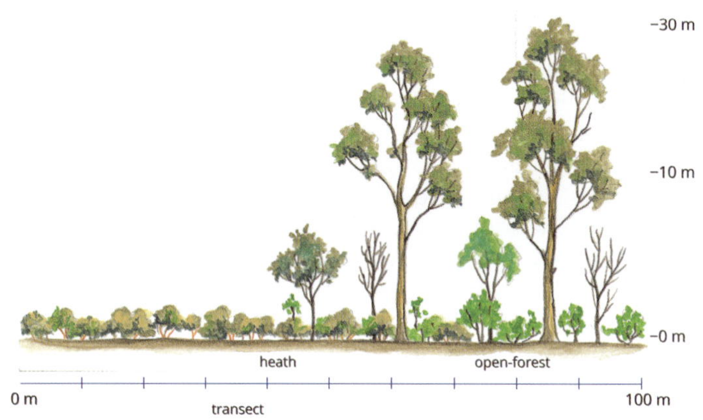

FIGURE 1.3.7 A transect is a straight line along which vegetation is sampled.

Techniques for sampling aquatic habitats

Marine and freshwater habitats can be sampled in many ways. Open water habitats are sampled by pulling a net through the water to collect swimming and floating organisms. Reefs and other underwater habitats are sampled using techniques similar to those used on land, such as quadrats and transects. You are most likely to sample a freshwater habitat during your studies (Figure 1.3.8). Water is collected in a bucket and tipped into a white tray so that free-floating organisms can be easily spotted. Loose rocks are also collected to search for organisms such as stoneflies. Mud might also be collected and sampled to search for organisms such as nematodes and snails. Larger animals such as fish and yabbies can be collected using a net, then quickly identified and returned to the water.

FIGURE 1.3.8 Sampling organisms in a freshwater stream. The researchers have placed a sample of water and rocks in a white tray and are searching for different types of organisms. After sampling, the water and rocks are returned to the stream.

Mark–recapture

In a **mark–recapture** study, animals are captured, marked and then released. When they are recaptured or observed again, their mark is used to identify them (Figure 1.3.9). Mark–recapture is used to determine the total population of a mobile species such as birds or turtles. It can also be used to track the movements of individual animals. However, it is very time-consuming and requires a lot of expertise to be done properly.

FIGURE 1.3.9 Bird banding is a common mark–recapture technique. The leg bands on these pied currawongs (*Strepera graculina*) uniquely identify each individual and include information about where and when the birds were captured.

Table 1.3.3 summarises these common fieldwork techniques.

TABLE 1.3.3 Summary of common fieldwork techniques

Methodology	Method	Uses	Considerations
point sampling	Individual points are chosen on a map and the organisms at those points are counted.	Determining the range of organisms that live in an area and how common they are.	Time efficient, and disturbance to the environment is minimised. Rare organisms may be missed. Sampling size may be small and therefore reduce the accuracy of the data collected.
transect sampling	Lines are drawn across a map. Organisms occurring along the line are sampled.	Determining how the community changes in an area and how common organisms are.	Time efficient, disturbance to the environment is minimised. Rare organisms may be missed. Only suitable for sampling stationary or slow-moving organisms. A large sample size can be obtained for stationary organisms as multiple samples can be taken along the transect line.
quadrat sampling	Sampling squares (quadrats) are placed in a grid pattern on the sample area and the occurrence of organisms in each quadrat is noted.	Determining the range of organisms that live in an area and how common they are. Gives very good data over a large area.	Disturbance to the environment is minimised. Time-consuming to do well. Only suitable for sampling stationary or slow-moving organisms.
mark–recapture	Animals are captured, marked and then released. After a suitable time period, the population is resampled using the same method.	Determining the total population of highly mobile species such as birds or possums. Movements of individuals can be tracked.	Time-consuming to do well. Not suitable for sampling stationary or slow-moving organisms. The marking of animals should not affect their behaviour or movement.

● *You will now be able to answer key questions 3–4 and 7.*

CASE STUDY

Intertidal molluscs

If you wander along rocky seashores when the tide is out, you will see many different species of mollusc (soft-bodied animals which often have shells) on the rocks. Limpets, mussels, periwinkles and tiny blue littorinids are some of the molluscs commonly found along the Victorian coastline. We can learn something about the dependence of the different mollusc species on water, and their ability to withstand drying out when the water recedes to the low tide mark, by observing the pattern of distribution of the molluscs along the shoreline (Figure 1.3.10). Rocky shorelines can be divided into a subtidal zone (always covered with water), an intertidal zone (between low and high tide marks), a spray zone (splashed by water at high tide) and a supratidal zone (out of range of sea water). Few molluscs are found in the subtidal and supratidal zones. In the intertidal region, limpets are scattered randomly, mussels are clumped together, and periwinkles are often found under rocks and in crevices. Littorinids are found mainly in the spray zone.

Some possible explanations for these observations are as follows:

- Each of these molluscs needs some contact with sea water.
- Limpets can tolerate a wide range of environmental conditions.
- Periwinkles cannot withstand long periods out of water.
- Mussels maintain higher moisture levels by clumping together.
- Littorinids need very little water.
- Periwinkles are eaten by predatory birds and only survive where they are hidden from view.

These explanations lead to further questions. How do molluscs protect themselves from drying out when they are out of water? Do different molluscs have different levels of resistance to drying out? Water is not likely to be the only factor affecting distribution. What other factors might be involved? You should think about whether your observations tell you anything about the importance of food sources, predation, temperature variation, the ability of the mollusc to withstand wave action, and how tightly each species of mollusc can hold onto rocks.

FIGURE 1.3.10 (a) Black mussels clump together along rock crevices in the intertidal region of the shoreline. (b) Limpets are distributed randomly over an exposed rock surface in the intertidal zone. (c) In the spray zone, tiny blue littorinids can be seen among the larger cream-coloured barnacles (which are small crustaceans in shells) and striped siphon-limpets.

COLLATING SECONDARY DATA FROM DATABASES

There are a number of open-access databases that provide a large body of information for investigating the living world. These databases are used to record and share information about species and may be created and managed by organisations such as museums and research institutions (Table 1.3.4). The information in the databases may come from specimen collections, field observations or laboratory analyses and include records such as information about the physical characteristics of specimens, photographs, DNA sequences, collection locations and geographic distribution (Figure 1.3.11).

TABLE 1.3.4 Useful databases for investigating the living world

Database	Type of data, information or applications
Encyclopedia of Life Tree of Life Web Project	species information, biodiversity, taxonomy, phylogeny
Museums Victoria	species data, classification, geographic distribution over time, skull image databases, biological data, fossils
American Museum of Natural History Smithsonian Museum of Natural History	research and collections with links to various resources, e.g. palaeobiology, bioinformatics
The Paleobiology Database Fossilworks	databases of fossils, geological distribution, timescales, analysis tools, construct maps

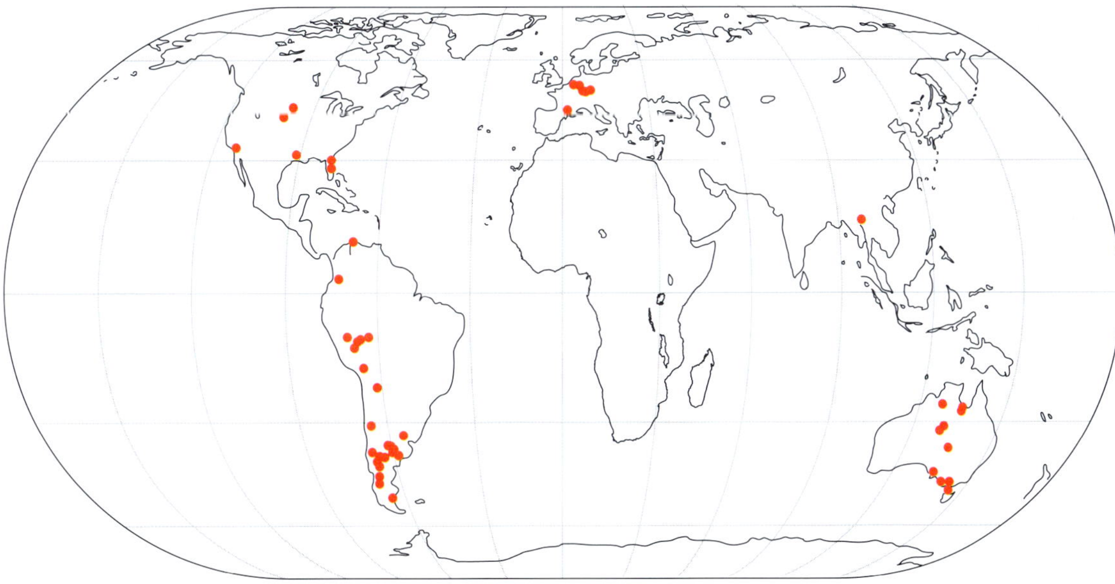

FIGURE 1.3.11 Map showing the distribution of marsupials in the Miocene geological period, constructed using a palaeontology database with search and mapping tools

CASE STUDY **ANALYSIS**

The rise of bioinformatics

Bioinformatics is the use of mathematics, statistics and computer science to analyse and understand biological data. Bioinformatics computer programs can be used to organise raw biological data to visualise patterns, identify genes, model protein structures (Figure 1.3.12), compare DNA sequences (Figure 1.3.13), predict evolutionary relationships and discover and design drugs, along with many other applications.

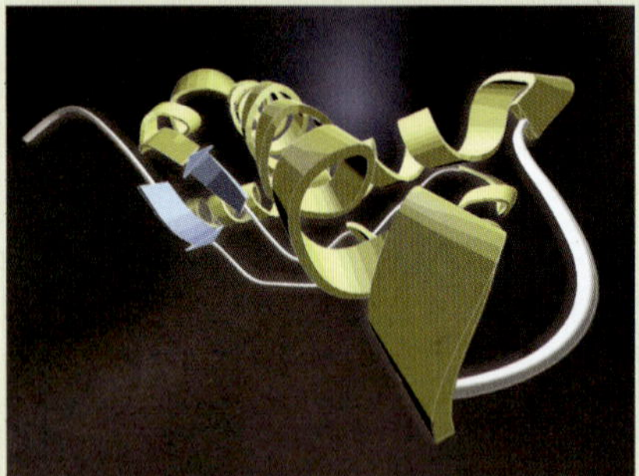

FIGURE 1.3.12 Bioinformatics tools enable you to view the secondary structure of proteins such as this bovine prion protein, which misfolds and forms clumps in the brains of animals with 'mad cow disease'.

Bioinformatics is one of the fastest growing areas of biological science and is now integral to most research and development in biology. The global bioinformatics market is predicted to grow from US$4.1 billion in 2014 to US$18.9 billion by 2026. This rapid growth is primarily driven by the medical biotechnology sector, with demand for more comprehensive and efficient storage and access to personal medical data, and the development of personalised medicine.

One of the best known applications of bioinformatics is whole genome sequencing. The Human Genome Project is still known as the world's largest collaborative biological project. It began in 1990 and by 2003 the three billion nucleotide bases of the human genome had been sequenced. With the technology available at the time, this was an enormous undertaking, costing approximately US$3 billion dollars. With rapid advances in sequencing technology, the output of genome sequencing has skyrocketed, while the cost to sequence a genome has plummeted. It now costs just over $1000 to have your entire genome sequenced, making it affordable for many people.

Although sequencing technology is becoming more accessible, the pool of raw data for analysis is growing, requiring efficient data storage and management systems and the processing power to analyse it. The potential applications of whole genome data are vast but are limited by the bioinformatics tools, computational power and specialised knowledge of bioinformatics currently available to most biologists. As the demand for and capabilities of this technology grows, the scope of biological research is also shifting. Biologists are increasingly required to hone interdisciplinary skills in computer science, mathematics and statistics in order to keep pace with the rapid rise of bioinformatics.

A sample of online bioinformatics resources is listed in Table 1.3.5.

TABLE 1.3.5 Bioinformatics resources

Centres providing bioinformatics databases	Type of data or information; applications
Biology Workbench, San Diego Supercomputer Center	search DNA and protein sequences; sequence alignment, construct evolutionary trees
US National Center for Biotechnology Information (NCBI) (GenBank) European Bioinformatics Institute (EMBL)	nucleotide (gene) sequences; protein sequences and protein structures, chromosome maps, genome maps, SNPs, epigenetics, molecular homology
NCBI–Cn3D OpenScience—Jmol	3D protein structure viewing—free downloads
Sanger Institute	bacterial, protozoan, virus and helminth (worm) genomes

```
B4F917.1   13 SIKLWPPSESTRIMLVDRMTNNLST..ESIFSRK..YRLLGKQEAHENAKTIEELCFALADE.....HFREEPDGDGSSAVQLYAKETSKMMLEVLK 100
A9S1V2.1   23 VFKLWPPSQGTREAVRQKMALKLSS..ACFESQS..FARIELADAQEHARAIEEVAFGAAQE......ADSGGDKTGSAVVMVYAKHASKLMLETLR 109
B9GSN7.1   13 SVKLWPPGQSTRLMLVERMTKNFIT..PSFISRK..YGLLSKEEAEEDAKKIEEVAFAAANQ.....HYEKQPDGDGSSAVQIYAKESSRLMLEVLK 100
Q8H056.1   30 SFSIWPPTQRTRDAVVRRLVDTLGG..DTILCKR..YGAVPAADAEPAARGIEAEAFDAAAA..SGEAAATASVEEGIKALQLYSKEVSRRLLDFVK 120
Q0D4Z3.2   44 SLSIWPPSQRTRDAVVRRLVQTLVA..PSILSQR..YGAVPEAEAGRAAAAVEAEAYAAVTES.SSAAAAPASVEDGIEVLQAYSKEVSRRLLELAK 135
B9MVW8.1   56 SFSIWPPTQRTRDAIISRLIETLST..TSVLSKR..YGTIPKEEASEASRRIEEEAFSGAST.......VASSEKDGLEVLQLYSKEISKRMLETVK 141
Q0IYC5.1   29 SFAVWPPTRRTRDAVVRRLVAVLSGDTTTALRKRYRYGAVPAADAERAARAVEAQAFDAASA....SSSSSSSVEDGIETLQLYSREVSNRLLAFVR 121
A9NW46.1   13 SIKLWPPSESTRLMLVERMTDNLSS..VSFFSRK..YGLLSKEEAAENAKRIEETAFLAAND.....HEAKEPNLDDSSVVQFYAREASKLMLEALK 100
Q9C500.1   57 SLRIWPPTQKTRDAVLNRLIETLST..ESILSKR..YGTLKSDDATTVAKLIEEEAYGVASN.......AVSSDDDGIKILELYSKEISKRMLESVK 142
Q2HRI7.1   25 NYSIWPPKQRTRDAVKNRLIETLST..PSVLTKR..YGTMSADEASAAAIQIEDEAFSVANA.......SSSTSNDNVTILEVYSKEISKRMIETVK 110
Q9M7N3.1   28 SFKIWPPTQRTREAVVRRLVETLTS..QSVLSKR..YGVIPEEDATSAARIIEEEAFSVASV.ASAASTGGRPEDEWIEVLHIYSQEIXQRVVESAK 119
Q9M7N6.1   25 SFSIWPPTQRTRDAVINRLIESLST..PSILSKR..YGTLPQDEASETARLIEEEAFAAAGS.......TASDADDGIEILQVYSKEISKRMIDTVK 110
Q9LE82.1   14 SVKMWPPSKSTRLMLVERMTKNITT..PSIFSRK..YGLLSVEEAEQDAKRIEDLAFATANK.....HFQNEPDGDGTSAVHVYAKESSKLMLDVIK 101
```

FIGURE 1.3.13 Bioinformatics tools allow comparison of many gene or protein sequences at once.

Analysis

1 A student was aiming to compare the genomes of four different organisms to determine their evolutionary relationships. The more closely related two organisms are, the more similar their DNA will be. A small section of the gene for cytochrome c (a protein involved in cellular respiration) is shown below for the organisms (labelled A to D).

Organism A	AGCCTATTTACGCAGTACGTAAACCCTATATACTATGCA	
Organism B	AGCCTATTTACGGACTACGTTAACACTATATACTATGCA	A/B = 4
Organism C	ACCCTATTTACGCAGTACGTAAACACTATATACTATGCA	A/C = 1, B/C = 3
Organism D	ACCCTATTTACGCAGTACGTAAACCCTATATACTATGGA	A/D = 2, B/D = 6, C/D = 2

a Explain how DNA sequences are used to infer evolutionary relationships.

b Compare the four sequences—which two organisms seem to be most closely related? Explain your answer.

2 Explain the reason the medical biotechnology sector is expected to expand significantly.

1.3 Review

SUMMARY

- Biologists use a range of techniques in scientific investigations.
- Fieldwork is commonly used to study the number and distribution of species in a particular area.
- When sampling in the field you should always consider the time and equipment available, the organisms involved and the impact the sampling may have on the environment. Some common sampling techniques are:
  - point sampling
  - quadrat sampling
  - transect sampling
  - mark–recapture.
- Observational studies involve observing the behaviours of organisms.
- Bioinformatics is the use of mathematics, statistics and computer science to analyse and understand biological data.
- Online databases such as the Encyclopedia of Life and Fossilworks are useful for investigations of the living world.

KEY QUESTIONS

Knowledge and understanding

1 Explain why it is important to choose appropriate equipment and instruments to conduct experiments.

2 Convert 2.5 mm (millimetres) into μm (micrometres).

3 Explain the difference between a quadrat and a transect.

4 Why is it important for quadrats to be placed randomly across the area being studied?

Analysis

5 a The total magnification is the product of the ocular lens (eyepiece) and objective lens. Complete the following table:

Ocular lens	Objective lens	Total magnification	Field of view
10×	4×		
	10×	100×	
	40×	400×	
10×	100×		

b Which magnification and field of view would be best for viewing the following?

i a *Paramecium* about 200 μm long

ii yeast cells about 7 μm in size

continued over page

1.3 Review *continued*

6 Which materials or methods from the following list could you use for each of the experiments listed in the table? Copy and complete the table by writing the appropriate letters into the right-hand column. More than one letter may be correct for each experiment.

A biochemical test
B bacterial culture
C glucose test strip
D pH meter, indicator or pH stick
E data logger—temperature probe
F plant tissue culture
G data logger—oxygen probe
H staining and microscopy
I spectrophotometer/colorimeter

Experiment	Materials or methods
measuring oxygen released in photosynthesis	
testing the effectiveness of antibiotics on the rate of bacterial growth	
quantitatively measuring protein concentration in an enzymatic reaction	
identifying phagocytosis in ciliate protozoa	
measuring glucose in an enzyme experiment	

7 Identify a sampling technique that would be suitable for each of the following investigations:

a the changes in a coastal community across sand dunes
b the number of turtles in a pond population
c the number of clover plants in a lawn.

1.4 Data collection and quality

In this section you will learn about data collection, and how to identify and reduce sources of error that can affect data quality. You will learn about generating primary data and how to record both qualitative and quantitative data. You will also learn about the various factors that contribute to data quality, and the importance of controlled experiments in producing valid results.

FIGURE 1.4.1 A student recording the results of a photosynthesis experiment directly into a logbook

KEEPING A LOGBOOK

Throughout Units 1 and 2, and during your investigation for Unit 1 Area of Study 3, you must keep a logbook that includes every detail of your research (Figure 1.4.1). The following checklist will help you remember what to include in your logbook:

- ☐ your ideas when planning the research
- ☐ clear protocols for each stage of your investigation (e.g. what standard procedures you will use and follow exactly each time)
- ☐ instructions noting exactly what needs to be recorded
- ☐ tables ready for data entry
- ☐ records of all materials, methods, experiments and raw data
- ☐ all notes, sketches, photographs and results; these should be recorded directly into your logbook, not on loose paper
- ☐ records of any incidents or errors that may influence the results.

DATA COLLECTION

The measurements or observations that *you* collect during *your* investigation are your **primary data**. Keep in mind there are different types of data that can be collected in a scientific investigation, including **secondary data** (data you have not collected yourself), so when planning your investigation, consider the type of data you will collect and how best to record it. Data can be raw or processed, and qualitative or quantitative.

Primary data is data you collect yourself. Secondary data is data that someone else has collected.

Raw and processed data

The data you record in your logbook is **raw data**. This data often needs to be processed or analysed before it can be presented. **Processed data** is raw data that has been organised, altered or analysed to produce meaningful information. If an error occurs in processing the data, or you decide to present the data in a different format, you will always have the recorded raw data to refer back to.

Raw data that should be recorded includes:

- ☐ tables of results
- ☐ all observations and other notes
- ☐ diagrams and/or photographs of results.

Raw data is the data you record directly into your logbook.

Processed data is data obtained by applying a calculation or formula to raw data.

For example, you might want to study the effect of glucose concentration on cellular respiration in yeast. To do this you might record two sets of raw data: the concentration of glucose added to each yeast culture flask and the amount of carbon dioxide produced by each culture (Figure 1.4.2).

Table 1 Carbon dioxide released by yeast cells in different glucose concentrations

Culture flask	Glucose concentration (g/L)	Amount of CO_2 released (ppm)	Amount of CO_2 released (ppm per 10^6 cells)
1	0.0	5	0.5
2	1.0	50	5.0
3	5.0	210	21.0
4	10.0	250	25.0

FIGURE 1.4.2 An example of a table that you might include in your logbook for primary data collection. Data tables should have a title and headings for each column and row, including units as required.

You can then process this data further. For example, to compare across different experiments in which the cell number may vary, it is useful to perform a cell count and express the result per cell (or per million cells). This value, shown in the last column in Figure 1.4.2, is processed data.

● *You will now be able to answer key question 3.*

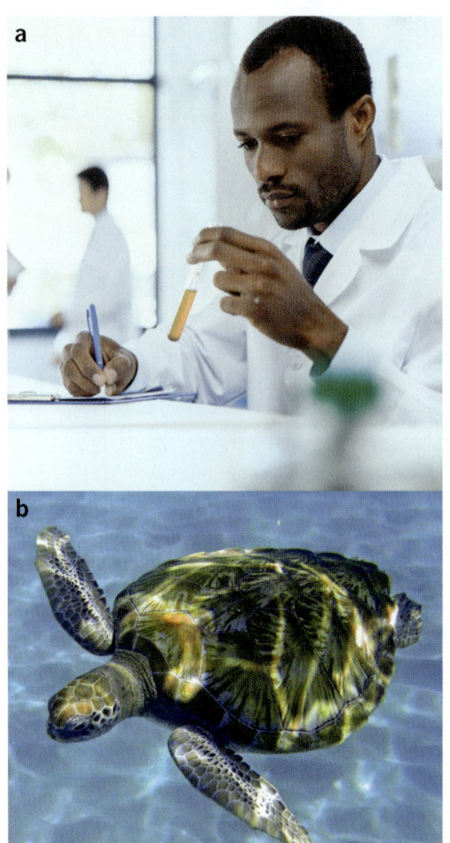

FIGURE 1.4.3 (a) Colour-based biochemical reactions and (b) structural features are examples of qualitative data. For example, different patterns on the top part of a turtle's shell (or carapace) can help distinguish one species from another. Reference keys and classification criteria are used to maintain consistency in recording qualitative data.

Qualitative data

Data collected about categorical variables is known as **qualitative data**. Categorical variables can be counted but not measured, and relate to a type or category, such as colour, or states such as on/off or wet/dry.

Recording qualitative data

Qualitative data can be represented as names, symbols or numbers. Observations of categorical variables can be descriptions or images. For example, dog breeds can be shown in a diagram, and textures of materials can be described using words such as brittle, coarse, crumbly, dense, flexible, rocky, rough, silky, slimy, smooth, spongy or velvety.

When you have to record qualitative data, think carefully about how each categorical variable will be defined. Creating a referencing system, such as assigning codes to different colours, allows you to quickly and easily record your data. For example, a photograph of reference colours with a scale (such as +++, ++, + for the colour reactions in Figure 1.4.3a) is a good way of maintaining consistency across experiments.

If you are recording details of structural features, such as when comparing variations in the patterns on turtle shells (Figure 1.4.3b), make a key with diagrams to define your criteria for recording each feature. Samples may have both qualitative data, such as a particular pattern on the top of the shell, and quantitative data, such as the number of clearly defined sections on the shell.

Quantitative data

Data collected about numeric variables is **quantitative data**. Like categorical variables, numeric variables can be counted. Unlike categorical variables, numeric variables can also be measured, because they have a measurable quantity, such as length, mass or time. Numeric variables can be discrete or continuous:

- **Discrete variables** are values that can be counted or measured, but which can only have certain values. Examples are number of chromosomes in a karyotype, number of white blood cells on a slide, or number of times a lever is pressed.
- **Continuous variables** may be any number value within a given range that can be measured. Examples are age, temperature, length, mass and wavelength.

Recording quantitative data

When you record quantitative data, remember to use SI (International System of Units) units, such as grams, centimetres or millimetres.

Sometimes qualitative data can become quantitative if accurate and consistent measurement is applied. For example, biochemical reactions based on a colour change, as in Figure 1.4.3a, can be prepared with known concentrations and a detailed grading system used to indicate colour intensity (such as ++, +, –; this might be considered semi-quantitative). If a colorimeter or spectrophotometer is available to read absorbance values, then you can obtain quantitative data. A calibration curve or standard curve can then be prepared for reading the experimental values. You will learn about standard curves in Section 1.5.

Figure 1.4.4 summarises the different types of data and their variables.

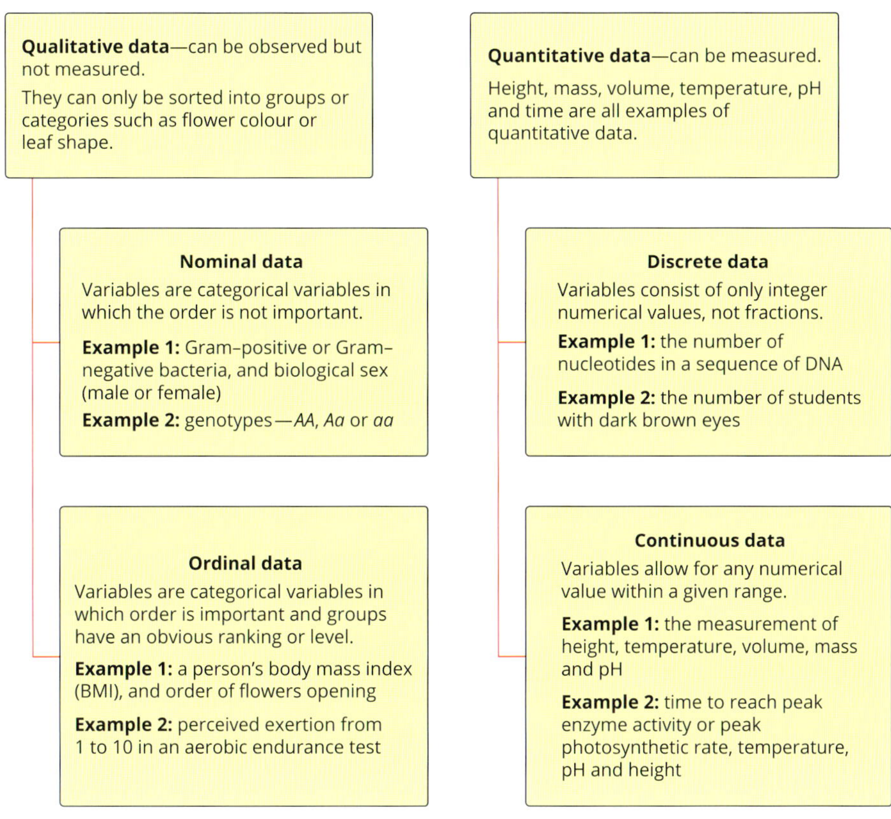

FIGURE 1.4.4 Qualitative and quantitative variables

IDENTIFYING AND REDUCING ERRORS

When an instrument is used to measure a physical quantity and obtain a numerical value, the aim is to determine the true value. The **true value** is the value, or range of values, that would be obtained if the variable could be measured perfectly. However, for a number of reasons the measured value is often not the true value. The difference between the true value and the measured value is called the error. This error in the measured value is the result of errors in the experiment. **Personal errors** are often mistakes or miscalculations. If you have made a personal error the data from this trial should be ignored and the trial should be repeated. The two types of experimental errors are systematic errors and random errors.

Systematic errors

A **systematic error** (or bias) is a consistent error that occurs every time you take a measurement and affects the accuracy of a measurement. Systematic errors are not easy to spot, because they do not appear as a single difference in the data set. Instead, repeated measurements give results that differ by the same amount from the true value. There are many different types of systematic errors, but the most common types are selection bias and measurement bias.

> A systematic error is an error that affects every result in the data set by the same amount. An example is a temperature probe that measures 0.2°C higher than the actual value.

Selection bias

Selection bias occurs when your sample is not representative of the population being studied. This can have a number of different causes, including sampling bias, which is when your sample has not been selected randomly, and time-interval bias, which is when you stop your study too early because the results support your hypothesis.

Measurement bias

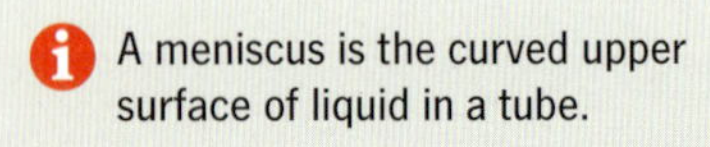

Measurement bias is usually a result of instruments that are faulty or not calibrated, or the incorrect use of instruments, which produces inaccurate results. For example, if a scale under-reads by 1%, a measurement of 99 mm will actually be 100 mm. Another example would be if you repeatedly used a piece of equipment incorrectly throughout your investigation, such as reading from the top of the **meniscus** instead of the bottom when using a measuring cylinder or graduated pipette (Figure 1.4.5).

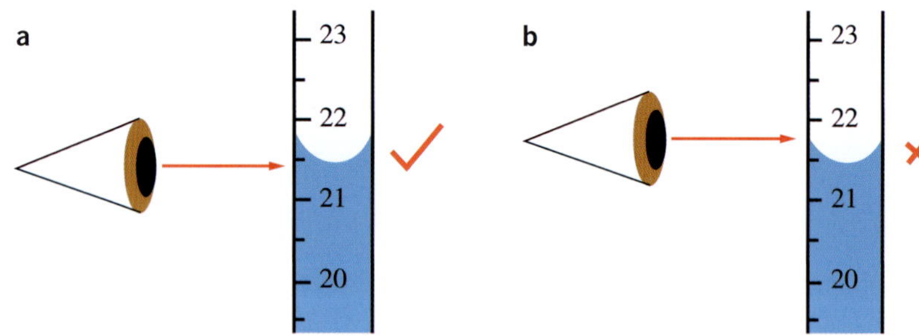

FIGURE 1.4.5 When measuring liquid levels in cylinders and pipettes, measure the value (a) at the bottom of the meniscus of the liquid, not (b) at the top.

Reducing systematic errors

The appropriate selection and correct use of calibrated equipment will help you reduce systematic errors. Because systematic errors are difficult to identify, it is also a good idea (if you have time) to repeat your measurements using different equipment.

Appropriate equipment

Use the equipment best suited to the data you need to collect. Determining the units and scale of the data you are collecting will help you to select the correct equipment. For example, if you need to measure 10 mL of a liquid, using a 10 mL graduated pipette or a 20 mL measuring cylinder will give more accurate readings than using a 200 mL measuring cylinder, because the pipette or 20 mL cylinder will have a finer scale.

Calibrated equipment

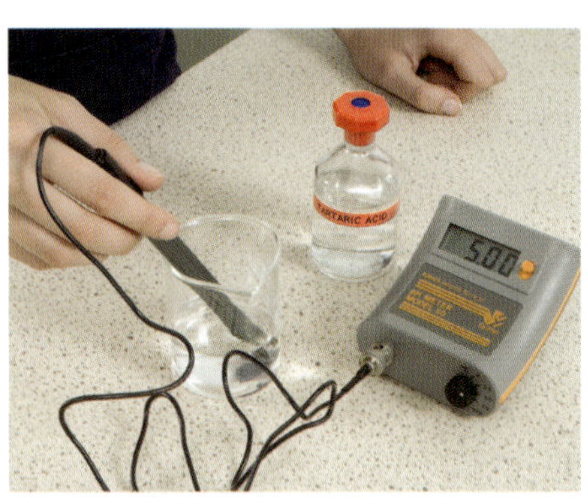

FIGURE 1.4.6 Measuring the pH level of tartaric acid with a pH meter. To ensure an accurate reading, the student would first have calibrated the meter using standard solutions of known pH.

Accurate measurement requires properly calibrated equipment. Before you conduct your investigation, make sure your instruments or measuring devices are properly calibrated and functioning correctly (Figure 1.4.6). Your school laboratory may have a set of standard masses that can be used to calibrate a balance or scale. A pH meter should have a set of standard pH solutions (e.g. at pH 4, pH 7 and pH 9) to check the meter readings and adjust the meter if necessary.

Random errors

Random errors (also called variability) are unpredictable variations that can occur with each measurement. Random errors affect precision and can occur because instruments are affected by small variations in their surroundings, such as changes in temperature. All instruments have a limited precision, so the results they produce will always fall within a range of values.

Reducing random errors

To reduce random errors you need to take more measurements or increase your sample size. You can then calculate the average (the mean), which should be close to the true value.

More measurements

The impact of random errors can be minimised by taking more measurements and then calculating the average value. In general, more measurements will improve the accuracy of the processed data (calculated values). The minimum number of measurements you should make is three. If one reading differs greatly from the rest, mention this in your results and discuss possible reasons for the difference.

Sample size

Increasing the sample size reduces the effect of random errors, which in turn makes your data more reliable. For example, if you are conducting an investigation into the effects of light intensity on the rate of photosynthesis in *Elodea*, do not test your hypothesis on just one stem. Test several stems (minimum three). If two stems photosynthesise and one does not, it is reasonable to conclude that one stem was unhealthy or the conditions incorrect. **Provisional data** is data that is subject to revision. If there are significant errors present, or results that you identify as outliers, you may wish to conduct another measurement under the same conditions. Using a large number of samples will reduce the likelihood of your results being skewed.

- *You will now be able to answer key questions 6–7.*

DATA QUALITY

The results of your data analysis will only be as good as the quality of the data. A well-designed scientific investigation should produce accurate, precise, reliable and valid results. You should consider all of these factors when collecting primary data in your investigations, and also when you evaluate the quality of secondary data from other sources. Being able to discuss systematic and random errors, and their effect on accuracy and precision, strengthens your written evaluation once your results have been obtained and analysed.

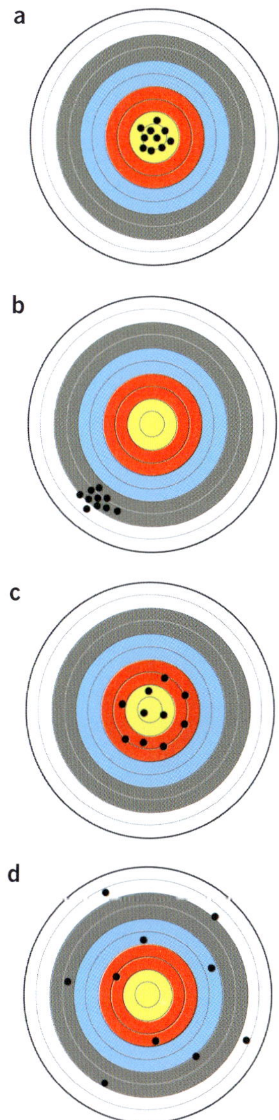

FIGURE 1.4.7 Examples of accuracy and precision: (a) both accurate and precise, (b) precise but not accurate, (c) accurate but not precise, and (d) neither accurate nor precise

Accuracy and precision

In science and statistics the terms 'accuracy' and 'precision' have very specific and different meanings:

- **Accuracy** is the ability to obtain the true value of the variable being measured. To obtain accurate results, you must minimise systematic errors.
- **Precision** is how closely a set of measurements agree with each other. Precision is different to accuracy in that it does not indicate how close the measurements are to the true value. To obtain precise results, you must minimise random errors.

To understand more clearly the difference between accuracy and precision, think about firing arrows at an archery target (Figure 1.4.7). Accuracy is being able to hit the bullseye, whereas precision is being able to hit the same spot every time you shoot. If you hit the bullseye every time you shoot, you are both accurate and precise (Figure 1.4.7a). If you hit the same area of the target every time but not the bullseye, you are precise but not accurate (Figure 1.4.7b). If you hit the area around the bullseye each time but don't always hit the bullseye, you are accurate but not precise (Figure 1.4.7c). If you hit a different part of the target every time you shoot, you are neither accurate nor precise (Figure 1.4.7d).

- *You will now be able to answer key questions 1–2 and 4*

Recording numerical data

When using measuring instruments, the number of significant figures (or digits) and decimal places you use is determined by how precise your measurements are.

This depends on the scale, accuracy and precision of the instrument and technique you are using (Figure 1.4.8). For example, a beaker is used to measure volumes approximately and has limited accuracy, for example ±5%, (meaning the value could actually be above or below the value shown by 5%). A graduated pipette is more accurate, with accuracies of ±0.1% or ±0.2%. Your pipette may be accurate but if your technique using the pipette is variable, the overall accuracy and precision will be limited.

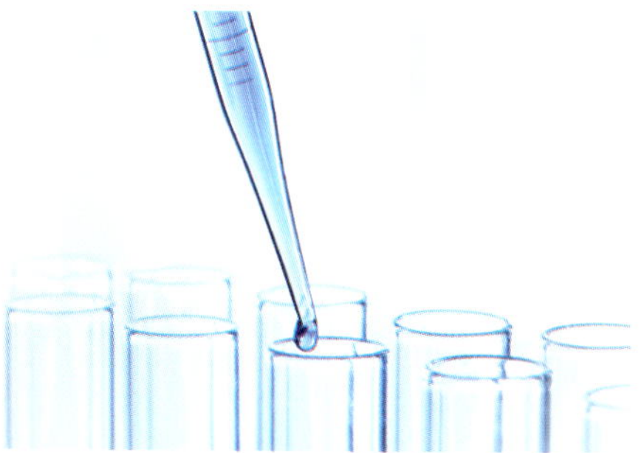

FIGURE 1.4.8 A 5 mL graduated pipette can measure volumes to an accuracy of one hundredth of a millilitre, or 5 mL ± 0.01 mL. The pipette has major divisions of 1 mL and minor divisions of 0.1 mL. You can estimate to 0.01 mL and record volumes to 2 decimal places, for example 3.80 mL or 4.52 mL.

When you record raw data and report processed data, use the number of significant figures available from your equipment or observation. Using either a greater or smaller number of significant figures can be misleading. For example, Table 1.4.1 shows measurements of five tissue samples taken using an electronic balance accurate to two decimal places. The data was entered into a spreadsheet to calculate the mean, which was presented with four decimal places. You would record the mean as 20.83 g, not 20.8260 g, because two decimal places is the precision limit of the instrument. Recording 20.8260 g would be an example of false precision.

TABLE 1.4.1 An example of false precision in a data calculation

Sample	1	2	3	4	5	Mean
Mass (g)	20.13	20.62	21.22	20.99	21.17	20.8260

Repeatability

Repeatability (sometimes called reliability) is the ability to obtain the same results if an experiment is repeated under the exact same set of experimental conditions (Figure 1.4.9). Because a single measurement or experimental result could be affected by errors, **replication** of samples within an experiment and **repeat trials** are key components of repeatability. To improve repeatability you should:

- specify the materials and methods in detail
- include replicate (several) samples within each experiment
- take repeat readings of each sample
- run the experiment or trial more than once.

FIGURE 1.4.9 If you can reproduce your results, they are reliable.

Conducting the experiment more than once allows the researcher to determine if their results are reproducible. **Reproducibility** is the ability to obtain the same results if an experiment is repeated under different conditions. Different conditions might include a different researcher conducting the experiment, the use of different equipment or instruments, or conducting the experiment at a different time or location.

Validity

Validity refers to whether your results measure what the investigation set out to measure. Results are invalid, for example, if you think you have measured a variable but have actually measured something else. Factors influencing validity include:

- whether your experiment measures what it claims to measure. In other words, your experiment should test your hypothesis.
- the certainty that something observed in your experiment was the result of your experimental conditions and not some other cause that you did not consider. In other words, whether the independent variable influenced the dependent variable in the way you have concluded.
- the degree to which your findings can be generalised to the wider population from which your sample is taken, or to a different population, place or time.

Controls

To ensure your results are valid, carefully determine:

- the independent variable (the variable that you will change) and how you will change it
- the dependent variable (the variable that you will measure)
- the controlled variables (the variables that must remain constant) and how you will maintain them (see Section 1.2).

Your experiment should be designed so that only one independent variable is changed at a time. The remaining variables must be kept constant (or controlled) so that meaningful conclusions can be drawn in turn about the effect of each independent variable on the dependent variable you are measuring.

A control group is a comparison group. To have a control group you need to set up two groups side by side within an experiment. Both groups are the same, except for the variable you are testing. This is the independent variable in the hypothesis, and is applied to your experimental group but not your control group. All the other variables have to stay the same. You do not want them to change, as these may affect the result of your experiment. For example, when testing a new medication (the variable being tested—the independent variable), two groups of patients are involved. The control group of patients is given a **placebo** (a blank capsule). The other group is given the actual medication, and the data collected from this group is compared to the data from the control group to see the effects of the medication.

A placebo simulates the same treatment but contains no active ingredient. In this way, it acts as a control group.

Randomisation

Random selection of your sample reduces selection bias and improves validity. Selection bias occurs when your sample doesn't reflect the wider population you wish to generalise your results to. For example, if you were scoring phenotypes in large trials of genetically selected or genetically modified crop plants, choosing locations at random throughout the plot, rather than choosing locations only at the edge of a plot, would reduce selection bias and improve the validity of the investigation.

- *You will now be able to answer key question 5.*

USING AND EVALUATING SECONDARY DATA

In researching your topic for various investigations (e.g. case study, correlational study, controlled experiment, literature review, modelling) you will find a range of sources of information. Not all will have reliable information suitable for a scientific investigation. Determine whether the information source is reputable, such as a university, research and education organisation, or peer-reviewed journal. Other sources of information, such as interest groups or companies, may have a specific bias (Table 1.4.2 on page 40). Current secondary school, university and specialist area textbooks are good places to start. The best source of experimental data and up-to-date information comes from peer-reviewed scientific journals that have been recently published.

Peer-reviewed means that other scientists have checked the information and have agreed that it is appropriate for publication.

As many peer-reviewed journals require a subscription, you may not have access to the original articles in full, but you can probably find the abstract (a summary of the study). Also, these original articles can be very complex and hard to interpret if you are not an expert in the field, so an alternative way to access current information about a topic is through print and online science magazines, such as *New Scientist* and *The Scientist*. Good science magazines and journalists provide the background, the results and the relevance of the study in a way that non-experts can understand. However, the methods will be in the original peer-reviewed report.

Your investigation may use scientific data such as protein structures, DNA sequences, or fossil and biogeographic data from open-access databases. Most of these databases are linked to large research centres and are usually reliable sources of information.

DATA QUALITY SUMMARY

Table 1.4.2 summarises factors to consider when evaluating and using primary and secondary data. Make sure you consider all the factors that might affect the quality of the data when you are doing your research and when you write a report of your investigation.

TABLE 1.4.2 Summary of factors impacting quality of primary and secondary data

	Primary data	Secondary data
Accuracy	• Use appropriate and calibrated instruments. • Address systematic errors.	• Use reputable sources such as peer-reviewed journals and books. • Check that systematic errors have been addressed.
Precision	• Use an appropriate number of significant figures. • Address random errors.	• Check that random errors were addressed. • Check that any data analysis was appropriate.
Repeatability	• Specify the materials and methods in detail. • Use replicates within the experiment. • Perform repeat readings. • Repeat your experiment.	• Check that the experimental method was relevant. • Check if the results were analysed and if they were found to be statistically significant. • Check that information is consistent with other reputable sources.
Validity	• Ensure your experiment tests your hypothesis. • Randomise your sample and use one or more controls.	• Check the study and information is current. • Check the information is based on scientific investigation, controlled trials or research. • Determine if the source is unbiased, or from a particular interest group, e.g. pharmaceutical company, religious group. • Check that the results relate to the hypothesis and aims.

You will now be able to answer key question 8.

1.4 Review

SUMMARY

- Record all information objectively in your logbook, including your data and method of investigation.
- Raw data is the data you collect in your logbook.
- Processed data is raw data that has been mathematically manipulated.
- Beware of potential errors when conducting an investigation:
  - Systematic errors are consistent errors that reduce accuracy.
  - Random errors are unpredictable errors that reduce precision.
- Reduce systematic errors by:
  - selecting appropriate equipment
  - properly calibrating equipment
  - repeating experiments.
- Reduce random errors by:
  - having a large sample size
  - repeating measurements.
- Accuracy is the ability to obtain the true value of the variable being measured.
- Precision is how closely a set of measurements agree with each other.
- Repeatability is the ability to reproduce your results under the exact same set of experimental conditions.
- Reproducibility is the ability to obtain the same results if an experiment is repeated under different conditions.
- Validity refers to whether your results measure what the investigation set out to measure.
- Controlled experiments are important for obtaining valid results.

KEY QUESTIONS

Knowledge and understanding

1 A journal article reported the materials and methods used to conduct an experiment. The experiment was repeated three times, and all values were reported in the results section of the article. Repeating an experiment and reporting results supports:
   A precision
   B reliability
   C accuracy
   D systematic errors

2 Explain the difference between accuracy and precision.

3 Explain the difference between raw and processed data.

4 What type of error is associated with the following?
   a inaccurate measurements
   b imprecise measurements

Analysis

5 A student investigated the effect of changing temperature on the rate of transpiration (water uptake) in a *Eucalyptus* species. The student measured the movement of water through a graduated pipette over the course of ten minutes. Describe other factors that would need to be controlled in this experiment.

6 Identify whether each of the following errors is a personal error, a systematic error or a random error.
   a A pipette that should have dispensed volumes of 25.00 ± 0.03 mL actually dispensed volumes of 24.92 ± 0.03 mL.
   b A student miscounts the number of cells across the field of view of a microscope.
   c A sample of glucose powder was weighed three times, and on the third weigh there was a fluctuation in power supply, giving an unexpected value.

7 Identify the type of error that is described in each scenario below, and how it could be avoided.
   a The electronic scale was not zeroed/tared before samples were weighed.
   b 1.0 mol/L hydrochloric acid was used rather than 0.1 mol/L hydrochloric acid.
   c One result was significantly lower than the rest.
   d A student incorrectly reads the measuring cylinder from the top of the meniscus rather than from the bottom.

8 Consider the following experiment.

Hypothesis
If seedlings are watered with mineral water, then they will grow more leaves than seedlings watered with tap water.

Experiment
Set up two identical trays of 20 seedlings. They should have the same type of plant, age of plant, type of potting mix, drainage and amount of sunlight and water. Everything should be the same except the type of water given to the plants.

Variables
Anything that could be different in the experiments must be kept the same. This includes everything listed above and even the height of the plants, the depth of potting mix and the intensity of the sunlight. These variables are kept the same—they are the controlled variables.

Only one variable is changed—the type of water. It is the effect of this variable that we are measuring. It is the independent variable. Its measurement should be objective (be able to be measured quantitatively).

The independent variable—the type of water—may change the number of leaves. The number of leaves is the dependent variable. The number of leaves depends on the type of water.

Results
Measure or count the number of leaves on each plant. This will give you objective results. Your friends could replicate the experiment at their houses. When you and your peers have repeated the experiment many times on different plants, the results can become a generalisation.

   a What is the sample size?
   b Identify the controlled variables.
   c Identify the independent and dependent variables.
   d Will the results be objective or subjective? Explain your answer.
   e Will the results be valid for all species of plants? Explain.

1.5 Data analysis and presentation

In Section 1.4 you learnt about different types of data and the factors that affect data quality. In this section you will learn about different descriptive statistics you can use to analyse your data, as well as how you can present your data in tables and graphs.

PREPARING DIAGRAMS

It is important to learn how to draw and label diagrams of equipment and biological specimens in your studies of biology. There are certain rules you should follow in order to produce a diagram that will be acceptable in your reports and exams.

When drawing scientific equipment, diagrams should:

- ☐ be large, simple, two-dimensional pencil drawings
- ☐ have ruled lines where possible
- ☐ keep proportions realistic (Figure 1.5.1).

When drawing biological specimens follow these guidelines:

- ☐ Draw the whole diagram (including labels, lines, magnification, heading and scale if possible) in pencil.
- ☐ A diagram of microscopic objects does not require a circle representing the field of view.
- ☐ Draw one or a few cells to represent a sample; there is no need to draw every cell in a field of view.
- ☐ Draw your diagram with simple and clear lines (do not sketch).
- ☐ Use stippling (small dots) rather than shading to indicate depth.
- ☐ Make your diagram as large as possible (at least 10 × 10 cm).
- ☐ Draw only the structures that you see, not things you think you should see, such as mitochondria (Figure 1.5.2).
- ☐ If there are many features to show, it is useful to pair a photo with a detailed supporting diagram that shows cellular detail (Figure 1.5.3).
- ☐ Include clear labels for the features you want to highlight.
- ☐ Place labels outside the drawing.
- ☐ Make sure label pointers do not cross over each other.
- ☐ Line up labels on either side of the diagram where possible.
- ☐ Use straight lines without arrowheads that meet the features being labelled.
- ☐ Include a scale bar or scale (e.g. 1:100) in the diagram, or state the magnification (e.g. 400×) in the caption.

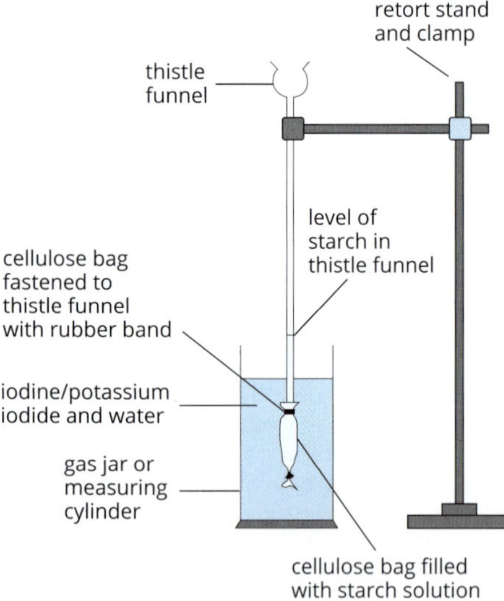

FIGURE 1.5.1 Diagram showing a dialysis tubing arrangement. Note the straight lines for labels that are horizontal where possible, and the realistic proportions of different parts in relation to each other.

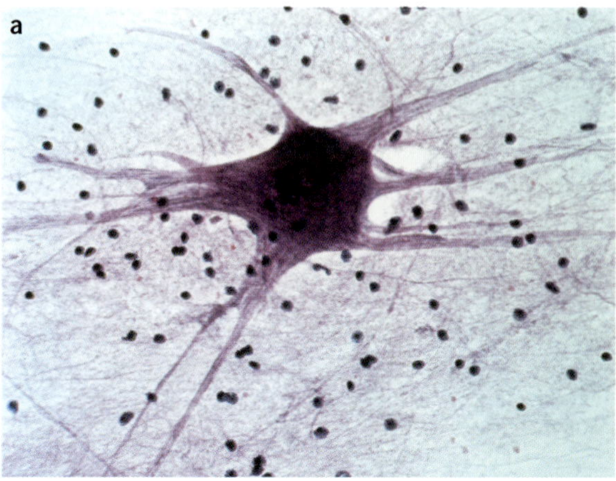

FIGURE 1.5.2 (a) A photomicrograph and (b) a scientific diagram showing a motor neuron in a spinal cord smear

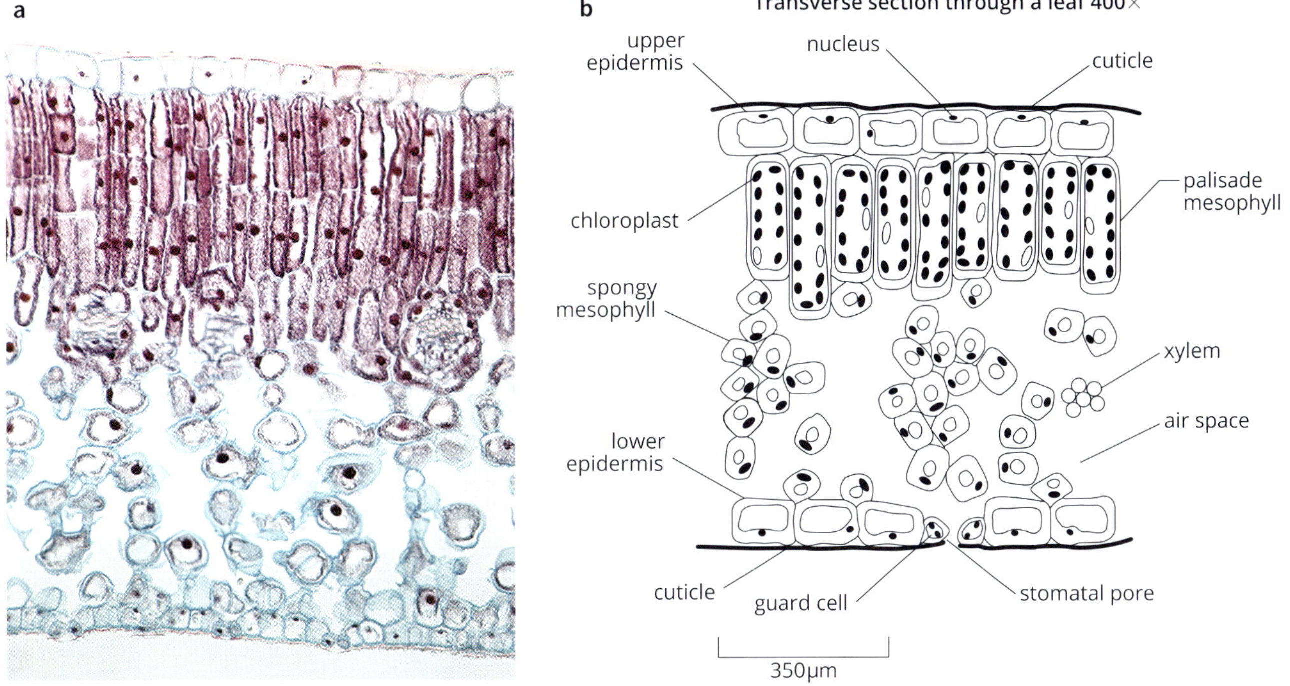

FIGURE 1.5.3 (a) A photomicrograph and (b) a diagram of a transverse section through a leaf

DESCRIPTIVE STATISTICS

Descriptive statistics can be used to analyse both quantitative and qualitative data. An important type of descriptive statistic is the measure of central tendency. It is good practice to use a measure of central tendency to provide a clearer understanding of the data.

Measures of central tendency

Measures of central tendency are single values that allow you to describe the central position in a set of data. Measures of central tendency are sometimes also called measures of central location. The mean, median and mode are all measures of central tendency.

Consider the following data set: 3, 5, 7, 8, 8, 8, 10.

- The **mean** (or average) is the sum of the values divided by the number of values, which in this case is $\frac{(3 + 5 + 7 + 8 + 8 + 8 + 10)}{7} = 7$.
- The **median** is the 'middle' value in an ordered list of values, which in this case is the fourth value, which is 8.
- The **mode** is the value that occurs most often in a list of values, which in this case is 8. This measure is particularly useful for describing qualitative or discrete data.

The mean, median and mode are all measures of central tendency. The mean can be used for nominal, ordinal, and discrete or continuous data so it is common to see the mean used in a variety of investigations.

The appropriate measure of central tendency to use depends on the type of data you are working with (Table 1.5.1).

TABLE 1.5.1 When to use the different measures of central tendency

Type of data	Mode	Median	Mean
nominal (qualitative)	✔	✘	✘
ordinal (qualitative)	✔	✔	maybe
discrete or continuous (quantitative)	✔	✔	✔

Percentage change

Calculating the change in a variable is a helpful statistic because it provides a general trend or pattern, rather than listing a specific value, which will vary depending on the sample being studied. Percentage change applies to increases and decreases relative to the control or the starting point of the measurement.

For example, Table 1.5.2 shows data collected in an experiment that investigated the osmotic strength of different solutions. Four sets of dialysis tubing (a semi-permeable membrane), each containing a different solution, were suspended in a beaker of physiological saline solution. The mass was measured at the start and after 24 hours.

TABLE 1.5.2 Percentage change in mass of dialysis tubing over 24 h

	Mass (g) at 0 h	Mass (g) at 24 h	% change
sample 1	20.55	20.89	1.65
sample 2	20.01	21.94	9.64
sample 3	21.25	22.09	3.95
sample 4	20.55	20.32	-1.12

The percentage change in mass is calculated using the equation:

$$\text{percentage mass change} = \frac{\text{final mass} - \text{original mass}}{\text{original mass}} \times 100$$

Calculating percentage change accounts for variation and/or errors in the replicates within your experiment, or for the same experiment repeated by others. In Table 1.5.2, the starting mass is not identical in each sample, perhaps due to errors in measuring the volume put into the tubing. Although the final mass for sample 3 is the greatest, the percentage change is less than for sample 2 because the original mass was higher. Calculating percentage mass change shows that sample 2 has the greatest osmotic effect.

Percentage difference

The percentage difference (also often expressed as a fraction) is a measure of the precision of two measurements. It is calculated by working out the difference between the two measurements and dividing by the average of the two measurements:

$$\text{percentage difference} = \frac{\text{measurement 1} - \text{measurement 2}}{\text{average of measurements}} \times 100$$

For example, if your two measurements were 25 cm and 24 cm, you would calculate percentage difference as follows:

$$\text{percentage difference} = \frac{(25-24)}{\frac{(25+24)}{2}} \times 100 = \frac{1}{24.5} = 0.041 \times 100 = 4.1\%$$

Range

The **range** is simply the difference between the highest and lowest values in a data set. Table 1.5.3 shows the measurements taken for five different plants after treatment with a plant hormone.

The range of a set of values can be found by subtracting the lowest value in the data set from the highest value.

TABLE 1.5.3 Plant height in a hormone treatment experiment

Plant	1	2	3	4	5	Mean	Range
hormone-treated plants (mm)	158	378	320	377	363	319.2	378 – 158 = 220
untreated control plants (mm)	140	135	170	171	193	161.8	193 – 135 = 58

To determine the range for values in Table 1.5.3 you would subtract the smallest value from the largest value. Notice how an abnormally large or abnormally small value in the data set makes the variability appear high. If one value appears way out of range, such as plant 1 in the hormone-treated group, it is considered an **outlier** and can be deleted from the calculations. The range for the hormone-treated plants would then be 378 – 320 = 58. This illustrates the importance of having a sample size that is large enough to limit the impact of anomalies in the data set.

Uncertainty in measurement

When averaging repeat measurements, the **uncertainty** should be reported alongside your average. Uncertainty results from errors and represents a realistic range within which the true value is likely to be. A simple way to calculate the uncertainty is the range divided by 2:

$$\text{uncertainty} = \pm\frac{(\text{maximum value} - \text{minimum value})}{2}$$

For example, if an experiment were conducted to measure the length of time it takes to convert a substrate to a product in an enzymatic reaction, and three replications of the experiment produced the times 2.50, 3.47 and 2.81 seconds, the average time taken would be 2.93 seconds. The uncertainty would be calculated as follows:

$$\text{uncertainty} = \pm\frac{(3.47 - 2.50)}{2} = \pm 0.48$$

The result showing the mean and uncertainty is expressed as mean = 2.93 ± 0.48 seconds.

For the data set in Table 1.5.3, in which the range was calculated, the uncertainties are as follows:

- control plants 161.8 ± 29.0 mm
- hormone-treated plants 359.5 ± 29.0 mm (with the outlier removed).

The higher the uncertainty, the less reliable your data may be (see Section 1.4).

You will now be able to answer key questions 2 and 4.

PRESENTING DATA

When you have completed your experiment, you will need to organise and present the data. This makes it much easier to identify trends or patterns in the data. It also helps to identify any relationships that result from cause and effect between the independent and dependent variables, and helps you see if one variable has had any effect on another variable.

Data can be presented in different ways, including tables, graphs, flow charts and diagrams. The best way of visualising your data depends on its nature. Try several formats before you make a final decision to create the best possible presentation.

Presenting data in tables

Tables record number values and allow you to organise your data.

Presenting raw data in tables

Tables organise data into rows and columns, and vary in complexity according to the nature of your data. Tables can be used to organise raw data and processed data, or to summarise results.

The simplest form of a table is a two-column chart. The first column should contain the independent variable (the one you manipulate) and the second column should contain the dependent variable (the one that may change in response to a change in the independent variable).

As you can see in Figure 1.5.4, tables should have the following features:

- a descriptive title
- column headings (including the units)
- aligned figures (align the decimal points)
- the independent variable placed in the left column
- the dependent variable/s placed in the right column/s.

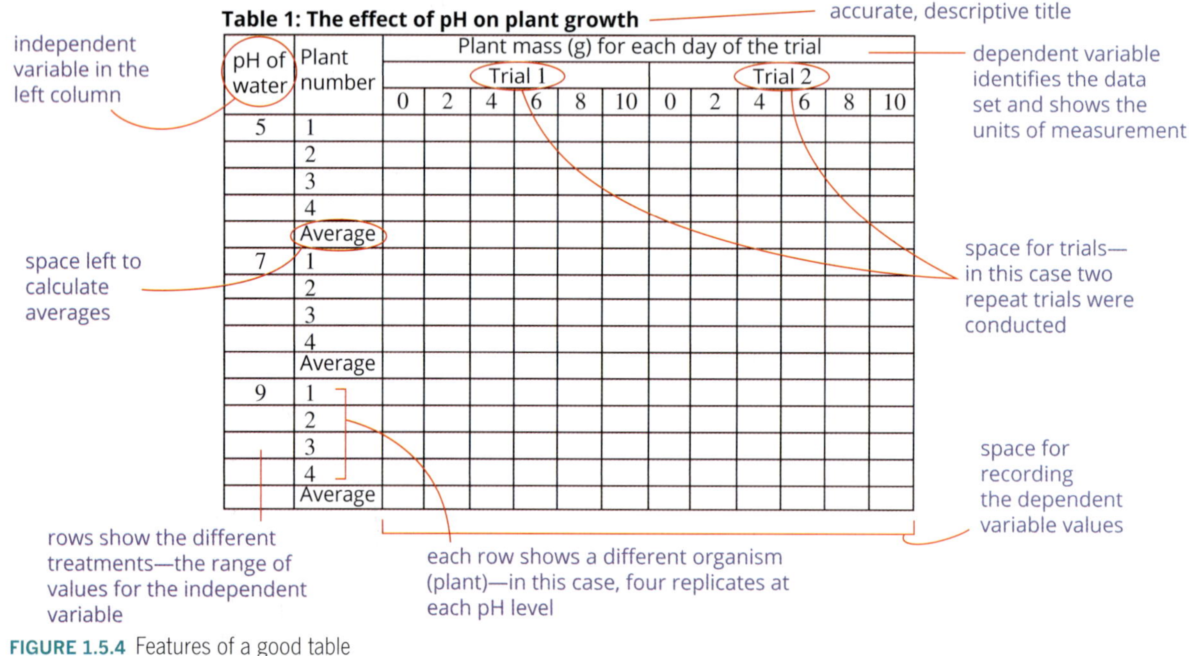

FIGURE 1.5.4 Features of a good table

You should tailor the layout of your data table to suit your experiment. Table 1.5.4 is an example of a raw data table. It contains data from an experiment on the effect of temperature on the activity of enzyme X. A reaction between the enzyme and substrate was conducted for 10 minutes and the reaction product was measured. Three trials were performed.

TABLE 1.5.4 Raw data table for the effect of temperature on reaction rate of enzyme X; measurement of reaction product

	Product released (μg)		
Temperature (°C)	**Trial 1**	**Trial 2**	**Trial 3**
10	100	120	120
20	850	790	820
40	1350	1420	1390
60	1250	1210	1150
80	200	220	230

Presenting processed data in tables

Table 1.5.5 also contains data on the relationship between temperature and mean enzyme reaction rate. It presents the data in a processed format; that is, the replicate values from Table 1.5.4 have been averaged to calculate the mean. The mean reaction rate per minute was also calculated using the equation $\frac{\text{mean}}{10\text{ min}}$. The mean of the reaction rate per minute and its uncertainty are listed in Table 1.5.6.

TABLE 1.5.5 Processed data table for the effect of temperature on enzyme X reaction rate; calculation of the mean product (μg) and rate (μg/min)

Temperature (°C)	Mean (μg)	Mean rate (μg/min)
10	113.3	11.3
20	820.0	82.0
40	1386.7	138.7
60	1203.3	120.3
80	216.7	21.7

TABLE 1.5.6 Processed data table for the effect of temperature on enzyme X reaction rate; calculation of mean and uncertainty

Temperature (°C)	Mean rate (μg/min)	Uncertainty
10	11.3	± 1.0
20	82.0	± 3.0
40	138.7	± 3.5
60	120.3	± 5.0
80	21.7	± 1.5

Presenting data in graphs

In general, tables provide more detailed data than graphs, but it is easier to observe trends and patterns in data in graph form than in table form. Graphs are used when two variables are being considered and one variable is dependent on the other. Table 1.5.7 on pages 51–52 summarises suitable graphs for qualitative and quantitative data.

There are several types of graphs, including line graphs, bar graphs, scatterplots and pie charts. The best one to use will depend on the nature of the data.

General rules to follow when preparing a graph include the following:

- ☐ Keep the graph simple and uncluttered.
- ☐ Use a descriptive title.
- ☐ Represent the independent variable on the *x*-axis.
- ☐ Represent the dependent variable on the *y*-axis.
- ☐ Start each axis at zero.
- ☐ Match the length of the axes to the data.
- ☐ Clearly label axes with both the variable and the unit in which it is measured.
- ☐ Use small symbols such as circles or squares for data points.
- ☐ Use different symbols for different data sets.

Line graphs

A **line graph** is a good way of presenting continuous quantitative data. In a line graph, the values are plotted as a series of points on the graph. A line can then be drawn from each point to the next. The independent variable, which is set by the experimenter, is always shown on the *x*-axis. The dependent variable, which is the variable measured in the experiment, is always shown on the *y*-axis. Each point should be drawn in pencil as a small symbol, such as a circle, square or cross. Alternatively you can use a computer program to generate your graphs.

The data from the enzyme example in Table 1.5.6 is presented in Figure 1.5.5.

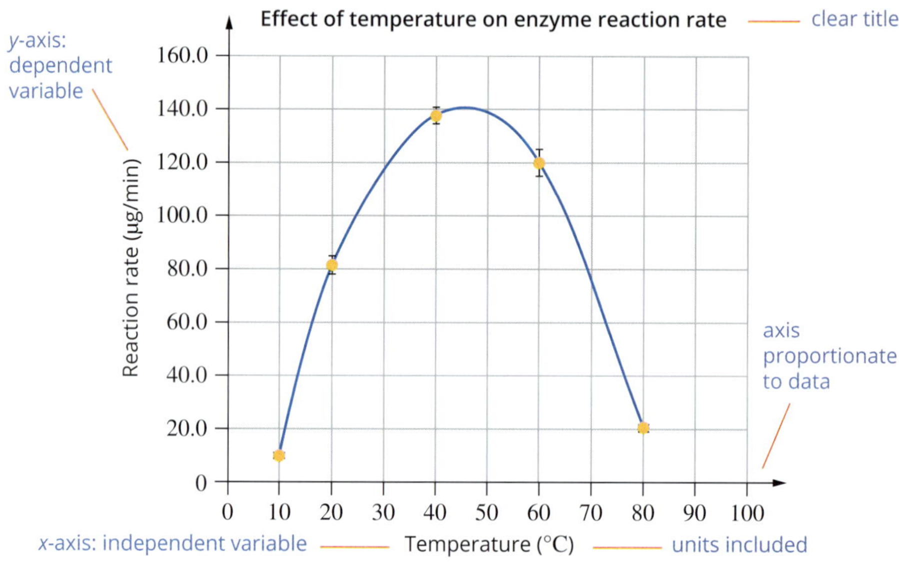

FIGURE 1.5.5 A line graph showing the relationship between two variables: temperature (independent variable) and reaction rate (dependent variable). The uncertainty values are represented as vertical bars above and below the mean at each data point.

For continuous data, the line does not need to join each data point. Rather a straight or curved line can be drawn to represent the overall trend of the data, as shown in the different graphs in Figure 1.5.6. This line is called a trend line or a line of best fit, and can be used to predict values between the data points. Its position can be estimated by eye or calculated mathematically from the data.

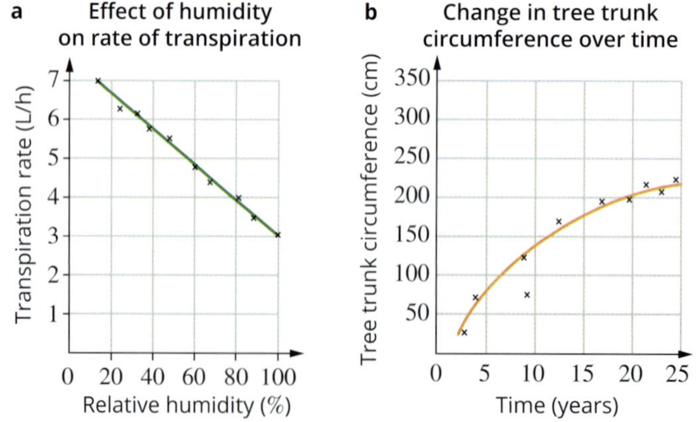

FIGURE 1.5.6 Graphs showing (a) straight and (b) curved trend lines or lines of best fit

When graphing discrete quantitative data, a line shows the change in data from one point to the next, but does not predict the value of a point between the plotted data. An example is plotting the incidence of disease in a population (Figure 1.5.7). This type of data could also be represented as a bar graph, but lines may be better when multiple data sets are being compared on the same graph, such as when plotting the incidence of several different diseases.

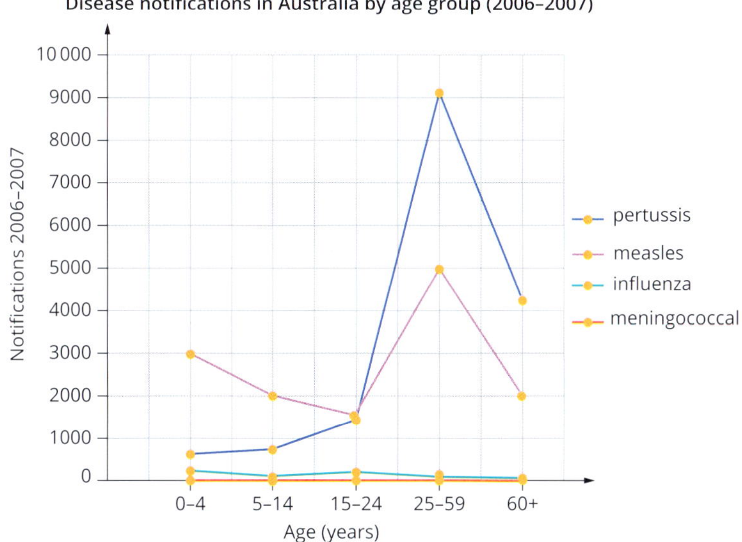

FIGURE 1.5.7 The Department of Health is notified of the number of new cases (or incidence) of a disease. The graph presents Australian Government Department of Health data for notifications of four infectious diseases (pertussis, measles, influenza and meningococcal) for the period 2006–2007. Vaccines are available for all of these infectious diseases.

You will now be able to answer key question 3.

Scatterplots

Scatterplots are commonly used to present data when looking to see if there is a correlation or relationship between two variables. For example, in human evolution there is extensive interest in the relationship between brain size and time since early human ancestors, hominins, first appeared (Figure 1.5.8), as well as other changes over time, such as body size, bipedalism and tool use.

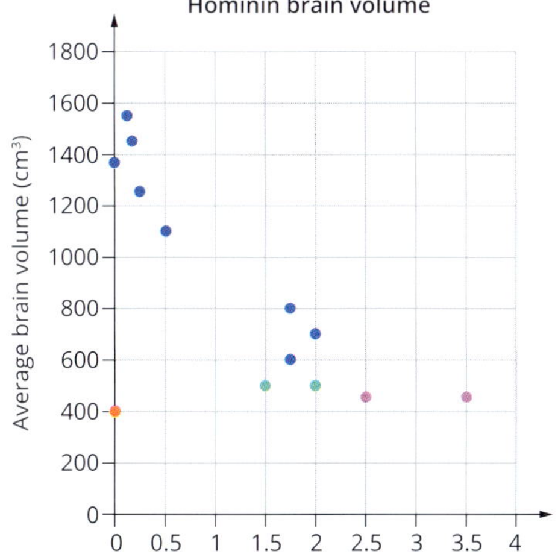

FIGURE 1.5.8 Scatterplot of brain volume in a range of hominin species. Modern *Homo sapiens* is the only *Homo* species alive now (0 million years ago on the graph). Colour code for different genera: *Homo*, blue; *Paranthropus*, green; *Australopithecus*, pink; a modern chimpanzee included for comparison, red.

Bar and column graphs

Bar and column graphs are used to show categories of data that have been counted.

- A **column graph** shows the value of the dependent variable by the height of the column; the categories are labelled across the x-axis.
- A **bar graph** shows the value of the dependent variable by the length of the horizontal bar; the categories are labelled up the y-axis.

Bar and column graphs are commonly used when the independent variable is categorical rather than numerical. The bars or columns are always the same width and the same distance apart. Bar and column graphs are very useful for graphing discrete quantitative data, such as the number of base pairs and number of genes on each human chromosome (Figure 1.5.9).

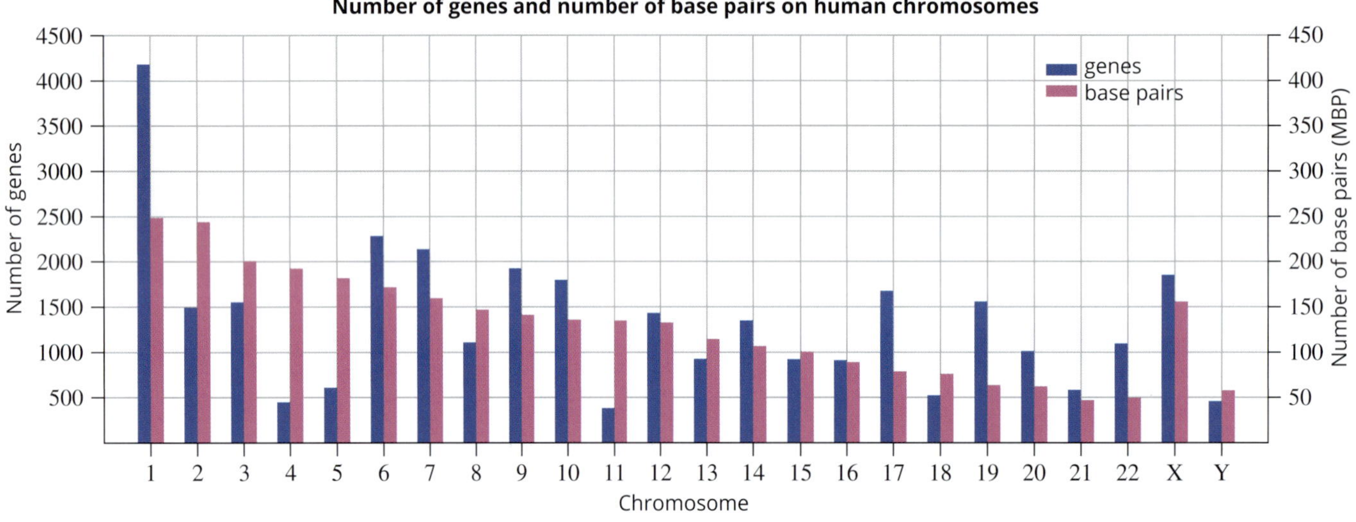

FIGURE 1.5.9 A column graph comparing the number of genes (blue, left axis) and the number of base pairs (pink, right axis) on human chromosomes. Note that two different vertical axes are used for the different data sets, which have very different scales.

When the labels of the variables are long, horizontal bar graphs can be used. Bar graphs are also used when the data ranges are variable and overlapping, such as genome sizes (Figure 1.5.10).

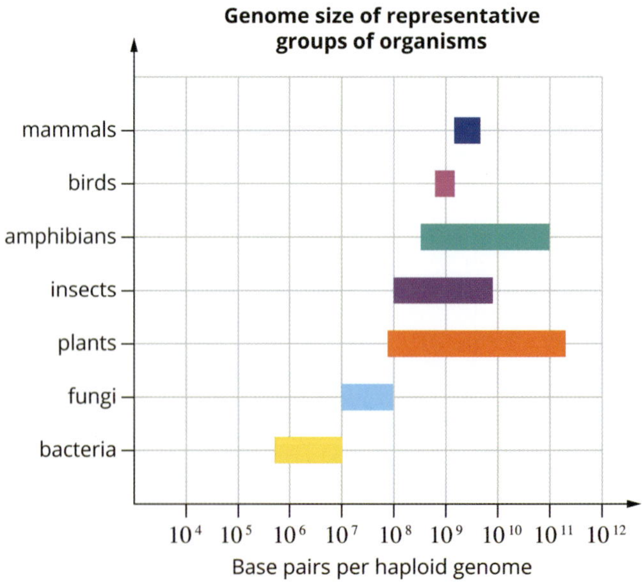

FIGURE 1.5.10 A horizontal bar graph comparing genome size of representative organisms from the different kingdoms of life

Pie charts

A **pie chart** is a way of presenting qualitative and categorical data. It shows each category of data as a proportion of the total data. The chart is a circle divided into sections according to the proportions of each category, like slices of a pie (Figure 1.5.11). Each category is coloured or shaded differently so that it can be distinguished clearly from the other categories. Pie charts should only be used when there are few categories.

To draw a pie chart you must find how many degrees are needed for each category to fit within the 360° circle. This can be done as follows:

- ☐ Add the amounts in each category to find the total.
- ☐ Divide 360° by the total (this will tell you how many degrees of the circle one value is worth).
- ☐ Multiply this value by the amount in each category.

This gives the degrees for each category that can then be marked, using a protractor, on the circle.

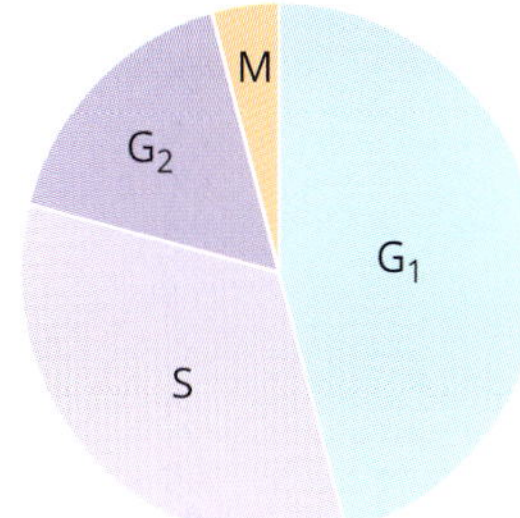

FIGURE 1.5.11 Pie chart presenting data on the length of time a population of mammalian cells spends in each stage of the cell cycle. Total cell cycle time is 24 h, G_1 is 11 h, S (DNA synthesis) is 8 h, G_2 is 4 h and M (mitosis) is 1 h.

TABLE 1.5.7 A summary of suitable graphs for discrete and continuous data

Type of data	Appropriate type of graph	Examples
discrete	bar graph, histogram or pie chart	Bar graph showing the turbidity of river water at four locations: **Water turbidity at various locations along the Murray River** Histogram showing the height distribution of students in class 11A: **Student heights in Class 11A** Pie chart representing the length of time a population of mammalian cells spends in each stage of the cell cycle: **Proportion of time spent in each stage of the cell cycle**

continued over page

TABLE 1.5.7 *Continued*

Type of data	Appropriate type of graph	Examples
continuous	line graph or scatterplot, including a trend line	Line graph showing absorbance of sodium in a sports drink: **Calibration curve of absorbance of standard solutions of sodium in a sports drink** Absorbance: 0.100, 0.200, 0.300, 0.400, 0.500 Concentration of sodium (mg/L): 1.00, 2.00, 3.00, 4.00, 5.00, 6.00, 7.00 Scatterplot of brain volume in a range of hominin species. Colour code for different genera: *Homo*, blue; *Paranthropus*, green; *Australopithecus*, pink; a modern chimpanzee included for comparison, red: **Hominin brain volume** Average brain volume (cm^3): 200, 400, 600, 800, 1000, 1200, 1400, 1600, 1800 Period the hominin species lived (million years ago): 0.5, 1, 1.5, 2, 2.5, 3, 3.5, 4

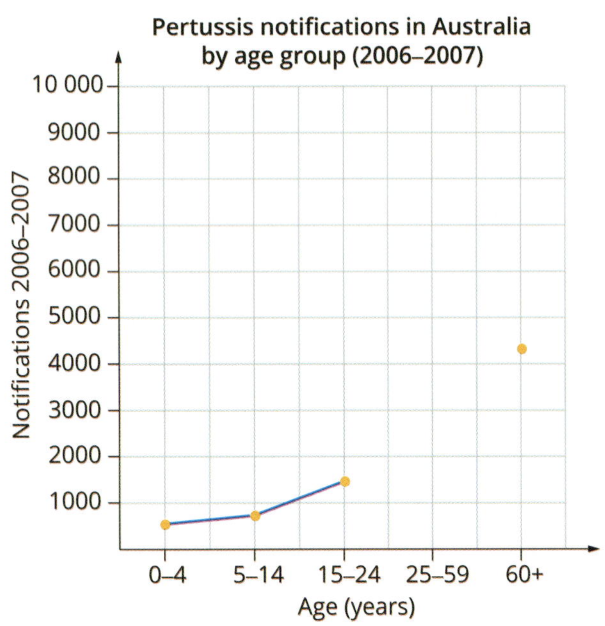

FIGURE 1.5.12 Line graph with missing data

Missing data

When you have missing data, leave a gap for it, as shown in Figure 1.5.12. Ensure that the axes are complete (do not skip values) and do not join data points that have data missing between them. Joining points could be misleading. For example, if the data in Figure 1.5.12 was collected to determine the need for a pertussis (whooping cough) booster vaccination program, it is important to know which age groups in the population are most at risk, so that the right age groups are targeted and public health funds are well-directed. Try to predict the result for the missing data in the 25–59 year old age group. The actual number of pertussis notifications recorded by the Australian Government Department of Health was 9000. This is significantly higher than the graph suggests, so it indicates the need for a significant change in the approach to vaccination programs for individuals within this age group.

Outliers

Sometimes when you plot data, there may be one point that does not fit the trend and is clearly an error. This is called an outlier. An outlier is often caused by a mistake made in measuring or recording data, or from a random error in the measuring equipment. You should plot all of the data points on your graph, but if a data point in a continuous data series is clearly outside the trend line (an outlier) you can ignore it when drawing the line of best fit (Figure 1.5.13). Outliers can also be calculated mathematically.

● *You will now be able to answer key questions 1 and 5–6.*

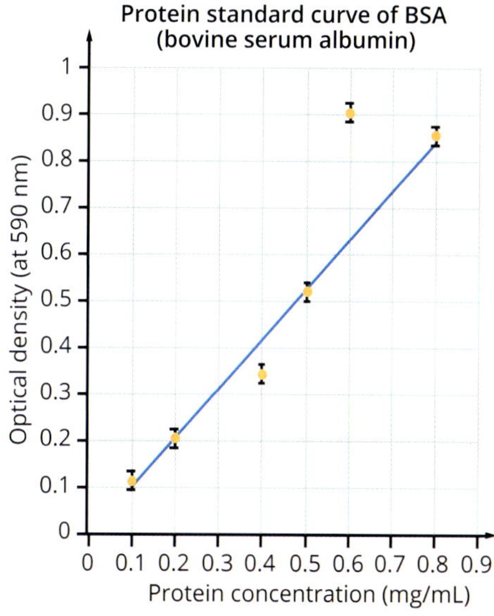

FIGURE 1.5.13 Line graph showing an outlier, which has been ignored when adding the line of best fit

Distorting the truth

Poorly constructed graphs can distort the truth. For example, in Figure 1.5.14 you can see two graphs that show the same data—the test results of two groups of students. One group of students did not eat breakfast before doing the test, and scored an average of 42 marks out of 50. The other group of students did eat breakfast and scored an average of 48 marks out of 50. One graph distorts the difference in marks between the two groups by using a scale from 40 to 50 marks on the *y*-axis. It is important to make sure the graphs you create do not distort your data. You should also be wary of distorted data when interpreting graphs in other publications.

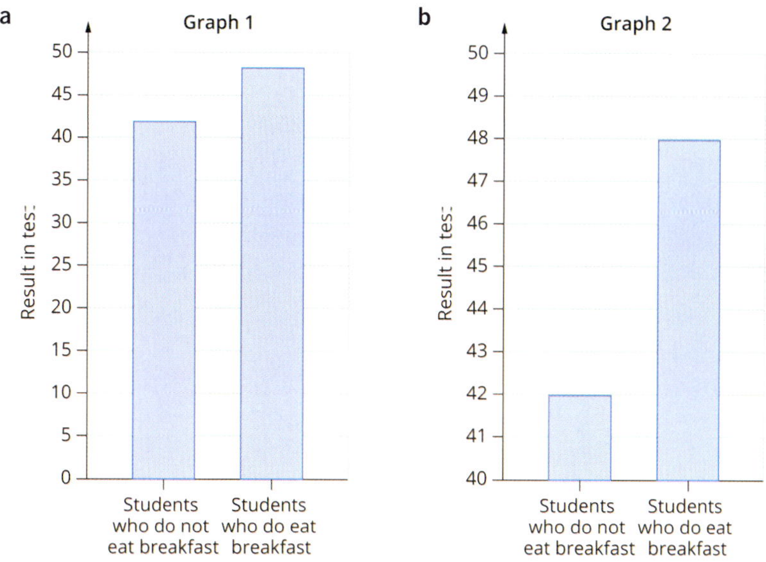

FIGURE 1.5.14 (a) Graph showing the difference between the test marks of two groups of students out of the total 50 marks on the *y*-axis. (b) Graph showing the difference between the test marks of the two groups with only a narrow range of marks on the *y*-axis, which distorts the difference and makes it appear larger than it really is

● *You will now be able to answer key question 7.*

1.5 Review

SUMMARY

- Descriptive statistics can be used for qualitative and quantitative data.
- Descriptive statistics include three measures of central tendency:
  - the mean, which is the sum of the values divided by the number of values
  - the median, which is the 'middle' value in an ordered list of values
  - the mode, which is the value that occurs most often in a list of values.
- Other helpful descriptive statistics include:
  - percentage change, which applies to increases and decreases relative to the control or the starting point of the measurement
  - percentage difference, which is a measure of the precision of two measurements
  - range, which is simply the difference between the highest and lowest values in a data set
  - uncertainty, which results from errors and represents a realistic range within which the true value is likely to be.
- Tables are used to record raw and processed data.
- Tables allow the presentation of more detail, while graphs allow trends to be shown more clearly.
- When presenting the results of an investigation, do not distort the truth—for example, this means you must select appropriate scales on graph axes, include outliers in graphs, and include and explain all errors.

KEY QUESTIONS

Knowledge and understanding

1 What are outliers, and what statistical measurement is most affected by them?

2 Describe the advantages of calculating percentage change for the results of an experiment repeated by different groups of scientists.

3 Explain when a line of best fit graph should be used and when a ruled graph line from point to point is more appropriate.

Analysis

4 For the following data set, calculate and record the following. Show your calculations.
Data: 21, 14, 15, 18, 14, 20, 16, 17, 19, 20, 14.
  a the median
  b the mode
  c the mean and uncertainty

5 A student monitors the temperature at different depths of two mines. Using the student's results below, draw an appropriate graph.

	Mining			
Mine A				
Depth	surface	300 m	700 m	1.00 km
Temperature	15°C	28°C	49°C	62°C
Mine B				
Depth	surface	600 m	1.5 km	3.5 km
Temperature	20°C	25°C	37°C	55°C

6 Describe at least four ways the graph below could be improved.

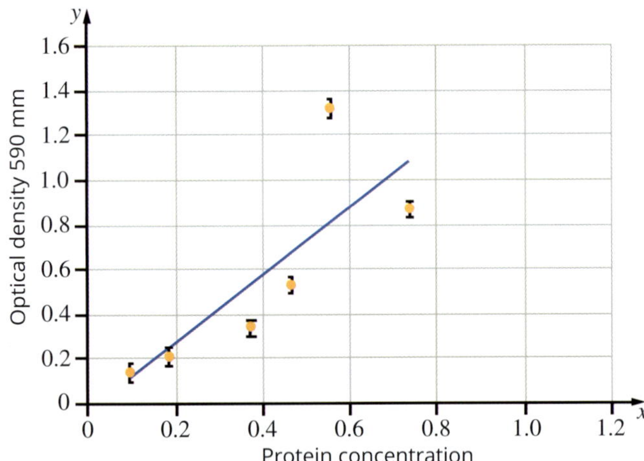

7 a Give two reasons why is it important to accurately represent data from your own investigations.
  b Explain why it is important to be able to interpret and understand representations of scientific data.

1.6 Reporting investigations

Now that you have thoroughly researched your topic, formulated a research question and hypothesis, conducted experiments, and collected data, it is time to bring it all together. The final part of an investigation involves summarising the findings in an objective, clear and concise manner for your audience.

Scientists report their findings in a number of ways: written peer-reviewed journal articles, on web pages, and at scientific conferences with short oral presentations or scientific posters (Figure 1.6.1). Regardless of the reporting and presentation method, the same key information is presented in the same order.

The coursework for Units 1 and 2 will include your completed logbook (which you should be completing as you conduct practical activities) and either written reports or multimodal presentations of scientific investigations. Upon completion of the scientific investigation for Unit 1, Area of Study 3, you are required to present your methodology, methods, results and conclusions. The final presentation for this report could take on a variety of forms, many of which are outlined in Table 1.6.1. In this section you will learn how to present your findings effectively, discuss your investigation, and draw evidence-based conclusions in relation to your hypothesis and research question.

FIGURE 1.6.1 Posters at a scientific conference

TABLE 1.6.1 Characteristics of the main formats for presenting research work

Format	Characteristics	General guidelines for the presentation format
poster presentation	• concise visual display of information • suitable for presenting information to many people • summary of ideas	• title that attracts attention • large headings that stand out • subheadings of a smaller size • attractive presentation • balance of written material and visual material such as diagrams, photographs, tables, graphs • writing large enough to be read from a distance
written report of a practical activity	• presents clear and detailed information on a topic • suitable for providing detailed and more comprehensive background information	• appropriate written style for introduction, materials, methodology, methods, results, discussion and conclusion • use subheadings to organise sections • text should be supported by tables, graphs, diagrams or photographs
oral presentation with supporting slides and/or handouts	• easy-to-follow format • good for presenting to a large audience • supporting slides can be printed as notes to be given to the audience • opportunity to answer questions from the audience	• brief oral descriptions • use clear visuals that complement what is spoken • minimal text on each slide • consistent format on all slides—background, colours and text • images, diagrams and graphs are clear and large
online presentation e.g. website, blog	• can present visual and written information • accessible to a worldwide audience • easy to follow • easy to update with new information	• include hyperlinks to related information • include multimedia, such as video clips and audio, if appropriate • use the same format throughout—font, background, colours • use clear headings • list all hyperlinks on the main page • include your name, credentials and date of publication

PRESENTATION FORMATS

All modes of presenting scientific investigations have the same elements, but with different emphases on visual or textual components depending on the mode of delivery. Table 1.6.1 on page 55 provides some guidelines for different presentation formats. Figure 1.6.2 is an example of a scientific poster.

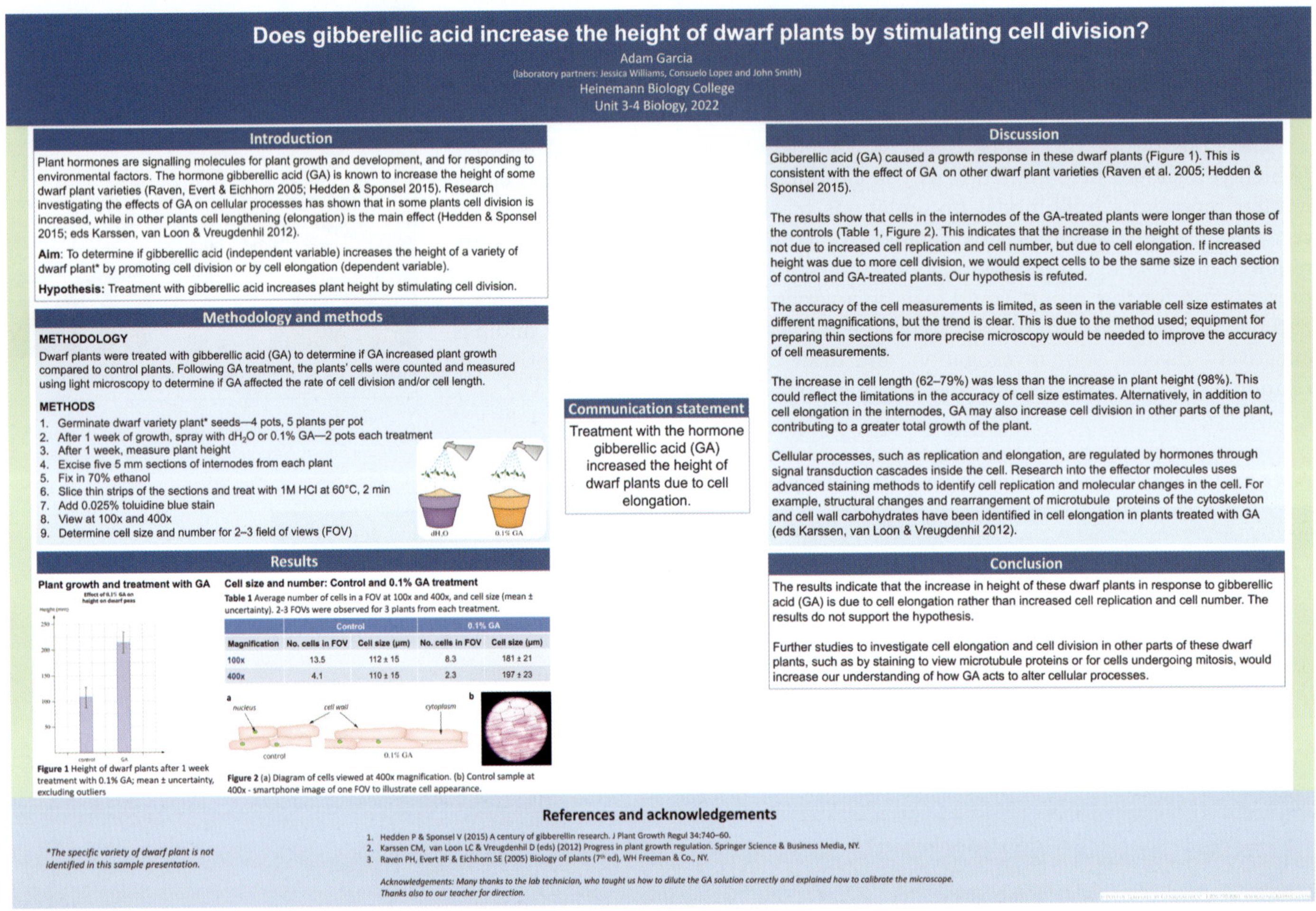

FIGURE 1.6.2 An example of a scientific poster

EFFECTIVE SCIENCE WRITING

Effective science writing is objective, clear and concise, and has a consistent narrative and visual support. If you have time, it is a good idea to put your finished writing aside for a few days and then go back and read it over again, fixing anything that is incorrect or poorly written. Checking the spelling is also an essential part of editing your writing. Do not rely only on computer programs to check spelling; they can make mistakes too, and often do not recognise scientific words. Make sure the spellchecker is set to Australian English; the default setting is usually American English.

Objective writing

Scientific reports should be written in an objective (unbiased) style. This is in contrast to literary writing, which often uses subjective (biased) techniques of persuasion (Table 1.6.2).

TABLE 1.6.2 Examples of unscientific and scientific writing

Unscientific writing examples	Scientific writing examples
Examples of biased and subjective language: • The results were weird/bad/atrocious/wonderful ... • This produced a disgusting odour ... • This is a major health crisis ... • This breathtakingly beautiful golden bowerbird ...	Examples of unbiased and objective language: • The results showed ... • This produced a pungent odour ... • This is a serious health issue ... • The golden bowerbird ...
Examples of exaggeration: • The object weighed a colossal amount ... • No one has ever seen this phenomenon ... • The magnesium exploded into flames ... • Millions of ants swarmed over the nest ...	Examples of accurate language: • The object weighed 250 kg ... • This phenomenon has not been reported previously ... • The magnesium burnt vigorously ... • Ants swarmed over the nest ...
Examples of everyday language: • The bacteria passed away ... • The results don't ... • We guessed that ... • Previous researchers were slack and missed ...	Examples of formal language: • The bacteria died ... • The results do not ... • It was predicted/hypothesised ... • Previous researchers did not report ...

Qualified writing

It is best to avoid words that are absolute, such as always, never, shall, will, or proven. Instead, qualify your writing using words such as may, might, possible, probably, likely, suggests, indicates, appears, tends, can and could.

Concise writing

To be concise use short sentences with a simple structure. The opposite of being concise is being verbose (wordy). When editing your writing consider how you could say the same thing using fewer words (Table 1.6.3).

TABLE 1.6.3 Examples of verbose writing and concise alternatives

Verbose	Concise	Verbose	Concise
due to the fact that	because	is well known to be	is
Carlos undertook an investigation into	Carlos investigated	on an annual basis	yearly
It is possible that the cause could be	the cause may be	until such time as	until
a total of five experiments	five experiments	in the vicinity of	near
the end result	the result	while in the process of preparation	while preparing
in the event that	if	I am of the opinion that	I think that
at the time of writing	today	we took measurement readings	we measured

It is common practice in scientific report writing to write in the passive voice and in past tense.

Voice

'Voice' means whether the subject of the sentence is the 'doer' or 'receiver' of the action. In the active voice the subject is the doer; for example, 'We added 20 mL of sodium chloride to the beaker.' In the passive voice the subject is the receiver; for example, '20 mL of water was added to the solution.' Scientific writing regularly avoids using the active voice and use of personal pronouns (Table 1.6.4).

TABLE 1.6.4 Examples of active and passive voice

Active voice	Passive voice
We recorded oxygen levels hourly.	The oxygen concentration was recorded every 60 minutes.
We used a pH meter to measure pH.	The pH was recorded with a pH meter.
A thermostat controlled the temperature in the water bath.	The temperature in the water bath was controlled by a thermostat.
We placed 50 g of solute in a conical flask containing distilled water and then slowly added 1 mol L^{-1} hydrochloric acid drop by drop.	Fifty grams of solute was placed in a conical flask containing distilled water, and then 1 mol L^{-1} hydrochloric acid was added dropwise.

Tense

Use the past tense when describing your research, including the planning, the experiments and the results, as well as the work of previous researchers. For everything else (including describing facts and theories) you should use the present tense. Avoid using conditional verbs (could or would) and the future tense (unless you are talking about something that has not yet happened). Table 1.6.5 shows some examples of the incorrect and correct use of tenses in scientific writing.

TABLE 1.6.5 Examples of correct and incorrect use of tense

Incorrect tense	Correct tense
Zhu (2013) describes a similar phenomenon.	Zhu (2013) described a similar phenomenon.
Hormone will then be added to the tips of coleoptiles.	Hormone was added to the tips of coleoptiles.
Enzyme B reacts best at pH 9.	Enzyme B reacted best at pH 9.
The DNA sequence comparison supports the conclusion that species A and B share a common ancestor.	The DNA sequence comparison supported the conclusion that species A and B share a common ancestor.

● *You will now be able to answer key question 3.*

Visual support

Use graphs or diagrams to present complex concepts or information. This will reduce the number of words you need, and also make your research more accessible for your audience. Details of experimental methods can be presented as a diagram or flow chart. This can make it easier to see the methods than to read through a series of steps. Flow charts use simple diagrams, small text boxes and connecting lines to represent the methods and sequence of steps in a scientific method. Diagrams should use clear outlines and labels—they are not works of art.

WRITING A SCIENTIFIC REPORT

Whether the investigation is presented as a poster, written report or oral presentation, the same key elements are included in the same sequence, as summarised in Figure 1.6.3.

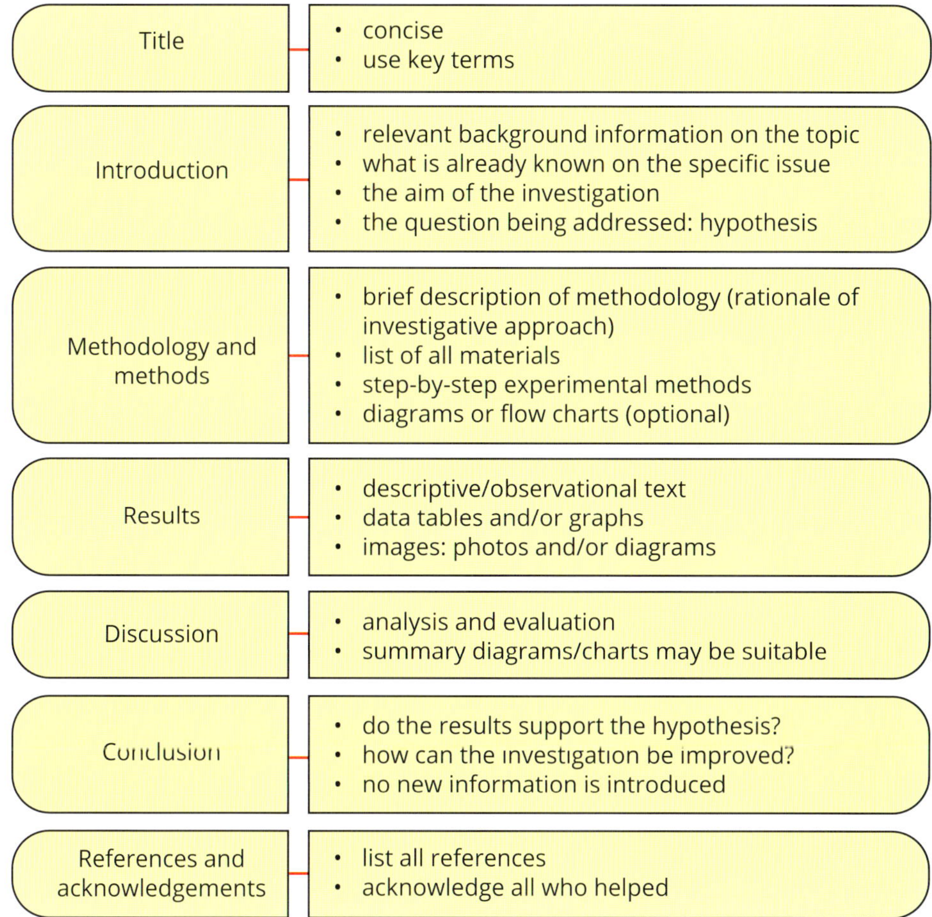

FIGURE 1.6.3 Elements of a scientific report or presentation

You will now be able to answer key question 1.

Title

The title should give a clear idea of what the report is about, without being too long. It should include key terms that tell the reader what your study is about.

Introduction

The introduction sets the context of your report. It should outline relevant biological ideas, concepts, theories and models, and how they relate to your specific research question and hypothesis. It introduces the key terms, the specific question to be addressed, and states your hypothesis. Any references used in the introduction should be correctly cited. This section should also identify the independent, dependent and controlled variables (see Section 1.2).

For example, consider a student investigating the cellular processes affected by a growth-promoting plant hormone. The research and introduction for this investigation might include the following points:

- the name and chemical nature of the hormone
- where the hormone is found (natural or synthetic)
- what is currently known about the actions of the hormone

- the specific question being addressed, identifying the independent and dependent variables
- the hypothesis.

While researching this topic, the student found prior evidence that suggests this hormone increases the height of some dwarf plants, but the mechanism for this effect was not clear. There were some reports of increased cell division, while other studies reported a change in cell length. The student's hypothesis was that the hormone would increase the growth of dwarf peas by increasing the cell number.

Methodology and methods

The methodology and methods section outlines the rationale of the investigative approach and describes in detail all the steps that were undertaken during the investigation, including a list of the materials used. For a poster presentation use step-wise lists, diagrams of specific methods, and/or flow charts of the overall experimental design. There should be enough detail for someone else to replicate your experiments. Therefore, your method needs to be in the correct sequence and include how you observed, measured, recorded and analysed the results.

Here is an example of a methodology and methods section for an experiment on plant hormone action as it might be presented in a written report. For a poster presentation, the methods may be easier to follow in a step-wise list accompanied by large, clearly labelled diagrams. Alternatively, flow charts are a good way to clearly present experimental designs.

Materials:

- 20 dwarf pea seeds
- 3 pots and potting mix
- plant hormone—gibberellic acid (GA) solutions, 0.01% and 0.1%, diluted from 1% stock solution in distilled water (dH_2O)
- small spray bottles
- scalpel blade and forceps
- toluidine blue stain (0.025%)
- 1 mol L^{-1} HCl
- microscope slides and coverslips
- compound light microscope

Methodology:

A controlled experiment was conducted in which dwarf plants were treated with gibberellic acid (GA) to determine if GA increased plant growth compared to control plants. Following GA treatment, the plants' cells were counted and measured using light microscopy to determine if GA affected the rate of cell division and/or cell length.

Methods:

Example experiment 1: Plant growth and treatment with GA

Dwarf pea seeds were germinated and transferred into 3 pots with potting mix, 5 plants per pot. After 1 week, when the seedlings were approximately 20 mm tall, plants were sprayed with either dH_2O, 0.01% GA or 0.1% GA (Figure 1). Plant height was measured 1 and 2 weeks later.

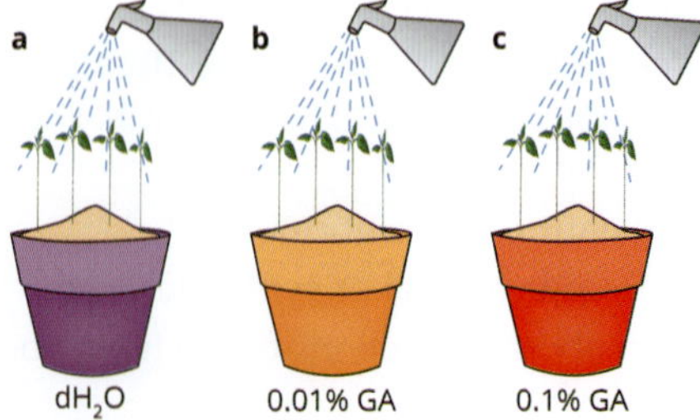

Figure 1: Seedlings were sprayed with (a) dH_2O, (b) 0.01% GA, or (c) 0.1% GA.

Example experiment 2: Microscopic analysis of internode cells

At week 3, a 5 mm section of internode was cut from the stem, placed on a microscope slide and sliced lengthwise. Three drops of 1 mol L^{-1} HCl were added to the tissue and the slide placed on a 60°C hotplate for 2 minutes. Excess HCl was soaked up with paper towel; 2 drops of toluidine blue stain were added for 2 minutes, then a coverslip was placed on the tissue and gently pressed down (Figure 2). The slide was viewed under the microscope at 100× magnification. Two stems from each pot were stained and viewed in this manner.

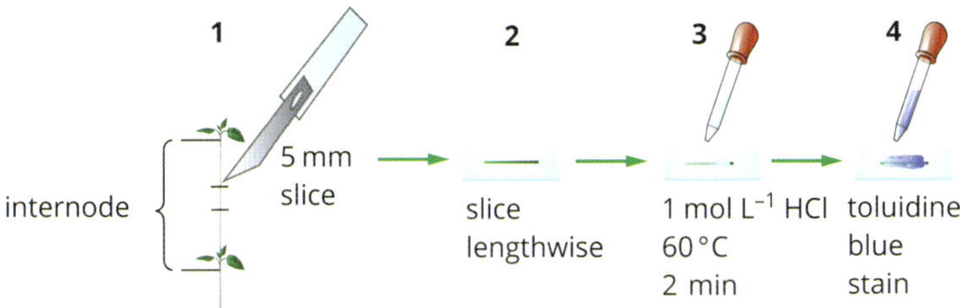

Figure 2: (1) A 5 mm section of internode was cut from the stem and (2) placed lengthwise on a microscope slide, then (3) 3 drops of 1 M HCl were added and the slide heated at 60°C for 2 minutes. (4) Finally, 2 drops of toluidine blue stain were added for 2 minutes, then a coverslip was placed on top before viewing by light microscopy.

Results

The results section is a record of your observations. It is where you present your data using graphs, diagrams, tables or photographs. In Section 1.5 you learnt tips on using graphs and tables appropriately.

For the plant hormone experiment described above, the results section might include the following table and figures.

Results:

Example experiment 1: Effect of hormone on plant growth

Table 1: Results of plant height (mm) at week 3 of GA treatment

	Plant height (mm) at different GA concentrations		
Plant no.	**0**	**0.01%**	**0.10%**
1	23	117	158
2	20	210	378
3	22	240	320
4	30	211	377
5	31	198	363
mean	25	195	319

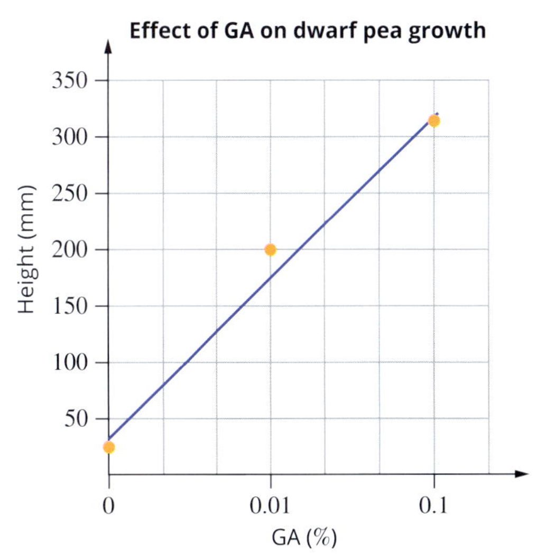

Figure 3: Graph of average plant height at week 3 (2 weeks after GA treatment)

Example experiment 2: Microscopy. The effect of GA on cell growth

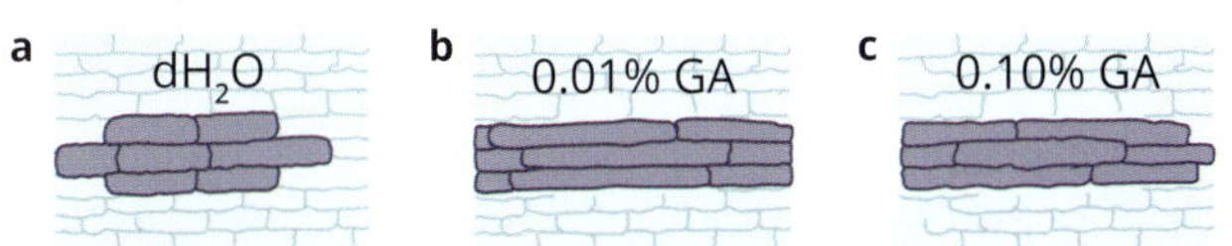

Figure 4: Diagrams of representative samples at week 3 (2 weeks after GA treatment), viewed at 100× magnification. Estimated average cell length in (a) control dH_2O, 60 µm; (b) 0.01% GA, 100 µm; (c) 0.10% GA, 100 µm.

a Direct or linear proportional relationship
- Variables change at the same rate (graph line is straight, slope is constant)
- Positive relationship—as *x* increases, *y* increases

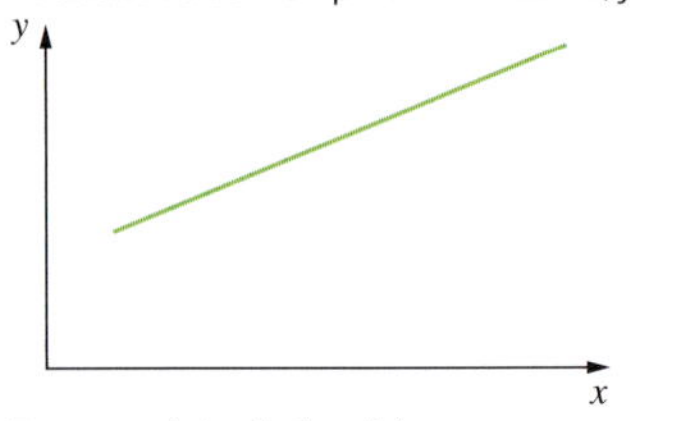

b Exponential relationship
- As *x* increases, *y* increases slowly, then more rapidly

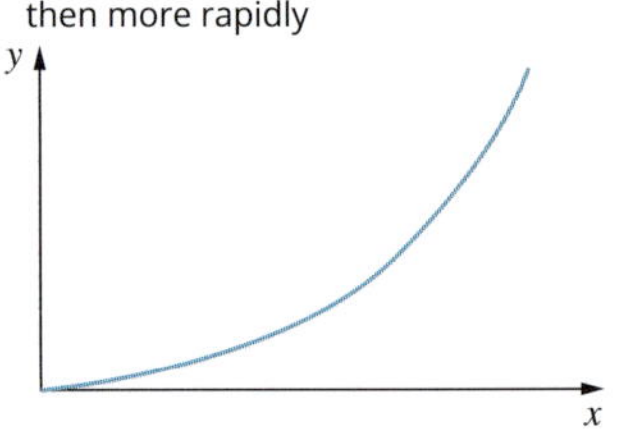

c Exponential rise, then levels off or plateaus (stops rising)
- As *x* increases, *y* increases rapidly at first, then slows, then finally does not increase at all—*y* reaches a maximum value

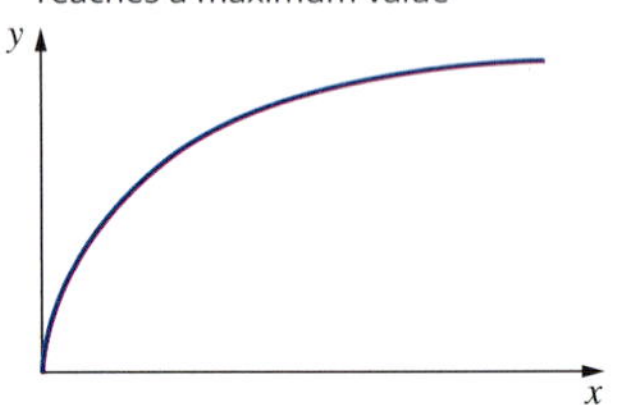

d Inverse direct or linear proportional relationship
- Variables change at the same rate (graph line is straight, slope is constant)
- Negative relationship—as *x* increases, *y* decreases

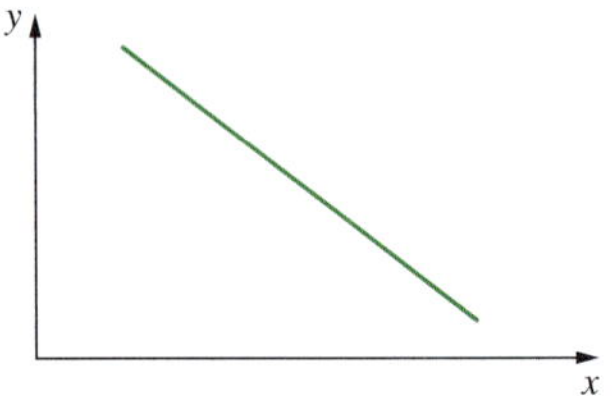

e Inverse exponential relationship
- As *x* increases, *y* decreases rapidly, then more slowly, until a minimum *y* value is reached

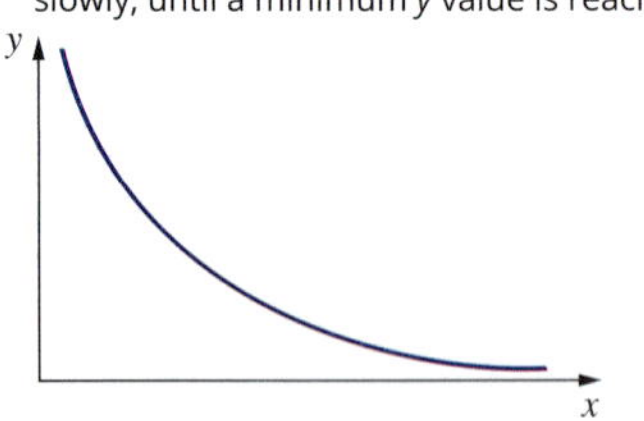

f No relationship between *x* and *y*
- As *x* increases, *y* remains the same

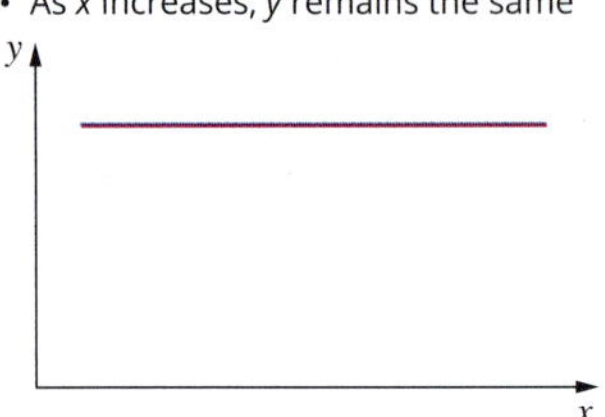

FIGURE 1.6.4 Line graphs illustrating common relationships between variables: (a) direct linear relationship, (b, c) exponential relationships, (d, e) inverse relationships, (f) no relationship

Discussion

In the discussion you interpret your results, and discuss how your findings relate to your initial question and hypothesis, the research of others, and the biological concepts outlined in your introduction. It is also important to evaluate the methods used and the impact of any errors on the results and conclusions that can be formed.

Interpret the results

When you interpret your results, you need to state clearly whether a pattern, trend or relationship was observed between the independent and dependent variables, describe what kind of pattern it was, and specify under what conditions it was observed.

In experiments with continuous variables, such as a range of concentrations, temperatures or pH, the types of relationships that may occur between variables are:

- **linear relationship**—variables that change in linear or direct proportion to each other produce a straight trend line (Figure 1.6.4a)
- **exponential relationship**—variables that change exponentially in proportion to each other produce a curved trend line (Figure 1.6.4b, c)
- **inverse relationship**—when there is an inverse relationship, one variable increases as the other variable decreases; this relationship may be linear or exponential (Figure 1.6.4d, e)
- none—when there is no relationship between two variables, one variable will not change even if the other does (Figure 1.6.4f).

More complex relationships might have to be evaluated mathematically to obtain a formula that describes the trend line.

Interpreting the result for the plant hormone experiment previously described, Figure 3 on page 61 shows that as the GA concentration increases, the height of the peas increases.

Evaluate investigative methodology and methods

Your discussion should evaluate your investigative methodology and methods and identify any issues that could have affected the validity, reliability, accuracy or precision of the data. Any possible sources of error in your experiment should be stated. Remember that controls are essential to the reliability and validity of your investigation, so if you have overlooked or were unable to control a variable that should have been controlled, this may explain unexpected results.

Make recommendations for modifying or extending the investigation. In the example of the plant hormone experiment, the sources of error the experimenter should consider include whether there were enough replicates to obtain reliable data, whether microscopy was an appropriate method for determining cell number and cell length, whether the microscope was calibrated, and whether enough cells were viewed. When writing your report, provide specific suggestions for improvements to the methodology based on what you have learnt.

It is also important to acknowledge contradictions in data and information. Again consider the example of the plant hormone experiment, in which the results of experiment 2 indicated an increase in cell length in the GA-treated plants compared to the controls, but both GA concentrations had the same effect. This is not consistent with the concentration effect on plant height. So this raises several questions. Is it a limitation of the experimental design or methods? Are there more biological effects that are not being detected or measured? In your discussion, acknowledge these sorts of issues and make suggestions for further experiments to address them.

Some experimental findings may lead you to formulate new research questions and develop new hypotheses. An extension of the experiment may be to make an alteration that will enable further investigation. For example, if the effect of temperature has been investigated, further understanding of temperature could be determined by using a different temperature range in a modification of the original method.

Some experimental findings may raise questions about what to do with the new information. This is of particular concern if animal studies have been undertaken. As stated in Section 1.2, research that involves animals, including humans, needs to have obtained approval from an ethics committee. If there are questions raised about the application of any findings from an experiment to animal or human models, the ethical implications of reporting on this data must include recognising the potential sociocultural, economic and political impact these results may have.

Relate findings to biological concepts

In your introduction you established a context. Now you have a framework in which to discuss whether your data supports or refutes your hypothesis. Providing context also enables you to compare your results with existing research and knowledge. Use the points in the Figure 1.6.5 to help frame your discussion.

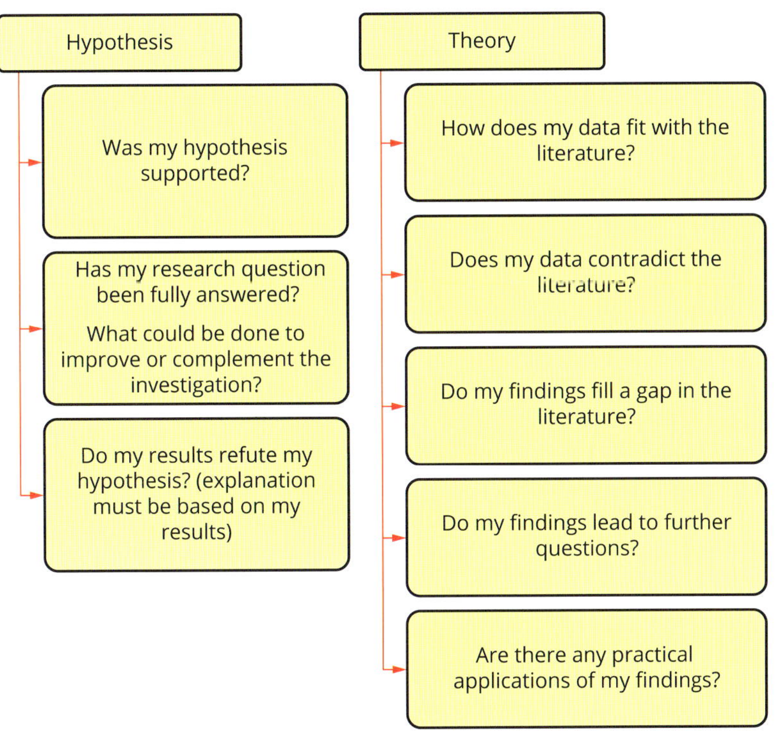

FIGURE 1.6.5 Points to help frame your discussion

- *You will now be able to answer key question 2.*

CASE STUDY

Testing the POLS hypothesis

The pace of life syndrome (POLS) hypothesis suggests that differences between individuals' life-history events, such as timing of birth, weaning, maturation and reproduction, are associated with individual differences in behavioural traits.

To test this hypothesis, a group of researchers from the University of Melbourne, the Max Planck Institute for Ornithology and the University of Munich studied the superb fairy-wren (*Malurus cyaneus*) in a wetland near Melbourne (Figure 1.6.6).

FIGURE 1.6.6 A female superb fairy-wren (*Malurus cyaneus*). Coloured bands are attached around the legs to identify individual birds.

Their study was titled 'Animal personality and pace-of-life syndromes: do fast-exploring fairy-wrens die young?'

The report on the research, which investigated long-term risk-related behaviours and looked at the impact of these behaviours on survival, was published in the journal *Frontiers in Ecology and Evolution* in 2015.

In their introduction, the authors explained their hypothesis and the relevant biological concepts. In their methods section they detailed the experimental design, including the number of fairy-wrens that were studied, how they were trapped and handled, the exact location of the population being studied, and the ethics approvals that permitted them to conduct their research. They also described how they had quantified the behaviour of adult fairy-wrens by exposing them to a new environment test in which the birds were temporarily removed from the wild and placed in a room with a water dispenser, perches, and a tray with 10 mealworms (Figure 1.6.7).

FIGURE 1.6.7 A superb fairy-wren in the personality assay room

The variables they quantified were the time it took for the bird to emerge from its cage into the room once the door was raised (emergence), the number of different areas the bird perched in the 5 minutes after entering the room (exploration), the total number of areas the bird perched in the 2 minutes starting from 6 minutes after entering the room and becoming familiar with it (activity), and how they responded to a mirror (mirror responsiveness). To test mirror responsiveness, after the bird had been in the room for 8 minutes a mirror was revealed that could only be seen by birds on the upper perches. The researchers scored their response to the mirror on a scale of 1 to 3, where 1 was swooping at the mirror, 2 was perching in front of the mirror, and 3 was pecking at the mirror. Based on their recapture rates, the same individuals were retested in this way up to five times.

The results, which included uncertainties, found the behaviour of individual fairy-wrens was consistent over several years and that risky individuals, like those with greater exploratory behaviour, were less likely to be in the population 12 months later. These results support the POLS hypothesis that consistent individual differences in risk-related behaviours are associated with variation in survival.

Conclusion

Your conclusion should be one or two paragraphs that uses your evidence (data) to support or refute the hypothesis. It should provide a carefully considered response to your research question based on your results and discussion. You should clearly state whether your hypothesis was supported or not. Draw your conclusions by identifying trends, patterns and relationships in the data.

It is important to recognise the limitations of your data and the limitations of the scientific method. Be careful not to overstate your conclusion. Your results will support or refute the hypothesis. They will not 'prove' something is true, as you can only ever provide evidence that indicates the probability of something being true.

Do not provide irrelevant information or introduce new information in your conclusion. Refer to the specifics of your hypothesis and research question, and do not make generalisations.

References

All the scientific papers and other sources that are mentioned in the report are to be listed at the end of your report. Cite the source of any information you obtained from secondary sources in the text of your report whenever it is used and referred to, and provide a list of references at the end of your report. This demonstrates that you are aware of previous work in the area, and allows readers to locate sources of information if they want to study them further.

The usual approach is to give a short reference in the text, such as 'Hedden and Sponsel (2015)', and give the full reference in the reference list. If you are stating factual information from another source, you can either quote it word-for-word, or rewrite it in your own words. However, if you rewrite it you must make it clear that the information is not your own. Plagiarism (claiming that another person's work is your own) is not tolerated in scientific research.

Table 1.6.6 shows examples of ways to reference the three most common sources of information: journal articles, books and web pages in American Psychological Association (APA) seventh edition style. This is only one of several referencing systems that you might be required to use throughout your career. Use a consistent format for all references.

TABLE 1.6.6 Examples of references in APA seventh edition style for three common information sources

Source of information and example of reference in text	Format for listing references and example of a reference as written in the reference list
Research article or review article in a scientific journal GA is well established as a naturally occurring plant growth regulator with effects on ... (Hedden & Sponsel, 2015, p. 743).	**Author, initials. (year). Title of article. Journal title, volume number(issue number), page numbers. Digital object identifier (doi) or URL** Hedden, P. & Sponsel, V. (2015). A century of gibberellin research. *Journal of Plant Growth Regulation, 34,* 740–760. doi:10.1007/s00344-015-9546-1
Book The molecular mechanisms of plant growth regulators are being studied by many groups (Karssen et al., 2012, p. 28).	**Author, initials. (year). Title of book (edition, if not first). Publisher.** Karssen, C. M., van Loon, L. C., & Vreugdenhil, D. (Eds). (2012). *Progress in Plant Growth Regulation.* Springer Science & Business Media.
Online article or page Plant hormones play many roles in plant growth and development, and sensing and responding to environment, possibly even by 'hearing' (Coghlan, 1998).	**Author, initials/name of organisation. (year). Title of webpage or web document. URL** Coghlan, A. (1998). Sensitive flower. *New Scientist,* (2153). https://www.newscientist.com/article/mg15921534-900-sensitive-flower/

Acknowledgements

Finally, it is important to acknowledge anyone who has assisted you in your investigation. This includes people who helped you find appropriate literature and references, learn to use equipment, prepare solutions, set up the experiments, find and navigate online databases, edit your report or prepare graphs and images. For example, statements similar to the following could be used:

- This research was supported by the staff of the Science Faculty at Western High School in Melbourne.
- Special thanks to Ms Smith for preparing stock solutions.

● *You will now be able to answer key question 4.*

1.6 Review

SUMMARY

- Your reports should include the following sections:
  - title
  - introduction
  - methodology and methods
  - results
  - discussion
  - conclusion
  - references
  - acknowledgements.
- The title should give a clear idea of what the report is about, without being too long.
- The introduction sets the context of your report. It should outline relevant biological ideas, concepts, theories and models, and how they relate to your specific question and hypothesis.
- The methodology and methods section should:
  - outline the methodology used and the rationale for using this approach
  - clearly state the materials required and the methods used to collect data during your investigation
  - be presented in a clear, logical order that accurately reflects how you conducted your study.
- The results section should state your results and present them using graphs, figures and tables, but not interpret the results.
- The discussion should:
  - interpret data
  - evaluate the investigative method and make recommendations for improving the method
  - explain the link between investigation findings and relevant biological concepts.
- The conclusion should succinctly link the evidence collected to the hypothesis and research question, indicating whether the hypothesis was supported or refuted.
- References and acknowledgements should be presented in an appropriate format.

KEY QUESTIONS

Knowledge and understanding

1 List the elements of a scientific report.

2 Describe the purpose of the discussion section of a scientific report.

3 Which of the following statements is written in third-person narrative? (More than one response may be correct.)

A The researchers reported ...
B Samples were analysed using ...
C The experiment was repeated three times ...
D I reported ...

Analysis

4 A scientist designed and completed an experiment to test the following hypothesis: 'Plants in a humid environment will have a lower rate of transpiration than plants in a dry environment'.

a Write a possible aim for this scientist's experiment.
b What would be the independent, dependent and controlled variables in this investigation?
c Would the scientist collect qualitative or quantitative data?
d List the equipment that could be used and describe the factors that might influence the precision of the scientist's measurements.
e What would you expect the graph of the results to look like if the scientist's hypothesis were correct?

Chapter review

accuracy
aim
bar graph
column graph
conclusion
continuous variable
control group
controlled variable
dependent variable
discrete variable
experimental group
exponential relationship
fieldwork
hypothesis
independent variable
inverse relationship
line graph
linear relationship
mark–recapture
mean
median
method
methodology
meniscus
mode
observation
outlier
peer-reviewed
personal error
pie chart
placebo
point sampling
precision
primary data
primary source
principle
processed data
provisional data
quadrat
qualitative data
quantitative data
random error
random selection
range
raw data
repeatability
repeat trial
replication
reproducibility
research question
risk assessment
safety data sheet (SDS)
scatterplot
scientific method
secondary data
secondary source
systematic error
theory
transect
true value
uncertainty
validity
variable

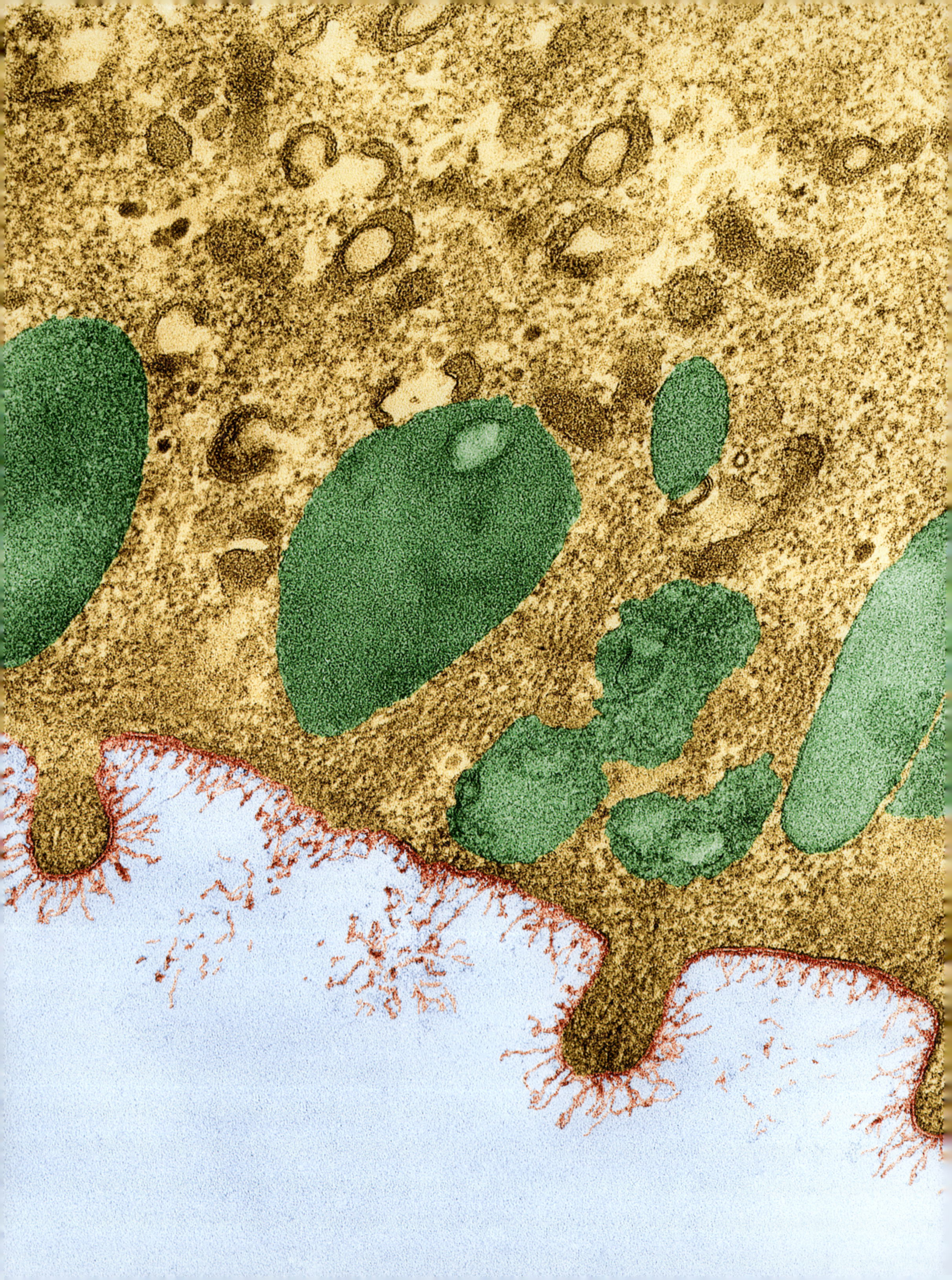

UNIT 1 How do organisms regulate their functions?

To achieve the outcomes in Unit 1, you will draw on key knowledge outlined in each area of study and the related key science skills on pages 7–9 of the study design. The key science skills are discussed in Chapter 1 of this book.

AREA OF STUDY 1

How do cells function?

Outcome 1: On completion of this unit the student should be able to explain and compare cellular structure and function and analyse the cell cycle and cell growth, death and differentiation.

AREA OF STUDY 2

How do plant and animal systems function?

Outcome 2: On completion of this unit the student should be able to explain and compare how cells are specialised and organised in plants and animals, and analyse how specific systems in plants and animals are regulated.

AREA OF STUDY 3

How do scientific investigations develop understanding of how organisms regulate their functions?

Outcome 3: On completion of this unit the student should be able to adapt or design and then conduct a scientific investigation related to function and/or regulation of cells or systems, and draw a conclusion based on evidence from generated primary data.

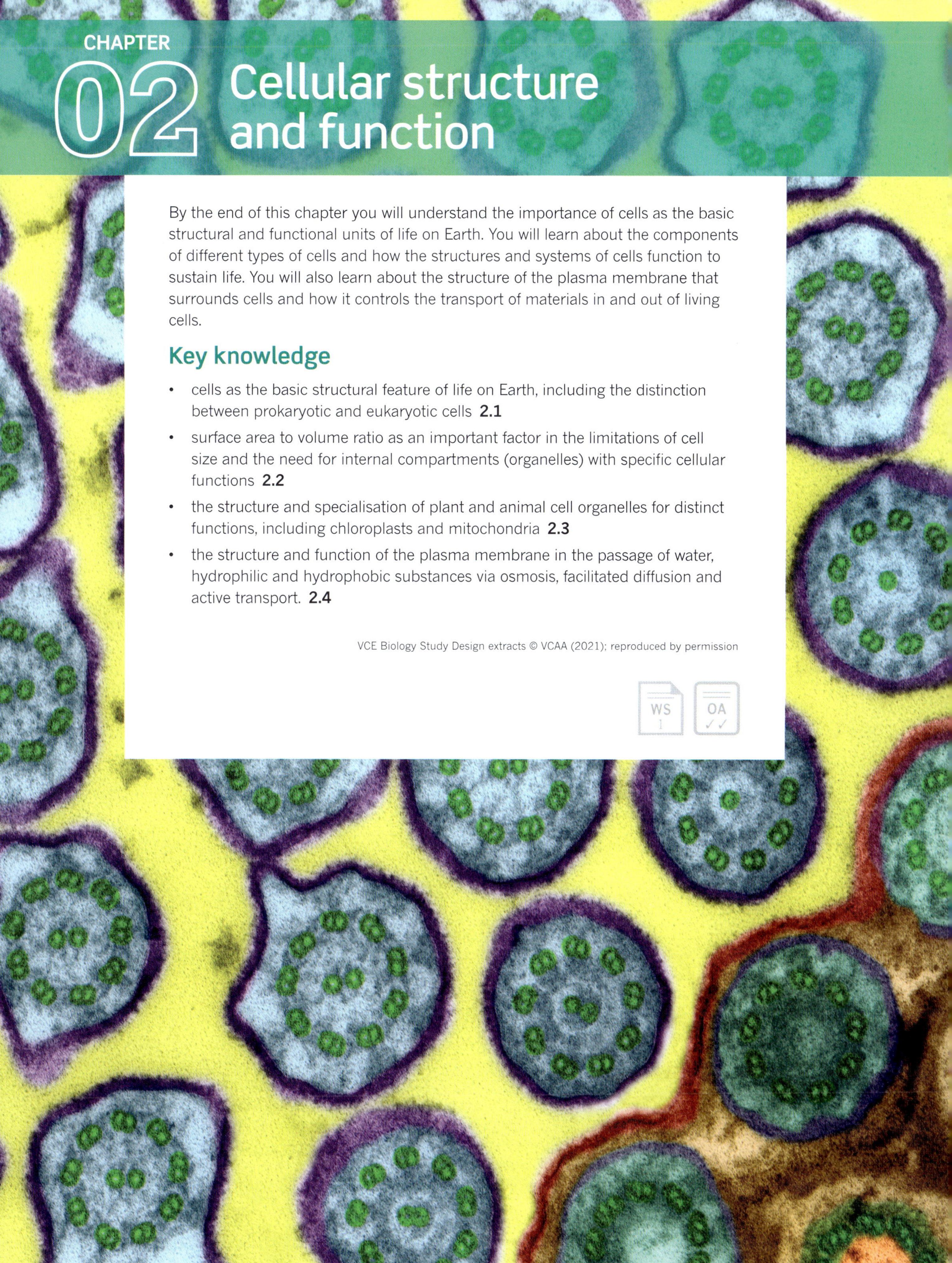

CHAPTER 02

Cellular structure and function

By the end of this chapter you will understand the importance of cells as the basic structural and functional units of life on Earth. You will learn about the components of different types of cells and how the structures and systems of cells function to sustain life. You will also learn about the structure of the plasma membrane that surrounds cells and how it controls the transport of materials in and out of living cells.

Key knowledge

- cells as the basic structural feature of life on Earth, including the distinction between prokaryotic and eukaryotic cells **2.1**
- surface area to volume ratio as an important factor in the limitations of cell size and the need for internal compartments (organelles) with specific cellular functions **2.2**
- the structure and specialisation of plant and animal cell organelles for distinct functions, including chloroplasts and mitochondria **2.3**
- the structure and function of the plasma membrane in the passage of water, hydrophilic and hydrophobic substances via osmosis, facilitated diffusion and active transport. **2.4**

WS 1

OA

2.1 Cell types

Cells are the basic structural units of all living things. The cell theory is one of the fundamental principles of biology. It is based on microscopic and experimental studies of tissues, from all types of organisms, carried out over the last 300 years.

In this section you will learn about cell theory, the features that are common to all cells and the differences between the two cell types, prokaryotic and eukaryotic cells.

CELL THEORY

Cells are the basic structural units of living organisms. The cell theory states that:

- all organisms are composed of cells
- all cells come from pre-existing cells
- the cell is the smallest living organisational unit.

Biogenesis

The cell theory states that all cells arise from pre-existing cells. This is known as **biogenesis**.

Until the 1850s, the idea of spontaneous generation was accepted as the origin of small organisms such as maggots. In other words, maggots could suddenly be formed from anything, even a grain of sand.

Experiments by Francesco Redi on maggots in the 17th century and Lazzaro Spallanzani on microorganisms in the 18th century suggested that 'spontaneous generation' was caused by contamination.

In 1859 Louis Pasteur finally disproved the theory of spontaneous generation. He did so by boiling beef broth in two flasks. Each flask had a glass 'swan-neck' to prevent contaminants in the air from reaching the broth (Figure 2.1.1). At this point no microorganisms grew in either of the flasks. When the swan-neck was broken on one flask and the broth was exposed to the air, microorganisms began to grow in the broth. But the unbroken flask remained free of microorganisms.

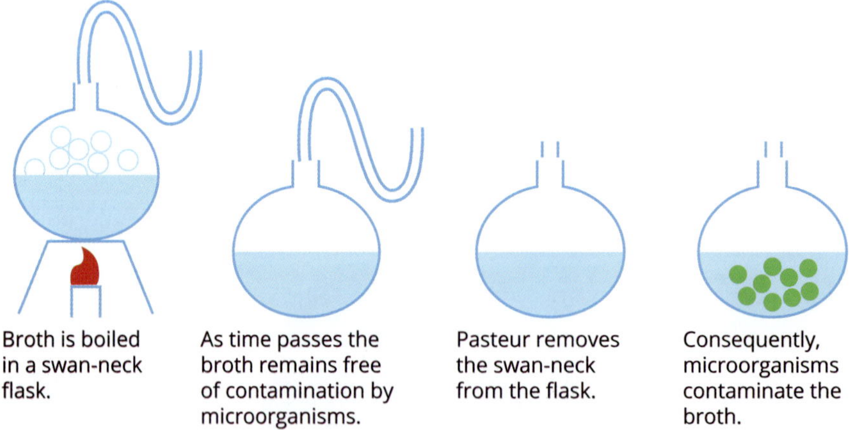

FIGURE 2.1.1 Pasteur's experiment disproved the theory of spontaneous generation.

Pasteur also showed that boiling and cooling wine or milk killed any microorganisms in them. This process is called pasteurisation—named after Pasteur.

An important implication of Pasteur's experiment is that it provided the scientific basis for the germ theory of infection. This theory states that microorganisms are widely present in the environment and are the cause of many diseases. Understanding germ theory eventually led to the development of antiseptic procedures in medicine.

CASE STUDY

History of cell theory

Hooke: the discovery of cells

The first description of cells was made by Robert Hooke in his book *Micrographia*, published in 1665. Hooke made a thin slice of cork from the bark of a tree and examined it under a microscope he had made himself (Figure 2.1.2). He saw that the bark was made up of hundreds of little 'empty boxes' that gave it a honeycomb appearance. He called the boxes 'cells'.

Hooke was actually looking at empty dead cells. When he later looked at fresh plant tissue, he noted the cells appeared to contain water. A few years later, Marcello Malpighi produced more detailed descriptions of plant cells.

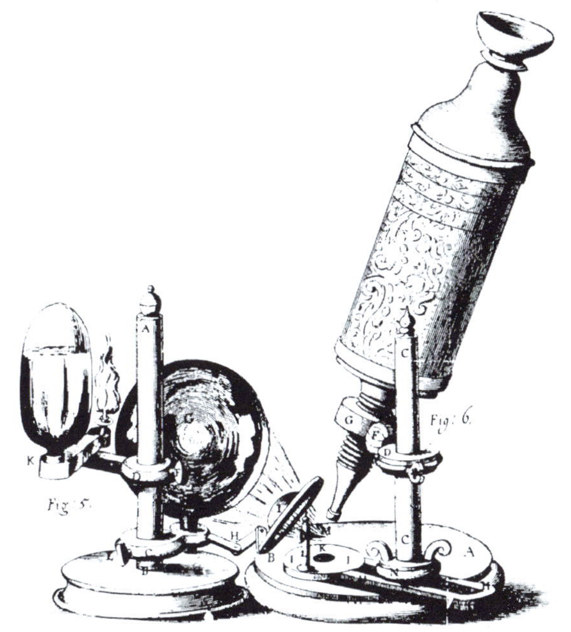

FIGURE 2.1.2 Robert Hooke's drawing of his light microscope in *Micrographia*, published in 1665

Leeuwenhoek: first observations of living cells

In 1676 Anton van Leeuwenhoek observed many living cells under the microscope, including bacteria, blood cells and sperm. He was the first to describe the reproduction of single-celled organisms, which he called 'animalcules'.

Lamarck and Dutrochet: all living things are composed of cells

By the early 19th century the microscope had become a standard tool of biologists, and living animal and plant cells were easy to observe. In the early 19th century Jean-Baptiste Lamarck stated that all living things are a mass of cells, and that complex solutions move in and out of cells. Henri Dutrochet supported this idea, stating: 'plants are composed entirely of cells, or of organs that are obviously derived from cells...the same is true for animals'.

Schleiden and Schwann: cells are organised into tissues

By the middle of the 19th century the fundamental principle that entire organisms are composed of highly organised groups of cells was broadly accepted. This was largely because of the work of Matthias Schleiden on plant tissues, and Theodor Schwann on animal tissues.

Remak and Virchow: the theory of biogenesis

Until the 1840s most biologists still believed that cells formed spontaneously from body fluids or from the nucleus, which they thought was the embryo of a new cell. Then Robert Remak discovered that new cells were formed by a single cell dividing in two, with the nucleus dividing at the same time (Figure 2.1.3). In the 1850s Rudolf Virchow used Remak's discovery to popularise the theory of biogenesis: that all cells come from pre-existing cells. Because of Virchow's great popularity, this theory was quickly accepted in Europe and then the rest of the world.

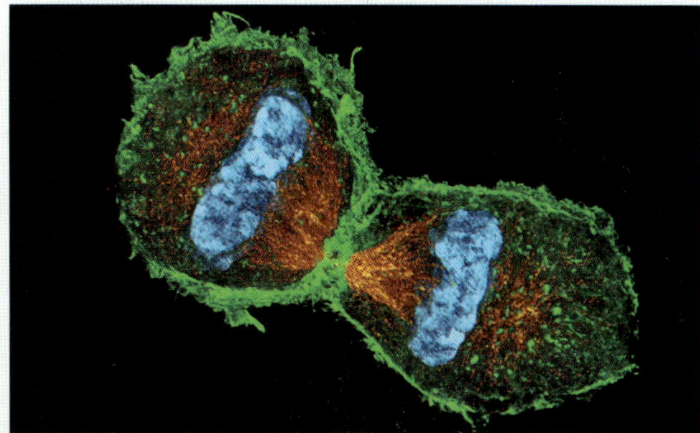

FIGURE 2.1.3 A cell dividing to form a new cell

COMMON CELL FEATURES

Cells are the basic structural unit of all living things. Although there are different types of cells, the cells of plants, animals and bacteria share a number of common features (Figure 2.1.4). These common features include:

- a **plasma membrane** (also called a cell membrane)—separates the interior of the cell from the outside environment
- **cytoplasm**—consists of the cytosol and specialised structures called **organelles**. **Cytosol** is a gel-like substance. It is made up of more than 80% water and contains ions, salts and organic molecules
- **deoxyribonucleic acid (DNA)**—carries hereditary information, directs the cell's activities and is passed from generation to generation
- **ribosomes**—organelles responsible for the synthesis of **proteins**.

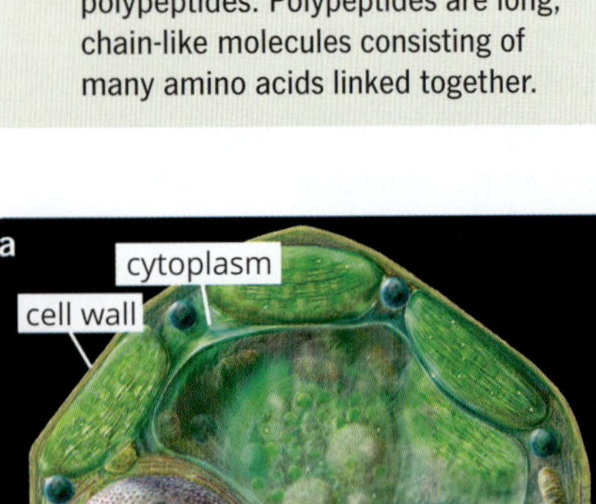
Proteins are large molecules composed of one or more polypeptides. Polypeptides are long, chain-like molecules consisting of many amino acids linked together.

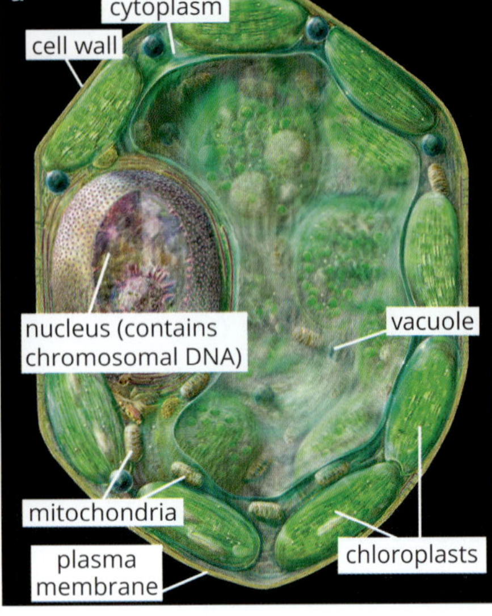

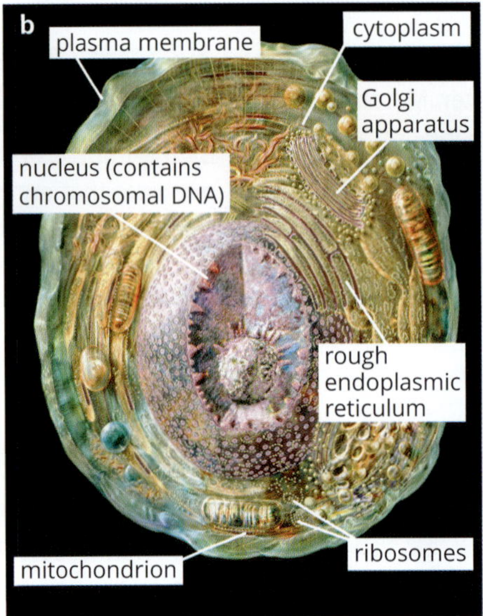

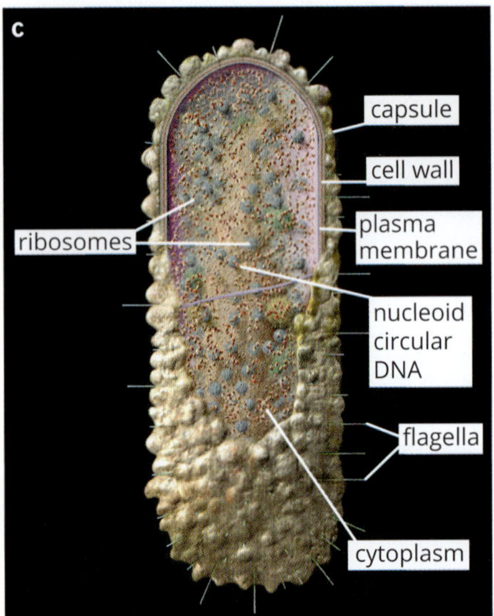

FIGURE 2.1.4 The cells in (a) a plant, (b) an animal and (c) a bacterium share common features, including a plasma membrane, cytoplasm, DNA and ribosomes.

CELL TYPES

The two fundamentally different cell types are prokaryotic cells and eukaryotic cells. Organisms are classified according to which cell type they have. Protists, fungi, plants and animals are composed of **eukaryotic cells** and are classified as eukaryotes. Bacteria and archaea are composed of single **prokaryotic cells** and are classified as prokaryotes. Prokaryotic cells are small and lack membrane-bound organelles, but they still have a number of features in common with eukaryotic cells (Figure 2.1.5).

FIGURE 2.1.5 (a) A typical prokaryotic cell and (b) eukaryotic cell. Note the different membrane-bound organelles in the eukaryotic cell and the lack of such organelles in the prokaryotic cell.

PROKARYOTES

Prokaryotic organisms are single-celled (**unicellular**) and have a simple cell structure. Bacteria, cyanobacteria (photosynthetic bacteria) and archaea such as methanogens are examples of prokaryotes. Prokaryotic organisms can be found everywhere, even in extreme environments such as volcanoes.

Most prokaryotic cells are small and therefore have a large surface area relative to their volume. This allows the cells to take in and release materials efficiently and replicate quickly. You will learn more about surface area to volume ratio in Section 2.2.

The structure of a typical prokaryotic cell is shown in Figure 2.1.6. Prokaryote cells lack membrane-bound organelles, and their cytoplasm contains scattered ribosomes that are involved in the synthesis of proteins. The genetic material of prokaryotic cells is usually a single, circular DNA **chromosome**, which is contained in an irregularly shaped region called the nucleoid. The nucleoid does not have a nuclear membrane, unlike the membrane-bound nucleus of eukaryotes.

In addition to this chromosomal DNA, many prokaryotic cells also contain small rings of double-stranded DNA called **plasmids**.

The plasma membrane of prokaryotic cells is surrounded by an outer cell wall. Many bacteria also have a capsule outside the cell wall, which protects the cell from damage and dehydration. Some prokaryotes have flagella that enable them to move freely.

Many prokaryotes also have small hair-like projections called pili, which are involved in the transfer of DNA between organisms and can also help generate movement. Specialised pili that can attach to surfaces are called fimbriae.

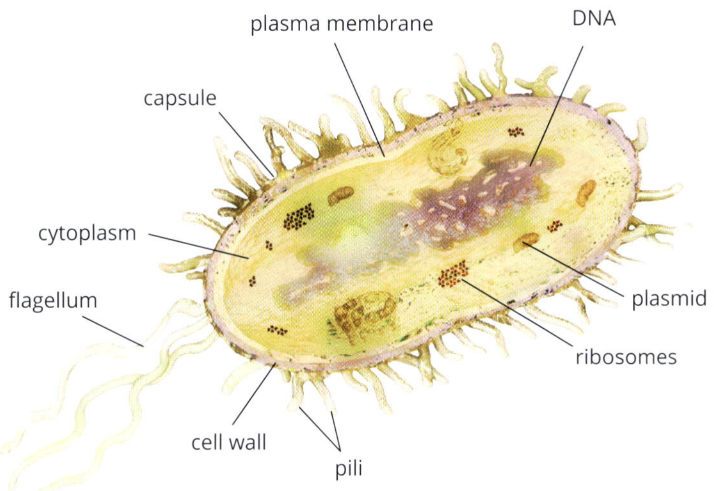

FIGURE 2.1.6 A typical prokaryotic bacterial cell

Bacteria

Most prokaryotes in the domain Bacteria are microscopic single-celled organisms. Fossil evidence dated at 3.5 billion years old confirms that bacteria were the first type of living organism on Earth. Today they are still the most numerous type of organism in the biosphere.

Bacteria have very diverse metabolisms, and they can survive in a great range of habitats and conditions. They are most common in environments of moderate temperature that are moist and low in salt, where sunlight or **organic compounds** are plentiful, and in or on plants and animals.

Bacteria need little oxygen to survive, as they have many ways of extracting energy and fixing carbon. Bacteria are able to obtain energy from sunlight (photosynthesis) or by reducing **inorganic compounds** such as sulfides or ferrous ions (chemosynthesis).

Because they break down many kinds of substances, including plant and animal remains and wastes, bacteria play an important role in ecosystems. They are also widely used in industry to manufacture foods, such as cheeses and yoghurt, and in medicine, to produce antibiotics, drugs and even human insulin. Some bacteria can break down oils and plastics, which makes them useful for pollution control.

Gram-positive and Gram-negative bacteria

The cell walls of prokaryotes are distinctive for containing **murein** (also known as peptidoglycan), which is a giant molecule consisting of sugar molecules linked by **amino acids**. In most bacteria the murein forms a cell wall in a mesh-like layer outside the plasma membrane. Prokaryotic bacteria are commonly identified as either Gram-negative or Gram-positive. A purple stain called crystal violet is used for this purpose.

BIOFILE

Classification

In older classification systems all organisms were divided into five ranks, called kingdoms. Prokaryotic organisms were placed in the kingdom Monera and eukaryotic organisms were placed in the kingdoms Protista, Plantae, Fungi and Animalia. These systems were based on the morphology (appearance and structure) of organisms.

However, in the late 1970s the use of DNA techniques in the emerging field of evolutionary genetics led to the discovery of two different types of prokaryotic cells. This resulted in the development of a system with three domains and six kingdoms (see figure below). Domains are now the highest rank in taxonomy, instead of kingdoms. Prokaryotes are divided into two domains: Bacteria and Archaea. All eukaryotic organisms are placed in a third domain called Eukarya. The four kingdoms within the Eukarya domain remain the same: Protista, Plantae, Fungi and Animalia.

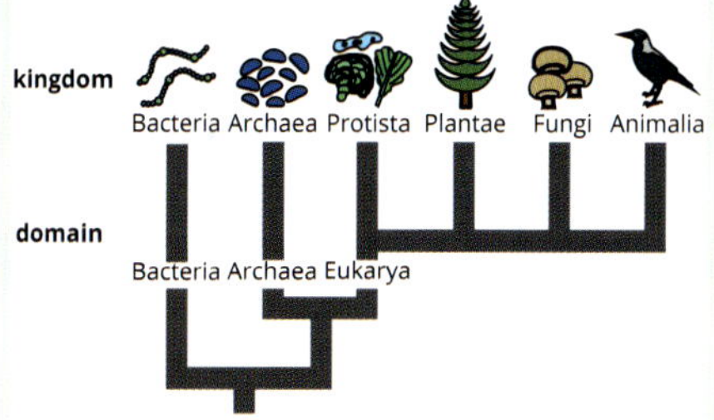

The classification of living things, showing the three domains based on cell types, and the six kingdoms. Bacteria and Archaea have prokaryotic cells. Protista, Plantae, Fungi and Animalia have eukaryotic cells.

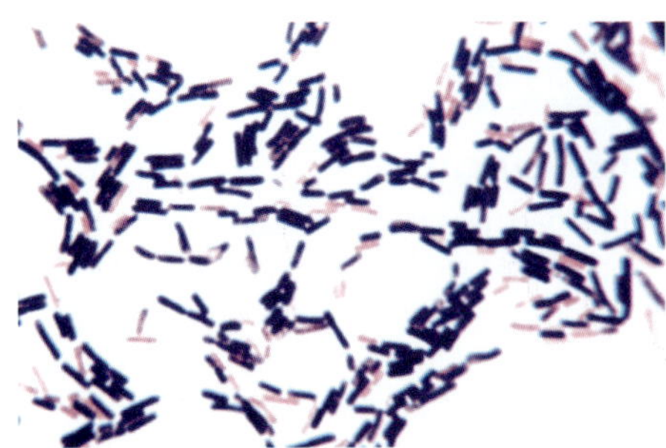

FIGURE 2.1.7 Light micrograph (LM) showing Gram-positive (stained purple) and Gram-negative (stained pink) bacteria

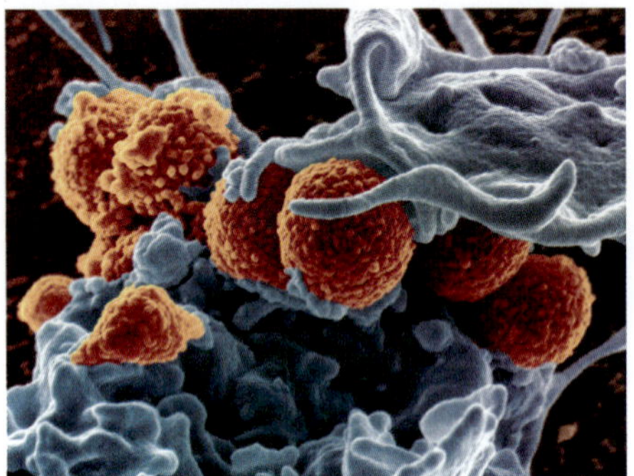

FIGURE 2.1.8 A scanning electron micrograph (SEM) of *Staphylococcus aureus* (commonly called 'golden staph') being engulfed by a white blood cell. The cocci are coloured orange in this image to represent their actual colour.

Gram-positive bacteria have a thicker layer of murein that absorbs and holds the stain, so they give a purple or 'positive' result. Gram-negative bacteria have a much thinner layer of murein that does not retain the stain as well, so they give a pink or 'negative' result (Figure 2.1.7).

There are numerous types of Gram-negative and Gram-positive bacteria. For example, Gram-positive cocci are spherical bacteria and include *Staphylococcus* and *Streptococcus*, which can cause serious diseases or death in humans (Figure 2.1.8).

An example of a Gram-negative bacterium is a cyanobacterium (Figure 2.1.9). Cyanobacteria were once called blue-green algae because they contain chlorophyll, but they were later found to be prokaryotes and placed in the Bacteria domain. They often form dense colonies in shallow estuaries or fresh water. Some species can form large colonies ('blooms') that produce toxins capable of killing fish and other aquatic life and causing illness in humans.

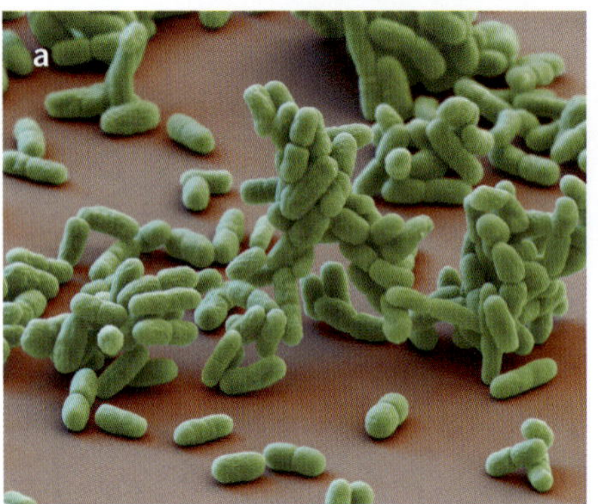

FIGURE 2.1.9 (a) SEM of *Synechococcus* cyanobacterium. (b) A colony of *Synechococcus* cyanobacteria in a lake

Archaea

The prokaryotes in the domain Archaea include **extremophiles**. These are organisms that can live in extreme conditions, such as:

- areas of high temperatures (thermophiles)
- areas of low temperatures (cryophiles)
- the upper atmosphere
- very alkaline environments
- very acidic environments (acidophiles)
- very salty environments (halophiles)
- environments with little or no oxygen
- very arid environments (xerophiles)
- areas without light
- petroleum deposits deep underground.

Archaea hold records for living in the hottest places (121°C), the most acidic environments (pH 0), and the saltiest water (about 30% salt). However, some archaea live in less extreme environments, such as the open seas.

The ability of archaea to live in extreme environments is due in part to their unique membranes. Like other living organisms, archaea possess a membrane composed mainly of **lipids**. Plasma membranes need to be fluid to respond to external deformations and damage and allow proteins to move around.

Lipids are 'fatty' organic compounds, including fats and oils, composed mainly of carbon, hydrogen and oxygen.

The lipids in eukaryotic plasma membranes have fluidity and selective permeability, but only in a narrow range of temperatures. The lipids that compose archaean membranes are different. They form a unique plasma membrane structure that remains fluid and permeable over a wide range of temperatures, from freezing cold to boiling hot.

There are many different types of extremophiles. Hyperthermophiles like *Pyrococcus furiosus* are extremophiles that can survive in very hot environments such as undersea vents, where temperatures are often above 100°C (Figure 2.1.10). They can also withstand extremely high pressures. *Sulfolobus* bacteria, which live in volcanic springs, are thermophiles as well as acidophiles: they can survive both high temperatures and high acidity (Figure 2.1.11).

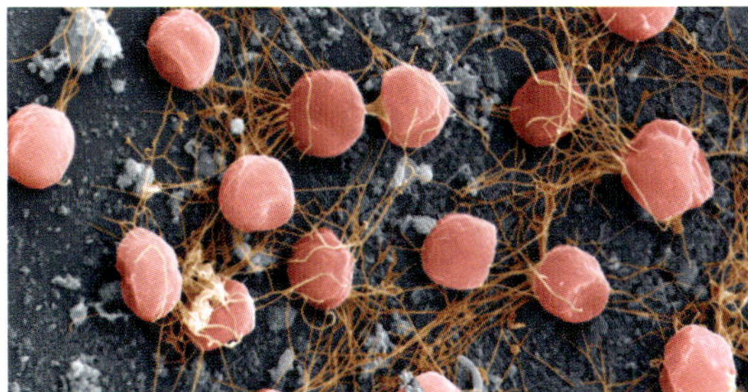

FIGURE 2.1.10 SEM of hyperthermophile *Pyrococcus furiosus*. These bacteria can only exist in very hot environments such as hot undersea vents.

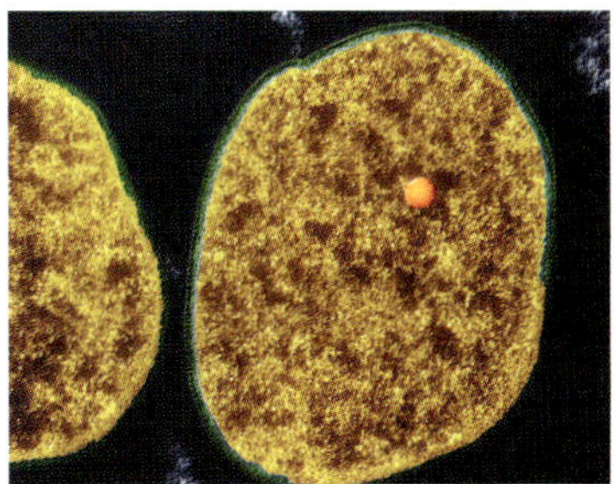

FIGURE 2.1.11 *Sulfolobus* bacteria are thermophiles as well as acidophiles. They thrive in hot, acidic environments.

COMPARISON OF PROKARYOTIC AND EUKARYOTIC CELLS

There are a number of differences between prokaryotic and eukaryotic cells (Table 2.1.1). Eukaryotic cells have their DNA in the nucleus, in the form of linear chromosomes, while prokaryotes have a single-stranded loop of DNA. Prokaryotic cells lack membrane-bound organelles, while eukaryotic cells contain many different membrane-bound organelles. You will learn more about eukaryotic cells and organelles in Section 2.2.

TABLE 2.1.1 Comparison of prokaryotic and eukaryotic cells

Feature	Prokaryotic cells	Eukaryotic cells
size	• very small	• larger, with large variation in size
membrane-bound organelles	• absent, no membrane-bound organelles	• many organelles bound by membranes, forming a compartmentalised internal structure
chromosomal DNA	• DNA chromosome in the form of a single-stranded loop • located in a region of cytoplasm called the nucleoid • DNA not enclosed in a membrane	• DNA in the form of linear, thread-like chromosomes • located in the nucleus, which is separated from the cytoplasm by a double-layered membrane
ribosomes	• many tiny ribosomes scattered in the cytoplasm	• many ribosomes, either attached to the endoplasmic reticulum (ER) or free in the cytoplasm
plasma membrane	• bilayer of phospholipid molecules enclosing the cytoplasm in bacteria	• bilayer of phospholipid molecules enclosing the cytoplasm
cell wall	• in bacteria, consists of a protein/carbohydrate compound called murein	• present in fungi, plants and some protists • consists mainly of carbohydrates: chitin in fungi and cellulose in plants
flagella	• may have flagella to provide movement • consists of three protein fibrils coiled in a helix and protruding through the plasma membrane and cell wall	• may have flagella or cilia for motility (but not in fungi) • consists of a highly organised array of microtubules (hollow protein tubes) enclosed by an extended plasma membrane

2.1 Review

OA ✓✓

SUMMARY

- The cell theory states that:
  - all organisms are composed of cells
  - all cells come from pre-existing cells
  - the cell is the smallest living organisational unit.
- All cells have the following key features
  - a plasma membrane
  - cytoplasm
  - genetic material in the form of DNA
  - ribosomes.
- There are two fundamentally different types of cells—prokaryotic and eukaryotic.
- Prokaryotic cells have a simple structure, with a nucleoid lacking a membrane, scattered ribosomes, and DNA mainly in a single-stranded loop in the nucleoid. Examples are bacteria and archaea.
- Archaea (the prokaryotic extremophiles) are often found in very harsh environments where their unique plasma membrane structure protects them.
- Archaea have a unique plasma membrane structure that remains fluid and permeable over a wide range of temperatures, from freezing cold to boiling hot.
- Eukaryotic cells have a complex structure, membrane-bound nucleus, many organelles in the cell cytoplasm, and DNA mainly in chromosomes in the nucleus. Examples are plants, animals and fungi.

KEY QUESTIONS

Knowledge and understanding

1 State the cell theory.

2 Describe the main differences between prokaryotic and eukaryotic cells.

3 All cells have a plasma membrane. Explain the difference between animal plasma membranes and the membranes of the prokaryotes called archaea.

4 a Which cells have a cell wall?
  b What is the main component of each cell type's cell wall?

Analysis

5 A cell was viewed through a microscope under a magnification of 1000×. It was observed to have large internal compartments and a cell wall. Determine what type of cell it might be from this brief observation.

6 The domains Bacteria and Archaea are believed to have the oldest lineage of any cell types on Earth. Outline at least four reasons, based on the structure and function of bacteria and archaea, to explain why they have survived so successfully.

2.2 Cell size and compartmentalisation

In Section 2.1, you learnt that the two main types of cells are prokaryotic and eukaryotic cells:

- prokaryotic cells are relatively small and lack membrane-bound organelles. Bacteria and archaea are called prokaryotes.
- eukaryotic cells are relatively large and more complex. They possess membrane-bound organelles, such as a nucleus and mitochondria. Protists, fungi, plants and animals are called eukaryotes because they are composed of eukaryotic cells.

In this section you will learn about cell size, surface area to volume ratio and the importance of cell compartmentalisation and membrane-bound structures in eukaryotes.

CELL SIZE

Cells vary greatly in size (Figure 2.2.1). Most cells are only visible under a light microscope, and their size is usually measured in micrometres (μm: there are 1000 μm in 1 mm). There are a few exceptions, though; the egg cell of some bird species can be many centimetres in diameter. Some typical cell sizes are as follows:

- bacterium: 0.1–1.5 μm long
- human: 8–60 μm long
- paramecium (a single-celled aquatic eukaryote): about 150 μm long.

The thickness of plasma membranes also differs between cells, and can be between 0.004 μm and 0.1 μm thick.

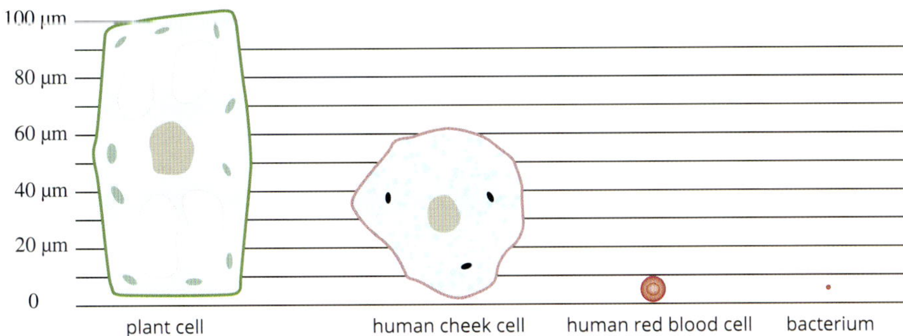

FIGURE 2.2.1 A typical plant cell, human cheek cell, human red blood cell and bacterium. Note the great difference in size between the eukaryotic cells and the prokaryotic cell (bacterium).

BIOFILE

Giant bacteria

In 1985 a huge single-celled organism was found in the gut of a surgeonfish (Figure 2.2.2) caught in the Red Sea. It measured up to 600 μm (0.6 mm) long and 80 μm (0.08 mm) wide, making it visible to the naked eye. It was first thought to be a protist, but further examination showed it to be a giant bacterium, now called *Epulopiscium fishelsoni* (Figure 2.2.3). The largest bacterium known is *Thiomargarita namibiensis*, which grows up to 750 μm long.

FIGURE 2.2.2 The brown surgeonfish (*Acanthurus nigrofuscus*), the species in which *Epulopiscium fishelsoni* was discovered

FIGURE 2.2.3 The genus *Epulopiscium* is an unusual group of cigar-shaped Gram-positive organisms that live in the guts of fish. The bacteria grow up to nearly a millimetre in length, big enough to be seen with the naked eye.

CASE STUDY

Investigating cells

Cytology is the study of cells. Cytologists use a variety of tools and techniques to study cells, including several microscopy techniques. Modern microscopy techniques, including light and electron microscopy, have greatly advanced our understanding of the structure and function of cells.

Light microscopy

Most cells are so small that they can only be seen with a microscope (Figure 2.2.4). The light microscope uses light and a system of lenses to magnify the image. One lens is called the objective and the other is the eyepiece or ocular lens. The total magnification of a microscope is calculated by multiplying the magnifying powers of the objective lens and the eyepiece. For example, a 10× objective lens used with a 4× eyepiece gives a total magnification of 40×.

One of the main advantages of light microscopy is that it can be used to view living cells in colour. A thin specimen is mounted on a glass slide and placed on the stage under the lenses. Light travels through the specimen and into the lens system, and the image is viewed by eye or with a digital camera.

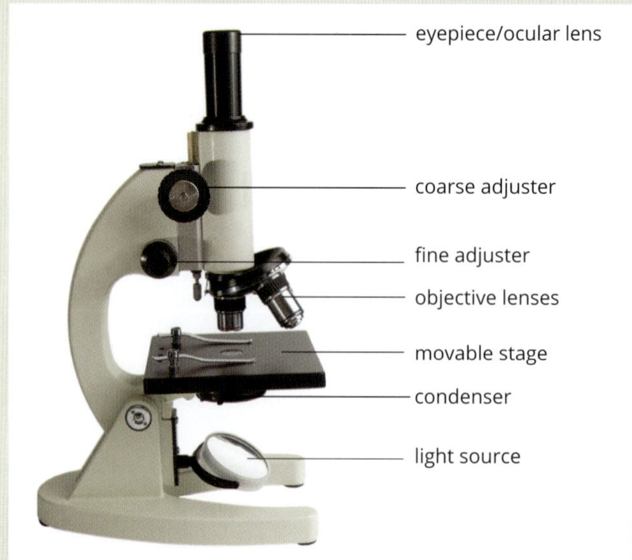

FIGURE 2.2.4 A light microscope and its parts

Electron microscopy

In electron microscopy, an object is viewed using an electron beam instead of light. This allows us to see structures in far more detail than is possible using light microscopy (Figure 2.2.5). An electron microscope produces a narrow beam of electrons that is maintained by electromagnetic lenses, which are coils that surround the tube and emit an electromagnetic field. Electrons striking the specimen are absorbed or scattered, or pass through it. The image obtained with an electron microscope has a much higher resolution and a greater depth of field than an image from a light microscope. Electron microscopy produces only black and white images, but these are often coloured later to highlight important features. Two types of electron microscopy commonly used are transmission electron microscopy and scanning electron microscopy.

Images taken using microscopes are called micrographs. The accepted abbreviations are:

- LM—light micrograph
- TEM—transmission electron micrograph
- SEM—scanning electron micrograph.

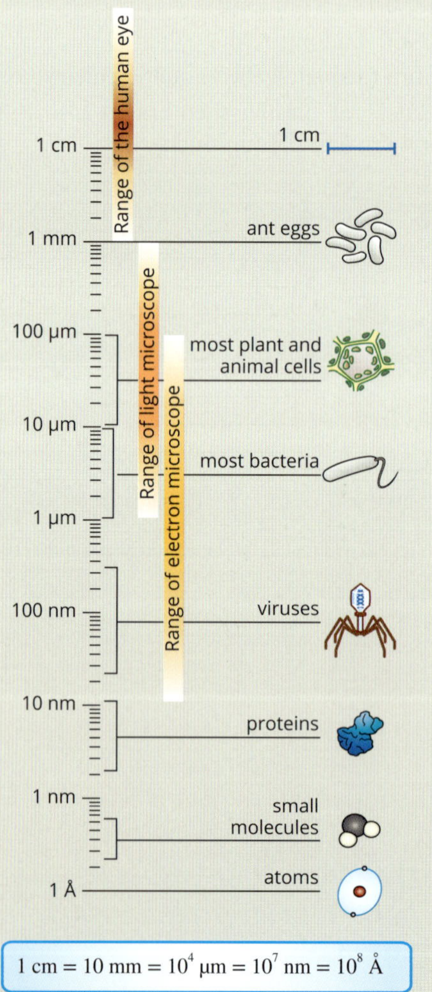

FIGURE 2.2.5 Comparison of the ranges of the light and electron microscopes. (Note that the scale is logarithmic, i.e.increases by powers of 10.)

SURFACE AREA TO VOLUME RATIO

All cells must exchange nutrients and wastes with their environment via the plasma membrane. In addition, organelles enclosed by their own plasma membrane also exchange materials with the cytosol.

The surface area of the plasma membrane around a cell and cellular compartments affects the rate of exchange that is possible between the cell organelle and its environment.

Larger cells have greater metabolic needs, so they need to exchange more nutrients and waste with their environment. However, because of the surface area to volume relationship, they do not have a proportionally larger surface area of plasma membrane for this exchange to take place. By compartmentalising specific areas of the cell into the organelles mentioned earlier in this section, the cell can maximise its efficiency in exchanging matter with its environment and its ability to undertake a variety of different life processes, such as cellular respiration or photosynthesis.

A high surface area to volume ratio increases the efficiency of cellular processes.

Surface area versus volume

The relationship between surface area and volume is known as **surface area to volume ratio** and can be explained using cubes (Figure 2.2.6). A cube with a side length of 1 cm has a surface area of 6 cm^2, a volume of 1 cm^3 and a surface area to volume ratio of 6. A cube with a side length of 10 cm has a surface area of 600 cm^2, a volume of 1000 cm^3, and a surface area to volume ratio of 0.6. Comparing these two cubes, it can be observed that, while the volume of the bigger cube is 1000 times larger than the volume of the smaller cube, its surface area is only 100 times larger.

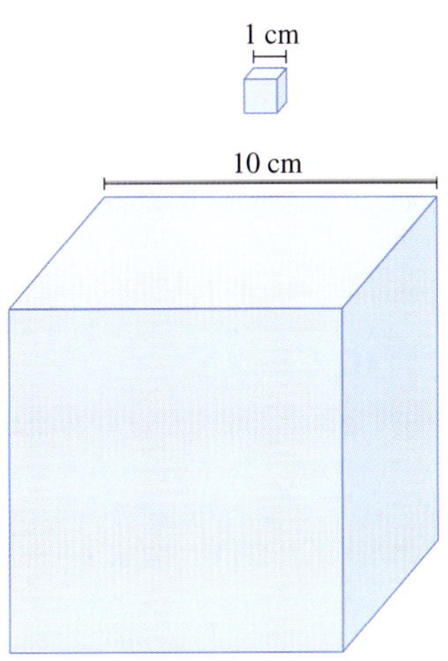

Side length	1 cm	10 cm
Surface area	6 cm^2	600 cm^2
Volume	1 cm^3	1000 cm^3
Surface area to volume ratio	6	0.6

FIGURE 2.2.6 Two cubes, showing the relationship between surface area and volume

Increasing the cell surface area to volume ratio

Three ways of increasing the membrane surface area of cells without changing cell volume are:

- cell compartmentalisation
- a flattened shape
- plasma membrane extensions.

Cell compartmentalisation

Cell compartmentalisation allows organelles to have the right conditions and concentration of **enzymes** and reactants for a particular function, making the processes in the organelles, and in turn the whole cell, highly efficient.

Cell compartmentalisation also allows eukaryotic cells to be much bigger than prokaryotic cells, because:

- it reduces the amount of exchange that needs to occur across the plasma membrane to maintain an environment suitable for all cell functions
- it creates more space for membrane-bound enzymes, allowing increased activity in the cell.

Enzymes are biological molecules that speed up the rate of biochemical reactions.

A flattened shape

As a cell increases in volume, the distance from the centre of the cell to the plasma membrane also increases. The rate of chemical exchange (or rate of diffusion) from the centre of the cell to the surrounding environment may then become too low to maintain the cell.

One way to counteract this effect is to be flatter. For example, flattening a cube while keeping the volume constant results in a larger surface area, and therefore a larger surface area to volume ratio (Figure 2.2.7). This larger surface area to volume ratio allows a higher rate of exchange through the plasma membrane, and reduces the distance that substances need to be transported to and from the plasma membrane.

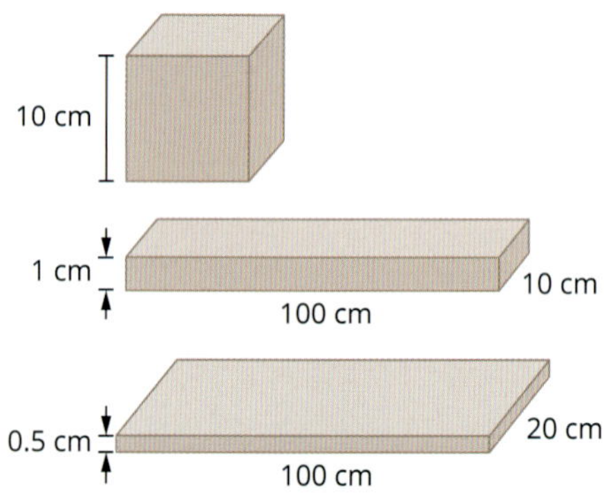

Surface area	Volume	Surface area to volume ratio
60 cm^2	1000 cm^3	0.06
2220 cm^2	1000 cm^3	2.22
4120 cm^2	1000 cm^3	4.12

FIGURE 2.2.7 The effect of changing shape on the surface area to volume ratio

This flattened shape solution is observed in nature in many types of cells, especially those involved in the rapid transport of substances, such as red blood cells and lung epithelium. These cells usually do not have a high metabolism and so do not contain many organelles, allowing their flattened shape.

Plasma membrane extensions

Instead of being larger or flatter, cells involved in absorbing nutrients or secreting wastes counteract the surface area to volume ratio problem by extending the surface area of their plasma membranes. For example, some animal cells have finger-like extensions of the plasma membrane called microvilli (singular microvillus), which increase the surface area (Figure 2.2.8). Another example is root hairs (lateral extensions of root cells) in plants, which increase the surface area of the root to allow the plant to absorb more water and nutrients from the soil.

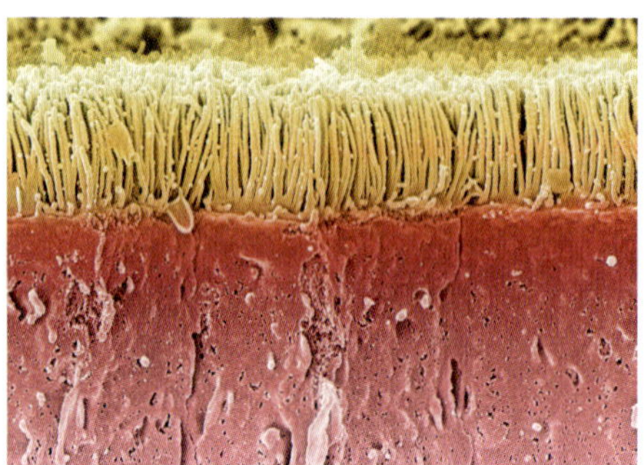

FIGURE 2.2.8 SEM of microvilli in the small intestine, where digested food is absorbed

A flattened shape would not be useful for these cells because they require an increased surface area in particular regions of the cell. For example, in cells of the small intestine an increased surface area for exchange is only required on the inside of the intestinal tube from which the cells absorb nutrients. In addition, these cells have a high metabolism and possess many organelles. If they were flattened, the distance between the different organelles of the cell would affect the movement of substances within the cell and reduce its functionality.

COMPARTMENTALISATION IN EUKARYOTIC CELLS

As well as a plasma membrane surrounding the cytoplasm, eukaryotes have internal membranes that form specialised membrane-bound compartments within the cell (Figure 2.2.9). This is known as cell compartmentalisation. The membrane-bound compartments are organelles (note that not all organelles have membranes).

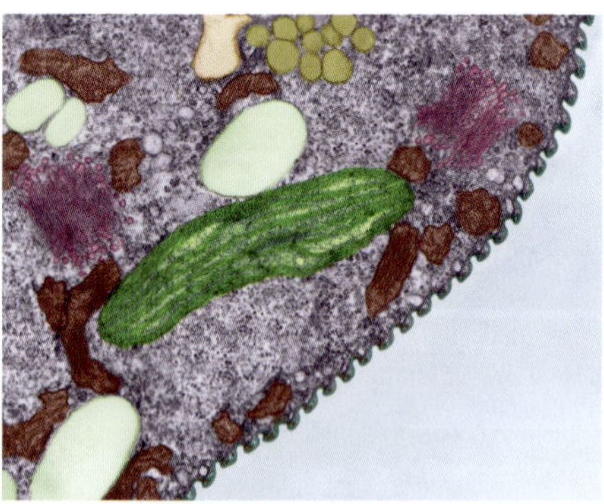

FIGURE 2.2.9 Transmission electron micrograph (TEM) of the single-celled eukaryote, *Euglena gracilis*. Membrane-bound organelles can be seen inside the cell—chloroplast (green), Golgi apparatus (pink) and mitochondria (dark red)—as well as non membrane–bound ribosomes (purple).

Each membrane-bound organelle has a different function. For this reason, each organelle requires a different internal composition, including a high concentration of enzymes and reactants that are needed for the organelle's particular function.

Role of organelle membranes

The membranes surrounding organelles control the movement of substances between the organelle and the cell's cytosol. Just as the plasma membrane of a cell enables the cytosol to have a different composition from the cell's external environment, the membranes of membrane-bound organelles enable each organelle to have a different composition from the surrounding cytosol and other organelles.

Benefits of compartmentalisation

Cellular compartmentalisation benefits the cell in several ways:

- it allows enzymes and reactants for a particular function to be close together in high concentrations and at the right conditions, such as at optimum pH levels, so that the processes within the organelles are very efficient
- it allows processes that require different environments to occur at the same time, in the same cell
- it makes the cell less vulnerable to changes in its external environment.

CASE STUDY ANALYSIS

Investigating surface area to volume ratio

All cells rely on the movement of dissolved solutes and water to be able to function effectively. The movement of solutes can be modelled using agar (a jelly-like substance) infused with a pH indicator such as phenolphthalein. When agar is submerged in an acidic solution, the movement of the solute can be observed as the indicator changes colour. The effect of surface area and volume can be investigated using different-sized agar cubes.

For example, a student wished to investigate the effect of surface area and volume on the movement of hydrochloric acid into agar containing phenolphthalein indicator. At high pH values, the phenolphthalein appears pink. When the indicator is exposed to lower pH values, i.e. those seen in acidic solutions, the indicator becomes clear.

The student cut three different-sized cubes in the dimensions shown in Table 2.2.1.

The cubes were then submerged in 100 mL of dilute hydrochloric acid for 20 minutes. The student gently stirred the agar cubes in each beaker every 5 minutes to ensure all sides of each cube were always exposed to the hydrochloric acid (Figure 2.2.10).

After 20 minutes, the agar cubes were removed from the beakers and cut in half. The height and width of the agar that remained pink was measured using a ruler as shown in Figure 2.2.11, and recorded in Table 2.2.2.

FIGURE 2.2.10 Agar cubes submerged in dilute hydrochloric acid

TABLE 2.2.1 Size of phenolphthalein cubes

Cube dimensions (cm)	1 × 1 × 1	2 × 2 × 2	4 × 4 × 4
Surface area (cm^2)	6	24	96
Volume (cm^3)	1	8	64
Surface area to volume ratio	6 : 1	24 : 8 or 3 : 1	96 : 64 or 1.5 : 1

FIGURE 2.2.11 Agar cubes cut in half

TABLE 2.2.2 Resulting volume of cubes that remained pink after exposure to hydrochloric acid

Cube dimensions (cm)	Dimensions of cube that remained pink (cm)	Volume of cube that remained pink (cm^3)	Volume of whole cube (cm^3)	Volume of uncoloured section (cm^3)	Proportion of cube uncoloured (%)
1 × 1 × 1	0.25 × 0.25 × 0.25	0.0156	1	0.9844	98.44
2 × 2 × 2	1.0 × 0.95 × 1.1	1.045	8	6.955	86.94
4 × 4 × 4	3.1 × 3.2 × 3.0	29.76	64	34.24	53.50

Analysis

1. Name the independent and dependent variables the student investigated in this experiment.
2. Describe some of the controlled variables the student needed to consider.
3. Predict what would have happened if the student had left the agar sitting in the hydrochloric acid solution for another 10 minutes.
4. The student cut a piece of agar into a rectangular prism with the dimensions 1 cm × 4 cm × 2 cm. Calculate the following for this piece of agar.
   a i surface area
   ii volume
   iii surface area to volume ratio.
   b Hypothesise about the degree of discolouration that would be seen in this prism if it were exposed to the same hydrochloric acid for 20 minutes. Justify your prediction.

CASE STUDY

An artificial cell with working organelles

Cells are extremely complex, yet incredibly efficient, processing multiple reactions simultaneously in a very small space. The efficiency of eukaryotic cells is largely a result of compartmentalisation. Chemical signals send messages between the compartments to ensure that the cell is functioning optimally as a unit.

Biochemists are interested in understanding how a cell can carry out such efficient chemistry on such a small scale, but the complexity of cells makes it extremely difficult for scientists to mimic these structures and their functions in the laboratory.

In 2014 a team of researchers from the University of Bordeaux in France and Radboud University in the Netherlands built the world's first artificial eukaryotic cell. The scientists created the cell's organelles from tiny enzyme-filled spheres. The organelles were placed inside a droplet of water, which was then coated with a polymer layer to create a plasma membrane (Figure 2.2.12). Using this method, the researchers had built a compartmentalised structure that mimicked a eukaryotic cell. To test whether the cell was functional, the researchers used fluorescent dyes to observe the series of chemical reactions within it. Just like the cells in our bodies, chemical reactions took place in the organelles and moved into the plasma membrane for processing elsewhere. The scientists had successfully created an artificial cell with working organelles! In the future, researchers hope to create cells that can produce their own energy.

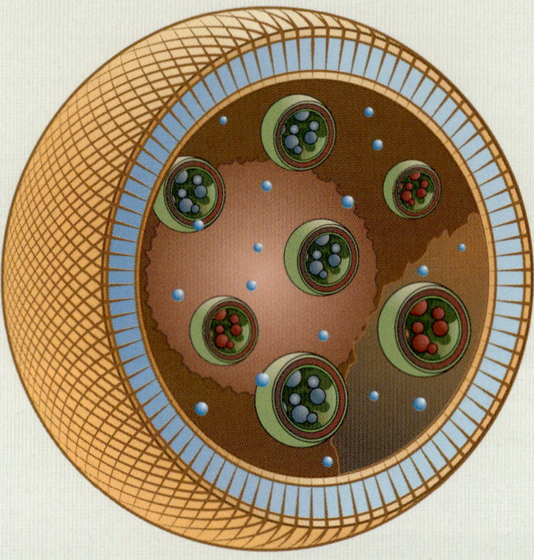

FIGURE 2.2.12 Cutaway diagram of an artificial cell. The cell consists of a polymer membrane surrounding a water droplet containing enzyme-filled spheres that function as organelles.

PA 2

2.2 Review

SUMMARY

- Cells vary greatly in size, and a microscope is needed to see most cells.
- Surface area to volume ratio is an important factor in limiting cell size.
- A large object has a smaller surface area to volume ratio than a small object with the same shape.
- The surface area of the plasma membrane around a cell and cellular compartments affects the rate of exchange that is possible between a cell, an organelle and their environments.
- Compartmentalising specific areas of the cell allows a cell to maximise its efficiency in exchanging matter with its environment and its ability to undertake a variety of different life processes.
- Three ways of increasing the membrane surface area of cells without changing cell volume are:
  - compartmentalisation (e.g. organelles)
  - a flattened shape (e.g. red blood cells)
  - plasma membrane extensions (e.g. microvilli and root hairs in plants).
- Compartmentalisation in eukaryotic cells allows them to:
  - reduce the amount of exchange that needs to occur across the plasma membrane to maintain an environment suitable for all cell functions
  - create more space for membrane-bound enzymes, allowing increased activity in the cell
- Membrane-bound compartments in eukaryotic cells are called organelles.
- Each membrane-bound organelle can have a different composition from the surrounding cytosol and other organelles.
- Organelles benefit the cell by:
  - allowing enzymes and reactants for a particular function to be close together in high concentrations and at the right conditions
  - allowing processes that require different environments to occur at the same time, in the same cell
  - making the cell less vulnerable to changes in its external environment.

KEY QUESTIONS

Knowledge and understanding

1 Identify three ways in which cells have evolved to overcome issues with a small surface area to volume ratio.

2 Organelles are surrounded by a plasma membrane. Explain the purpose of this membrane.

3 Cells of the small intestine have multiple extensions of their plasma membrane called microvilli. Explain the purpose of these microvilli and their importance in the plasma membrane.

Analysis

4 Suggest why, in contrast to eukaryotic cells, prokaryotic cells do not contain membrane-bound organelles.

5 **a** Explain what is meant by 'surface area to volume ratio'.

b Consider the two objects shown, which have the same volume of 8 cm^3. Which shape has the greater surface area?

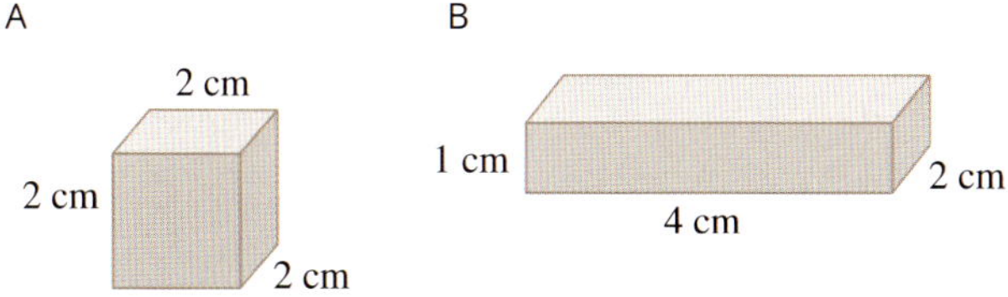

6 Explain why eukaryotic cells are often compartmentalised. Refer to surface area and volume in your answer.

2.3 Organelle structure and function

In Section 2.2 you learnt about the importance of the surface area to volume ratio as a limiting factor for cell size.

Eukaryotic cells are relatively large and complex. Eukaryotes are able to maintain their cell size and efficiency of cellular processes using compartmentalisation. Most of their internal compartments are membrane-bound organelles.

In this section you will learn about the structure and specialisation of plant and animal cell organelles for distinct functions, including chloroplasts and mitochondria.

MEMBRANE-BOUND AND NON-MEMBRANE-BOUND ORGANELLES

Organelles are subcellular structures that each have a specific function (Table 2.3.1). Some organelles, as mentioned earlier, are membrane-bound compartments within the cytoplasm. Membrane-bound organelles are only present in eukaryotic cells.

Prokaryotic cells have some non-membrane-bound organelles, such as ribosomes, a cell wall and sometimes flagella, although the structure and composition of these are usually different to those of eukaryotic cells.

TABLE 2.3.1 Organelle structure and function

Organelle	Structure	Function
nucleus	• membrane-bound: double membrane • contains DNA	• contains genetic information, which controls the activities of the cell • site of DNA transcription
rough endoplasmic reticulum	• membrane-bound: network of cisternae • ribosomes bind to its membranes	• processes and modifies proteins
ribosome	• non-membrane-bound • made of proteins and ribosomal RNA (rRNA)	• synthesises proteins
Golgi apparatus	• membrane-bound: stack of cisternae that are not connected to each other	• processes and packages proteins
lysosome	• membrane-bound: vesicle containing digestive enzymes	• digests cellular waste material and foreign matter
smooth endoplasmic reticulum	• membrane-bound: network of cisternae	• synthesises lipids
mitochondrion	• membrane-bound: double membrane. The inner membrane is highly folded • contains its own DNA molecule	• obtains energy from organic compounds
chloroplast	• spherical or ellipsoidal, with double membrane • contains its own DNA molecule • contains thylakoid sacs	• uses light energy, carbon dioxide and water to produce glucose
centriole	• small structure in the cytoplasm, consisting of microtubules	• involved in cell division and the formation of cell structures such as flagella and cilia
cilium or flagellum	• external structure consisting of microtubules	• motility; movement of substances across cell surface
vacuole	• membrane-bound: fluid-filled vesicle	• stores substances; also involved in cell structure in plant cells
plastid	• small, with double membrane • contains its own DNA molecule	• synthesises and stores various organic molecules
cell wall	• external structure surrounding plasma membrane • composition depends on type of cell	• cell structure and protection

FUNCTION AND ULTRASTRUCTURE OF ORGANELLES

Cellular organelles are involved in a number of different functions (Table 2.3.2). Their functions include the synthesis and processing of proteins and lipids, energy transformations, storage, and maintaining the structure of the cell.

TABLE 2.3.2 Organelles and their functions

Function	Organelle	Present in plants	Present in animals
Synthesis and processing of proteins and lipids	nucleus	✓	✓
	ribosome	✓	✓
	rough endoplasmic reticulum	✓	✓
	Golgi apparatus	✓	✓
	lysosome	✗	✓
	smooth endoplasmic reticulum	✓	✓
Energy transformations	mitochondrion	✓	✓
	chloroplast	✓	✗
Storage and cell structure	centriole	sometimes	✓
	flagellum or cilium	✓	✓
	vacuole	✓	small
	cell wall	✓	✗

SYNTHESIS AND PROCESSING OF PROTEINS AND LIPIDS

The following organelles are involved in the synthesis and processing of proteins and lipids in eukaryotic cells.

Nucleus

In eukaryotes most of the DNA (genetic material) is contained in the **nucleus**, which is a large organelle surrounded by a double-layered nuclear membrane. The genetic material in the nucleus takes the form of linear chromosomes composed of DNA and proteins. Chromosomes are usually not clearly visible, except during cell division when they can be seen in prepared slides under a light microscope. The nuclear membrane contains pores that link it with the cytoplasm (Figure 2.3.1).

The information for the synthesis of new proteins is present in the DNA. Genes in the DNA are transcribed (copied) into a type of **ribonucleic acid (RNA)** called **messenger RNA (mRNA)**. The mRNA leaves the nucleus and moves into the cytoplasm. The most visible structure inside the nucleus of a non-dividing cell is the **nucleolus**. The nucleolus is composed of proteins, DNA and RNA, and is where ribosomes are assembled before being released into the cytoplasm.

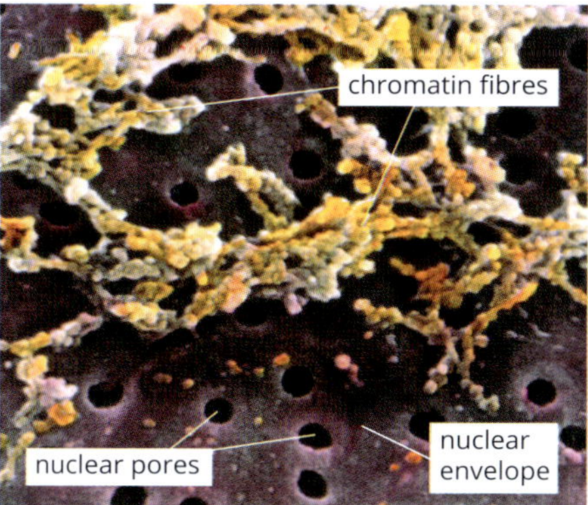

FIGURE 2.3.1 SEM of the external surface of a nuclear envelope in an onion root tip cell. The envelope consists of a double membrane (purple) which encloses the nuclear DNA, seen here as chromatin fibres.

Ribosomes

Cells contain many thousands of ribosomes, which are only about 30 nanometres in diameter and therefore only visible using an electron microscope. Ribosomes are composed of proteins and **ribosomal RNA (rRNA)**, and are sites of protein synthesis. They consist of two subunits joined together (Figure 2.3.2). The subunits in eukaryote ribosomes are different to those in prokaryote ribosomes. Ribosomes are non-membrane bound organelles and occur either free in the cytoplasm or bound to rough endoplasmic reticulum.

Ribosomes translate mRNA into proteins. The mRNA specifies the sequence of amino acids in the protein. Proteins produced in free ribosomes will function in the cell's cytoplasm, while proteins synthesised in ribosomes bound to endoplasmic reticulum are secreted out of the cell, packaged into organelles or inserted into plasma membranes.

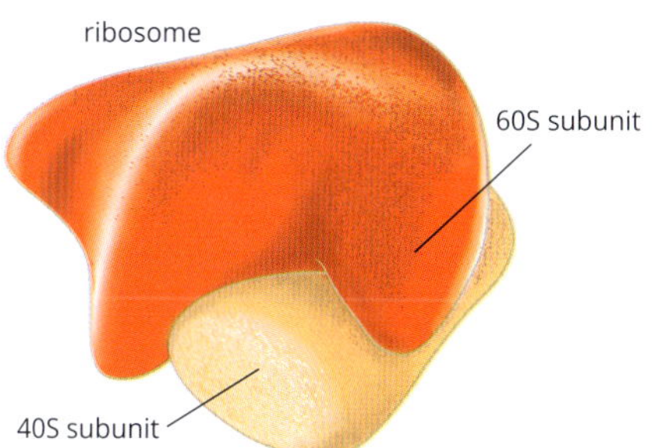

FIGURE 2.3.2 A single eukaryote ribosome consists of a larger subunit and a smaller subunit.

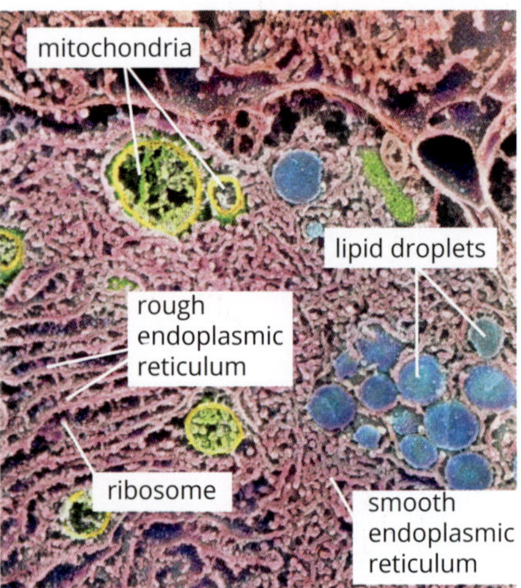

FIGURE 2.3.3 SEM showing smooth (right) and rough (left) endoplasmic reticulum (light pink) inside a Leydig cell of a 14-week-old human fetus. Leydig cells synthesise steroid hormones in the male testis.

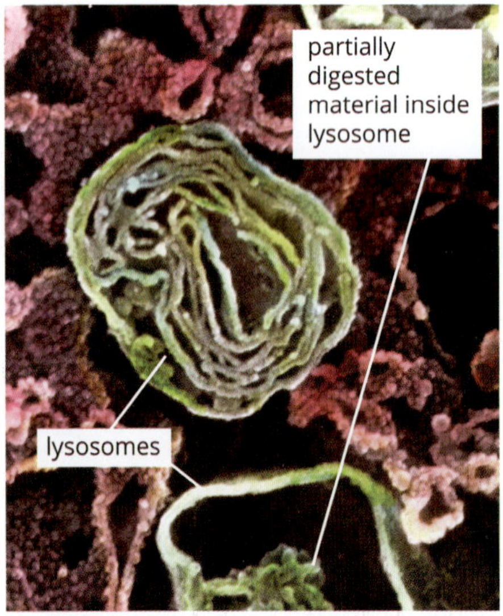

FIGURE 2.3.4 SEM of two lysosomes in a pancreatic cell

The Golgi apparatus has a *cis* and *trans* face. The *cis* face faces the endoplasmic reticulum and the *trans* face faces the plasma membrane.

Endoplasmic reticulum

Endoplasmic reticulum is a network of intracellular membranous sacs (cisternae) and tubules that link with the plasma membrane and other membranous organelles, including the nucleus. The endoplasmic reticulum can be rough or smooth.

Rough endoplasmic reticulum has ribosomes attached, which synthesise proteins. These ribosomes are bound to the membrane of the rough endoplasmic reticulum (Figure 2.3.3). After the proteins are made, they pass into the endoplasmic reticulum cavity, which contains enzymes. The enzymes add sugar molecules to the proteins to form glycoproteins.

Rough endoplasmic reticulum is abundant in cells that actively produce and export proteins, such as pancreatic cells that secrete digestive enzymes. From the rough endoplasmic reticulum, proteins move into the Golgi apparatus for export from the cell.

Smooth endoplasmic reticulum contains the enzymes involved in the synthesis of molecules other than proteins, such as phospholipids and steroids. It is abundant in steroid–hormone secreting cells in the testes, ovaries, kidneys and adrenal glands (Figure 2.3.3).

Lysosomes

Lysosomes are specialised **vesicles** that digest (break down) unwanted matter such as damaged organelles or foreign material (Figure 2.3.4). They are the recycling units of the cells. They are found only in animal cells. Lysosomes are formed when a transport vesicle containing enzymes is released from the Golgi apparatus and fuses with another vesicle called an endosome. The endosome contains material brought into the cell by endocytosis.

Lysosomes fuse with vesicles containing unwanted matter such as damaged organelles or foreign matter. The enzymes in the lysosome then digest the unwanted matter and any small molecules produced that the cell can re-use may diffuse back into the cytoplasm. The rest is retained in the lysosome or released from the cell by exocytosis. You will learn more about these processes in Section 2.4.

Golgi apparatus

The **Golgi apparatus** (also called Golgi body, Golgi complex or simply the Golgi), is a stack of flattened smooth membrane sacs called **cisternae** (Figure 2.3.5). Unlike the rough endoplasmic reticulum, the cisternae in the Golgi apparatus are not connected. When proteins formed in the rough endoplasmic reticulum reach the Golgi apparatus, vesicles are formed from each cisterna to transport the proteins from one cisterna to the next. These vesicles can then be modified for use by the cell, or for transport out of the cell. The cisternae form transport vesicles to move these materials into the cytosol or out of the cell, as is the case with secreted hormones. Vesicles budding from the Golgi apparatus also carry membrane-bound proteins to the plasma membrane and digestive enzymes into lysosomes.

Secretory cells have a well-developed Golgi apparatus, but in other cells the Golgi apparatus is small. Some products packaged by the Golgi apparatus, such as the enzymes found in lysosomes, are not released from the cell.

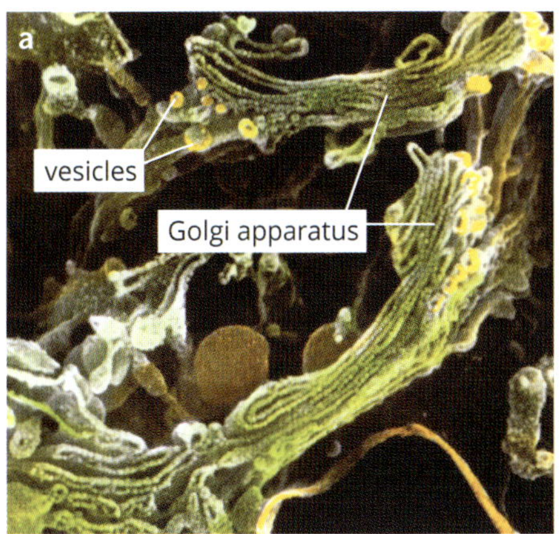

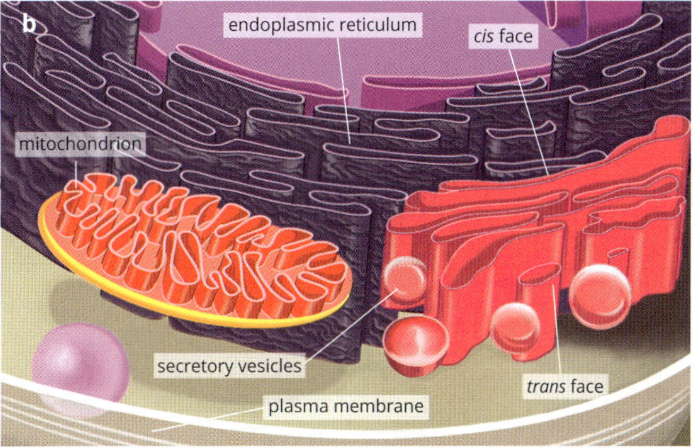

FIGURE 2.3.5 (a) SEM of the Golgi apparatus, the site of synthesis of biochemicals that are packaged into swellings at the margins of the sacs, which are then pinched off as vesicles (small yellow spheres). (b) The Golgi apparatus has a *cis* face, which faces the endoplasmic reticulum, and a *trans* face, which faces the plasma membrane. Here the vesicles are coloured red.

Summary: Synthesis and processing of proteins and lipids

Protein and lipid synthesis and processing is shown in Figure 2.3.6. DNA is transcribed inside the nucleus into RNA. RNA moves out of the nucleus and binds to ribosomes where proteins are synthesised from the RNA information. Proteins that will be secreted out of the cell are made in the ribosomes bound to the rough endoplasmic reticulum. These proteins are modified and packaged in the Golgi apparatus. Vesicles arising from the Golgi apparatus fuse with the outer plasma membrane, releasing their contents from the cell (exocytosis). They also insert membrane-bound proteins into the plasma membrane. Lipids are synthesised and processed in the smooth endoplasmic reticulum.

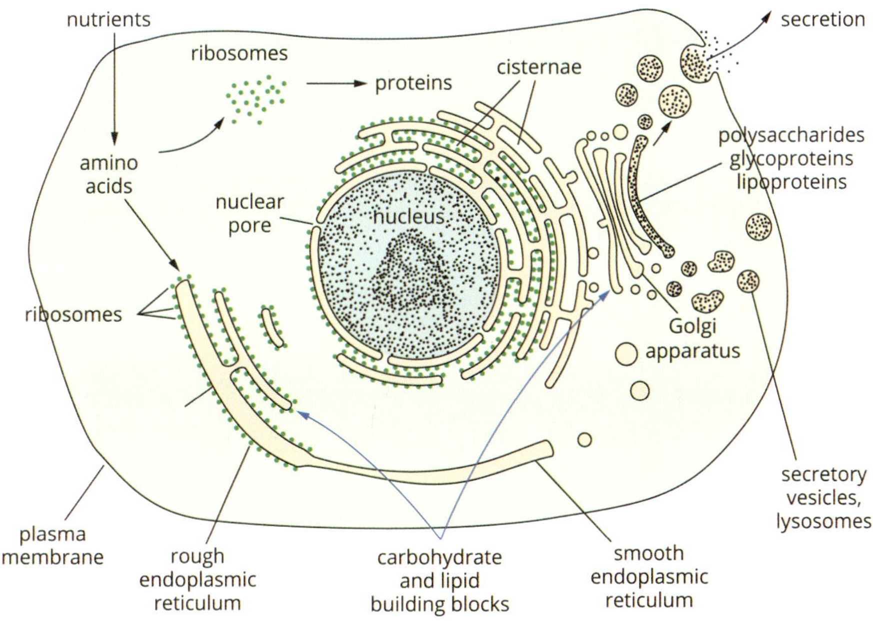

FIGURE 2.3.6 A typical animal cell, showing the organelles involved in synthesising and processing proteins and lipids

BIOFILE

Camillo Golgi (1844–1926)

Camillo Golgi was an Italian physician, anatomist and histologist. Golgi developed a method of staining tissues with silver nitrate, which he called the 'black reaction'. He was the first person to describe the membranous structure in the cell that is now known as the Golgi apparatus or Golgi complex.

Golgi won the Nobel Prize in Physiology or Medicine in 1906, but it was for his work on the structure of the nervous system, not his discovery of the Golgi apparatus. In fact, many biologists did not think the Golgi apparatus existed, and it was not until the 1950s that its existence was confirmed using electron microscopy. Today scientists are increasingly referring to the Golgi apparatus as simply 'the Golgi'.

Camillo Golgi

ENERGY TRANSFORMATIONS

Cells need energy to do work. The energy used by organisms and their cells is stored in organic compounds. Organic compounds are used by all organisms for growth and repair, as well as to perform everyday functions. Mitochondria and chloroplasts are the organelles involved in converting organic compounds into energy within eukaryotic cells.

Mitochondria

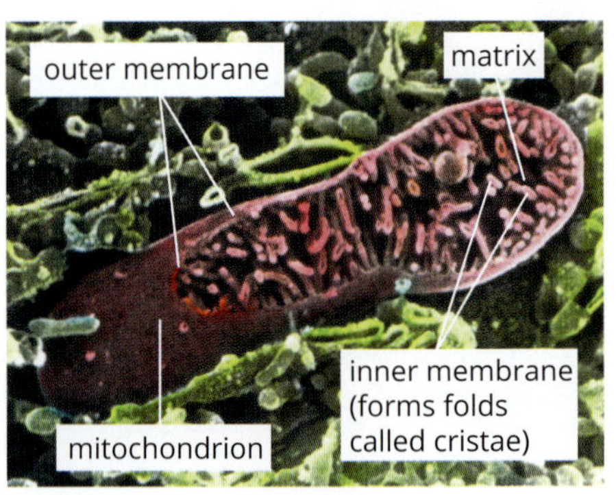

FIGURE 2.3.7 SEM of a single mitochondrion (pink) in the cytoplasm of an intestinal epithelial cell. The highly folded internal membrane provides a large surface area for cellular respiration. The spaces between the folds are filled with fluid.

Mitochondria (singular, mitochondrion) are known as the powerhouses of cells because they convert chemical energy from food into a form that is readily usable by the cell (adenosine triphosphate (ATP)) using a series of biochemical reactions called **cellular respiration**. Mitochondria are composed of two membranes and are filled with fluid. The inner membrane of each mitochondrion has folds called cristae (Figure 2.3.7). There are two different compartments inside mitochondria: an intermembrane space and the matrix. The matrix is the fluid-filled space enclosed by the inner membrane and contains a double-stranded DNA molecule. Different enzymes which control cellular respiration are found inside each compartment and on each membrane.

The number of mitochondria in a cell is related to the cell's energy requirements. Very active cells, such as heart muscle cells and neurons, have many thousands of mitochondria.

CASE STUDY

Mitochondria and cartilage injury

Dr Michelle Delco at Cornell University in the USA is a veterinarian specialising in treatment of injuries to horses (Figure 2.3.8). Her research published in January 2020 established links between equine osteoarthritis and trauma injury to mitochondria in cartilage cells. A drug treatment devised by her collaborating team of biomedical engineers may also be useful for future protection and repair of mitochondria in human cartilage. Their experimental work compared large numbers of cartilage mitochondria during and after a high-speed impact to cartilage tissue samples in the laboratory. Special dyes were used to track healthy mitochondria compared to those that became dysfunctional.

Damaged mitochondria became swollen and lost their cristae structure, usually resulting in the death of their cells. Treated mitochondria were found to maintain a normal, healthy structure and their cells were much less likely to die.

FIGURE 2.3.8 High-performance horses like this prize show jumper are vulnerable to cartilage trauma injury, which leads to osteoarthritis. New research indicates the problem is caused by damage to mitochondria.

Chloroplasts

Chloroplasts are the specialised organelles that carry out **photosynthesis**. Photosynthesis is the series of biochemical reactions that converts solar energy from sunlight into chemical energy to be used by cells. Chloroplasts appear green because of the large amounts of chlorophyll they contain (Figure 2.3.9). They are present in plants and many protists, but never in animals or fungi.

Chloroplasts are composed of a system of three membranes: the outer membrane, the inner membrane and the thylakoid system. Thylakoids are disc-shaped sacs. This system of membranes forms compartments within the chloroplast that contain different enzymes. Chloroplasts also contain a double-stranded DNA molecule.

Chloroplasts trap light energy, which is used to split water molecules into hydrogen and oxygen. The hydrogen then combines with carbon dioxide to make glucose, and the oxygen is released into the atmosphere as a waste product.

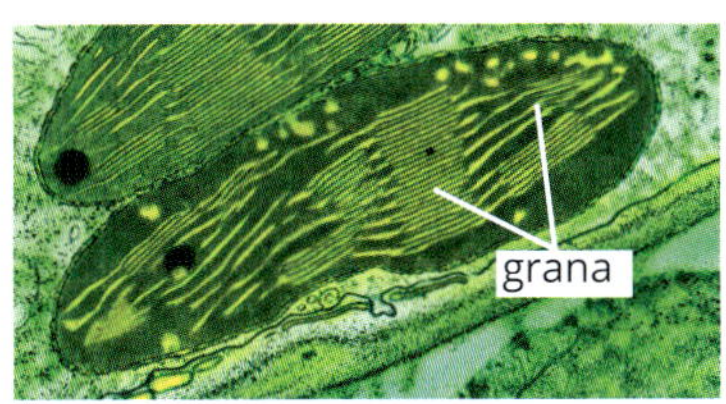

FIGURE 2.3.9 TEM of two chloroplasts seen in the leaf of a pea plant (*Pisum sativum*). Each is seen cut lengthways and contains stacks of grana (yellow) formed from the thylakoid membranes.

Thylakoids are disc-shaped sacs. Multiple thylakoids stack on top of each other and are known as grana (singular, granum).

CASE STUDY ANALYSIS

The endosymbiotic theory

Endosymbiosis is a type of symbiosis in which one organism lives inside the other. The endosymbiotic theory suggests that it is possible for a large cell to ingest a smaller bacterial cell and for the two to become dependent on each other. It was first suggested as the origin of large, complex cells by Konstantin Mereschkowsky in 1910, but he was ridiculed for the idea and it was largely forgotten.

Then in 1967 American biologist Lynn Margulis published a paper titled 'On the origin of mitosing cells'.

In this paper she argued that mitochondria and chloroplasts were both once free-living prokaryotic cells that came to live inside larger cells, eventually becoming specialised organelles that cannot survive outside the cell today.

Although most biologists were extremely sceptical when she first put forward her hypothesis, it is now widely accepted. Evidence supporting the theory includes that mitochondria and chloroplasts have double membranes and their own circular loop of DNA, which you would expect if they were once free-living prokaryotes (Figure 2.3.10). Mitochondria and chloroplasts also have ribosomes that are similar to those found in bacteria, with ribosomal subunits of different sizes to those in the cytosol of eukaryotic cells.

It is now thought that mitochondria came from aerobic bacteria (bacteria that can survive in the presence of oxygen) and that chloroplasts came from cyanobacteria (bacteria that obtain their energy by photosynthesis).

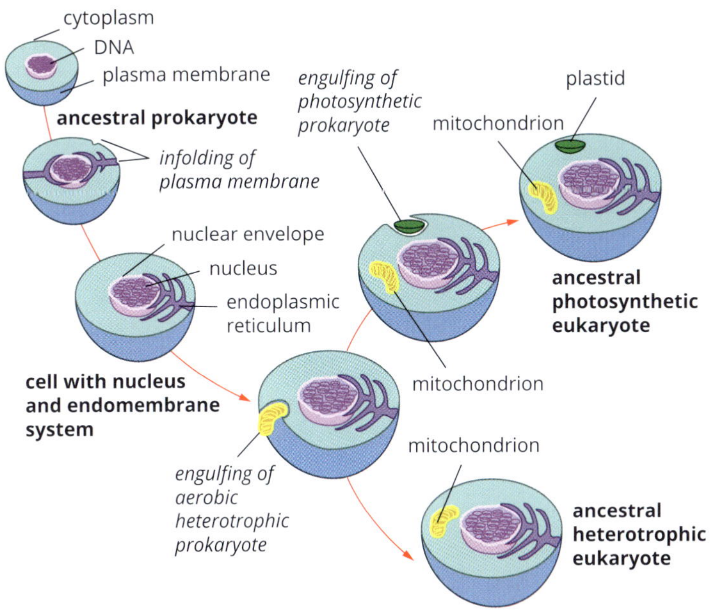

FIGURE 2.3.10 The theory of endosymbiosis explains how eukaryotes originated from the symbiosis of prokaryotic ancestors.

Analysis

1 Other evidence used to support the theory of endosymbiosis is the presence of ribosomes within the chloroplast and mitochondria. Infer the purpose of these ribosomes in what was believed to be the aerobic heterotrophic prokaryote and cyanobacteria from which they have theoretically evolved.

2 Given mitochondria and chloroplasts are believed to be a form of prokaryote, what would be the expected structure of their DNA, if it was to offer further support of the endosymbiotic theory?

STORAGE AND CELL STRUCTURE

The following organelles are involved in storage and also support cell structure in eukaryotic cells.

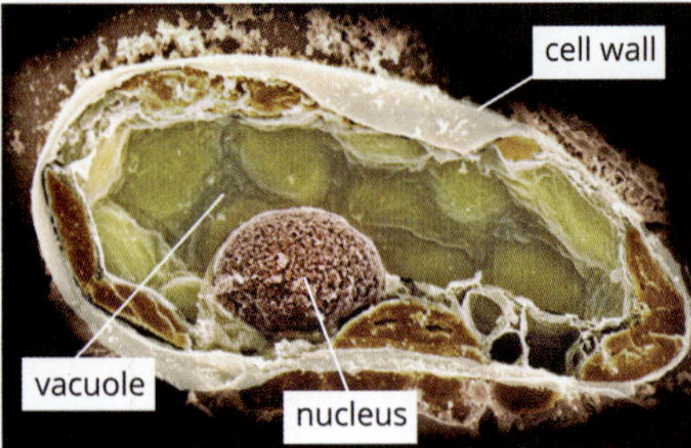

FIGURE 2.3.11 SEM of a sectioned plant cell revealing the large central vacuole

Vacuoles

Vacuoles are membrane-bound, liquid-filled spaces that store enzymes and other organic and inorganic molecules. They occur in most cells, but the number varies. Vacuoles in animal cells and plant cells are different. Animal cells contain many small temporary vacuoles, but most plant cells contain a single large permanent vacuole surrounded by a membrane (Figure 2.3.11). In plants the vacuole provides structural support by helping to maintain **turgor**. The presence of a large vacuole can displace the nucleus away from the centre of the plant cell, as seen in Figure 2.3.11.

Plastids

Plastids are organelles involved in the synthesis and storage of different chemical compounds. They contain a double-stranded DNA molecule and possess a double membrane. Plastids develop from simple organelles called proplasts. Animal cells lack plastids. The best-known plastids are chloroplasts, which are involved in photosynthesis and are found only in plants and some protists.

Cell wall

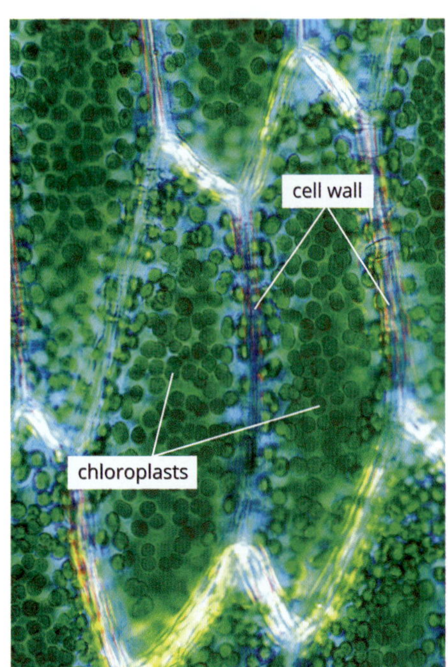

FIGURE 2.3.12 LM of cells in a leaf of shining Hookeria moss (*Hookeria lucens*). The leaf is made up of a single layer of cells. Each cell contains numerous green chloroplasts.

The cell wall is a rigid structure outside the plasma membrane of plant cells, fungal cells and some prokaryote cells (Figure 2.3.12). In plants the cell wall is composed mainly of cellulose. The fungal cell wall is made of chitin. In prokaryotic cells the cell wall is commonly formed from a mesh-like layer of murein, also called peptidoglycan.

The cell wall provides support, prevents expansion of the cell, and allows water and dissolved substances to pass freely through it. Lignin in the cell walls of woody plants, especially in the **xylem**, gives them additional strength.

> Xylem is the tissue in vascular plants that transports water and nutrients upwards from the roots.

Cytoskeleton

The cytoskeleton consists of microtubules of the protein tubulin and filaments of the protein actin (Figure 2.3.13). The cytoskeleton supports the cell's structure, allows the cell to move and assists in the transport of organelles and vesicles within the cell.

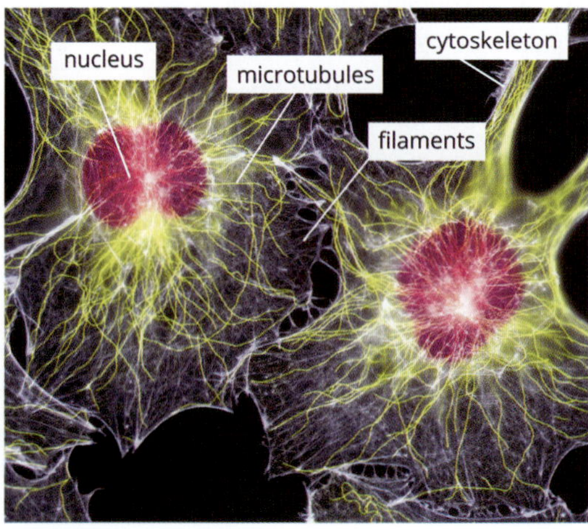

FIGURE 2.3.13 Fluorescent LM of two cells found in human connective tissue, showing their nuclei (pink) and cytoskeleton (yellow)

Centrioles

Centrioles are a pair of small cylindrical structures composed of microtubules (Figure 2.3.14). They are present in most eukaryotic cells, but many plant cells do not have centrioles. Centrioles are involved in cell division, in transport of substances within the cell and in the formation of cell structures such as cilia and flagella.

Cilia and flagella

Cilia and flagella (singular cilium and flagellum) are hair-like structures on the surface of cells (Figures 2.3.15 and 2.3.16). They each consist of an arrangement of microtubules enclosed by an extension of the plasma membrane. Cilia move with an oar-like rowing motion and are usually shorter and more numerous than flagella. Both structures are involved in the movement of the cell or of substances around the cell.

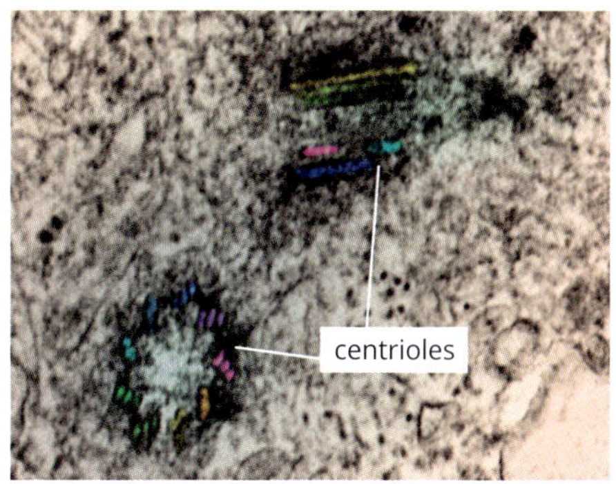

FIGURE 2.3.14 TEM of centrioles, the organelles that help certain cells to divide. These cylindrical structures are mainly composed of the protein tubulin and are involved in assembling the spindle that pulls chromosomes apart in animal cells during cell division by mitosis.

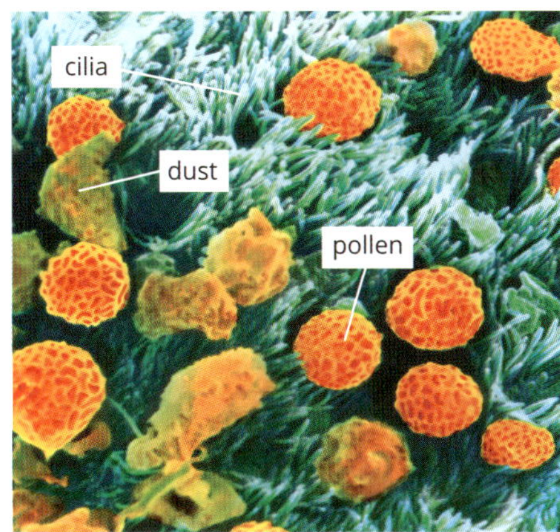

FIGURE 2.3.15 SEM of the internal surface of the trachea (windpipe) with inhaled pollen (orange) and dust (brown). The surface cells have hair-like cilia (green) which, together with mucus, trap airborne particles for removal of foreign matter from the airways and lungs.

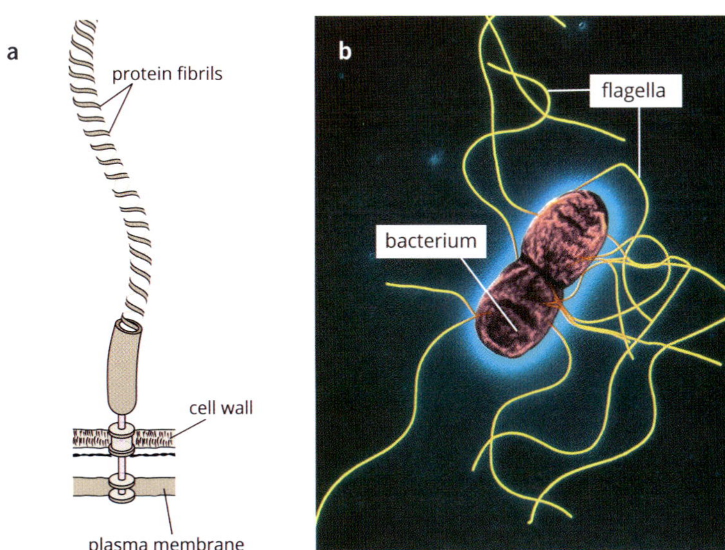

FIGURE 2.3.16 (a) Bacterial flagella consist of three protein fibrils coiled in a helical pattern. (b) SEM of a *Salmonella typhimurium* bacterium. This rodshaped, Gram-negative bacterium moves by using its long, hair-like flagella (yellow).

BIOFILE

The role of centrioles

Centrioles are a vital but often overlooked organelle in animal cells. They construct and anchor microtubules to form a transport network. Specialised motor proteins link microtubules with a substance or small unit (e.g. vesicles) to be transported. Like a freight train with a protein engine, the load is dragged along the microtubular track and delivered to its new location.

Centrioles are made of nine sets of microtubules, each in groups of three. These triplet microtubules are made of tubulin strongly bonded together to form hollow tubes. The structure is important to centriole function.

The evolution of animal cells from prokaryotes with small genomes demanded a mechanism to manipulate more complex genomes. During cell division microtubules and centrioles first align the chromosomes, then slowly and gently pull them apart. This prevents damage to the genetic material. In plants and other cells without centrioles, alternative mechanisms separate duplicated chromosomes.

Centrioles in a human cell with nine sets of triplet microtubules forming a hollow tube

ANIMAL AND PLANT CELLS

There are cellular differences not only between prokaryotes and eukaryotes but also between different groups of eukaryotic organisms, and even between tissues in one organism. Organelles are involved in specific functions, so their presence depends on the needs of the cells. A good way to understand this is to compare animal and plant cells.

Animal and plant cells are very similar. They both contain a nucleus with cytoplasm around it, all enclosed by the plasma membrane. They have mitochondria for cellular respiration and organelles such as the endoplasmic reticulum and Golgi apparatus where the synthesis and processing of organic molecules occurs. However, there are a number of differences between plant and animal cells (Figure 2.3.17).

The main differences between plant and animal cells are as follows.

- Plant cells have rigid cell walls made from cellulose and lignin outside the plasma membrane. The cell wall provides structural support and results in a fixed shape. Animal cells do not have cell walls.
- Plant cells have a large permanent vacuole that stores minerals and nutrients in a solution called cell sap. The vacuole also provides structure to plant cells by maintaining turgor pressure against the cell wall. Animal cells have many small temporary fluid-filled vacuoles called vesicles which do not provide structural support.
- Green plant cells have chloroplasts, the best-known type of plastid, in their cytoplasm. Chloroplasts are the site of photosynthesis. Animal cells do not contain chloroplasts and do not perform photosynthesis.
- Animal cells contain lysosomes for the digestion of unwanted cellular material. These are not present in plant cells and the large vacuole in plant cells often undertakes this role in plant cells.

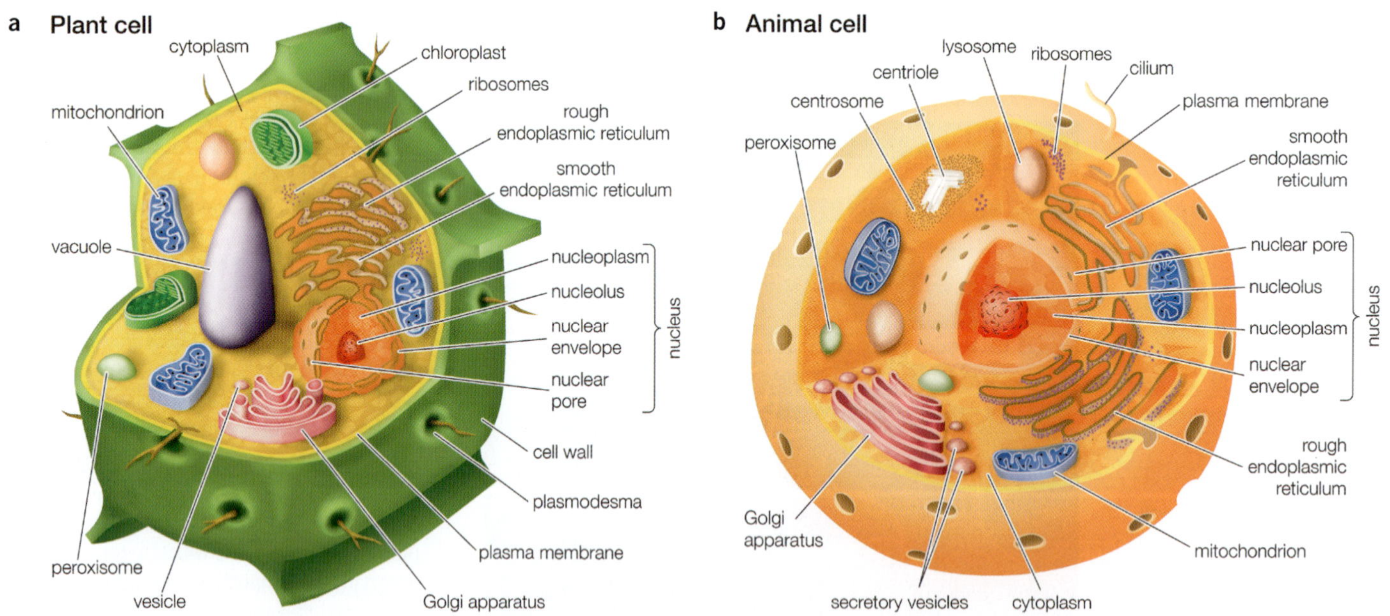

FIGURE 2.3.17 The many organelles of eukaryotic cells can be seen in these illustrations of (a) a plant and (b) an animal cell.

2.3 Review

SUMMARY

- Membrane-bound compartments in eukaryotic cells are called organelles.
- The nucleus contains the genetic material of a cell and therefore controls the activities of the cell.
- The ribosomes are responsible for manufacturing proteins in a cell.
- The rough endoplasmic reticulum is involved in protein synthesis and modification of proteins.
- The smooth endoplasmic reticulum is involved in the manufacture of other cellular molecules, such as lipids.
- The Golgi apparatus packages products, usually proteins, for export from the cell.
- Lysosomes are specialised vesicles that digest cellular debris.
- Mitochondria are found in all eukaryotic cells and are the site of cellular respiration and energy (ATP) production.
- Chloroplasts are only found in plants and protists and are the site of photosynthesis.
- Vacuoles store enzymes and other organic and inorganic molecules. In plants, vacuoles are particularly large as they maintain cell turgor and contain water.
- Plastids, such as chloroplasts, are only found in plants and often contain coloured pigments.
- The cell wall is the outer rigid structure of the cells of plants, fungi and some prokaryotes. In plants the cell wall is composed of cellulose and lignin.

KEY QUESTIONS

Knowledge and understanding

1 Which one of the following options lists only membrane-bound organelles?
  - A nuclei, mitochondria, vacuoles, ribosomes
  - B nuclei, mitochondria, centrioles, ribosomes
  - C nuclei, mitochondria, vacuoles, chloroplasts
  - D nuclei, plastids, centrioles, cilia

2 Recall the function that is shared by mitochondria and chloroplasts.

3 Label the parts of the plant cell in the diagram at right.

4 Identify where the DNA is contained in eukaryotic and prokaryotic cells.

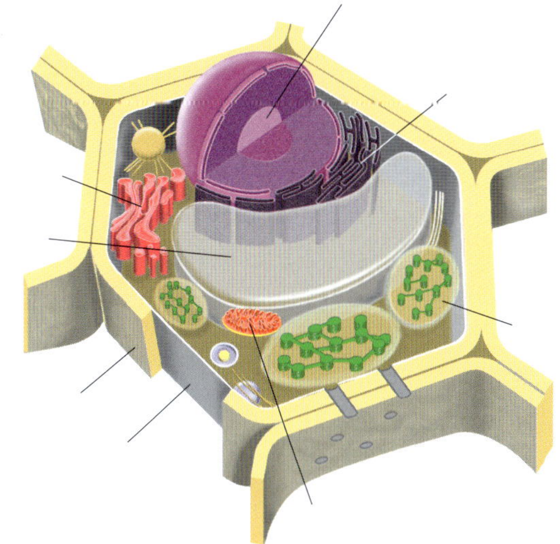

Analysis

5 Using the organelles seen in eukaryotic cells, distinguish between plant and animal cells.

6 The diagrams at right are of two cells observed with an electron microscope.
  - a Describe evidence that cell A is prokaryotic.
  - b Describe evidence that cell B is eukaryotic.

7 Explain why plant cells have a single large vacuole whereas animal cells are likely to contain multiple smaller vacuoles.

8 A pancreatic cell is responsible for secreting a high number of digestive enzymes into the small intestine. Suggest and justify which organelles would be very active in human pancreatic cells.

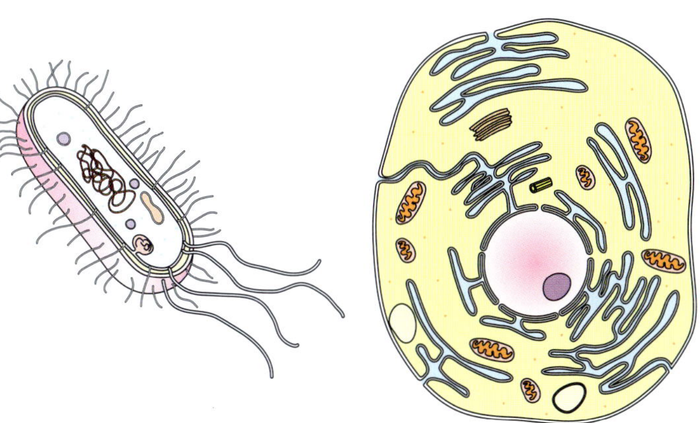

cell A (4500×) cell B (3000×)

2.4 Plasma membrane structure and function

Cells exist in a watery environment of **extracellular fluid**, which can be a large amount of fluid or a thin surface layer of fluid. In plants the cell wall is porous and does not obstruct the movement of molecules. So all cells are surrounded by a layer of fluid in contact with the outer plasma membrane. The composition of this fluid is critical to the stability of cells because it is from this environment that cells get the nutrients they need. The plasma membrane controls the movement of substances between the extracellular fluid outside and the **intracellular fluid** (or cytosol) inside the cell (Figure 2.4.1).

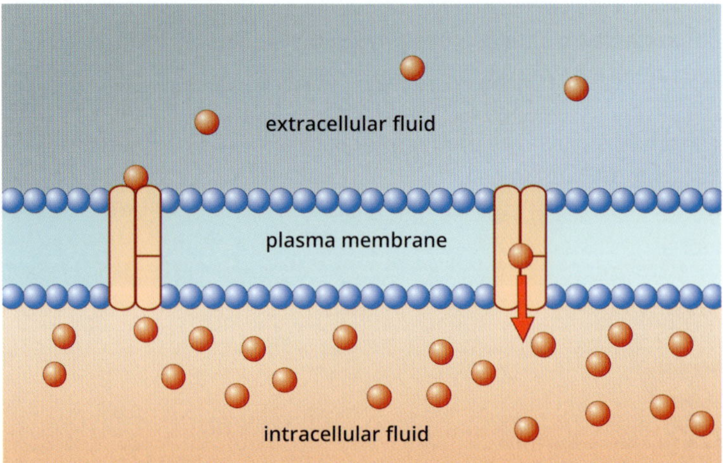

FIGURE 2.4.1 The plasma membrane regulates the movement of substances between the extracellular fluid and intracellular fluid.

In Section 2.2 you learnt about the importance of surface area to volume ratio and the ways in which the surface area of the plasma membrane is increased in some cells for more efficient function. In this section you will learn about the structure and function of the plasma membrane.

EXTRACELLULAR FLUID AND UNICELLULAR ORGANISMS

Extracellular fluid—body fluid outside the plasma membranes; includes blood plasma and interstitial fluid.

For unicellular organisms the extracellular fluid is simply the watery **external environment** in which they live. Unicellular organisms can do little to control their environment and may die if it changes significantly.

However, some unicellular organisms such as yeasts can become dormant until their environment returns to normal again. Others can move to a place where conditions are more suitable to their needs. For example, unicellular algae are able to move slowly towards light, and some bacteria can detect and move towards nutrients or away from toxic substances.

EXTRACELLULAR FLUID IN MULTICELLULAR ORGANISMS

Conditions for cells in multicellular organisms are better than those of unicellular organisms. The more complex the organism, the more control it will have over the environment in which its cells exist, and the more independent it will be from the external environment. Whether they live in water or on land, multicellular organisms have an outer layer that acts as a protective barrier (Figure 2.4.2). This outer layer creates an **internal environment** for the organism that is different from their external environment. Therefore, in multicellular organisms, the environment of the cells is the extracellular fluid that surrounds them.

FIGURE 2.4.2 Crabs have an external skeleton (carapace) that protects them from water loss when on land.

Most multicellular organisms can regulate the conditions of their internal environment, often very precisely. This allows them to provide the specific environments needed by specialised cells and tissues, and for their cells to function more efficiently. Commonly regulated aspects of the internal environment are:

- temperature
- concentration of oxygen
- concentration of carbon dioxide
- pH (acidity or alkalinity)
- osmotic pressure (concentrations of salts or ions)
- concentration of nitrogen wastes
- concentration of glucose.

Importantly, the way cells interact with the extracellular fluid of the internal environment is regulated by the plasma membrane.

PLASMA MEMBRANE STRUCTURE

This **fluid mosaic model** that describes the structure of the plasma membrane was first proposed by Jonathan Singer and Garth Nicholson in 1972. It is now widely accepted as the basic model of all biological membranes. According to this model, plasma membranes consist of two layers of **phospholipid** molecules, with other molecules including proteins, carbohydrates and cholesterol scattered throughout the membrane (Figure 2.4.3).

Plasma membranes are phospholipid bilayers that enclose the cytoplasm and subdivide a eukaryotic cell into compartments (organelles).

Plasma membranes have the same basic structure in all organisms, which serves to separate the interior of the cell (the cytoplasm) from its external environment. Most membranes are also asymmetrical, meaning one layer has different properties from the other. For example, the pattern of proteins and carbohydrate molecules in the external surface is different from the pattern in its internal surface.

The composition and characteristics of the plasma membrane are related to the needs and function of the cell. In addition to transporting molecules into and out of the cell, the plasma membrane performs other important functions, such as cell recognition and communication with other cells (Figure 2.4.6 on page 99).

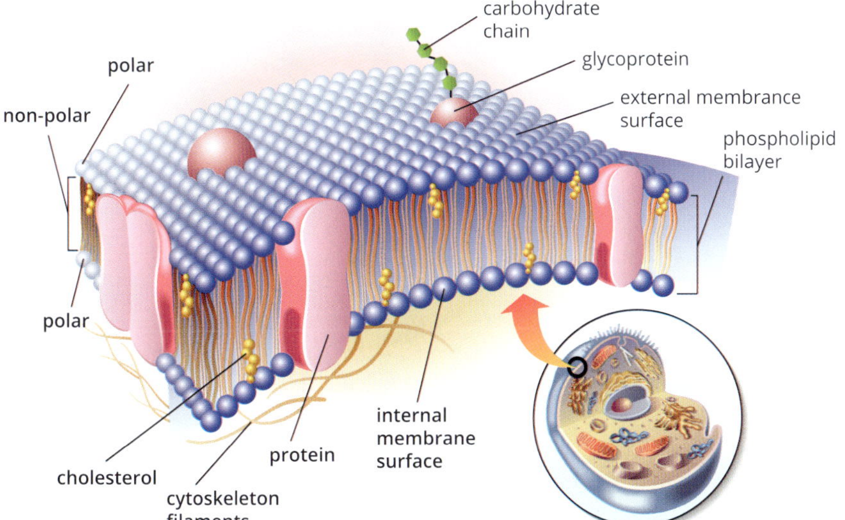

FIGURE 2.4.3 The fluid mosaic model of a plasma membrane, showing the phospholipid bilayer in which large protein molecules are embedded

Phospholipids

Phospholipid molecules have a **hydrophobic** (water-repelling) 'tail' and a **hydrophilic** (water-attracting) 'head'. The **phospholipid bilayer** of the plasma membrane is called a bilayer because it has two layers of phospholipids. The hydrophilic heads form the outside and inside lining of the plasma membrane, and the hydrophobic tails of the two layers of phospholipids meet in the middle (Figure 2.4.4).

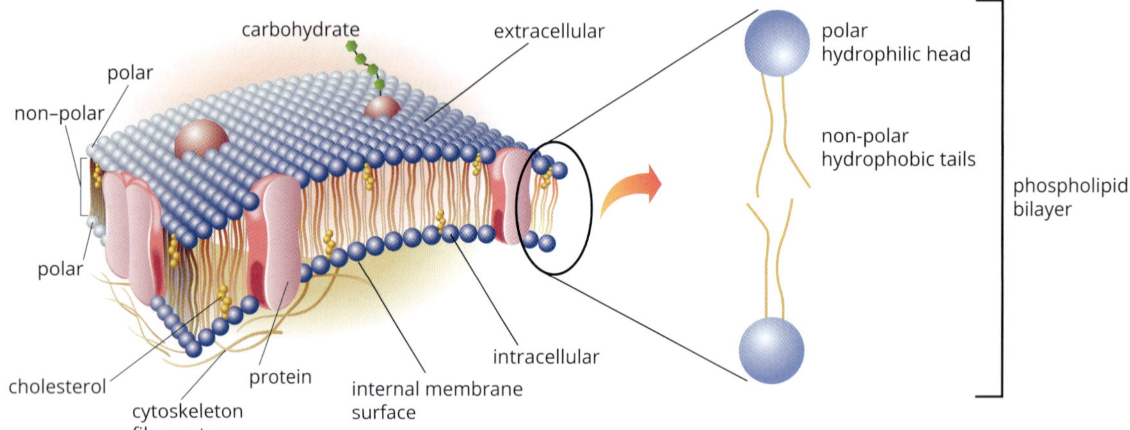

FIGURE 2.4.4 A phospholipid has a hydrophobic 'tail' and a hydrophilic 'head'.

> A phospholipid is a molecule that consists of long-chain fatty acids (which are hydrophobic) and a phosphate (which is hydrophillic). It is the major component of plasma membranes.

The phospholipids in plasma membranes make them impermeable to water-soluble particles, ions and polar molecules. The movement of these molecules across the membrane is controlled by protein channels, which allow the cell to regulate the exchange of molecules with the environment. Controlling the movement of substances into and out of the cell is essential to important processes that keep the cell alive, such as cellular respiration, digestion, and elimination of wastes. You will learn more about transport across plasma membranes later in this section.

Plasma membranes are fluid structures, which means that individual phospholipid molecules (and some proteins) are free to move about within the layers. However, they rarely cross from one layer of the membrane to the other. The level of fluidity depends on the percentage of unsaturated fatty acids in the phospholipid molecules—the greater the percentage, the more fluid the membrane.

Figure 2.4.5 shows the components of the plasma membrane that are discussed on the following page.

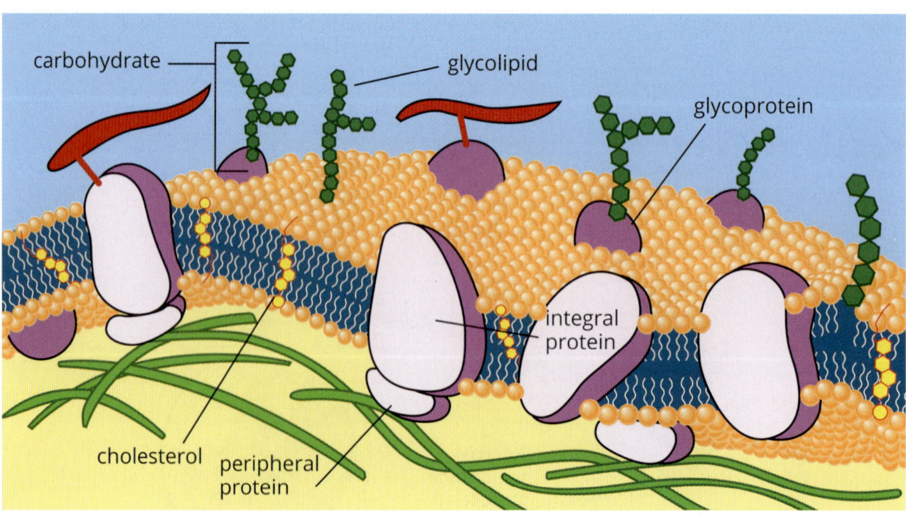

FIGURE 2.4.5 Components of the plasma membrane

Cholesterol

Plasma membranes contain many fatty molecules between the phospholipid molecules (Figure 2.4.5). The plasma membranes of eukaryotes contain **cholesterol**, a type of fatty molecule that gives stability to the membrane without affecting its fluidity, and reduces the permeability of the membrane to small water-soluble molecules.

Proteins

Like phospholipid molecules, proteins in the plasma membrane are able to move about to some extent, but this movement may be limited to particular regions.

Proteins that are a permanent part of the plasma membrane are called **integral proteins**. Proteins that are a temporary part of the plasma membrane are called **peripheral proteins**. Peripheral proteins bind to integral proteins or penetrate into one surface of the plasma membrane (Figure 2.4.5). When integral proteins span both phospholipid layers they are also called **transmembrane proteins**. Transmembrane proteins are involved in a number of important cellular and intercellular activities (Figure 2.4.6).

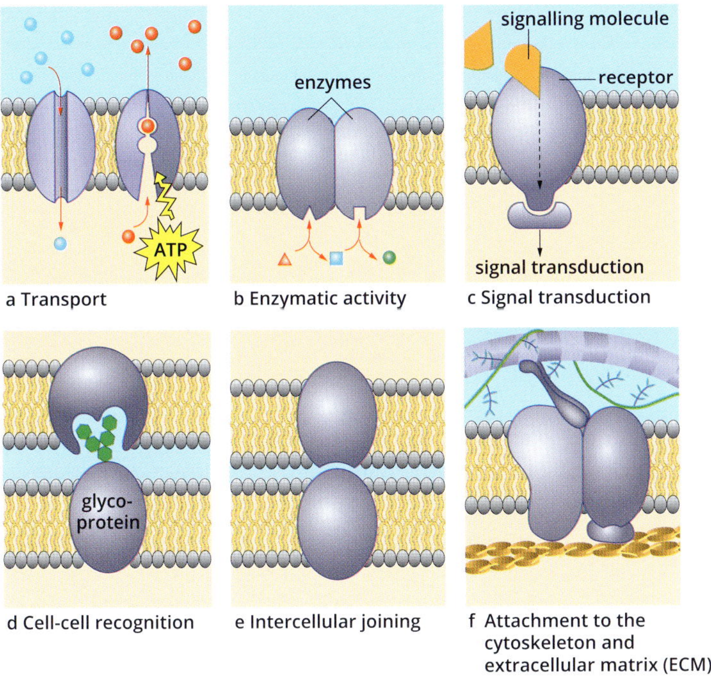

FIGURE 2.4.6 Different functions of plasma membrane proteins

Carbohydrates

Carbohydrates associated with plasma membranes are usually linked to protruding proteins (forming **glycoproteins**) or to lipids (forming **glycolipids**) on the outer surface of the membrane (Figure 2.4.5). They play a role in recognition and adhesion between cells, and in the recognition of antibodies, hormones and viruses by cells.

TRANSPORT ACROSS THE PLASMA MEMBRANE

A plasma membrane is described as semi-permeable (also partially or selectively permeable) because it only allows certain substances to cross it.

One of the main functions of the plasma membrane is to control the exchange of molecules between the cytoplasm and the external environment of the cell. The plasma membrane is described as a **semi-permeable membrane** because it allows some substances to cross it but acts as a barrier to others. The semi-permeable nature of the plasma membrane allows for various processes that control the exchange of molecules, including diffusion, osmosis, facilitated diffusion, active transport, endocytosis and exocytosis.

Plasma membrane permeability

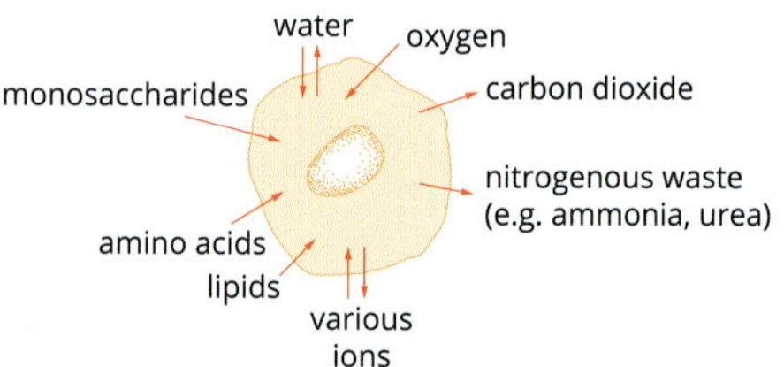

FIGURE 2.4.7 Cells exchange many substances with their environment across the plasma membrane.

The movement of different types of molecules across plasma membranes (Figure 2.4.7) depends on their properties, such as size and charge, if they are hydrophilic or hydrophobic, and whether or not the phospholipid bilayer is permeable to each substance (Table 2.4.1). Some substances such as glucose and amino acids have both hydrophilic and hydrophobic parts to their molecules.

TABLE 2.4.1 The plasma membrane's permeability to different molecules.

Molecule/Ion	Examples	Permeability of membrane to the molecule/ion
small uncharged molecule	oxygen, carbon dioxide	permeable
lipid-soluble, non-polar molecule	alcohol, chloroform, steroids	permeable
small polar molecule	water, urea	permeable or semi-permeable
small ion	potassium ion (K^+) sodium ion (Na^+), chloride ion (Cl^-)	non-permeable (molecule passes through protein channels)
large, polar, water-soluble molecule	amino acid, glucose	non-permeable (molecule passes through protein channels)

Because of their lipid nature, plasma membranes are permeable to small molecules and lipid-soluble molecules that can move freely through the phospholipid bilayer. However, their lipid nature also makes plasma membranes impermeable to:

- most water-soluble molecules
- ions (atoms or groups of atoms with an overall positive or negative charge)
- polar molecules (molecules with charged regions but no overall charge).

These substances must therefore pass through specific protein channels in the plasma membrane (Figure 2.4.8).

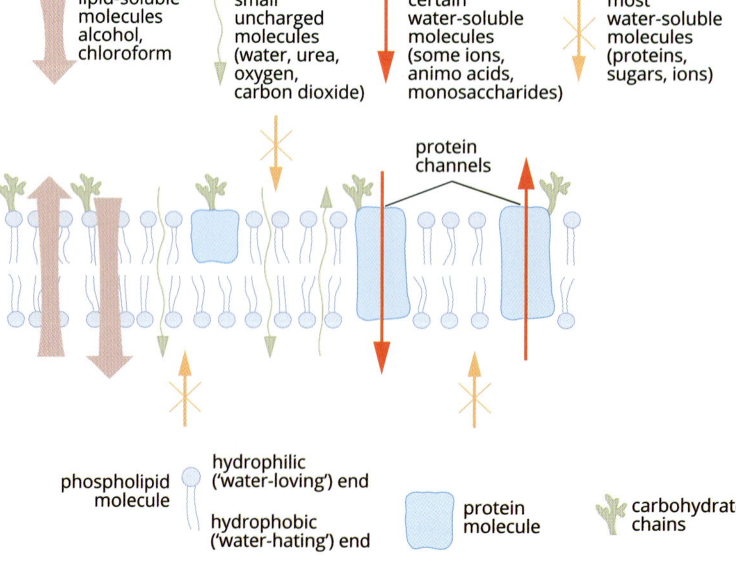

FIGURE 2.4.8 If the plasma membrane is not permeable to particular molecules, protein channels must assist these molecules to cross the membrane.

DIFFUSION

Particles in a solution move from an area of high concentration to an area of low concentration. This process is called **diffusion** (Figure 2.4.9). As there are many particles colliding with each other during this process, the overall movement of particles is very slow.

Diffusion can be seen when a drop of ink (the **solute**), is placed in a jar of still water (the **solvent**). The dye particles in the ink move randomly through the water until the colour is evenly spread. In other words, the solute particles move from an area of high solute concentration (the drop of ink) to the areas of low solute concentration (the water). The solute particles are said to have moved along the **concentration gradient**.

Diffusion is called a passive process because it does not require energy. It occurs only because there is a concentration gradient.

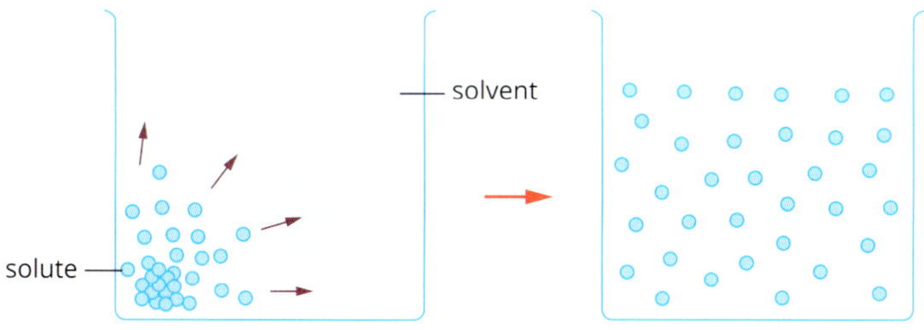

FIGURE 2.4.9 Diffusion results in the random dispersal of solute molecules throughout a solvent.

Diffusion across membranes

There are two types of diffusion that occur across plasma membranes: simple diffusion and facilitated diffusion. These are both passive types of diffusion, and both move molecules along the concentration gradient.

Simple diffusion

Simple diffusion is the movement of particles from an area of high concentration to an area of lower concentration without the use of transmembrane proteins. Small, non-polar molecules are transported through the plasma membrane by simple diffusion.

Solute molecules can diffuse across a membrane only if the membrane is permeable to them. There is a constant movement of solute molecules backwards and forwards across the membrane. If the concentration of solute molecules is the same on both sides of the membrane, there will always be about the same number moving across in either direction. That is, there will be no net movement from one side to the other.

However, if the concentration of the solute molecule is higher on one side of the membrane than the other, more solute molecules will cross from the area of higher concentration to the area of lower concentration (i.e. down its concentration gradient), as you can see on Figure 2.4.10a on page 102.

In Figure 2.4.10b on page 102, you can see what happens if the membrane is not permeable to the solute molecules. This happens if they are too large or are repelled by the hydrophobic nature of the lipids in the membrane. Instead the solvent molecules will move across to dilute the side with the solute.

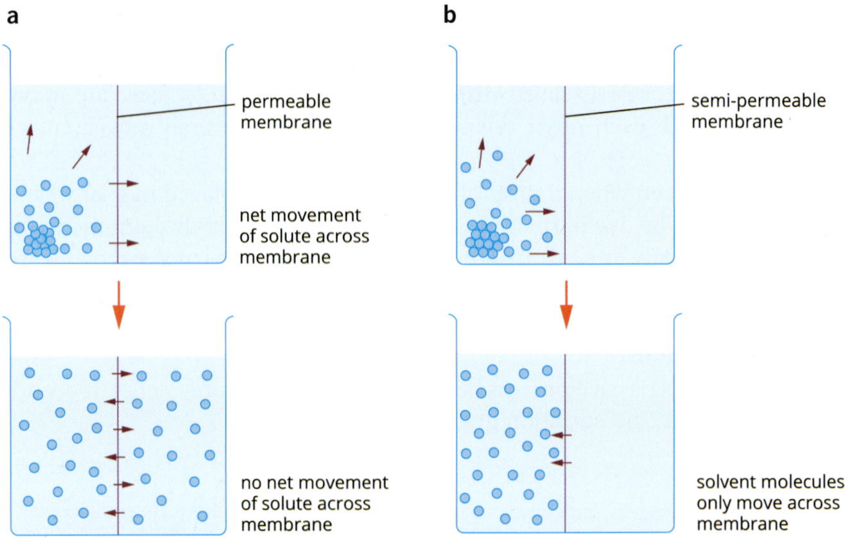

a

permeable membrane

net movement of solute across membrane

no net movement of solute across membrane

b

semi-permeable membrane

solvent molecules only move across membrane

FIGURE 2.4.10 (a) A membrane that is permeable to both the solute and the solvent will not affect the pattern of diffusion. (b) A semi-permeable membrane that allows the solvent molecules to pass through, but not the solute molecules, stops the solute from diffusing through the membrane.

Facilitated diffusion

You have seen that the phospholipid bilayer of the plasma membrane is impermeable to certain particles (ions or molecules). However, transport proteins in the plasma membrane allow for the movement of these particles. **Facilitated diffusion** is the movement of particles from an area of high concentration to an area of lower concentration using transport proteins in the plasma membrane. Large, polar molecules are transported through the plasma membrane by facilitated diffusion.

In facilitated diffusion:

- the membrane transport proteins are specific for particular types of particles, so transport is selective; some particles are transported and others are not (selectivity)
- transport is more rapid than by simple diffusion
- the transport proteins can become saturated (fully occupied) as the concentration of the transported substances increases
- the transport of one particle may be inhibited by the presence of another particle of a similar type that uses the same transport protein. This is known as **competitive inhibition**.
- no energy is required; the particles move down their own concentration gradient.

The two main types of membrane transport proteins in facilitated diffusion are **channel proteins** and **carrier proteins**. Membrane proteins provide channels for the passage of hydrophilic (that is, water-soluble, also referred to as polar) molecules and ions across the hydrophobic phospholipid bilayer. Channel proteins are specific for a substance. They do not usually bind with the molecules being transported. They function like pores that open and close to allow the passage of specific molecules. Channel proteins are mainly involved in the passage of water-soluble polar particles such as ions.

Carrier proteins bind the molecules being transported, causing the protein to undergo changes in shape (conformation) that allow specific molecules to be transported across the plasma membrane. After the molecule has crossed the plasma membrane the original shape of the protein is restored.

> A molecule with a partial positive or negative charge is called polar or hydrophilic (Greek for 'water-loving'). Polar molecules dissolve easily in water.

BIOFILE

Fast molecules

At body temperature, water molecules travel at about 2500 km/h. Glucose, a larger molecule, travels more slowly at only about 850 km/h!

Factors affecting the rate of diffusion

The three main factors that affect the rate of diffusion across a plasma membrane are:

- concentration—the greater the difference in concentration gradient, the faster the rate of diffusion. When the concentration is equal on both sides of the membrane the net diffusion is zero, even at high temperatures
- temperature—the higher the temperature, the higher the rate of diffusion. Increasing temperature increases the speed at which molecules move
- particle size—the smaller the particles, the faster the rate of diffusion through a membrane.

Osmosis

Osmosis is the net diffusion of water molecules across a semi-permeable membrane, from a solution of lower concentration to a solution of higher concentration.

If a diluted and a concentrated solution are separated by a semi-permeable membrane that allows the movement of free water molecules across the membrane, but not the movement of the solute molecules, the free water molecules will move across the membrane from the diluted to the concentrated solution.

In osmosis, net diffusion of water occurs through a semi-permeable membrane from a diluted to a concentrated solution along its own concentration gradient, which is known as the **osmotic gradient** (Figure 2.4.11). The pressure causing the water to move along this gradient is called **osmotic pressure**.

> Osmosis is the movement of water from a solution of lower concentration to a solution of higher concentration.

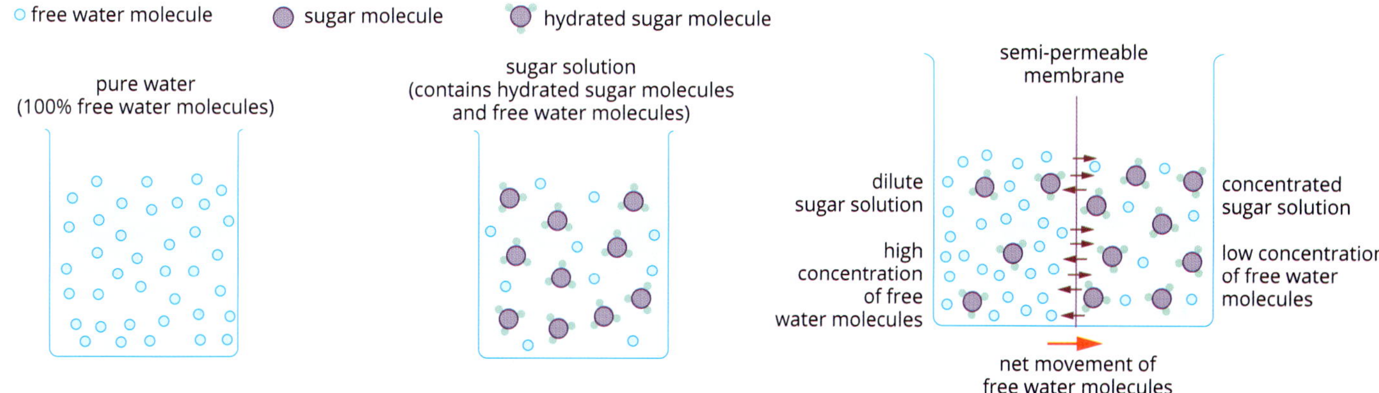

FIGURE 2.4.11 A net movement of water molecules from a dilute solution through a semi-permeable membrane into a concentrated solution is osmosis.

The plasma membrane is permeable to water, so when cells are placed in fresh water an osmotic gradient will draw water into the cells. This is because the cytosol is a concentrated solution containing many dissolved substances. For example, if red blood cells are placed in fresh water, the cells absorb so much water by osmosis that they swell and may eventually burst, releasing red pigment into the water (Figure 2.4.12c). Conversely, if red blood cells are placed in a solution that is more concentrated than their cytosol, water leaves the red blood cells by osmosis and causes them to shrink (Figure 2.4.12a).

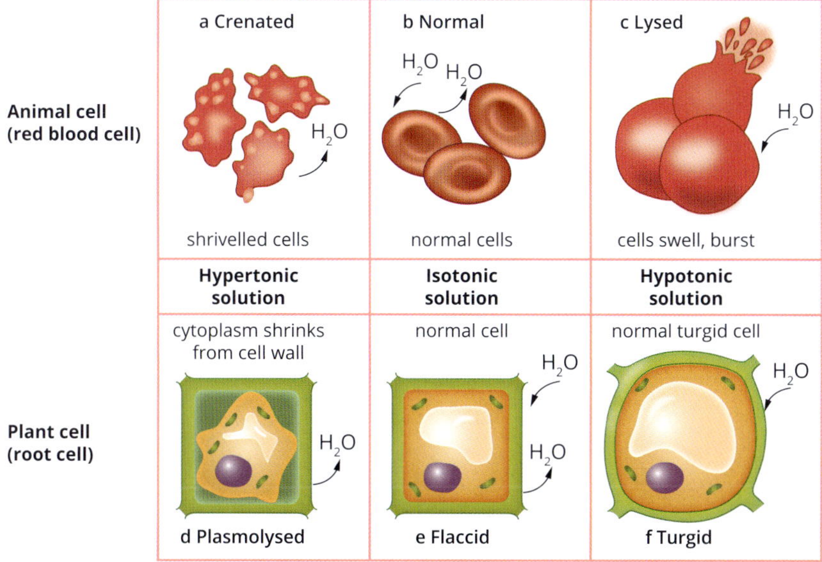

FIGURE 2.4.12 The effect of three different solution concentrations on an animal cell and plant cell

Plant cells tightly filled with fluid are described as turgid. When they lose water they become flaccid.

If a plant cell absorbs water it swells to some extent, but the cell wall prevents the cell from bursting (Figure 2.4.12f on page 103). Water will continue to enter the cell along an osmotic gradient until the internal fluid pressure equals the osmotic pressure drawing water in, at which point no more water will enter. Plant cells with high internal fluid pressures have a high turgor because they are full of fluid tightly pressing against the cell wall. Turgor in cells provides structural support for plants. Dehydrated plants droop because their cells are flaccid.

In osmosis, we are always comparing solute concentration between two solutions. The terms isotonic, hypertonic and hypotonic solution are often used to describe the differences:

- isotonic solutions—the solutions being compared have equal concentrations of solutes
- hypertonic solution—the solution with a higher concentration of solute (hence lower concentration of free water molecules)
- hypotonic solution—the solution with a lower concentration of solute (hence higher concentration of free water molecules).

BIOFILE

Osmosis in salty environments

There is no biological mechanism for actively transporting water molecules across plasma membranes. Net movement of water across membranes occurs only by passive osmosis. Organisms that are adapted to extremely salty environments survive by retaining much higher ion concentrations within their cells. They also produce small, osmotically active but otherwise inert molecules to reduce the osmotic gradient. This prevents the loss of water to their salty surroundings. Their proteins are specialised to function normally despite a high concentration of salts in the cytosol. Organisms such as these are called halophiles.

The salt lakes of Murray-Sunset National Park in Victoria owe their colour to the presence of red algae.
These algae, along with the solid salt bed of the lakes, create the distinctive pinkish red colour of the lake.

Active transport

Diffusion, facilitated diffusion and osmosis are examples of passive transport, because they do not require energy to move particles across the plasma membrane. **Active transport** involves the use of energy by the cell to transport particles across membranes (see Figure 2.4.13).

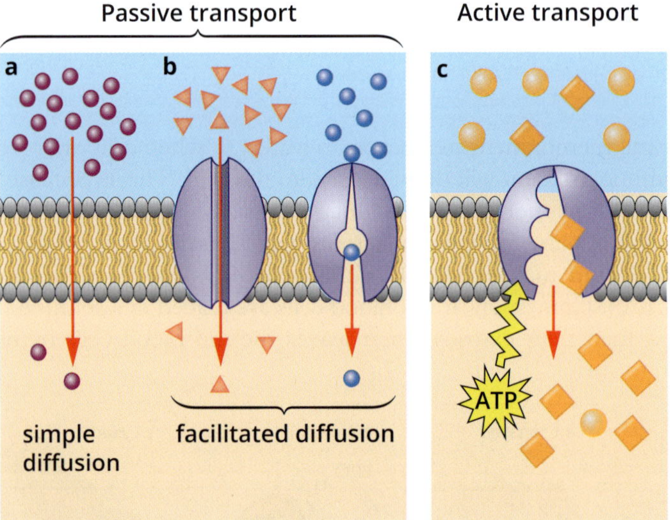

FIGURE 2.4.13 Passive transport does not require an energy source. Diffusion is a type of passive transport. (a) Simple diffusion is where substances move from high to low concentrations. (b) Facilitated diffusion is where substances move from high to low concentrations with help from a carrier protein. (c) Active transport requires an energy source. As a result, active transport usually moves substances from low to high concentrations. This shows a protein pump that assists the movement of substances.

Active transport and facilitated diffusion compared

Active transport has the same properties of selectivity, saturation and competitive inhibition as facilitated diffusion, because it also occurs through transport proteins (Figure 2.4.14). Selectivity means that some substances are transported but others are not. Saturation means that there is no increase in the rate of transfer when all transport proteins are open. Competitive inhibition means that one substance can inhibit the transport of another substance by using the same transport protein.

But unlike facilitated diffusion, which can occur through either channel or carrier proteins, active transport only occurs through carrier proteins. Because active transport uses energy, it can move substances against a concentration gradient (from low concentrations to high concentrations). In comparison, facilitated diffusion uses no energy, so it can only move substances down a concentration gradient in a passive way.

In different situations, either facilitated diffusion or active transport may be used to transport a particular molecule. Whether a cell uses facilitated diffusion or active transport depends on the specific needs of the cell.

For example, glucose is actively transported from the gut into epithelial cells lining the gut so it can enter the bloodstream. The regulation of this process is controlled by hormones, principally insulin and glucagon. If gut glucose levels are high, blood glucose levels will increase. If gut glucose levels are low, active transport makes sure that the little glucose that is in the gut gets pumped into the epithelium from where it can move to the bloodstream via facilitated diffusion.

In contrast, red blood cells move glucose by facilitated diffusion. This makes sense because glucose concentration in the blood is usually maintained within a narrow range. In addition, cells convert glucose into other chemicals as soon as it enters them, keeping the intracellular concentration of glucose lower than the blood concentration of glucose.

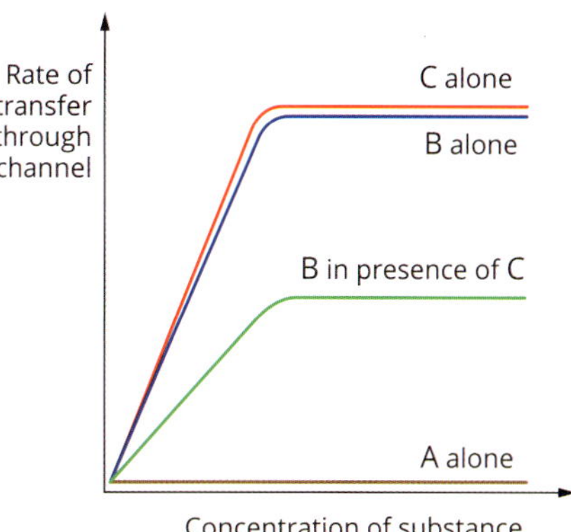

FIGURE 2.4.14 Theoretical transport rate vs concentration graph for the movement of three substances through a channel protein. Substances B and C are transported, but not substance A, demonstrating selectivity. The rate of transfer of substances B and C flattens out when their concentrations reach a certain level, demonstrating saturation. The rate of transport of B is less when C is present, demonstrating competitive inhibition.

Endocytosis and exocytosis

Some large molecules and other particles and fluids are moved into or out of the cell by exocytosis and endocytosis (Figure 2.4.15). These are both forms of active transport because they require energy.

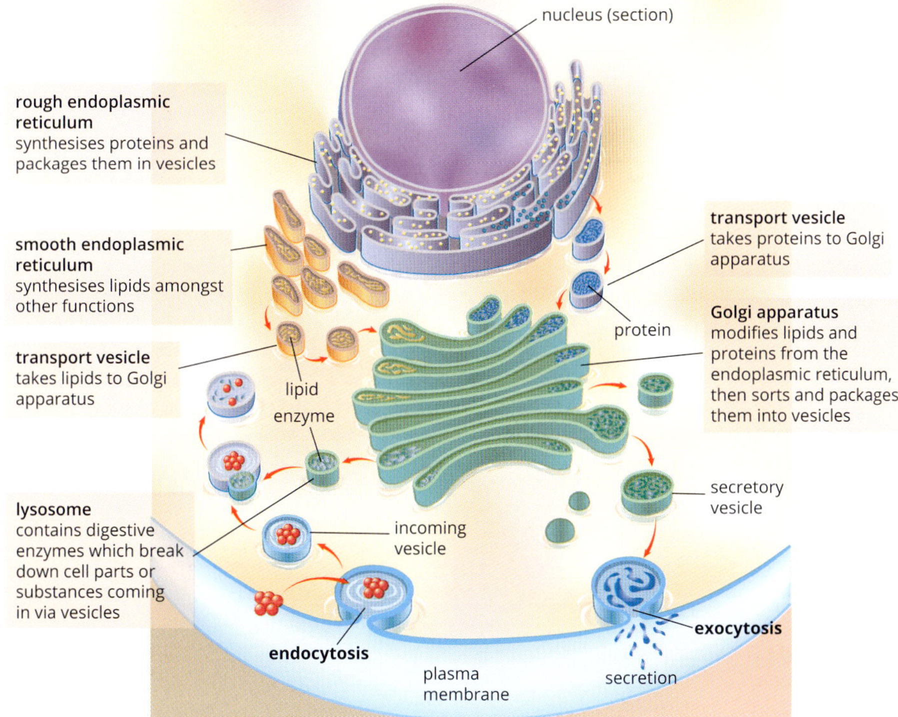

FIGURE 2.4.15 To transport large molecules that proteins or pumps cannot transport, cells use endocytosis and exocytosis, which require vesicles and energy.

Exocytosis is the movement of substances out of the cell, from the cytoplasm to the extracellular fluid. A transport vesicle, which may contain wastes or substances needed for secretion (e.g. digestive enzymes or hormones), fuses with the plasma membrane and the junction then breaks down, releasing the enclosed materials. Unicellular organisms such as amoebas remove digestive wastes in this way.

Endocytosis is the movement of substances into the cell, from the extracellular fluid into the cytoplasm. Particles near the plasma membrane are enclosed by the membrane, which then pinches off to form a vesicle enclosing the particle. In eukaryotes this vesicle may then become fused with a lysosome so that its contents can be digested for use by the cell. The two forms of endocytosis are pinocytosis and phagocytosis. **Pinocytosis** is the entry of extracellular fluid and substances such as proteins and sugars that are carried in it. **Phagocytosis** is the entry of large particles such as bacteria and cell debris, and is carried out by some white blood cells.

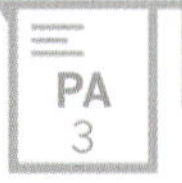

CASE STUDY

Ilya Mechnikov (1845–1916)

Ilya Mechnikov (Figure 2.4.16) was a Russian scientist who first described phagocytosis, a process that uses endocytosis. In 1882 he noticed free-moving cells (which he called 'wandering cells') in transparent sea star larvae. It occurred to him that these cells take in nutrients, and so perhaps they take in other microbes too. And if they did then the same might be true for humans, and perhaps our 'wandering cells' (white blood cells) also protect us from germs.

To test his hypothesis Mechnikov inserted thorns from garden plants into sea star larvae. The next day he found the free-moving cells were not moving around aimlessly, but rather they had surrounded the thorns as if attempting to drive them out. In that moment Mechnikov knew he was right about what he would later term phagocytosis, from the Greek words *phag* (eat) and *kytos* (vessel or cell), with the ending *-osis* (a process). In other words, the process of eating cells.

FIGURE 2.4.16 Ilya Mechnikov

Mechnikov also laid the foundation for the science of immunology with his concept of cell-mediated immunity, for which he won a Nobel Prize in 1908.

2.4 Review

OA ✓✓

SUMMARY

- Cells are surrounded by a plasma membrane that is in direct contact with a layer of extracellular fluid.
  - For unicellular organisms, such as prokaryotes, the extracellular fluid is their watery environment, which undergoes fluctuations largely out of their control.
  - For multicellular organisms, such as eukaryotes, the external environment of their cells is the extracellular fluid that surrounds the cell.
- The plasma membrane separates the cytoplasm inside the cell from the external environment and controls the movement of substances between the two.
- Plasma membranes consist of a double layer of phospholipid molecules. They contain protein molecules of various sizes, fatty molecules such as cholesterol and some other molecules, including carbohydrates.
- The phospholipid nature of the plasma membrane makes it hydrophobic and impermeable to water-soluble particles, ions and polar molecules. These substances can only pass through protein channels.
- Plasma membrane proteins:
  - can change position within their membrane layer
  - provide selective channels to transport water-soluble particles and ions
  - catalyse reactions associated with the plasma membrane
  - communicate with the external environment and other cells
  - bind with other cells.
- Diffusion is the passive movement of solute molecules along a concentration gradient, from a region of high solute concentration to a region of low solute concentration.
- There are two types of diffusion across plasma membranes: simple and facilitated.
- Passive transport has three forms and does not require energy to move particles across the plasma membrane.
  - Simple diffusion involves solutes that the membrane is permeable to, including lipid-soluble substances. The rate of diffusion is affected by concentration, temperature and particle size.
  - Facilitated diffusion is through selective channels in membranes that permit or enhance passive movement of particular ions and molecules down their own concentration gradient. Facilitated diffusion generally occurs more rapidly than simple diffusion.
  - Osmosis is the net diffusion of water across a semi-permeable membrane down its own concentration gradient, from a low solute concentration to a high solute concentration (called the osmotic gradient).
- Active transport requires energy to move substances across plasma membranes through protein channels against their concentration gradient.
- Exocytosis (moving substances out of the cell) and endocytosis (moving substances into the cell) are forms of active transport using vesicles that fuse with the plasma membrane. These forms of transport are generally used to transport larger molecules.

KEY QUESTIONS

Knowledge and understanding

1 List three functions of the plasma membrane around a cell.

2 Label the key components of the plasma membrane shown in the diagram at right.

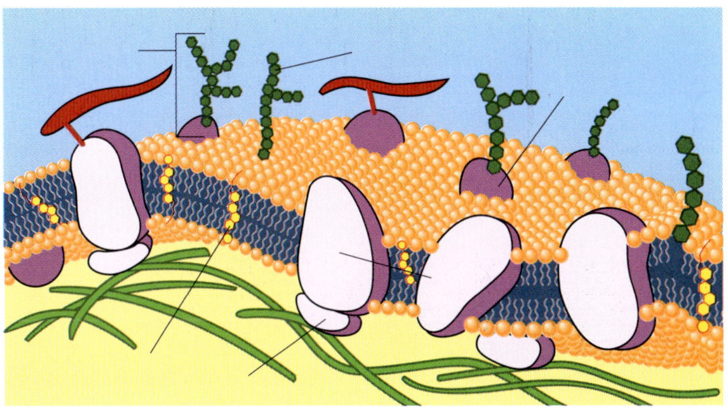

continued over page

2.4 Review *continued*

3 Complete the following table by stating whether the phospholipid bilayer is permeable, semi-permeable or not permeable to each substance described.

Substance	Examples	Permeability
small uncharged molecule	oxygen, carbon dioxide	
lipid-soluble, non-polar molecule	alcohol, chloroform, steroids	
small, polar molecule	water, urea	
small ion	potassium ion (K^+) sodium ion (Na^+), chloride ion (Cl^-)	
large, polar, water-soluble molecule	amino acid, glucose	

4 Recall the two types of proteins used in facilitated diffusion and explain how they are different.

5 Describe diffusion and explain the difference between simple and facilitated diffusion. Include an example of each.

6 a Define active transport.
b How is this process different from diffusion?

Analysis

7 Consider the images of red blood cells in the following figure. The arrows indicate the direction of net movement of water. Using your understanding of osmosis, explain why red blood cells, when moved from blood plasma into different solutions, would:
a shrink
b swell and burst.

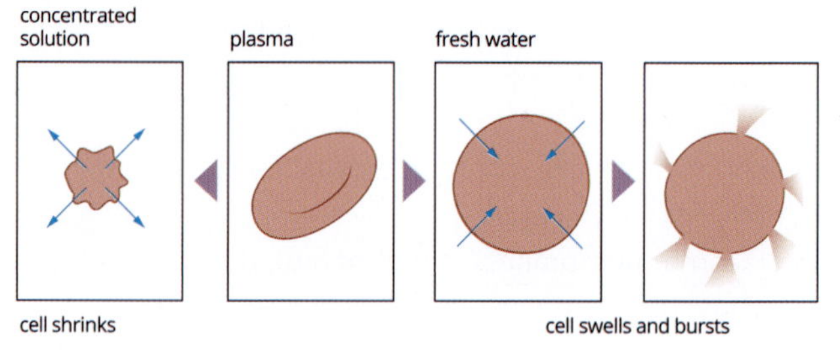

8 Membrane-bound proteins may have carbohydrates attached. Examine the role these membrane-bound proteins play in the plasma membrane.

Chapter review

KEY TERMS

active transport
amino acid
biogenesis
carbohydrate
carrier protein
cell
cell compartmentalisation
cellular respiration
channel protein
chloroplast
cholesterol
chromosome
cisternae
competitive inhibition
concentration gradient
cytoplasm
cytosol
deoxyribonucleic acid (DNA)
diffusion
endocytosis
enzyme
eukaryotic cell
exocytosis
external environment
extracellular fluid
extremophile
facilitated diffusion
fluid mosaic model
glycolipid
glycoprotein
Golgi apparatus
hydrophilic
hydrophobic
inorganic compound
integral protein
internal environment
intracellular fluid
lipid
lysosome
messenger RNA (mRNA)
mitochondria
murein
nucleolus
nucleus
organelle
organic compound
osmosis
osmotic gradient
osmotic pressure
peripheral protein
phagocytosis
phospholipid
phospholipid bilayer
photosynthesis
pinocytosis
plasma membrane
plasmid
prokaryotic cell
protein
ribonucleic acid (RNA)
ribosomal RNA (rRNA)
ribosome
rough endoplasmic reticulum
semi-permeable membrane
simple diffusion
smooth endoplasmic reticulum
solute
solvent
surface area to volume ratio
transmembrane protein
turgor
unicellular organism
vesicle
xylem

REVIEW QUESTIONS

Knowledge and understanding

1 Select the list containing structures that are common to all cells.
- **A** cytoplasm, chloroplasts, DNA, plasma membrane
- **B** DNA, ribosomes, plasma membrane, cytoplasm
- **C** ribosomes, DNA, mitochondria, cell wall
- **D** cell wall, plastids, DNA, ribosomes

2 Recall that prokaryotic cells:
- **A** lack membranes
- **B** lack internal membrane-bound organelles
- **C** have internal membrane-bound organelles
- **D** have only the nucleus that is membrane bound

3 Identify which of the following is/are never found in prokaryotic cells.
- **A** DNA
- **B** mitochondria
- **C** cytosol
- **D** cell wall

4 Select the statement that accurately describes eukaryotic cells.
- **A** Eukaryotic cells have circular chromosomes and membrane-bound organelles, and some also have cell walls.
- **B** Eukaryotic cells have linear chromosomes but not membrane-bound organelles, and some have cell walls.
- **C** Eukaryotic cells have linear chromosomes and membrane-bound organelles, and some also have cell walls.
- **D** Eukaryotic cells have linear chromosomes and membrane-bound organelles, but not cell walls.

5 Summarise the properties of archaean plasma membranes that allow them to be extremophiles.

6 **a** The micrometre (μm) is the unit used to measure cell size. There are 1000 μm in 1 millimetre (mm). Convert 1.6 mm into μm.

b Select the answer that is closest to the diameter of the animal cell shown in the following photograph.
- **A** 10 μm
- **B** 100 μm
- **C** 1 μm
- **D** 1000 μm

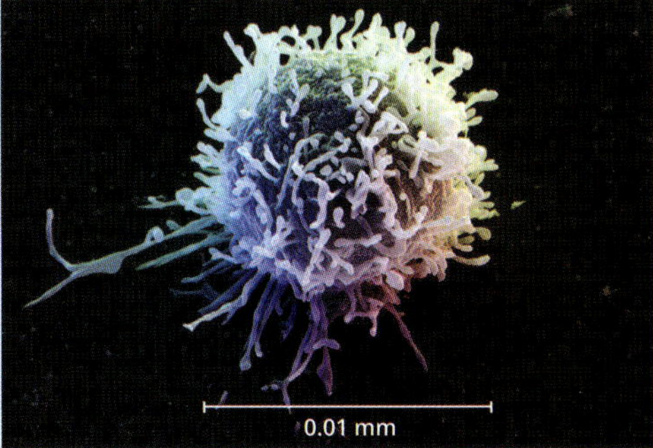

7 **a** What is cell compartmentalisation?

b Outline three ways that it benefits a cell. Use examples to support your answer.

8 A cell from an organism has a distinct nucleus, green organelles, and a plasma membrane within a cell wall. What cell type is this?

continued over page

9 The internal environment of eukaryotic cells is divided into compartments by membranes. This makes eukaryotic cells more efficient than prokaryotic cells because:

A it allows tasks to be merged together
B it decreases the surface area of eukaryotic cells
C each compartment is entirely independent of the others
D it allows different areas of the cell to maintain different conditions

10 Identify which organelles would be most abundant in each of the following cell types. Explain your reasoning in each case.

a enzyme-secreting cells
b muscle cells
c cells that carry out photosynthesis in a leaf

11 Draw and label a typical plant cell and and a typical animal cell.

12 Many single-celled organisms such as amoeba feed by a process in which the plasma membrane engulfs solid food particles to form a food vacuole. This process is called:

A phagocytosis
B active transport
C pinocytosis
D osmosis

13 According to the fluid mosaic model of plasma membrane structure, proteins of the membrane are mostly:

A spread in a continuous layer over the inner and outer surfaces of the membrane
B confined to the hydrophobic interior of the membrane
C embedded in a lipid bilayer
D randomly orientated in the membrane, with no fixed inside–outside polarity

14 Identify which factor would most increase the fluidity of a plasma membrane.

A greater proportion of saturated phospholipids
B greater proportion of unsaturated phospholipids
C lower temperature
D greater proportion of large glycolipids compared with lipids of smaller molecular masses

15 Create a table to summarise the major functions of phospholipids, cholesterol, glycoproteins, proteins and glycolipids in cell plasma membranes.

Application and analysis

16 You have been assigned the task of determining whether a sample contains plant or animal cells.

a How would you examine the sample?
b Identify what features of the sample would help you in this task.

17 An experiment was conducted to investigate the synthesis of a type of organic substance from β-cells in the pancreas. Radioactive amino acids (the subunits used to make proteins) were injected into the secretory tissue. The level of radioactivity in various organelles of this type of cell was then measured every 60 minutes. The results are summarised in the table below.

	Percentage of total radioactivity			
Time (mins)	**Rough endoplasmic reticulum**	**Golgi apparatus**	**Immature secretory vesicles**	**Mature secretory vesicles**
0	77	10	0	13
60	17	57	15	11
120	20	15	45	10
180	21	13	16	50
240	21	11	13	55
300	20	11	12	57

a Deduce the type of substance being synthesised.
b Identify the trend or pattern in radioactivity in the:
i endoplasmic reticulum
ii Golgi apparatus.

18 **a** Complete the table below of surface area to volume ratios. Show your working.
Note: each cube is 1 cm × 1 cm × 1 cm.
b What happens to the surface area to volume ratio as objects increase in size?
c In the table, objects (ii) and (iv) have the same volume, as do objects (iii) and (v). How does a change in shape affect the surface area to volume ratio of these objects?
d Explain the significance of the surface area to volume ratio for cells.

Object	**Surface area**	**Volume**	**SA : V ratio**
i		1 cm × 1 cm × 1 cm = 1 cm^3	
ii	4 cm^2 × 6 sides = 24 cm^2		
iii			
iv			
v			

19 Below are two cells observed under a scanning or a transmission electron microscope.

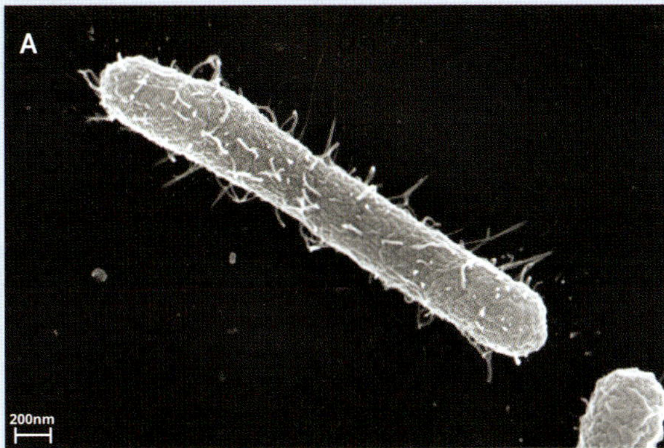

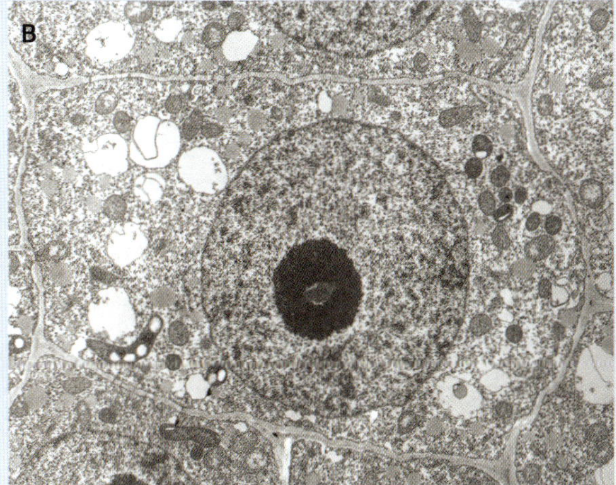

a One of the two cells is from a prokaryote. State which one, and explain your reasoning.

b Determine if the eukaryotic cell is from an animal or a plant.

20 You are given a microscope slide with a sample of cells smeared on it and are asked to identify the cell type. The cells are circular with a dark, round mass at their centre. You estimate that the cells are approximately 20 μm in diameter.

a Are the cells prokaryotic or eukaryotic? What feature(s) led you to your conclusion?

b What is the dark, round mass at the centre of the cells?

c What are three features you could look for to determine if the cells are from a plant or an animal?

21 Describe how prokaryotic cells are different from eukaryotic cells and outline the advantages and disadvantages of each cell type.

22 Two solutions are separated by a semi-permeable membrane as illustrated at the top of the next column. Deduce in which direction (if any) there would be a net movement of particles if water is the solvent and the membrane is impermeable to the solute.

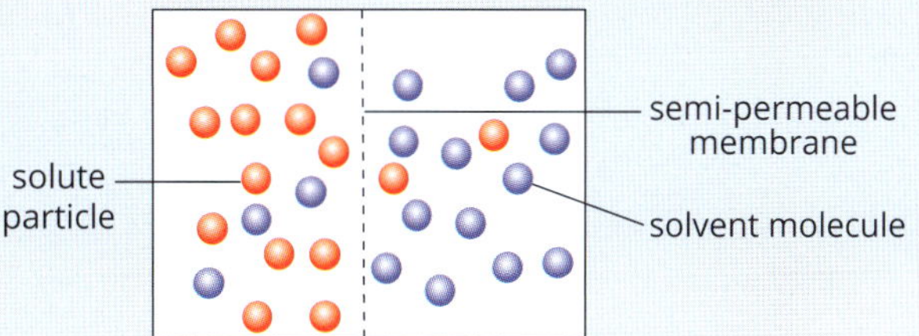

23 Two different solutions in water with the same volume are placed on either side of a semi-permeable membrane in a U-shaped glass tube, as shown in the following diagram. The membrane is permeable to salt but not glucose. The tube is left to stand for several days. Explain what you would expect to happen to:

a the salt concentration on each side of the membrane

b the glucose concentration on each side of the membrane

c the fluid levels on each side of the membrane.

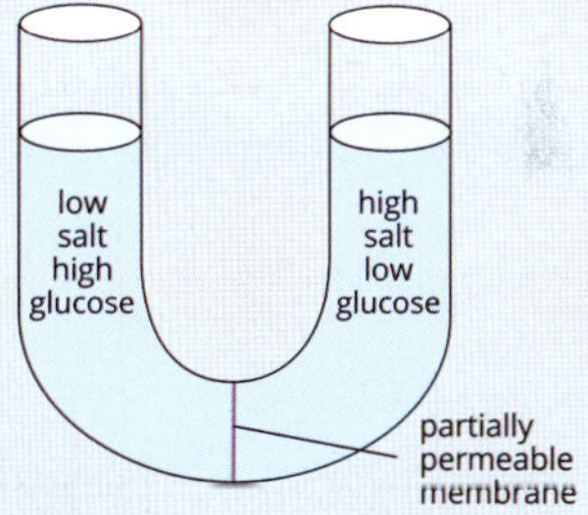

24 Root hair cells on the roots of plants use energy to take up some nutrients from the soil, but not others. State the circumstances in which energy is expended during nutrient uptake.

25 Solutions of different sugar concentrations were prepared and a rod of peeled potato tissue of the same known mass was put into each solution. After 1 hour, the potato was removed and its mass was measured again. Results are summarised in the following table.

Concentration of sugar (g/100 mL)	Change in mass (g)	
20	0.68	decrease
18	0.40	decrease
14	0.01	increase
12	0.18	increase
10	0.32	increase
6	0.59	increase
2	0.84	increase

a Use this data to plot a graph showing the change in the mass of the potato tissue with changes of the concentration of sugar.

b Use the graph to predict the mass change if a rod of potato tissue was placed in a sugar solution of 8 g/100 mL.

c Identify any trends and patterns in the data and explain these results, making specific reference to the data.

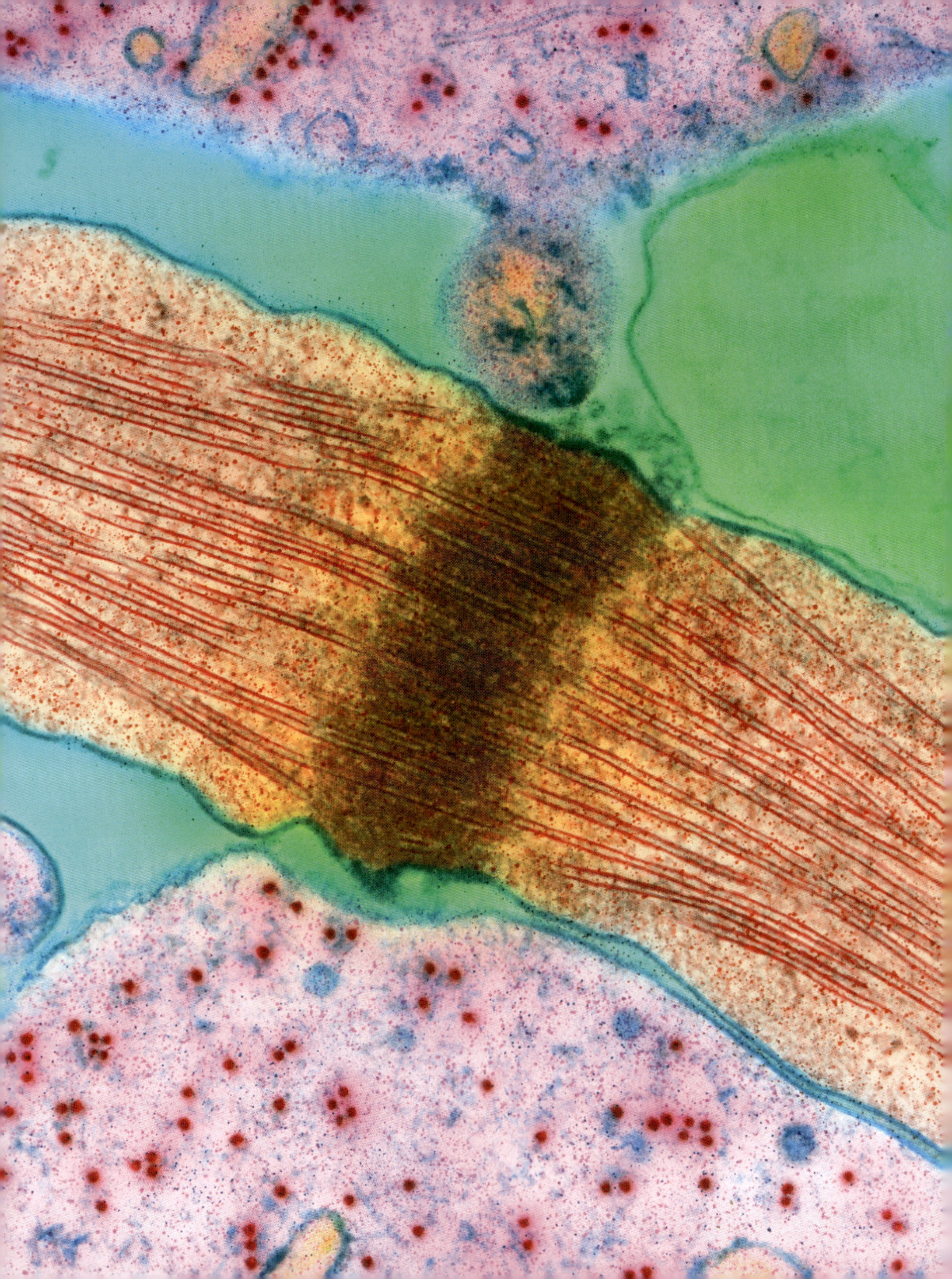

CHAPTER

03 The cell cycle and cell growth, death and differentiation

A characteristic of living things is their ability to reproduce to create more of their own kind. This idea was first proposed in 1855 by the German–Polish embryologist Robert Remak, who showed that new cells originate from existing cells by cell division. Three years later, the German physician Rudolf Virchow wrote about this finding in an essay using the Latin phrase '*omnis cellula e cellula*' (all cells originate from other cells). Cells need to replicate in order to maintain the continuity of life, and this is true for both prokaryotic and eukaryotic cells.

In this chapter you will explore how cells replicate and the purposes of replication. You will examine the mechanisms that control replication, and what happens if these mechanisms fail. You will also learn about the properties of stem cells and their role in the differentiation, specialisation and renewal of cells and tissues.

Key knowledge

- binary fission in prokaryotic cells **3.1**
- the eukaryotic cell cycle, including the characteristics of each of the sub-phases of mitosis and cytokinesis in plant and animal cells **3.2**
- apoptosis as a regulated process of programmed cell death **3.3**
- disruption to the regulation of the cell cycle and malfunctions in apoptosis that may result in deviant cell behaviour: cancer and the characteristics of cancer cells **3.3**
- properties of stem cells that allow for differentiation, specialisation and renewal of cells and tissues, including the concepts of pluripotency and totipotency. **3.4**

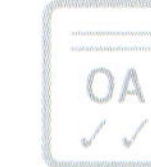

3.1 Cell replication

FIGURE 3.1.1 Coloured scanning electron micrograph (SEM) of a fertilised human egg (zygote). This one cell must replicate to become a fetus.

Once, you were a single cell—a fertilised egg known as a zygote (Figure 3.1.1). Now, your body is made up of about 37 trillion cells. In order to start creating the millions of cells that make you, that first single cell had to replicate itself. As the cells in your body wear out and die or are damaged, more cells are replicated to replace them. Replication results in genetically identical cells.

The cell theory, which you learnt about in Chapter 2, states that all cells arise from pre-existing cells. In order for this to occur, cells must be able to replicate. This process is essential to the life of all organisms. In this section, you will learn about the reasons for cell replication and the methods of replication for eukaryotes and prokaryotes.

PURPOSES OF CELL REPLICATION

Cell replication is a form of cell division in which a parent cell divides to produce two genetically identical **daughter cells**. It is essential that the genetic information is passed on accurately, because the activities of cells are controlled by the genetic information in the nucleus (in eukaryotes) or nucleoid (in prokaryotes).

In eukaryotes (protists, fungi, plants and animals), cells replicate by mitosis. Mitosis is part of the eukaryotic cell cycle, which you will learn about in Section 3.2.

Cells replicate for a number of reasons. For a multicellular organism, cells replicate for:

- restoring the nucleus-to-cytoplasm ratio
- growth and development
- maintenance and repair.

Unicellular organisms do not need to replicate for these purposes because they remain a single cell throughout their entire life cycle. Instead, cell replication in unicellular organisms (whether prokaryotes or eukaryotes) is a simple form of reproduction and creates a new, genetically identical individual.

Restoring the nucleus-to-cytoplasm ratio

Most cells in multicellular organisms are microscopic. The largest cells are egg cells (Figure 3.1.2). Some egg cells, such as bird and reptile eggs, are many millions of times larger than the body cells of the animals into which they will grow.

A consequence of the large size of cells such as egg cells is a very low nucleus-to-cytoplasm ratio. Large cells are therefore unable to balance nucleic acid and protein synthesis in relation to their increase in volume, which results in a decrease in the efficiency of cellular processes due to relatively large diffusion distances. The rate of cellular processes therefore limits the size of cells.

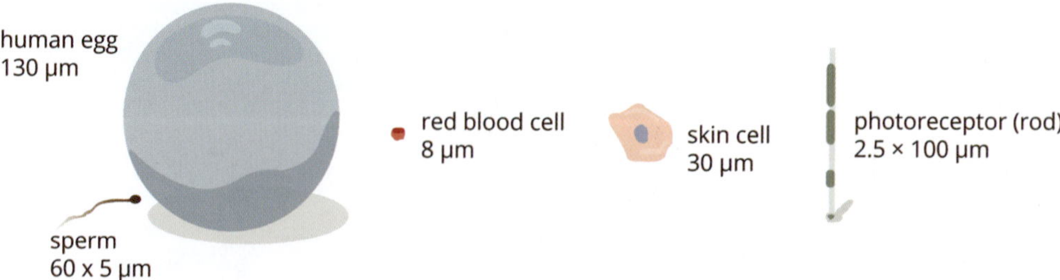

FIGURE 3.1.2 The human egg cell is many times larger than other human cells.

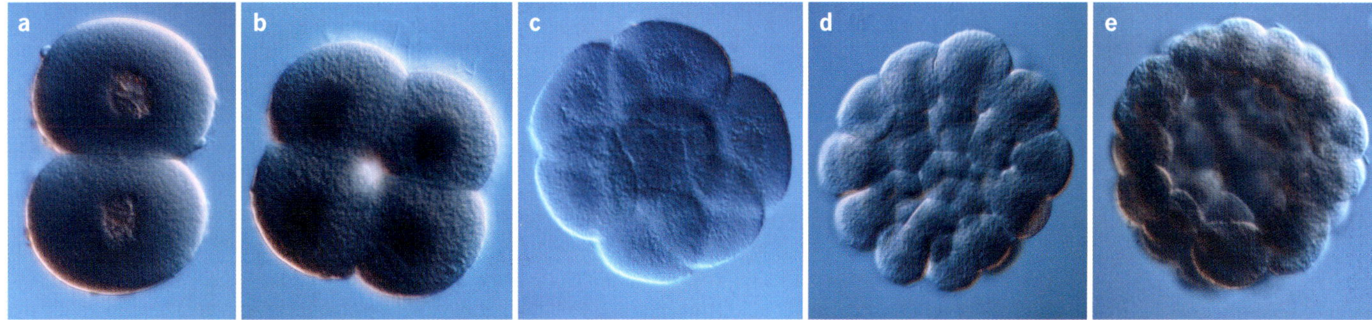

FIGURE 3.1.3 A series of light micrographs (LM) showing the change in the nucleus-to-cytoplasm ratio in the first five divisions of a zygote by mitotic cell division. (a) The first division results in 2 cells; (b) the second results in 4 cells; (c) the third results in 8 cells; (d) the fourth results in 16 cells, and (e) the fifth results in 32 cells.

After an egg is fertilised, the cell (now called a **zygote**) divides repeatedly by mitosis (mitotic division) to produce many smaller cells and form an **embryo**. These early rounds of mitotic division serve to restore the nucleus-to-cytoplasm ratio, so that each nucleus is located in a normal-sized cell.

You can see the cell division from 2 to 32 cells in the first stages of embryo development in Figure 3.1.3. You will notice that the overall size of the early embryo does not increase as the number of cells increases.

Growth and development

Multicellular organisms grow in size by increasing the number of their cells through repeated cell replications. The new cells then grow in size, increasing the size of the organism. As the new individual continues to develop, new cells become specialised for different purposes. Muscle cells in animals and root cells in plants are examples of specialised cells (Figure 3.1.4). More replications follow and the specialised cells become organised into tissues, which form the body of the organism.

Development in multicellular organisms involves a balance between cell replication and cell death. As development continues to the adult form, some cells are destined to die through 'programmed cell death', which is known as apoptosis. You will learn about apoptosis in Section 3.3. Other cells, such as nerve cells and red blood cells, may become highly specialised and no longer undergo replication.

FIGURE 3.1.4 As this soya bean plant grows, cells replicate and specialise (for example, into root or leaf cells). As growth and development occur, some cells will also undergo apoptosis.

BIOFILE

Nematodes

Growth in the size of an organism usually involves increasing the number of cells, but nematodes are unusual. Cell division ceases when a nematode hatches from its tiny egg. Nematodes hatch as miniature adults, and all subsequent growth is the result of the increasing size of cells present at hatching. Because there is a fixed number of mitotic divisions before hatching, all individuals of a species have exactly the same number of cells. This uniformity in number of cells makes the nematode, *Caenorhabditis elegans*, a good model organism for studies in cell biology.

SEM of nematode worm *Caenorhabditis elegans*

Tissue maintenance and repair

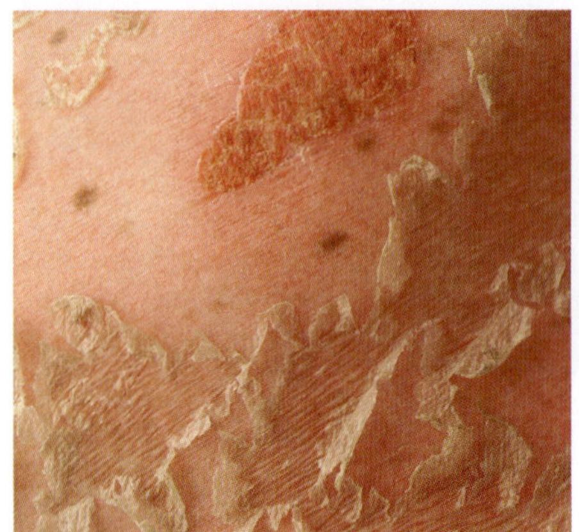

FIGURE 3.1.5 Sunburn kills skin cells. Layers of dead cells then peel off, while healthy skin cells underneath replicate to replace them.

Cells in the tissues of multicellular organisms, such as the skin cells covering the body's surface, become damaged or die as a result of normal functioning, and also as a result of injury, such as sunburn in the case of skin cells (Figure 3.1.5). In either case, maintaining and repairing tissues requires the production of new cells to replace those that die. These new cells are produced by cell replication.

The extent to which different organisms can carry out repair varies dramatically. Given the right conditions, many plants can grow from a fragment of a stem or leaf, and most mosses and liverworts can regrow from just a few cells. In most animals, stem cells (cells that retain some embryonic characteristics and can turn into specialised cells) are involved in the growth, repair, replacement and regeneration of tissues. For example, stem cells in bone marrow give rise to the different blood cells and replenish them as needed by the body (Figure 3.1.6).

a

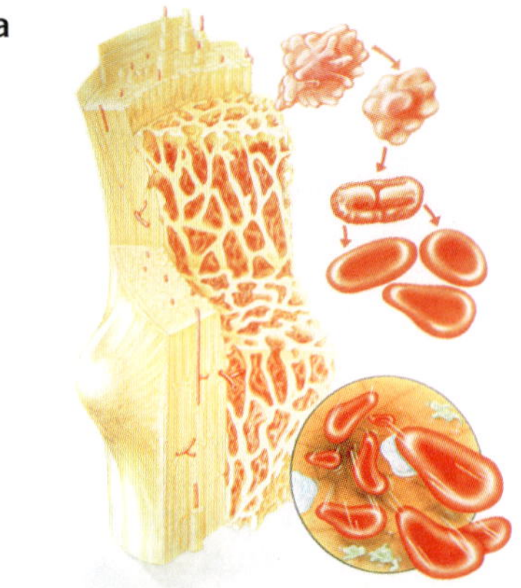

b

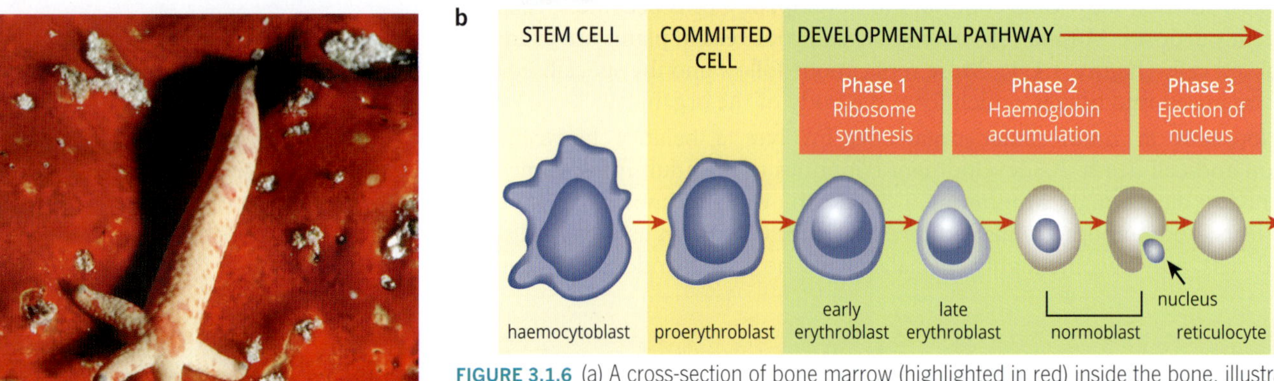

FIGURE 3.1.6 (a) A cross-section of bone marrow (highlighted in red) inside the bone, illustrating the formation of red blood cells from stem cells in the marrow. (b) The development of haemocytoblasts (stem cells from the bone marrow) into red blood cells. Note the nucleus exclusion during the later stages to create the red blood cell.

FIGURE 3.1.7 A sea star regenerating from a single arm (upper right) that was severed from the body of a mature sea star. The single arm is developing into a new sea star with four new arms.

A sea star can produce an entire new individual from a single arm by cell replication and subsequent specialisation (Figure 3.1.7). Humans can repair many tissues, but they cannot grow new limbs. A nematode, once hatched, cannot produce any new cells at all.

BIOFILE

Permanent cells: the vertebrate lens

In some tissues there is no cell replication. Sufficient numbers of specialised cells are produced during development to last the entire life of the organism. These permanent cells include the cells of the eye lens in vertebrates. Although permanent cells cannot be replaced, they can be repaired by the replacement of various organelles, a process that continues throughout the life of the cell.

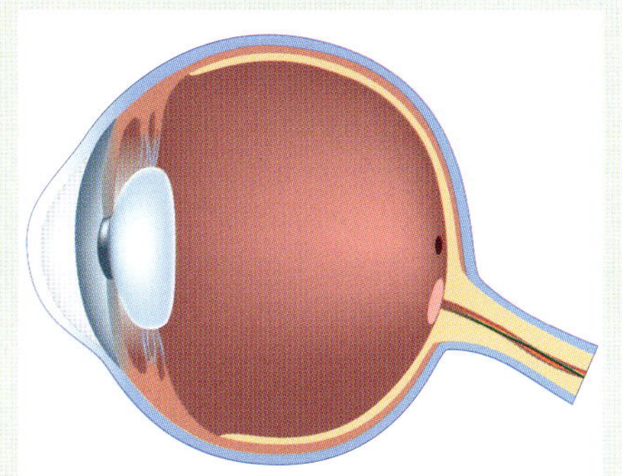

The highly specialised cells that make up the lens of a vertebrate eye must be transparent so that light can pass through the lens without being distorted. Once the lens has formed, individual cells cannot be replaced.

CELLS DIVIDE EXPONENTIALLY

As cells replicate, their numbers increase exponentially: 2 cells give rise to 4, 4 cells give rise to 8, 8 cells give rise to 16, and so on (Figure 3.1.8). The total number of cells doubles at every replication, because two cells are produced from each cell replication. This can be explained by the formula, $C = 2^n$ where C is the number of cells and n is the number of cell divisions that have occurred.

Exponential growth by cell replication:
$C = 2^n$
where
C = number of cells
n = number of cell divisions that have occurred

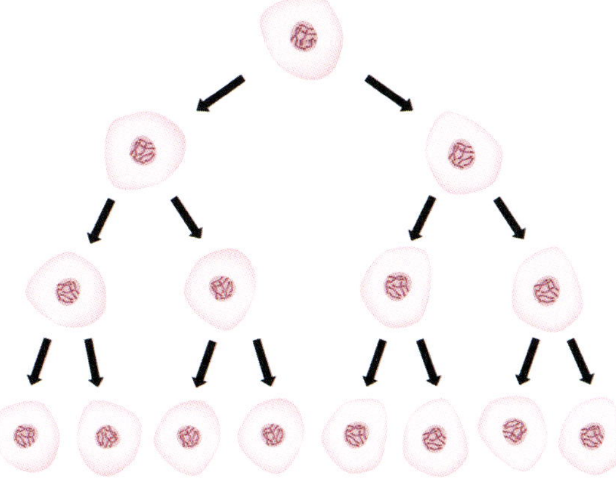

FIGURE 3.1.8 When cells replicate, they increase their numbers exponentially. This one cell becomes eight cells after only three divisions.

CELL REPLICATION IN EUKARYOTES

In eukaryotes (protists, fungi, plants and animals), cells replicate by mitosis followed by cytokinesis.

Mitosis is the division of the nucleus into two daughter nuclei. At the end of mitosis the cytoplasm also divides, separating the two nuclei and other organelles into two complete daughter cells. This separation of the cytoplasm is known as **cytokinesis**. Mitosis and cytokinesis are part of the cell cycle, which is covered in more detail in Section 3.2.

Another form of cell division in eukaryotic cells is **meiosis**. Meiosis is not a form of cell replication because the daughter cells are different from each other and also from the parent cell. Meiosis is an important cell division that is required for sexual reproduction; it produces four daughter cells that are genetically unique. You will learn more about meiosis in Chapter 6.

Replication and reproduction in multicellular organisms are different processes. Replication produces two genetically identical cells from one parent cell. Reproduction produces a new organism from one or two parent organisms.

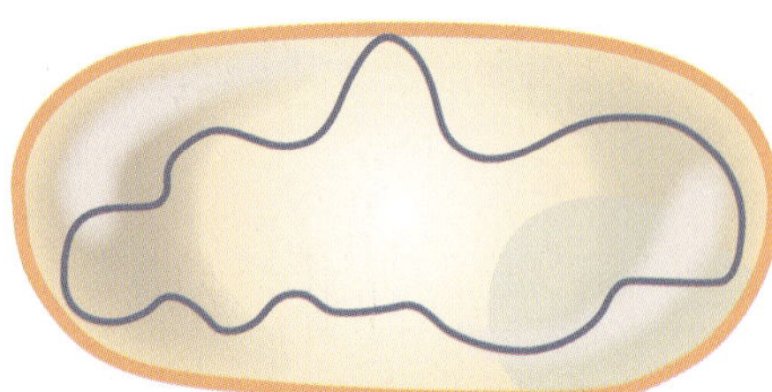

FIGURE 3.1.9 Prokaryotes are simple cells with a single continuous (or circular) DNA chromosome (shown in blue).

CELL REPLICATION IN PROKARYOTES

Prokaryotes such as bacteria and cyanobacteria are simple single-celled organisms. Prokaryotes have no nucleus or membrane-bound organelles; instead their genetic material is a single circular DNA chromosome attached to the plasma membrane at a point called the **origin**. (Figure 3.1.9). The circular DNA chromosome of prokaryotes is contained within an irregularly shaped region called the nucleoid. Because prokaryotes lack membrane-bound organelles and have a smaller amount of DNA, cell replication is much simpler and occurs more quickly than in eukaryotes.

Cell replication in prokaryotes is called **binary fission**. Like mitotic division, binary fission is an exponential process because (in ideal conditions) the population doubles after every cycle of division.

Some bacteria can undergo binary fission every 20 minutes. In other words, the number of cells can double every 20 minutes. This means that in six hours up to 18 cycles of binary fission could occur, and in this time one bacterium could have produced 2^{18} (262 144) individuals.

Binary fission

Binary fission occurs in a series of steps, as shown in Figure 3.1.10.

1 Before a prokaryotic cell undergoes binary fission, it has just one DNA molecule.

2–3 The DNA molecule (which in bacteria is in a circular chromosome) is duplicated within the nucleoid, resulting in two identical DNA molecules. Replication of the circular chromosome begins at the origin.

3–4 The cell grows until it has almost doubled in size.

5 The two DNA molecules are pulled to separate poles as the cell increases in size.

6 A new cell wall and plasma membrane form between the separating chromosomes, dividing the cell into two relatively equal halves. These halves eventually separate, forming two daughter cells from the single parent cell.

Figure 3.1.11 shows binary fission in a bacterium. At this stage the new cell wall and plasma membrane are beginning to form.

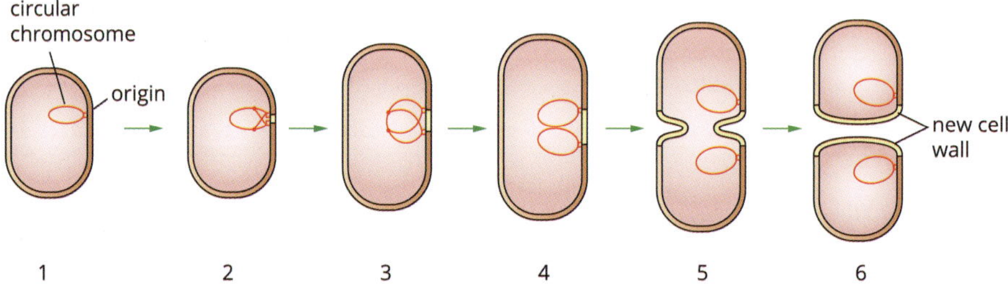

FIGURE 3.1.10 Binary fission of a prokaryotic cell. Replication of the chromosome begins at a point called the origin, after which new cell wall and plasma membrane is laid down to divide the cell in two.

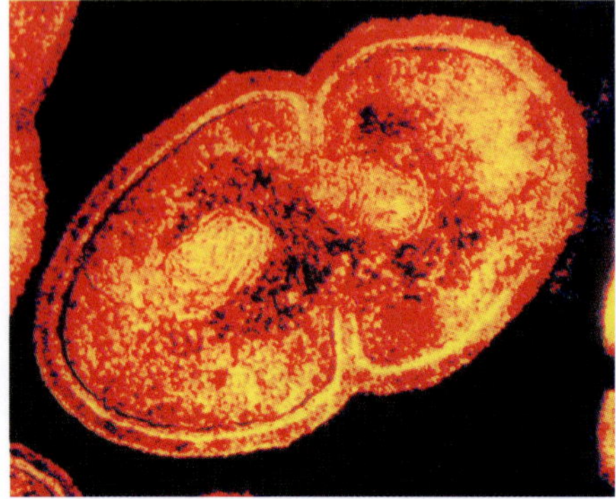

FIGURE 3.1.11 Transmission electron micrograph (TEM) of an *Enterococcus faecalis* bacterium undergoing binary fission. The regions containing the nucleic material are visible as round, yellow structures within the two daughter cells. The formation of new cell walls can also be seen as yellow indentations in the sides of the cell.

BIOFILE

Antibiotic resistance

Binary fission produces organisms that are genetically identical to the parent. One consequence of this method of reproduction is that all individuals in a colony (a population of bacteria originating from one parent cell) are vulnerable to the same environmental factors, such as antibiotics that are designed to target a particular strain of bacteria. Genetic mutations can introduce variation into a bacterial population, potentially creating resistance to antibiotics that the population was previously vulnerable to. The resistant individuals can survive antibiotic treatment and rapidly replicate using binary fission.

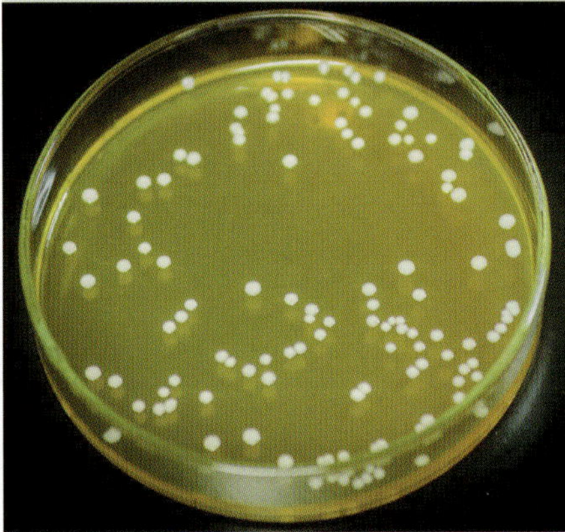

Colonies of bacteria growing on an agar plate. Each bacterium in a colony is genetically identical.

3.1 Review

OA ✓✓

SUMMARY

- Eukaryotic cells replicate by mitotic cell division.
- For multicellular organisms, the purposes of cell replication are to restore the nucleus-to-cytoplasm ratio, growth, development, maintenance and repair.
- Some cells, such as fertilised eggs, can replicate many times while keeping the same amount of cytoplasm. Therefore, replication can restore the nucleus-to-cytoplasm ratio in cases where there is too much cytoplasm for the nucleus to control.
- As cells multiply through replication, they can also enlarge and specialise, allowing the organism to grow and develop.
- As cells die or are damaged, cells produced through replication can replace those cells, allowing an organism to maintain and repair itself.
- As cells replicate, their numbers increase exponentially, according to the formula $C = 2^n$, where C is the total number of cells and n is the number of cycles of replication that have occurred.
- Mitotic cell division occurs in two distinct stages: mitosis (a division of the nucleus) followed by cytokinesis (the division of the cytoplasm).
- For unicellular organisms, the purpose of cell replication is reproduction.
- Prokaryotic cells replicate by binary fission.
- Binary fission is an efficient process wherein the DNA molecule replicates, then the cell grows larger and splits into two daughter cells.
- Cell replication is much simpler and occurs more quickly in prokaryotes than in eukaryotes because prokaryotes lack membrane-bound organelles and have less DNA.

KEY QUESTIONS

Knowledge and understanding

1 Which of the following is not a purpose of cell replication by mitosis in multicellular organisms?
  A growth
  B repair
  C reproduction
  D restoring the nucleus-to-cytoplasm ratio

2 Which of the following statements is true?
  A Cytokinesis is also called binary fission.
  B Cytokinesis involves the division of the nucleus.
  C Cytokinesis occurs during meiosis.
  D Cytokinesis occurs after mitosis.

3 **a** Outline the purposes of cell replication for multicellular organisms.
  b Give an example of a specific tissue or cell type in both plants and animals that illustrates each point in **a**.

4 Describe the key events in binary fission in bacteria.

Analysis

5 Classify the examples of cell replication into the correct purposes.
  - toddler's height increasing by 2 cm
  - cut healing
  - bacteria cell dividing
  - embryonic cell dividing
  - seed germinating
  - unicellular Protista organism dividing

Purpose	Example
reproduction	
repair and maintenance	
growth and development	
restoring nucleus-to-cytoplasm ratio	

6 If a bacterium undergoes binary fission in ideal conditions every 30 minutes, calculate how many bacteria there will be after four hours.

3.2 The cell cycle

FIGURE 3.2.1 LM of hyacinth root cells undergoing mitosis. Mitosis is one of the phases of the cell cycle.

> The cell cycle is the series of stages that a cell passes through, from its formation by cell division through its growth and function until it divides again.

Individual organisms have a life cycle. A life cycle is simply the phases through which an organism passes during its life, including reproduction. A butterfly's life cycle, for example, includes stages from egg, to caterpillar, to chrysalis, to butterfly. The cycle of life begins again when the adult butterfly lays an egg.

Just as individual organisms have a life cycle, so do cells. The phases in the life of a cell are known collectively as the **cell cycle**. The cell cycle begins with a single cell that grows and then divides into two daughter cells through cell replication. Figure 3.2.1 shows plant cells in different phases of the cell cycle.

In this section you will learn about the cell cycle and its sequential phases.

THE EUKARYOTIC CELL CYCLE

In eukaryotic cells the cell cycle has three main phases:

- interphase
- mitosis
- cytokinesis.

These phases always occur in this order, beginning with interphase. During interphase the contents of the cell are duplicated and the cell mass doubles. During mitosis the nucleus divides, and during cytokinesis the cytoplasm divides.

The dividing of the cytoplasm during cytokinesis marks the creation of the two new cells. So the cell cycle is the period between one cytokinesis and the next. In actively growing cells, mitosis and cytokinesis occupy only a small part of the cell cycle, and interphase occupies a large portion, as you can see in Figure 3.2.2.

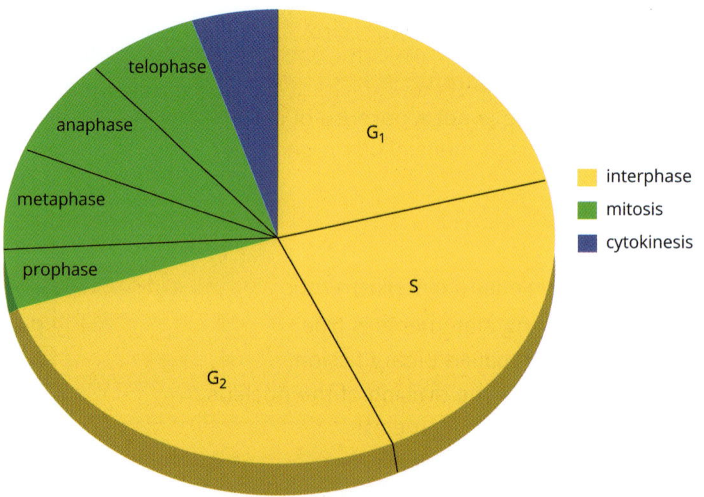

FIGURE 3.2.2 The eukaryotic cell cycle. Mitosis and cytokinesis occupy only a small portion of the time taken for one cell cycle, while interphase occupies a large portion.

Interphase

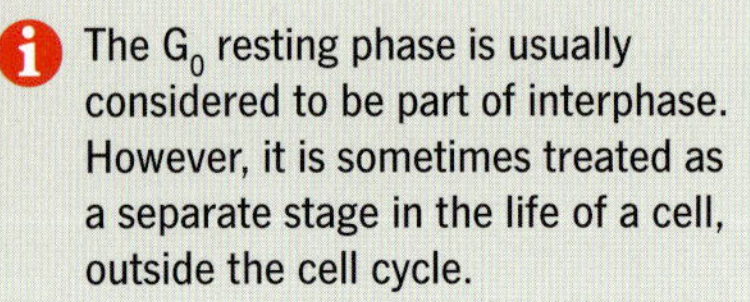

> The G_0 resting phase is usually considered to be part of interphase. However, it is sometimes treated as a separate stage in the life of a cell, outside the cell cycle.

The first stage of the cell cycle is **interphase**. It begins immediately after the end of cell division. During interphase, a cell that is about to divide grows larger, and copies its chromosomes in preparation for cell division. Interphase is divided into three main phases:

- G_1 (pre-DNA synthesis)
- S (DNA synthesis)
- G_2 (post-DNA synthesis).

During interphase there is sometimes a fourth phase called G_0 (G zero), also known as the 'resting phase'.

During interphase the cell grows by producing proteins and organelles such as mitochondria and Golgi apparatuses. Chromosomes, however, are only copied during the S phase. Thus a cell grows (in G_1 phase), continues to grow as it duplicates its chromosomes (in S phase), then grows more as it completes preparation for cell division (in G_2 phase). As the cell grows, the normal functions of the cell occur alongside the activities described above, including the synthesis of the many components required for G_1, S and G_2. Interphase always lasts much longer than mitosis, which lasts for about two hours in a human cell (Figure 3.2.3). The length of time a cell spends in interphase varies. Slow-growing cells, such as liver cells, spend many weeks or even years in interphase. Bone marrow cells may pass through interphase in less than a day as they generate many new blood cells. In an early embryo there is little growth and the cells rush from one round of mitosis to the next (Figure 3.2.4).

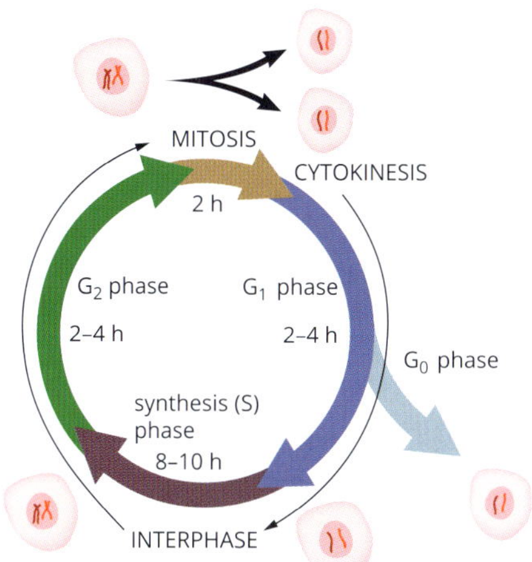

FIGURE 3.2.3 The cell cycle. Mitosis and cytokinesis (top) occupy only a small part of the whole cycle. A cell spends most of its time in interphase (G_1, G_0, S and G_2). The cell cycle shown here takes approximately 14–20 hours to complete, but other cell cycles can take much longer.

G_1 phase

After division, a daughter cell is quite small. During the **G_1 phase** the cell gains energy and undergoes metabolic processes such as protein and membrane synthesis, and almost doubles in size. This growth includes various structures within the cytoplasm, including a large increase in the number of organelles. The progress of G_1 is very variable, and if conditions are not right the cell cycle is stopped in G_1 or goes into G_0. In any group of dividing cells, most will be in interphase. Most cells are seen in the G_1 phase because it is the longest part of interphase.

G_0 phase

At the start of G_1, cells may enter the **G_0 phase** (also known as the resting phase). During G_0 they carry out the normal functions of the cell but do not change their internal structure or size. In most of these cells G_0 is temporary and ends when the cell re-enters the G_1 phase. However, some specialised cells such as nerve cells and red blood cells remain permanently in the G_0 phase and therefore usually do not replicate (Figure 3.2.5). For example, in skin cells, the surface keratinised cells are highly specialised and no longer able to divide because they have permanently entered the G_0 phase. Skin replacement is from precursor (partially differentiated) cells in lower layers that are still able to divide.

i The G_0 phase is the pathway to cell differentiation. Some cells remain permanently in the G_0 phase and do not replicate, while others rest temporarily and then return to the G_1 phase.

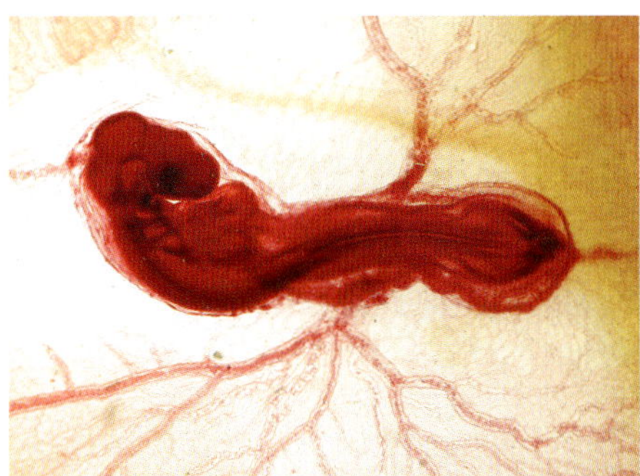

FIGURE 3.2.4 Early in the development of an embryo, such as this chicken embryo, cells replicate rapidly but little growth occurs.

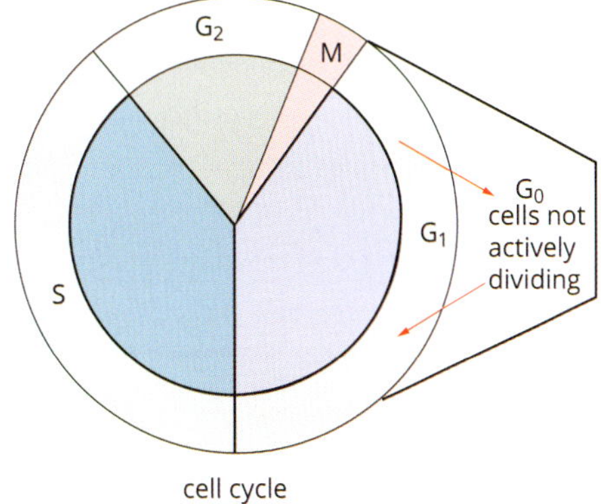

FIGURE 3.2.5 The cycle of a cell that has temporarily entered the G_0 phase and then returned to the G_1 phase. Not all cells enter the G_0 phase. Of the cells that do enter this phase, some enter it permanently and others temporarily.

BIOFILE

Cell cycle 'gaps'

In 1879, German physician and professor of anatomy Walther Flemming was the first to observe the behaviour of the chromosomes during cell division. He achieved this by using a stain he had developed that highlighted the nucleus of the cell during cell division. Because of the limited power of the microscopes available at the time, all the activity of the cell during interphase was not evident. For this reason the phases of interphase were named Gap 1 (G_1) and Gap 2 (G_2). We know now that interphase is a period of growth and activity in the cell cycle.

Walther Flemming (1843–1905), a pioneer in research on cell division

CASE STUDY ANALYSIS

The life span of cells

The life span of human cells has only been determined accurately in the last decade. The difficulty lay in being able to measure back in time to when each mature cell was born. Cell cycles measured in cells grown in laboratory cultures cannot be used to accurately represent those in living organisms.

In 2005, a team of Swedish researchers developed a dating technique for cells by applying carbon-14 (^{14}C) dating to the DNA of cells. Carbon-14 dating has long been used to date fossils, but the decay rate is too slow for it to be used for short time spans such as cell cycles.

The breakthrough came when the research team took advantage of the increased levels of ^{14}C in the atmosphere during the Cold War due to nuclear testing. By the time above-ground nuclear testing ended in 1963, the levels of atmospheric ^{14}C were double natural background levels. Since then, it has halved every 11 years. By taking this into account, there were detectable changes in levels of ^{14}C in modern DNA over short time spans. A scale for converting the ^{14}C enrichment into calendar dates was calculated from ^{14}C measurements of tree rings in Swedish pine trees exposed to these atmospheric changes (Figure 3.2.6). Since nuclear DNA fixes carbon atoms at the time of cell division it serves as a time capsule for measuring a cell's age.

Not surprisingly, the human cell cycle lengths were found to correlate in part to the function and position of each cell type. Cells that are worn away regularly need to be replaced regularly, meaning they have a shorter cell cycle (Figure 3.2.7). Some examples from the study are shown in Table 3.2.1.

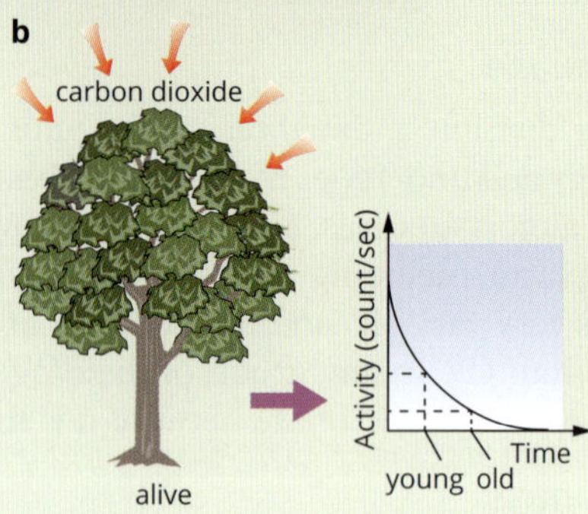

FIGURE 3.2.6 (a) The annual growth rings of pine trees containing carbon-14 from the atmosphere provided scientists with (b) a timeline for determining the age of human cells exposed to the same level of carbon-14.

TABLE 3.2.1 Approximate cell cycle lengths of different cell types in the human body

Cell type	Approximate age of cell (cell cycle length)
gut lining	2–5 days
epidermis	2–4 weeks
red blood cell	3–4 months
liver	10–17 months
bone	10–30 years
rib muscle	15 years
visual cortex	same age as the person
eye lens	same age as the person

FIGURE 3.2.7 Coloured SEM of overlapping cells on the outer layer of the human skin. These cells have a short life span as they are located at sites where they are continually worn away and need to be replaced rapidly. Cells to be shed undergo programmed cell death (apoptosis).

Analysis

1 Describe the correlation between growth rings of pine trees and ^{14}C from the atmosphere.
2 How can this correlation be used to calculate the length of the human cell cycle?
3 Do all cells in our body have the same age?
4 Using the information from Table 3.2.1, discuss why different cell types have different cell cycle lengths.

S phase

The 'S' in **S phase** stands for synthesis. During this phase, chromosomes are replicated in the nucleus. Figure 3.2.8 shows how each chromosome, which in eukaryotes are linear, makes an exact copy of itself in preparation for mitosis. Each strand of the replicated chromosomes is known as a **chromatid**. Chromatids are held together at the **centromere**. At the end of this phase the amount of DNA in the cell doubles, but the number of chromosomes (ploidy) of the cell remains the same.

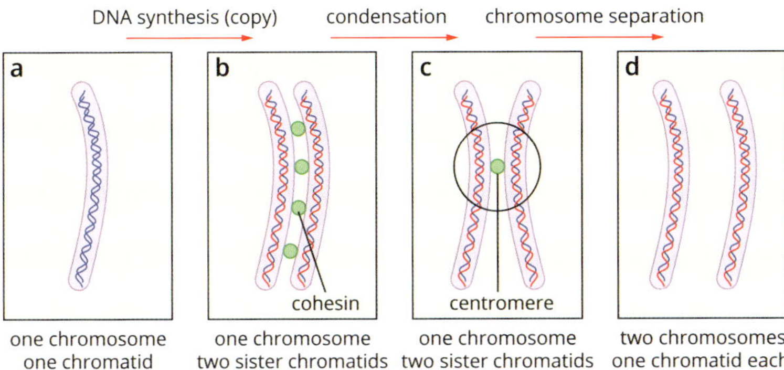

FIGURE 3.2.8 (a) In interphase chromosomes are present in the nucleus as a single unit (also called a chromatid). (b) A copy of each chromosome is synthesised. The copies are called sister chromatids, and they are identical. (c) The centromere represents the connection between the chromatids. This arrangement is still considered to be one chromosome, although it consists of two sister chromatids. (d) After the sister chromatids separate in anaphase, each sister chromatid is then considered to be an individual chromosome again.

G_2 phase

During the **G_2 phase** the cell undergoes a secondary stage of growth, metabolism and energy acquisition. It prepares for mitosis by synthesising the materials needed for division, such as proteins.

Mitosis

Mitosis is the division of the nucleus. It is a continuous process but has four sub-phases:

1 prophase
2 metaphase
3 anaphase
4 telophase.

Each sub-phase can be distinguished by the appearance and the position of the chromosomes in the cell. During interphase in the cell cycle, the chromosomes are duplicated. However, they are not visible under a microscope because they have not condensed. Figure 3.2.9 shows the movement of the chromosomes during mitosis.

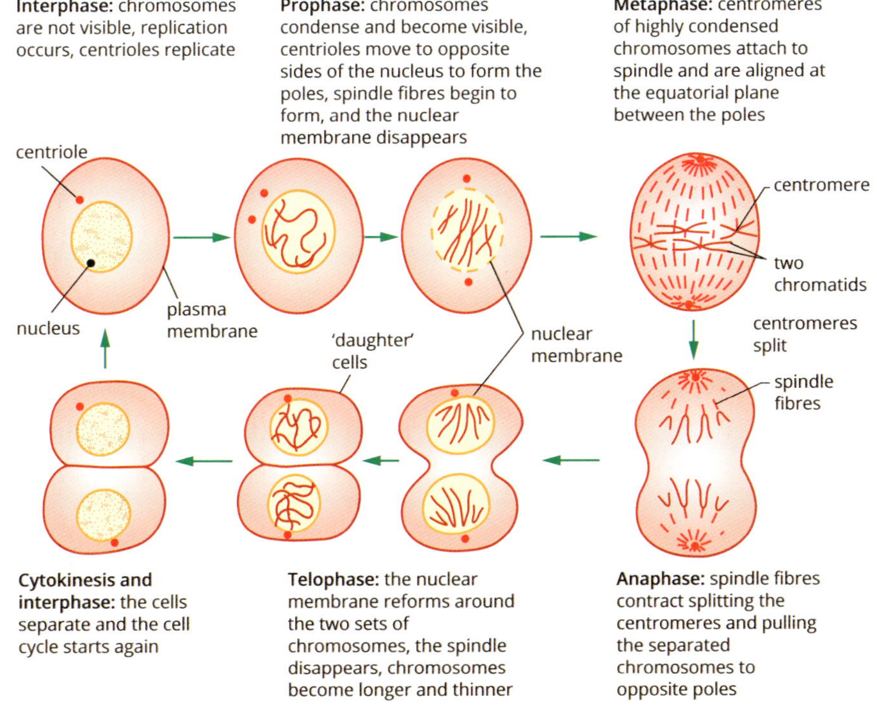

FIGURE 3.2.9 Major stages in the cell cycle, including mitosis

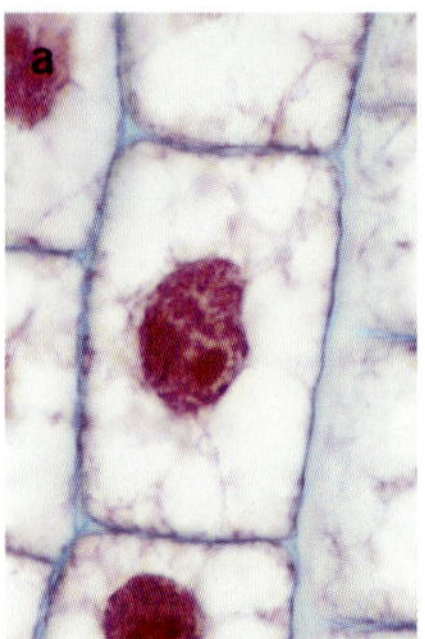

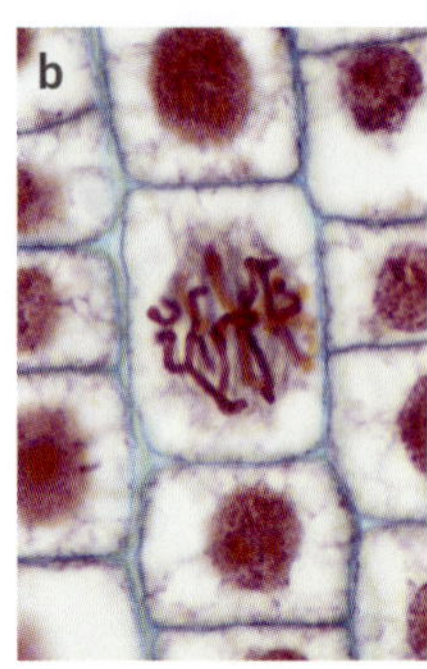

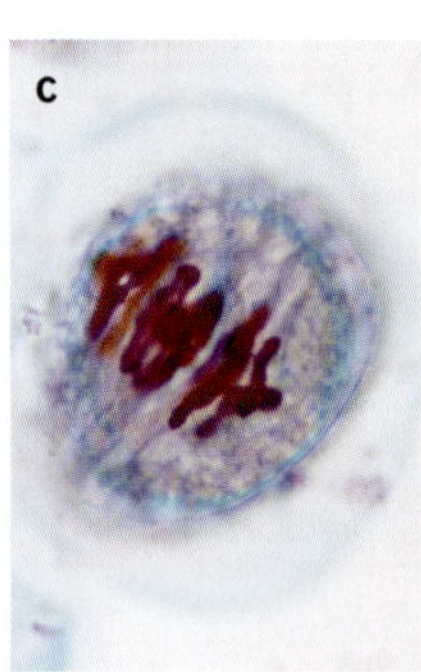

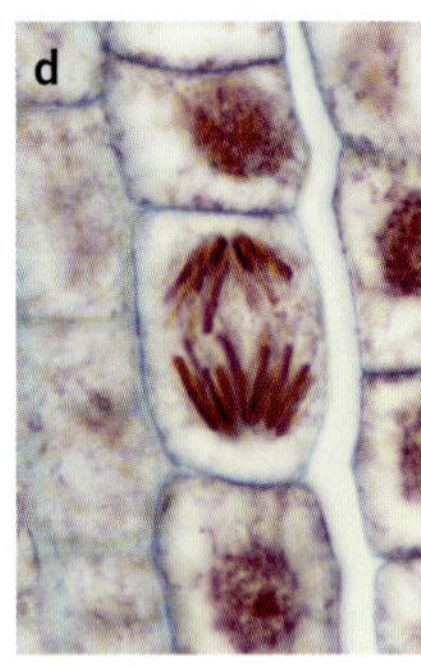

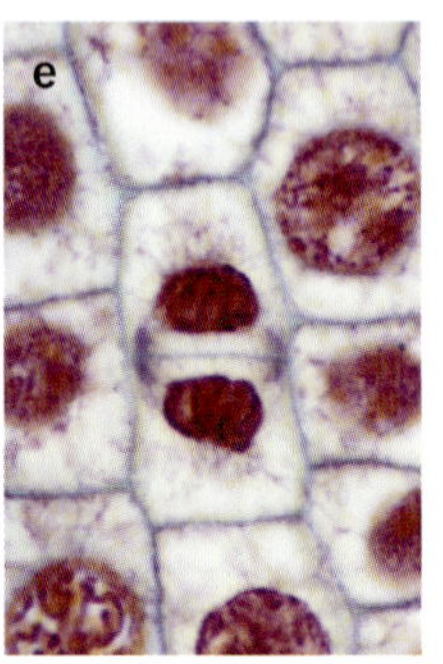

FIGURE 3.2.10 LMs of mitotic division in a garlic root tip cell: (a) interphase, (b) prophase, (c) metaphase, (d) anaphase and (e) telophase

Prophase

Early in **prophase**, chromosomes begin to condense (shorten and thicken) and become increasingly visible under the microscope. As they condense further, each chromosome can be seen as two chromatids held together at the centromere. At the same time the **centrioles**, which were replicated during interphase, move to opposite ends of the cell to form the poles.

Later in prophase the nuclear membrane breaks down. The centrioles begin to form a network of fibres, called the **mitotic spindle**, which extends between the two poles of the cell. Plant cells do not usually have centrioles; they use a different mechanism to produce the mitotic spindle.

Metaphase

During **metaphase** the centromere of each individual chromosome attaches to spindle fibres extending from each of the poles. The centromeres continue to be drawn by the spindle fibres so that the chromosomes are aligned in the middle of the cell. Chromosomes are most easily observed at this stage because they are highly condensed.

Anaphase

In **anaphase** the spindle fibres contract, pulling the two centromeres in opposite directions. The centromere splits, separating the two chromatids. Contraction of the spindle fibres continues and the separated chromatids are pulled to opposite poles. Thus, daughter cells receive the same genetic information—one copy of every chromosome that was in the original nucleus at interphase.

Telophase

The final stage of mitosis is called **telophase**. It is rather like prophase in reverse. A nuclear membrane forms around the chromosomes at each pole. The mitotic spindle is dismantled and disappears. The chromosomes become longer and thinner, and therefore less visible under the microscope. When mitosis is complete, each daughter nucleus moves into G_1 of interphase.

At the end of mitosis the cell contains one copy of every chromosome that was present in the parental cell. Before a cell can divide again, the chromosomes must again replicate into two identical 'sister' chromatids. This replication of the DNA occurs during interphase.

You can see the stages of interphase through to telophase in the light micrographs in Figure 3.2.10.

Cytokinesis

During cytokinesis the cytoplasm divides and the new nuclei separate (Figure 3.2.11). Cytokinesis in animal cells occurs in a different way to cytokinesis in plant and fungi cells.

In animal cells the plasma membrane moves inwards, pinching the two daughter cells apart. Plant and fungi cells lay down a new plasma membrane and cell wall between the two daughter nuclei to separate the daughter cells. Components of the new cell wall, called the cell plate, are initially deposited in the centre of the cell (Figure 3.2.12). The growth of the cell plate extends outwards until the two daughter cells are completely separated.

In some cells mitosis is not followed by cytokinesis, resulting in a large cell containing many nuclei. This type of cell is called a coenocyte. Examples of coenocytes include the endosperm of plant seeds, filamentous fungi and plasmodial slime moulds.

In contrast, a syncytium is a multinuclear cell formed from fusion of many separate cells, rather than a lack of cytokinesis. Skeletal muscle is an example of fusion of myocytes, which arose through normal mitosis.

FIGURE 3.2.11 Cytokinesis is as important as mitosis in eukaryotic cell replication because it results in the generation of two daughter cells. To form two cells, the cytoplasm of the parent cell must be divided into two by the plasma membrane.

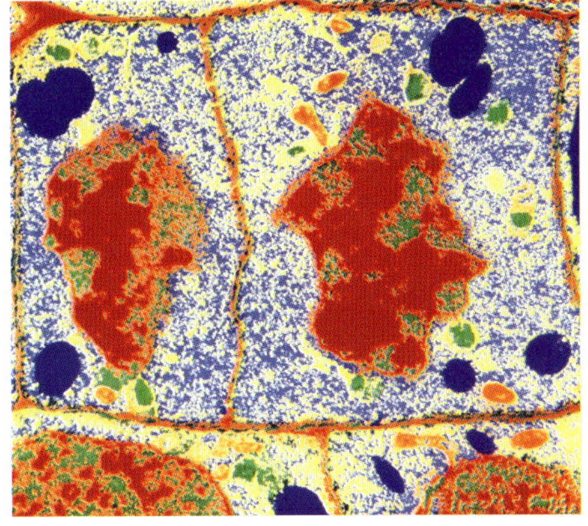

FIGURE 3.2.12 TEM of a late stage of cell division in a plant cell. The image has been coloured to show the organelles. The daughter nuclei (red and green bodies) are reforming into membrane-bound organelles. Between the two forming nuclei is the developing cell wall.

DIFFERENCES BETWEEN BINARY FISSION AND MITOSIS

As you learnt in the previous section, prokaryotic cells are much simpler than eukaryotic cells. As a result, the cell replication of prokaryotic cells (binary fission) is much simpler and quicker than that of eukaryotic cells (mitosis). Table 3.2.2 highlights some of the differences between binary fission and mitosis.

TABLE 3.2.2 Differences between binary fission and mitosis

Feature	Binary fission	Mitosis
cell type	prokaryotic	eukaryotic
division	a single organism divides into two daughter organisms	vegetative cell division
rate and complexity	efficient and rapid relatively simple	relatively slow relatively complex
primary function	reproduction	reproduction, repair and growth
structural changes	no nuclear membrane to break down and reform	nuclear membrane must break down and reform
cellular components	ribosomes and other cellular components are doubled before binary fission	organelles are doubled at interphase in order to separate into two cells
spindle apparatus	a spindle apparatus is not formed	a spindle apparatus is formed
DNA	single loop of double-stranded DNA	linear double-stranded DNA held together by centromeres
replication	DNA is replicated from origin	DNA is replicated in nucleus
DNA attachment	DNA is directly attached to plasma membrane	DNA is attached to the spindle apparatus
reliability of replication	replication errors sometimes result in unequal distribution of genetic material between daughter cells	high fidelity replication in which chromosome number is maintained through a checkpoint at metaphase. Errors occur, but more rarely than in binary fission.
cytokinesis	occurs directly after replication	cell division occurs long after replication is completed in the S phase

DNA REPLICATION

The DNA molecule is passed on from one cell to another when cells divide. In order to be able to transmit an exact copy of the DNA molecule without losing any instructions, cells must have a mechanism for accurately copying (replicating) and synthesising new DNA. This process is known as **DNA replication**.

DNA is a double-stranded molecule. Each chromosome is one long double strand of DNA. During chromosome replication, each strand of a parental DNA molecule acts as a template strand on which a new strand is synthesised. This involves the double-stranded DNA 'unzipping' and the enzyme, DNA polymerase, moving along the exposed template strands, adding new nucleotides. The new nucleotides pair up with the template, using the rules of complementary base pairing (A pairs with T, C pairs with G), to make new strands.

As each daughter DNA molecule consists of one old and one newly synthesised strand, DNA replication is described as semi-conservative replication. It is an extremely accurate process with three distinct phases (Figure 3.2.13).

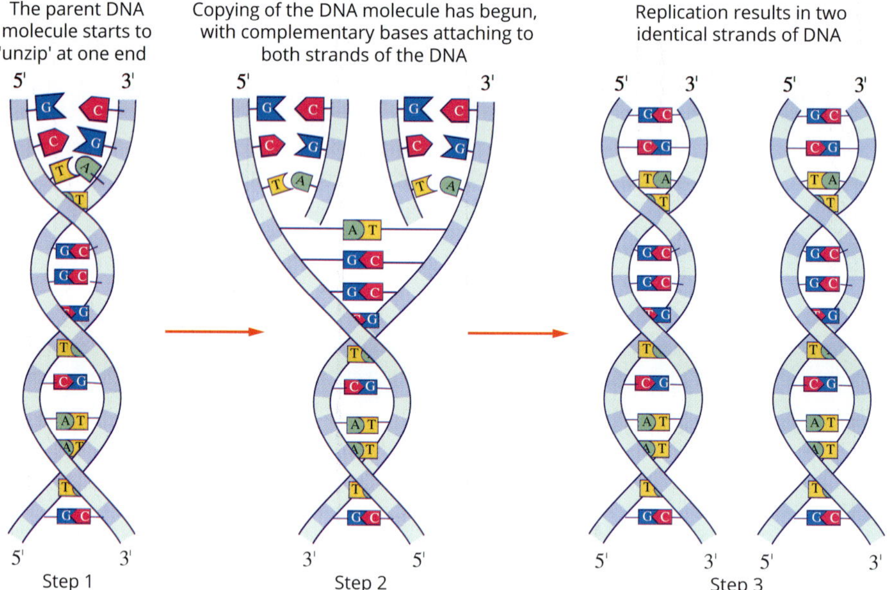

FIGURE 3.2.13 DNA replication involves three distinct phases.

3.2 Review

OA ✓✓

SUMMARY

- The cell cycle has three main stages: interphase, mitosis and cytokinesis.
- Interphase involves three phases: G_1 (pre-DNA synthesis), S (DNA synthesis) and G_2 (post-DNA synthesis). Some cells move out of the cell cycle into an additional phase called G_0 (resting phase).
- Cellular activities that usually occur during interphase include DNA synthesis (replication of chromosomes), increase in cell size, protein synthesis and other normal cellular processes, and preparation for mitosis.
- Mitosis is a continuous process that is described in sub-phases: prophase, metaphase, anaphase and telophase.
- During mitosis, identical copies of each chromosome are passed from the parent cell to two daughter cells.
- Cytokinesis is the division of the cytoplasm, and the separation of the new nuclei, to form two new daughter cells.
- Cytokinesis marks the beginning of two new cells, and the cell cycle is the period between one cytokinesis and the next.
- DNA replication involves synthesising new DNA by replicating parent strands of DNA. Daughter DNA molecules consist of one parent strand and one synthesised DNA strand.
- The characteristics of the sub-phases of mitosis are summarised in the following table.

Mitosis		
	Prophase	• Chromosomes condense and become visible. • Centrioles move to opposite sides of the nucleus and form poles. • Nuclear membrane breaks down. • Centrioles form spindle fibres between the two poles.
	Metaphase	• Chromosomes align at equatorial plane of cell. • Spindle fibres attach to centromeres of chromosomes.
	Anaphase	• Spindle fibres contract, splitting the centromeres and separating the sister chromatids. • The sister chromatids of chromosomes are separated. • Chromosomes are pulled to opposite poles.
	Telophase	• Nuclear membrane reforms around the two sets of chromosomes. • Spindle fibres disappear. • Chromosomes become longer and thinner.

KEY QUESTIONS

Knowledge and understanding

1 Which organelle divides into two during mitosis?

A nucleus
B vacuole
C mitochondrion
D chloroplast

2 Which of the following statements is not correct?

A Although it is divided into stages, mitosis is a continuous process.
B Cytokinesis marks the beginning of two new cells.
C DNA is replicated during interphase.
D Mitosis is the longest phase of the cell cycle.

3 What events occur during interphase of the cell cycle?

4 Identify the three phases of the eukaryotic cell cycle and describe what occurs in each of them.

5 Draw a diagram to represent the events that occur at each phase of mitosis.

6 Explain how cell division in plants is different from cell division in animals.

7 How do dividing parent cells ensure that the DNA is copied correctly to the daughter cells?

8 Distinguish between a chromatid and a chromosome. When are they visible?

9 **a** Does every cell go through the G_0 phase?

b What is the result if a cell does not move out of this phase?

10 During the growth of onion roots the cells in the roots divide with high frequency, and in each cell division the chromosomes are divided between two daughter cells. Overall the number of chromosomes in each cell does not change. What processes ensure that the normal number of chromosomes is maintained after each cell division?

Analysis

11 In cell replication it is necessary for the division of the nuclear material to be exact, but this is not necessary for the division of the cytoplasm. Infer why.

3.3 Controlling cell division

Cells replicate by dividing: a parent cell divides to create two new daughter cells. If all of the cells in your body replicated at the same rate as the first embryonic cells (2–3 times faster than adult cells), each day you would make up to 100 trillion cells—enough to make three bodies! This does not happen because the rate of cell replication is controlled, so that most cells are not replicating at any particular time. Some cells also take on specialist roles and do not replicate again. Cell division is a carefully controlled process.

In this section you will examine the factors that cause cells to divide and the factors that prevent them from dividing. You will also learn how cell division in eukaryotic cells is controlled.

REGULATING THE CELL CYCLE

The eukaryotic cell cycle is a highly regulated process. This regulation is critical to the proper development and function of organisms. A decrease or increase in the rate of cell replication can have negative effects. For example, if replication of bone marrow cells slowed down, this would also slow down the production of red blood cells, which takes place in the bone marrow. This would in turn have serious consequences for the supply of oxygen to cells.

The cell cycle is regulated by internal and external factors.

Internal checkpoints in the cell cycle

The purpose of the cell cycle is to produce two genetically identical daughter cells. If **mutations** (DNA changes) occur during the cycle, these need to be repaired or the cycle needs to stop.

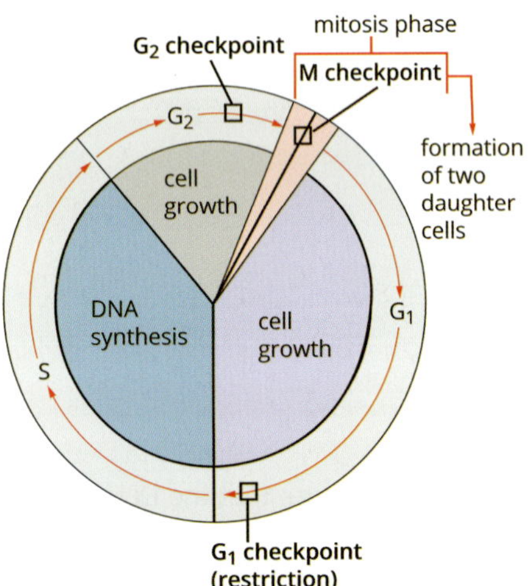

FIGURE 3.3.1 The cell cycle is controlled at three checkpoints: the G_1 checkpoint, G_2 checkpoint and M checkpoint.

An internal regulation system consisting of a group of regulatory proteins produced within the cell is responsible for determining whether the cell cycle continues or stops at each stage. It is known as the **cell cycle control system** (Figure 3.3.1). The regulatory proteins in this control system are active at three checkpoints in the cell cycle:

- G_1
- G_2
- metaphase (M).

G_1 checkpoint

The **G_1 checkpoint** occurs towards the end of G_1 in interphase. This checkpoint, also known as the restriction point, checks three things:

- There are adequate resources for the cell to divide, such as enough nucleotides and energy supply to copy the DNA.
- The cell is large enough to divide.
- The DNA in the nucleus has not been damaged.

If a cell passes this checkpoint, it is committed to the cell division process.

G_2 checkpoint

The **G_2 checkpoint** occurs towards the end of G_2 in interphase. As in the G_1 checkpoint, the cell is checked for adequate resources, such as proteins required for mitosis, and cell size. However, the most important check is of the DNA and chromosomes, to ensure that they have all been replicated without mistakes or damage. A cell cannot enter the mitotic phase if all of these requirements are not met.

M checkpoint

The **M checkpoint** occurs towards the end of metaphase in mitosis. This checkpoint, also known as the spindle checkpoint, determines if all of the spindle fibres have correctly attached to the sister chromatids and that the chromosomes are correctly aligned at the midline (equator) of the cell. The cell will not proceed to anaphase unless all sister chromatids are correctly attached and aligned.

External cell cycle controls

Signals for cell division can also originate from outside of the cell. Signals from outside the cell include contact inhibition, where crowding and contact with neighbouring cells cause a cell to slow its passage through the cell cycle or to stop in interphase (stop dividing). Lack of contact inhibition is typical of cancer cells, which keep growing rampantly regardless of the level of crowding and contact.

Molecules called mitogens act as signals to promote cell division. Mitogens include molecules such as growth factors and hormones. This is important during development and in repair. When cells die in a tissue, neighbouring cells release mitogens that stimulate cell division for tissue repair. These are the types of factors that can bring cells resting in G_0 back into the cell cycle. One example is the ability of liver cells to return to the cell cycle for tissue repair after liver damage. Tumours can also release factors that stimulate division in nearby capillary cells, directing the growth of their own personal blood supply (Figure 3.3.2).

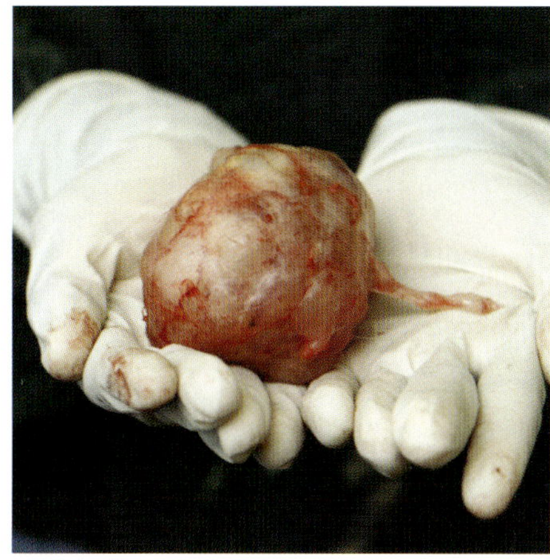

FIGURE 3.3.2 A tumour can release growth factors that encourage new blood vessels to grow into the tumour. This allows the tumour to grow larger and more quickly.

In addition to these molecular signals that promote the cell cycle, environmental conditions must be optimal for the cell cycle to progress. These include temperature, pH and the amount of nutrients.

CELL REPAIR AND CELL DEATH

During embryonic development and after birth, many cells are only temporarily useful. Once they have done their job they are removed. This also applies to cells that have been irreversibly damaged.

Cells cannot function properly if their genetic material is damaged. Damage to a cell's DNA can occur during DNA replication or it can be caused by environmental factors. Checkpoints throughout the cell cycle check for damage to the DNA. If the damage to the DNA cannot be repaired, the cell will undergo apoptosis (programmed cell death).

Repair

Cells have **enzymes** that can detect and repair damage to DNA. These enzymes run along strands of DNA like a zip, checking that the DNA is intact and has been replicated properly (Figure 3.3.3). If there is minor damage, such as DNA breakages, these will be corrected by enzymes before the cell continues its progress through the cell cycle.

FIGURE 3.3.3 The enzyme DNA ligase (shown in yellow) joining together a broken strand of DNA. Millions of DNA breaks occur during the normal course of a cell's life. Without molecules that can connect the pieces, cells can malfunction, die or become cancerous.

Apoptosis

Apoptosis, also called programmed cell death, is a highly regulated form of cell death that is vital for the normal functioning of every organism. If damage to the DNA of a cell is too great, enzymes will be activated that will cause the cell to die (Figure 3.3.4 on page 130).

Several specific alterations take place in apoptosis that result in cell shrinkage and separation from adjoining cells. The cell's plasma membrane buds to form what are known as **apoptotic bodies**. These contain the remnants of the cytoplasm, organelles, and DNA and protein fragments. **Phagocytic cells** (cells specialised to engulf and break down cellular debris) remove the apoptotic bodies. In this way, apoptosis has a role in maintaining the genetic integrity of cells, ensuring that only cells with intact DNA proceed through the cell cycle.

Apoptosis, also called programmed cell death, is a series of changes in a cell that results in the death of the cell. The changes are driven by biochemical changes.

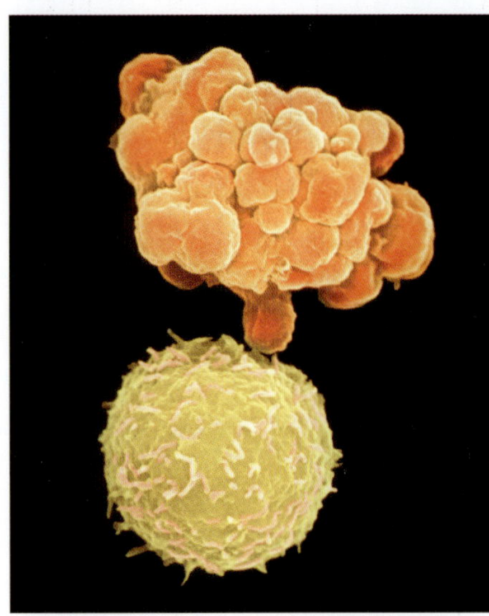

FIGURE 3.3.4 Coloured SEM of two human white blood cells. The top cell is undergoing programmed cell death, or apoptosis. Its cytoplasm is forming clusters. The cell will later break into vesicles, which will then be engulfed by phagocytes.

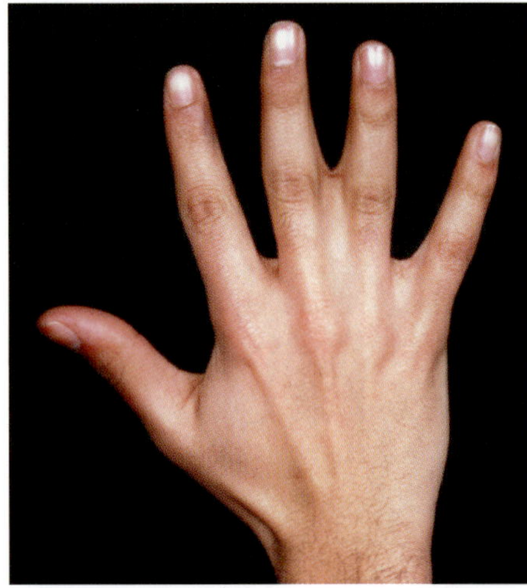

FIGURE 3.3.5 Syndactyly (excess webbing between the fingers or toes) can occur in humans if the apoptosis of the webbing between fingers or toes is incomplete during development.

Apoptosis is also a necessary part of normal development. An example is the programmed breakdown and death of the cells forming the initial webbing in the developing limbs of humans and some other vertebrates, which breaks down to allow the individual movement of fingers and toes. If the normal process of apoptosis does not occur or is incomplete, webbing remains between the fingers or toes. This condition is called syndactyly (Figure 3.3.5). Apoptosis occurs in many other tissues, too. In the developing brain, many more nerve cells are produced than will survive in the adult. Those nerve cells that do not make a functional connection with other parts of the nervous system will die by apoptosis.

Apoptosis is a five-step process that consists of:

1 activation of apoptotic enzymes, called caspases
2 digestion of cell content
3 cell shrinkage
4 cell blebbing and breakage
5 removal of fragments by phagocytic cells.

Apoptosis can be initiated through two different pathways: the mitochondrial and the death receptor pathway.

Mitochondrial pathway—The mitochondrial pathway is triggered when cell components (e.g. DNA) are damaged. Intracellular signalling proteins then act directly on the cell's mitochondria, which leads to the release of cytochrome c into the cytosol. This process is regulated by members of the BCL protein family bound to the mitochondrial membrane, including Bax and BCL-2, which act as pro- or anti-apoptotic regulatory proteins, respectively. Release of cytochrome c leads to the formation of an apoptosome and the start of a series of reactions that result in the activation of apoptotic enzymes called **caspases**.

Death receptor pathway—Extracellular signals (e.g. proteins released by immune cells) can be recognised by the death receptor molecule on the surface of cells that are under stress from factors such as lack of growth factors, hypoxia, DNA damage, viral infection or ultraviolet radiation. Similar to the mitochondrial pathway, this leads to the initiation of a cascade of reactions, which also result in caspase enzyme activation.

After activation by these pathways, caspases travel throughout the cell and digest specific proteins ultimately leading to the degradation of all organelles and of the cytoskeleton. As a consequence, the cell and the nucleus shrink, and the integrity of the cells weakens. The membrane forms **blebs**, detaches from the cell (Figure 3.3.4), and causes the cell to break up into vesicles (apoptotic bodies) which contain the digested intracellular material. The vesicles are recognised, engulfed and digested by phagocytes.

Necrosis

Not all cells die by apoptosis. Some cells die by necrosis (accidental cell death). This can occur as a result of physical damage, toxins, pathogens or a lack of oxygen. It often affects large, compact clusters of cells. In necrosis the plasma membrane becomes damaged, allowing water and ions to enter the cell, causing it to swell. This results in a messy 'explosion', as illustrated in Figure 3.3.6. Cell contents are released in an uncontrolled manner, resulting in inflammation and damage to surrounding cells.

Limited mitotic divisions

Another internal cell cycle control is rather like an internal clock that counts the number of divisions that have occurred in a line of cells. In **tissue culture**, cells from most multicellular organisms, including mammals, do not normally live indefinitely but die after a certain number of divisions. If the cells are taken from an embryo or young animal, they undergo more divisions than if they are taken from an adult.

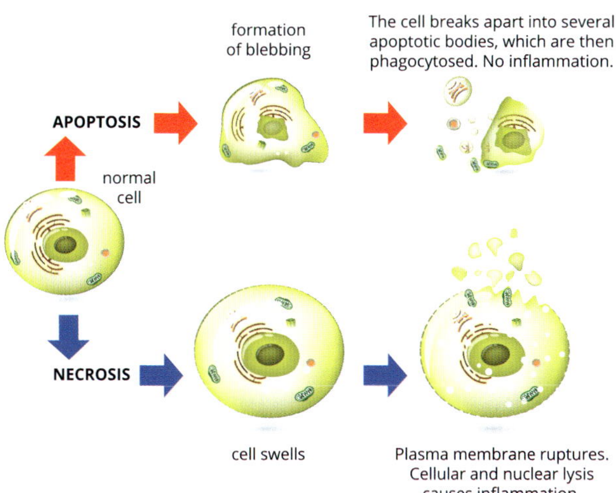

FIGURE 3.3.6 Illustration showing the difference in the structural changes of cells at cell death via apoptosis and necrosis. Apoptosis is triggered by normal, healthy processes in the body. Necrosis is cell death that is triggered by external factors or disease, such as trauma or infection.

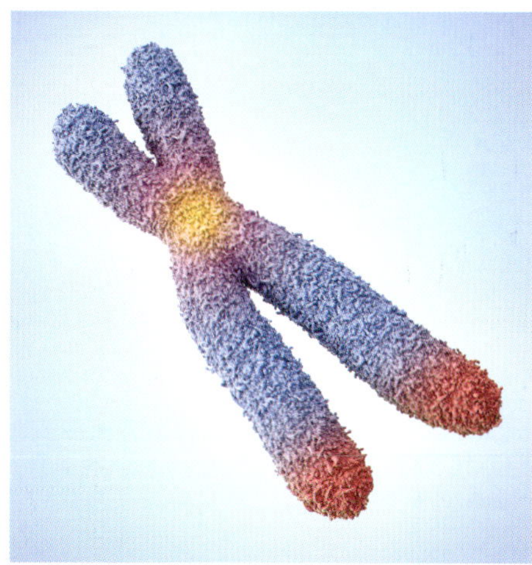

FIGURE 3.3.7 Illustration of a chromosome. At the tip of each chromosome arm are the telomeres (lower two shown in red), which protect the ends of the chromosome from damage. Telomeres get slightly shorter with each chromosome replication.

It is as though the cells can 'count' how many divisions they have made. It seems that they do this by losing a small amount of DNA from the tips of their chromosomes, known as **telomeres**, at each division (Figure 3.3.7). After about 50 or so divisions, the tips are lost and the cell either stops dividing or enters apoptosis.

Uncontrolled cell division

If uncontrolled cell division occurs during embryo development, the embryo will be abnormal and, in most circumstances, will abort. If uncontrolled cell division occurs in a mature organism, a neoplasm may form.

A **neoplasm** is an abnormal growth of tissue that usually, but not always, forms a mass. Neoplasms are more commonly referred to as tumours, but not all are cancerous. There are three types of neoplasm:

- benign—these form localised masses but do not transform into cancer
- potentially malignant—these form localised masses that will eventually invade other tissues and transform into cancer
- malignant—these form masses that invade other tissues and transform into cancer.

DISRUPTION TO CELL CYCLE REGULATION AND CANCER

The cell cycle is highly regulated, but its regulatory mechanisms can be disrupted by a number of factors. If this occurs, uncontrolled cell division, growth and death may lead to neoplasms, which can eventually transform into cancer.

Cancers are a group of diseases that commonly involve unregulated and abnormal cell growth and division. Cancer can be caused by genetic mutations in the cells that either increase the rate of cell division and/or result in the suppression of apoptosis. Either case can lead to the growth of cancer cells.

Characteristics of cancer cells

Cancer cells differ from normal cells in many ways:

- They divide at a faster rate than normal cells of the same type. Some divide very rapidly, others more slowly.
- They are not affected by the normal signals that control the cell cycle, such as contact inhibition.

BIOFILE

The HeLa cell line

A population of cells grown continuously by mitosis in a cell culture is called a cell line. There are now many cell lines that can be isolated and grown in culture indefinitely, especially cells derived from tumours. HeLa cells are a particular line of cultured cells that are used worldwide in experimental studies of cell functions. These cells were isolated from a human cervical carcinoma in 1951 and have been grown continuously ever since. They are named after the person from whom they were obtained, Henrietta Lacks.

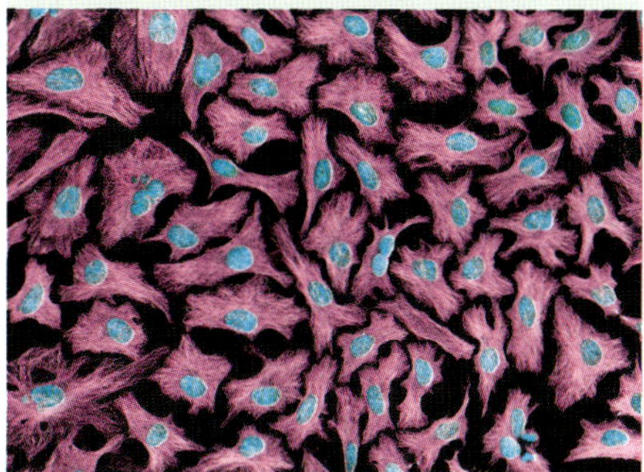

Fluorescence LM of a group of cultured HeLa cells, showing the cell nuclei (stained blue)

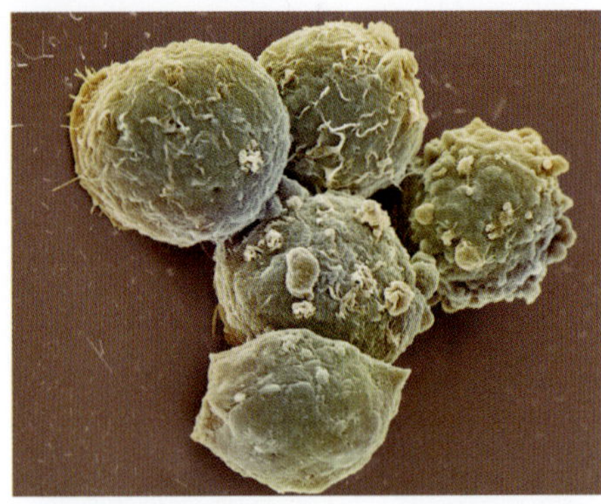

FIGURE 3.3.8 Coloured SEM of cancerous white blood cells

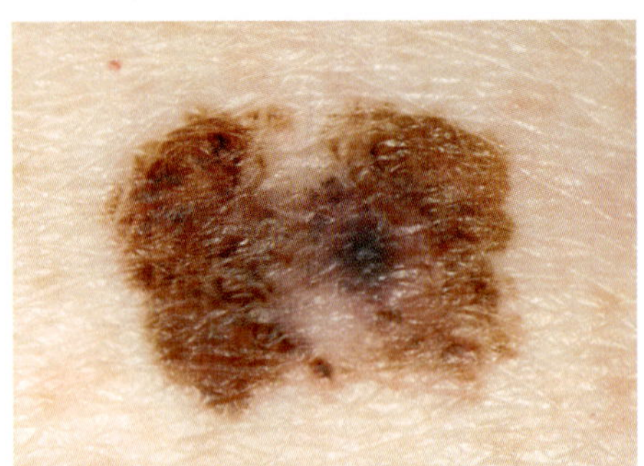

FIGURE 3.3.9 A melanoma is a malignant neoplasm that will continue to grow as a skin cancer unless removed.

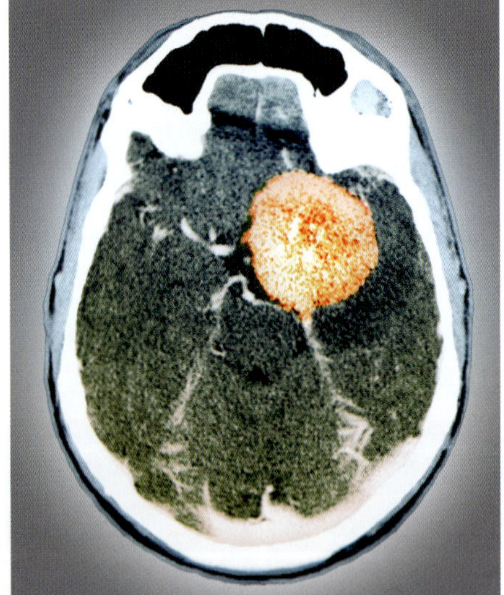

FIGURE 3.3.10 Coloured computed tomography scan of a section through the head of a 42-year-old patient with a benign (non-cancerous) meningioma (orange).

- They look different and may become less specialised (Figure 3.3.8).
- They release factors that stimulate the development of their own blood supply.
- Their DNA mutates, making them different and sometimes resistant to earlier successful treatments.
- They can 'colonise' new parts of the body and continue to grow unchecked.
- They can continue dividing endlessly, whereas normal cells undergo a limited number of cell cycles (Figure 3.3.9).
- They avoid proceeding to death by apoptosis.

Benign and malignant neoplasms can seem similar; Table 3.3.1 outlines the main differences between them.

TABLE 3.3.1 Comparison of benign and malignant neoplasms

Benign neoplasms	Malignant neoplasms
Cells divide uncontrollably, yet not as rapidly as those of a malignant neoplasm.	Cells divide uncontrollably.
The organism controls the growth of the neoplasm to a certain extent by encapsulation. Cells are contained and do not penetrate the blood and lymph vessels.	Growth of the neoplasm: uncontrolled cell growth breaks out of capsule. Neoplastic cells can spread to other tissues (i.e. they metastasise).
Because the neoplasm grows inside a capsule, it does not destroy the surrounding tissues.	The growing neoplasm destroys the surrounding tissues.

Malfunctions in apoptosis

Like many processes in the body, apoptosis is highly regulated. Too much cell death may result in a loss of vital tissue, while too little may result in tumours, cancer or other diseases. For example, meningioma is a tumour arising from the meninges, the membranes that surround the brain and spinal cord. The growth of the tumour is due to a breakdown in the apoptosis pathways, leading to uncontrolled proliferation of cells (Figure 3.3.10).

In the cancer B-cell lymphoma, excessive amounts of anti-apoptotic BCL-2 proteins are produced. The BCL-2 family of proteins is central to controlling the mitochondrial apoptotic pathway. Anti-apoptotic BCL-2 proteins inhibit the release of cytochrome c from mitochondria. Without cytochrome c, proteins needed for apoptosis cannot form and apoptosis cannot continue. As a result, mutated B cells that would normally be removed by apoptosis instead survive, replicate and develop into cancer.

Defects in other aspects of the apoptotic pathways have been identified in various other cancers. Some of these defects include mutations in the enzymes responsible for apoptosis (i.e. caspases) and defects in proteins essential to the apoptosis pathways. In each case, apoptosis does not occur when it should, and defective cells survive and replicate, potentially developing into cancer.

Many cancer treatments involve the use of radiation or chemicals, which aim to induce the mitochondrial pathway of apoptosis. However, the reduced likelihood of apoptosis in cancerous cells also explains why these treatments are not always effective.

Excessive apoptosis has been linked to neurodegeneration. This is seen, for example, in Alzheimer's disease, a neurodegenerative disease that is characterised by a shrinking of the brain due to a loss of neurons associated with excessive apoptosis. Other degenerative diseases such as Parkinson's disease and motor neuron disease are also influenced by excessive apoptosis.

Genetic factors

Genes code for the enzymes that regulate cell division or regulate apoptosis. When mutations to these genes occur, regulation of cell division or apoptosis can cease. **Oncogenes** are genes that have the potential to cause cancer. In cancer cells oncogenes are frequently expressed at high levels or are mutated.

> Oncogenes are genes that have the potential to cause cancer. Oncology is the medical field that studies and treats cancer.

Proto-oncogenes

Proto-oncogenes are a group of normal genes involved in the regulation of cell division. One of their functions is to stimulate cell growth. They are required for the normal growth and development of cells. However, mutations of these genes can change them into oncogenes, which induce uncontrolled cell division leading to the development of neoplasms.

One example is an important regulator of cellular proliferation called platelet derived growth factor (PDGF). Mutations of the *PDGF* gene are connected with the development of brain cancer. Mutations of the *PDGF* receptor gene are associated with thyroid cancer (Figure 3.3.11).

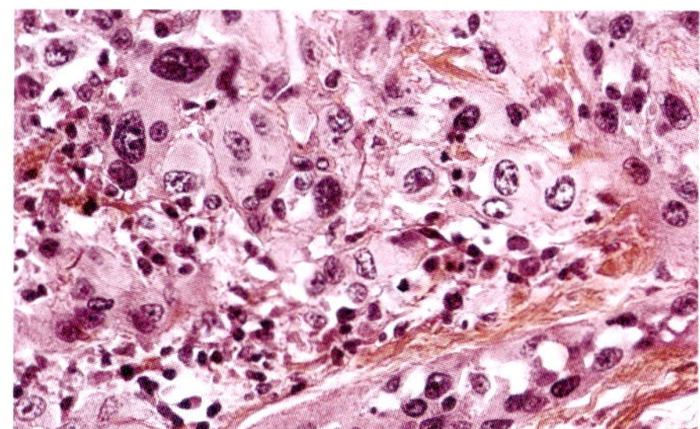

FIGURE 3.3.11 LM of thyroid cancer tissue

Tumour-suppressor genes

The group of genes that code for proteins involved in the slowing down of cell division, the repair of DNA or apoptosis are called **tumour-suppressor genes**. DNA changes (mutations) in tumor-suppressor genes can lead to cancer.

Inherited genes

Some people also have a genetic predisposition to certain forms of cancer. An individual inherits their genes from their parents, so they could receive an oncogene or mutated tumour-suppressor gene from one or both parents. This does not necessarily mean that all family members with that gene will develop the cancer. It appears that more than one gene mutation must occur before cancer develops. However, individuals who inherit an affected gene do have a higher chance of developing the cancer because they have one such mutation.

A family history of breast cancer, for example, is associated with mutations in the *BRCA1* or *BRCA2* genes, which are located on chromosome 17. A protein coded by these genes takes part in the repair of damaged DNA, but changes in its structure lead to uncontrolled growth. The growth occurs mainly in tissues whose functioning depends on oestrogens, for example, tissues in the mammary glands (Figure 3.3.12) and ovaries.

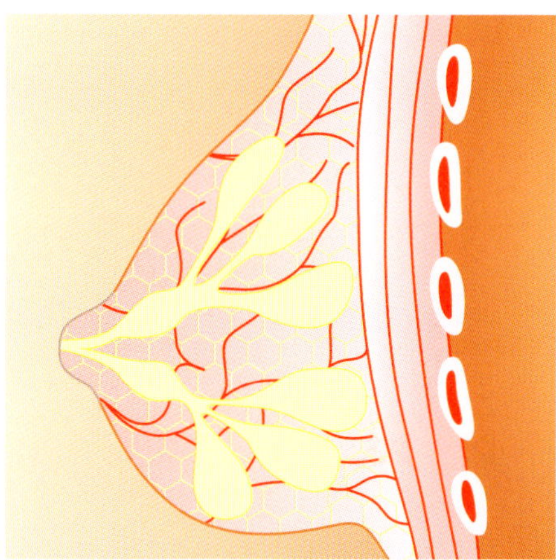

FIGURE 3.3.12 The mammary glands of the breast (highlighted in yellow) contain tissues that depend on oestrogens and are vulnerable to neoplasms caused by changes in the *BRCA1* or *BRCA2* genes.

CASE STUDY

Turning off BCL-2 to halt cancer

Venetoclax is a newly available anti-cancer drug, developed from research into the function of the protein BCL-2 at the Walter and Eliza Hall Institute of Medical Research in Melbourne, Australia. In patients with chronic lymphocytic leukaemia (CCL), infected cells are highly mobile through the lymph tissue and usually spread through the blood, bone marrow and then organs. This makes it difficult to target them with specific treatments. These infected cells have a disrupted BCL-2 protein that prevents apoptosis from occurring, resulting in cancer cells living for a long time.

Venetoclax binds to the BCL-2 protein receptor site that controls the intrinsic mitochondrial apoptosis pathway in these cancerous cells. This inhibits the BCL-2 from functioning (Figure 3.3.13). Damaged cells die through natural apoptotic pathways, stopping the cancer spreading, and also become more susceptible to other chemotherapy treatments.

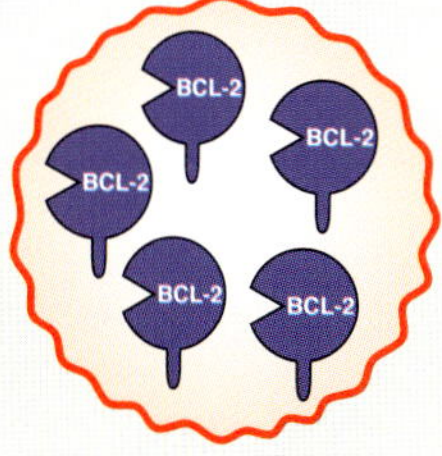

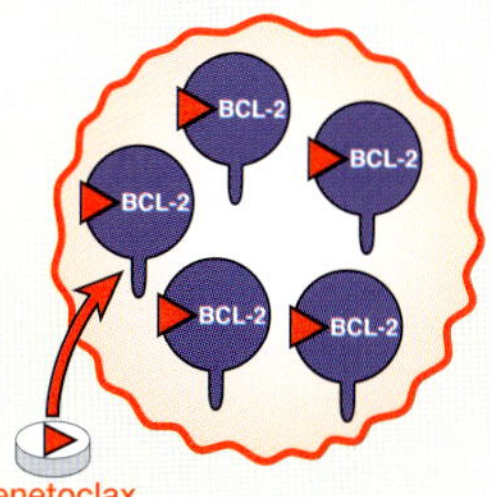

- Addicted to high levels of BCL-2
- Cell becomes long-lived
- Resistant to anti-cancer treatments

- BCL-2 inhibited
- Cancer cell dies, or responds to other anti-cancer treatments

FIGURE 3.3.13 The anti-cancer drug venetoclax works by binding to the BCL-2 protein receptor site in cancerous cells, allowing apoptosis to proceed as normal.

Environmental factors

Environmental factors such as exposure to radiation or certain chemicals can damage DNA. Such agents that cause damage in DNA are called **mutagens**. If the damage occurs to proto-oncogenes or tumour-suppressor genes, the control of cell division can be disrupted.

There are three types of carcinogens (cancer-causing agents) that can lead to the formation of neoplasms: chemical, physical and biological. Some examples are outlined in Table 3.3.2.

TABLE 3.3.2 The three types of carcinogens

Carcinogen type	Examples
chemical	Tobacco contains mutagenic and carcinogenic compounds. Smokers are much more likely to develop malignant neoplasms of the respiratory system, pharynx, larynx and lungs than non-smokers.
	Some chemicals in air pollution can be carcinogenic. Factories can emit smoke containing chemicals that increase the risk of developing a neoplasm.
physical	The ionising radiation in X-rays can have dangerous effects because it can cause cancerous mutations in cells.
	Ultraviolet light is a mutagen. Overexposure to sunlight or tanning beds, and therefore ultraviolet radiation, increases the risk of skin cancer or melanoma.
biological	Some viruses are oncogenic viruses, that is, they cause cancer. HTLV-1 is a retrovirus, a type of virus that inserts its genetic information into the host cell's chromosomes. Such viruses either carry oncogenes themselves or enhance the host cell's proto-oncogenes.

Loss of immunity

The immune system is usually able to detect and destroy abnormal cells, including those that are replicating in an uncontrolled manner. If an individual's immune system is weakened, cells that are dividing in an uncontrolled manner may not be detected and may continue to divide to form tumours.

CASE STUDY **ANALYSIS**

Mitotic index and cancer therapy

In cancer cells, cell cycle regulation can be disrupted, leading to uncontrolled cell growth. As a consequence, cancerous tissues are characterised by a high proportion of cells undergoing cell divisions compared to normal tissues (Figure 3.3.14).

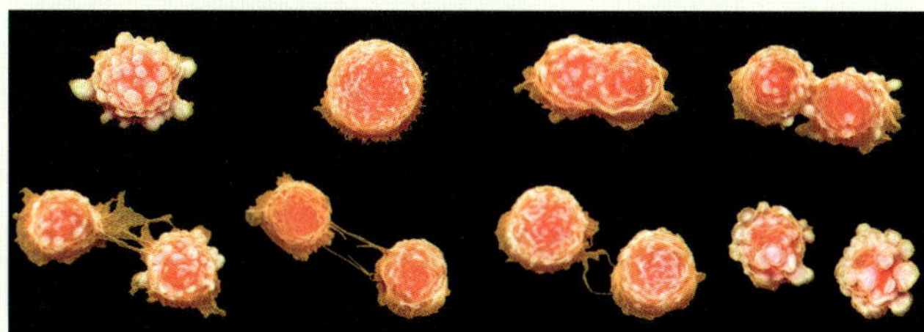

FIGURE 3.3.14 SEM image of dividing cancer cells. Cancerous cells ignore or override some of the factors that control cell division, often replicating a lot faster than the organism's non-cancerous cells.

The mitotic index (MI) is an indirect measure of cell proliferation and has been shown to be a reliable predictor of survival for several human cancers. MI is the ratio between the number of cells in mitosis and the total number of cells in a tissue sample. MI has also been used as a prognostic tool to predict response and overall survival to chemotherapy.

Ovarian cancer case study

Ovarian cancer is classified into three types—common epithelial (approximately 90% of all cases), germ cell (approximately 4% of cases) and stromal (approximately 6% of cases). It is the eighth most common cancer affecting women in Australia, with a five-year survival rate of 45%.

A pathologist evaluated sections of ovarian tissues via light microscopy to assess three patients' prognosis for cancer therapy. The region of the tumour samples with the highest overall mitotic activity was chosen for evaluation. A section of normal tissue was used as a control. Table 3.3.3 shows the data collected from the ovarian tissue samples.

TABLE 3.3.3 Cell counts in ovarian tissue samples

Tissue	Number of cells in interphase	Number of cells in mitosis	Mitotic index
normal ovarian tissue	372	71	
cancerous ovarian tissue	284	187	
cancerous ovarian tissue at beginning of therapy	323	156	
cancerous ovarian tissue after 6 months of therapy	407	110	

Analysis

1 Complete Table 3.3.3 by calculating the MI for the three different samples using the formula:

$$\text{mitotic index} = \frac{\text{number of cells in mitosis}}{\text{total number of cells}} \times 100$$

Give your answers to one decimal place.

2 Compare the mitotic index of the three tissues. What does this information tell you about the cells in the samples?

3 How does the mitotic index of the three tissue samples help researchers assess the prognosis of the patients for cancer therapy?

3.3 Review

SUMMARY

- The cell cycle is controlled by signals from inside and outside the cell.
- There are three internal checkpoints in the cell cycle: the G_1 checkpoint, G_2 checkpoint and metaphase (M) checkpoint.
- External cell cycle controls include:
  - contact inhibition
  - growth factors and hormones
  - environmental conditions such as temperature and pH.
- Apoptosis is programmed cell death during which cells respond to signals and die in a controlled way. Apoptosis is important during development.
- Necrosis involves a damaged cell bursting in an uncontrolled manner. It can cause inflammation and damage to surrounding cells.
- Cells usually have a limited number of mitotic divisions.
- Cancers are a group of diseases that commonly involve unregulated and abnormal cell growth and division.
- Uncontrolled cell division can result in neoplasms. Neoplasms can be benign, potentially malignant or malignant.
- Uncontrolled cell division can be due to genetic factors, environmental factors, pathogens and loss of immunity.
- Malfunctions in apoptosis can lead to defective cells surviving and replicating, potentially causing cancer and other diseases.

KEY QUESTIONS

Knowledge and understanding

1 Which of the following is not checked during the G_1 checkpoint of the cell cycle?

A DNA

B the attachment of the spindle fibres to the centromeres

C the amount of resources in the cell

D the size of the cell

2 Label the parts of the diagram below.

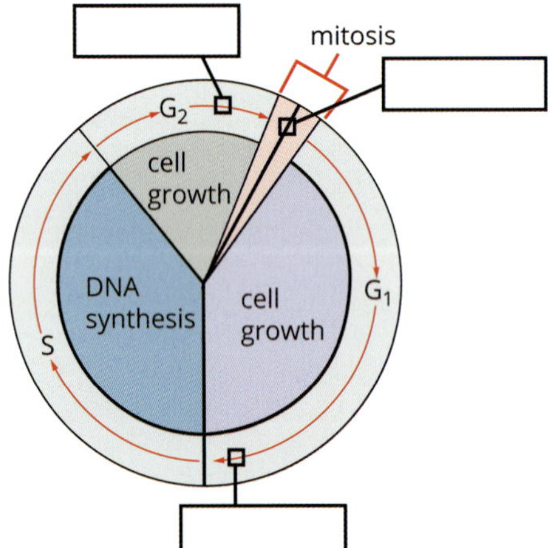

3 Classify each of the following as either internal or external controls of the cell cycle.

- temperature
- DNA quality
- telomere length
- contact inhibition
- size of cell

4 Which section of the chromosome is important for attachment to spindle fibres and thus progression of cells through the M checkpoint?

5 Which section of the chromosome do scientists believe has a function in controlling the number of mitotic divisions that a cell can undertake? Where is this section located?

6 When a tadpole develops into a frog, the cells in the tail die. What is the name given to this type of cell death?

7 Give an example for each of the chemical, physical and biological factors that cause neoplasms.

8 Identify the two types of cell death and explain how they differ.

Analysis

9 Infer how proto-oncogenes and tumour-suppressor genes can cause cancer.

3.4 Stem cells

Stem cells are cells that are yet to be specialised (Figure 3.4.1). Stem cells are also capable of self-renewal—they can replicate themselves to produce new stem cells. Stem cells allow the body to develop, grow and repair damaged cells and tissues. Certain stem cells are able to divide indefinitely in order to replace cells. For these reasons, stem cells are considered by many to be the basis of a potentially powerful technology in medicine, called stem cell therapy.

In this section you will learn about the properties of stem cells that allow for differentiation, specialisation and renewal of cells and tissues.

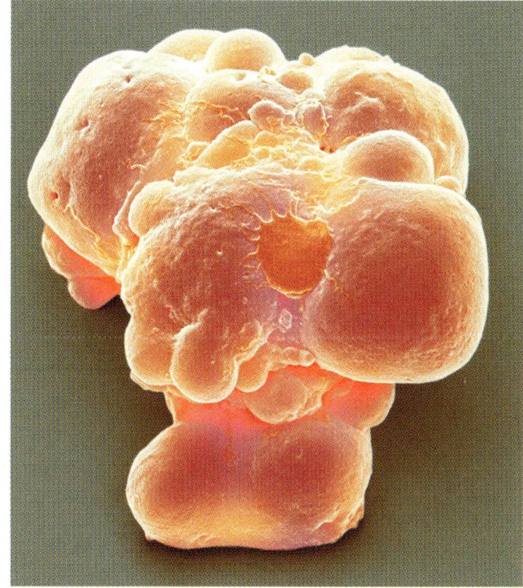

FIGURE 3.4.1 SEM of a clump of stem cells. The particular type of stem cell in this image is able to differentiate into any of the cell types in the human body.

TYPES OF STEM CELLS

Stem cells can be broadly classified as embryonic stem cells and adult stem cells, but can also be categorised based on their differentiation potential or potency.

Embryonic stem cells

Embryonic stem cells (also known as pluripotent stem cells) are the undifferentiated or relatively undifferentiated cells of embryos. They can be obtained from surplus three- to five-day-old embryos from IVF programs. Embryonic stem cells can become many types of cell and can replicate indefinitely.

Adult stem cells

Adult stem cells (also known as somatic stem cells) are present in small numbers in some adult tissues, such as hair follicles, bone marrow, the spinal cord and germ cells, and remain as stem cells throughout an individual's life. They can give rise only to a limited range of cells, such as bone marrow (haemopoietic) stem cells (Figure 3.4.2). They are therefore considered multipotent or unipotent. The biological purpose of adult stem cells is repair and regeneration of damaged and aged tissue, such as skin and liver cells. Adult stem cells cannot replicate indefinitely.

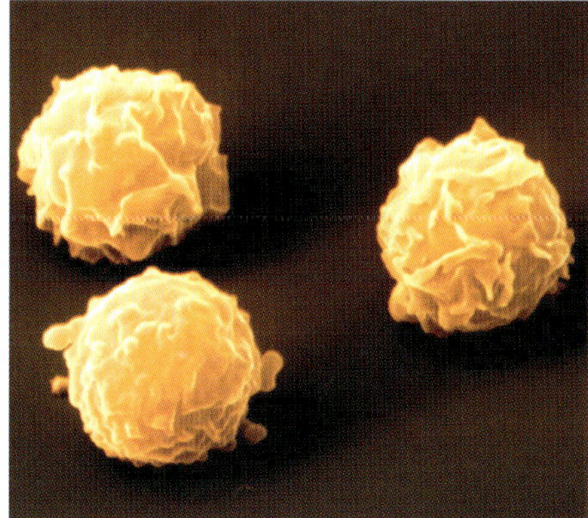

FIGURE 3.4.2 SEM of human bone marrow stem cells. These stem cells can give rise to different types of blood cells.

POTENCY

Scientists also characterise stem cells by cell potency. Cell **potency** is a cell's ability to differentiate into other cell types. The more cell types into which a cell can differentiate, the greater its potency. Stem cells are classified as totipotent, pluripotent, multipotent or unipotent, depending on what sort of cells they can become.

- **Totipotent** stem cells have the greatest potency as they are capable of giving rise to any cell type or even another embryo. The zygote formed at fertilisation is totipotent until three to four days after fertilisation, when it consists of 16 cells and is known as a morula. After the 16-cell stage, the embryo is no longer totipotent.
- **Pluripotent** stem cells have the ability to form a range of cell types, but not all cell types. The cells of the blastocyst (from five days after fertilisation until implantation in the uterus) are pluriopotent because they can differentiate into any of the three **germ layers**: endoderm (e.g. lungs and gut lining), mesoderm (e.g. muscle, bone, blood) or ectoderm (e.g. skin and nervous system). Once this germ layer differentiation has occurred, the cells are no longer pluripotent. The primordial germ cells (PGCs) that give rise to gametes are also pluripotent.

- **Multipotent** stem cells have the ability to give rise to multiple, but limited, cell types. Following implantation in the uterus, the blastocyst undergoes gastrulation and becomes a gastrula with three distinct layers of cells (the germ layers). The cells from the three germ layers are multipotent because they can only give rise to certain types of cells.

  Adult stem cells are more limited in their ability to differentiate than embryonic stem cells. Some adult stem cells, such as haematopoietic (blood-forming) stem cells in red bone marrow, are multipotent, as they can give rise to lymphocytes, macrophages, platelets and other blood cells (Figure 3.4.3).

- **Unipotent** stem cells can only form one cell type found in a specific tissue but can divide repeatedly. Skin epidermal stem cells are examples of unipotent cells that give rise only to new skin cells.

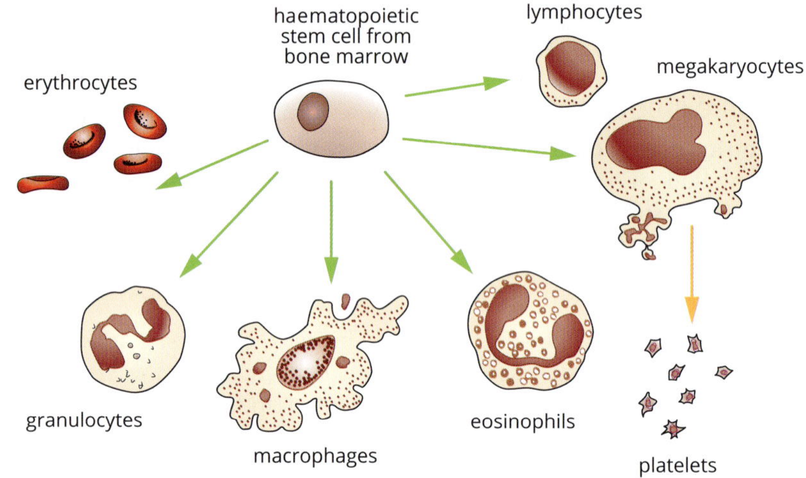

FIGURE 3.4.3 Haematopoietic cells (blood forming) stem cells are set apart very early in development. They are found in bone marrow and are responsible for the continued production of a number of distinct cell types.

Embryonic stem cells (ESC) are totipotent or pluripotent. Adult stem cells are multipotent or unipotent. The four potencies of stem cells and the life stages in which they are found are outlined in Figure 3.4.4.

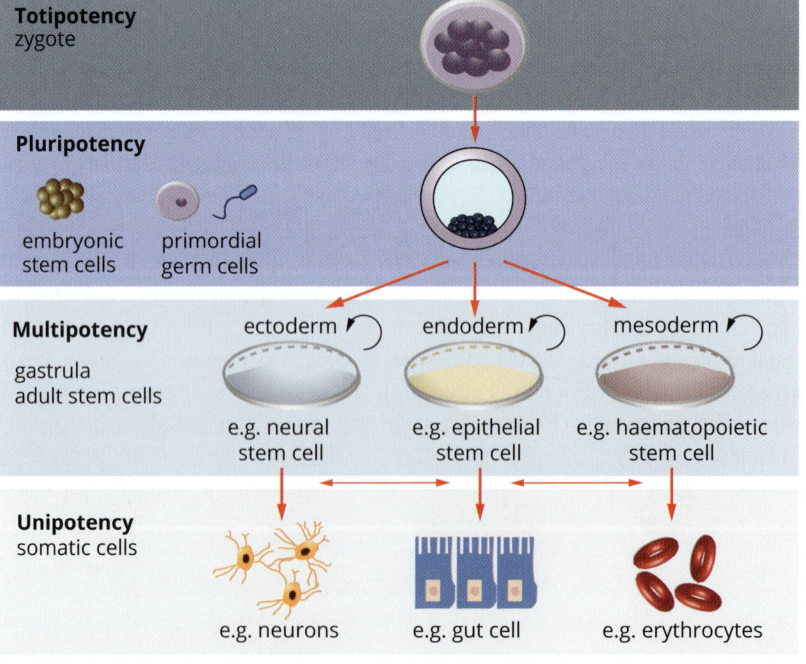

FIGURE 3.4.4 Cell potency is determined by how many cell types the stem cell can differentiate into. Totipotent cells have the greatest cell potency, followed by pluripotent cells, then multipotent cells.

BIOFILE

Stem cells in other organisms

Stem cells are found in most adult organisms. Some organisms, such as free-living planarians and sea stars, retain a population of stem cells throughout their life. These stem cells can develop into any cell type in the body, giving these organisms the remarkable ability to regenerate a body part that was completely lost through injury.

This marine planarian, or flatworm, has a population of stem cells that enables it to regenerate if a part of it is severed.

CASE STUDY

Growing 'beating hearts'

An important area of stem cell research is the development of induced pluripotent stem cells (iPSCs). iPSCs are adult cells that have been genetically reprogrammed to an embryonic stem cell–like state. This forces the cell to express genes and factors that are characteristic of embryonic stem cells.

In 2015, researchers from the University of California published the results of a study in which they used human iPSCs to grow miniature 'beating hearts'. These were nothing like the four-chambered heart inside your chest, but the cells did form hollow, pulsating chambers rather than simple layers of cells. Previously, stem cells had been grown in wells that were about 2.5 cm in diameter, such as those shown in Figure 3.4.5, and under these conditions they only grew into flat sheets of cells.

To create the beating chambers, the researchers created tiny wells only 200–600 μm wide—the thickness of a few strands of human hair—in the bottom of a petri dish. They then took iPSCs that had been genetically reprogrammed from human skin tissue and grew them in the tiny wells. The stem cells formed 'microchambers' because the mechanical pressure on the cells on the outside caused them to develop into cells that produced collagen, while the cells on the inside developed into heart muscle cells.

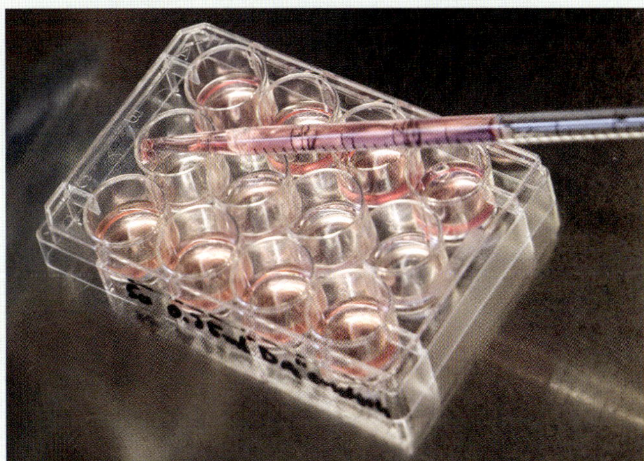

FIGURE 3.4.5 A standard 12-well culture dish containing human stem cells

Once they had created the beating microchambers, the researchers added thalidomide, which is a drug known to cause heart defects and deformities in fetuses. They found that the drug made it difficult for the microchambers to contract, which resulted in them beating slower than ones that had not been exposed to thalidomide.

Stem cells will not be used to create replacement hearts for people in need of heart transplants any time soon. But thanks to this new research, using them to create beating heart microchambers that can then be used to test drugs for any side effects on heart chamber formation is now a reality.

CASE STUDY

Ethical considerations of stem cell therapy

There is still much to learn about stem cells and the way that they normally work in maintenance and repair, such as how they find their way to damaged tissues and how they limit replication and avoid forming neoplasms. Stem cell research has the potential to revolutionise medicine but there are ethical issues associated with the application of this technology.

The main ethical considerations of stem cell therapy involve the use of embryonic stem cells. This is because the first harvesting techniques for embryonic stem cells involved the destruction of embryos at the blastocyst stage. However, new harvesting techniques that do not damage the embryo are being developed (Figure 3.4.6).

Adult stem cells are usually harvested from adults or from cord blood during a birth. In both cases, those who are donating the stem cells are able to give their consent and there are minimal adverse side effects.

Despite the ongoing ethical and moral debates about the use of embryonic stem cells, research into embryonic stem cell use is exciting, because these stem cells can theoretically be made to differentiate into any human tissue type and treat many more diseases and disorders than adult stem cells can.

Table 3.4.1 outlines the advantages and disadvantages of the use of the embryonic and adult stem cells in stem cell therapy.

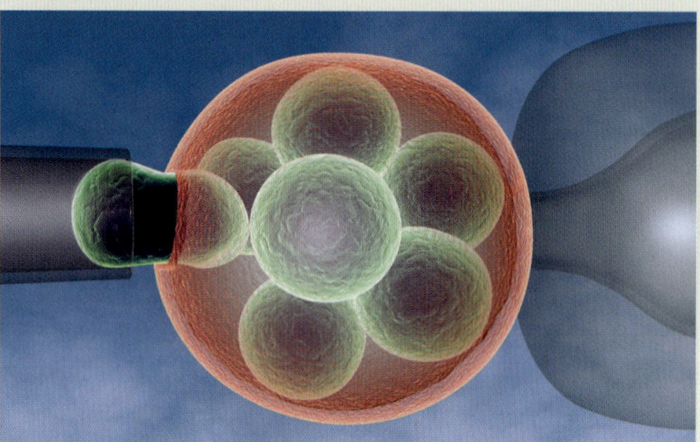

FIGURE 3.4.6 Illustration of an embryonic stem cell being removed from a blastocyst. This technique removes only one cell, so the blastocyst remains intact.

TABLE 3.4.1 Advantages and disadvantages of embryonic, adult and induced pluripotent stem cells

	Embryonic stem cells	Adult stem cells	Induced pluripotent stem cells
Advantages	• unlimited supply, e.g. left over from IVF • cell division can occur indefinitely • can grow into large quantities • can differentiate into any type of cell	• already programmed to make a particular cell type and therefore can be used for a specific treatment, e.g. brain stem cells make new neurons	• use normal somatic cells (many cells available) • use cells from the person who needs replacement cells • no immune rejection
Disadvantages	• not yet technically possible to train stem cells to become every particular cell type • ethical objections • potential for uncontrolled growth (including teratoma formation)	• no indefinite growth • limit on cell types into which they can differentiate • difficult to obtain • small number of cells in only some tissues • potential for uncontrolled growth	• time and technology required to reprogram the cells • may have limited differentiation capability • not enough knowledge yet about how to differentiate into all cell types • objections to any genetic reprogramming of human cells • potential for uncontrolled growth

3.4 Review

OA ✓✓

SUMMARY

- Stem cells can be broadly classified as embryonic stem cells and adult stem cells.
- Up to five days after fertilisation, embryonic stem cells can be thought of as 'all-purpose' cells that have the potential to develop into many different kinds of cells; stem cells are relatively undifferentiated cells.
- Stem cells be categorised based on their differentiation potential (potency). Stem cells are classified as totipotent, pluripotent, multipotent or unipotent, depending on what sort of cells they can become.
- The primary roles of adult stem cells in a living organism are to maintain and repair the tissue in which they are found. They have the ability to give rise to multiple, but limited, cell types.
- Stem cells allow organisms to differentiate, specialise and renew cells and tissues.

KEY QUESTIONS

Knowledge and understanding

1 What are pluripotent stem cells?

2 **a** What is the name given to cells that self-renew and remain undifferentiated when removed from an embryo?
b What advantages do these cells have for the body?

3 **a** What are adult stem cells?
b What is their role?
c How do they differ from embryonic stem cells?

4 State the four types of stem cell potency. How do the different types of stem cell potency differ?

5 Classify the following cell types with their correct potency.
- zygote/morula cell
- embryonic stem cell
- adult stem cell

6 Planarians (flatworms) retain a population of totipotent stem cells throughout their life. How does this observation explain why a planarian, when cut in half, can regenerate the missing part?

Analysis

7 Use the terms 'self-renewal' and 'potency' to explain how plant shoots grow.

Chapter review

OA ✓✓

03

KEY TERMS

adult stem cell
anaphase
apoptosis
apoptotic body
binary fission
bleb
cancer
caspase
cell cycle
cell cycle control system
centriole
centromere
chromatid
cytokinesis
daughter cell
death receptor pathway
DNA replication
embryo
embryonic stem cell
enzyme
G_0 phase
G_1 checkpoint
G_1 phase
G_2 checkpoint
G_2 phase
germ layer
interphase
M checkpoint
meiosis
metaphase
mitochondrial pathway
mitosis
mitotic spindle
multipotent
mutagen
mutation
neoplasm
oncogene
origin
phagocytic cell
pluripotent
potency
prophase
proto-oncogene
S phase
telomere
telophase
tissue culture
totipotent
tumour-suppressor gene
unipotent
zygote

REVIEW QUESTIONS

Knowledge and understanding

1 Which process is not associated with cell division?

A cytokinesis
B DNA replication
C pairing of homologous chromosomes
D formation of two diploid daughter cells

2 Consider the following diagram.

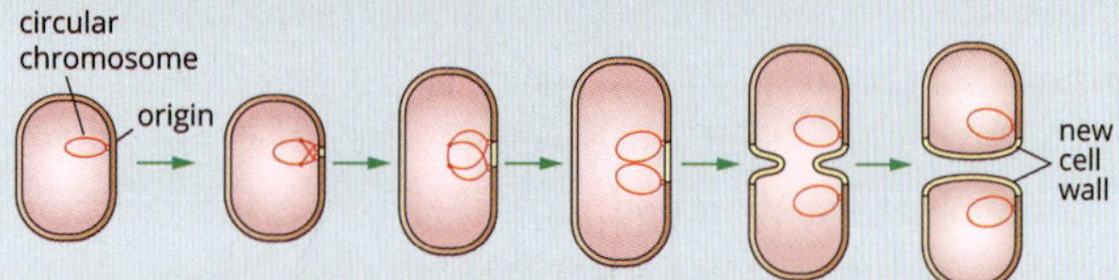

a What process does this diagram show?
b What is one advantage and one consequence of this process when it is used as a form of reproduction?

3 Which of the following statements shows a correct sequence of events for mitosis?

A chromatids separate, chromosomes duplicate, cytokinesis occurs
B cytokinesis occurs, chromosomes duplicate, chromosomes line up at the equator
C chromosomes line up at the equator, chromatids separate, cytokinesis occurs
D chromosomes duplicate, cytokinesis occurs, chromatids separate

4 Use a table to summarise the different phases of the cell cycle and what occurs during each phase.

5 The diagram below represents the stages of mitosis, but they are not in the correct order. Determine the correct sequence and add labels for the name of each stage of mitosis.

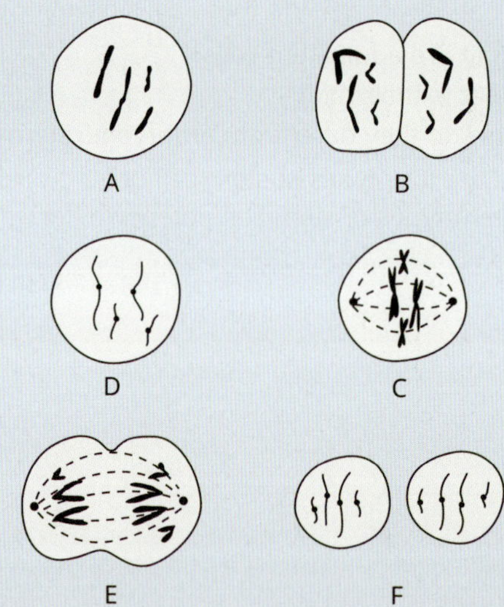

6 Mitosis in unicellular eukaryotes has a different function to mitosis in multicellular eukaryotes. Explain the purposes of mitosis in:

a unicellular eukaryotes
b multicellular eukaryotes.

7 Examine the following diagram of a chromosome.

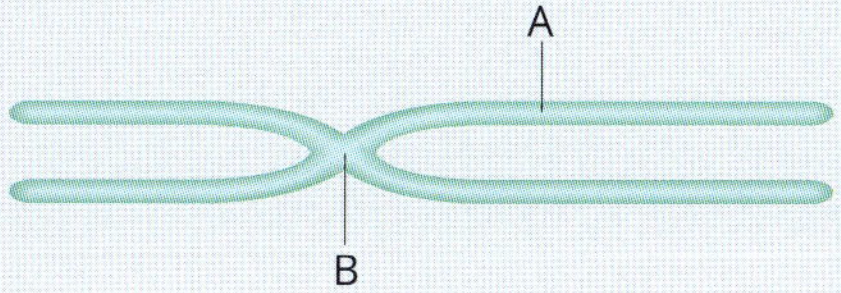

a Name structures A and B.
b During what phase of mitosis does the chromosome first appear in this state? Explain what happens to cause this appearance.
c Chromosomes do not always look like the one shown in the diagram. Describe the changes in the appearance of chromosomes during the different phases of the cell cycle.
d Draw a typical representation of an animal cell in interphase.

8 List three differences between malignant and benign neoplasms.

9 Use a table to summarise the different steps of apoptosis and what occurs during each step.

10 Give an example of each of the following:
a normal apoptosis in human embryonic development
b normal apoptosis in an adult human
c unregulated apoptosis associated with a human disorder or disease.

11 Defects in aspects of the apoptotic pathways have been identified in various cancers. Describe some of the defects that can potentially lead to the development of cancer.

12 Some cancers are at least partially hereditary; the genetic predisposition can be passed down in a family line.
a Give one example of a hereditary predisposition to cancer.
b If one family member has a particular type of hereditary cancer, do all other relatives develop the cancer? Why or why not?
c What other factors can influence the development of a neoplasm?

Application and analysis

13 Describe the importance of cell division to an organism.

14 Consider a defect in a multicellular organism resulting in the inability to complete mitosis correctly all the time. Outline some of the possible consequences for the organism.

15 Different organisms vary in how much repair their cells can carry out. Describe how three different organisms vary in their ability to carry out replication and, therefore, repair.

16 Contrast cytokinesis in plant and animal cells.

17 Why is programmed cell death important in embryo development?

18 Describe where mitosis would be occurring in a pregnant woman.

19 The cell cycle is regulated by the cell cycle control system, which ensures that no abnormalities occur.
a Name the checkpoint at which the process would cease if the following occurred:
i Some of the spindle fibres have not attached to the sister chromatids.
ii The cell is too small to divide.
iii The DNA has not been correctly copied.
b Name the phases of the cell cycle associated with the three checkpoints.

20 Suggest how the cell cycle may differ in stem cells and fully differentiated cells.

REVIEW QUESTIONS

How do cells function?

Multiple-choice questions

1 Which of the following is an example of a eukaryotic cell?
- A a fungal cell
- B a bacterium
- C an enzyme
- D a virus

2 Identify the process that most bacteria use to divide.
- A mitosis
- B meiosis
- C crossing over
- D binary fission

3 What are organelles?
- A small structures with multiple functions found in the cytosol of cells
- B membrane-bound structures that are found near the nucleus of all cells
- C specialised structures with a specific function found inside cells
- D membrane-bound structures only found in eukaryotic cells

4 Select the example that is not a specialised cell.
- A meristematic cell
- B nerve cell
- C red blood cell
- D mesophyll cell

5 Which of the following statements about chloroplasts in plants is true?
- A They only function at night.
- B They occur only in green leaves.
- C They function independently of mitochondria.
- D They use oxygen and produce carbon dioxide.

6 Which list contains organelles that are found in both animal and plant cells?
- A mitochondria, nucleus and chloroplasts
- B mitochondria, Golgi apparatus and chloroplasts
- C mitochondria, ribosomes and cell wall
- D mitochondria, Golgi apparatus and nucleus

7 Which of the following is required for osmosis to occur?
- A a fully permeable membrane
- B a semi-permeable membrane
- C ATP
- D an enzyme

8 Three cells (X, Y and Z) containing different internal solute concentrations were placed next to each other, as shown in the following diagram. (M is a unit of measurement of solute concentration. A higher value means a higher solute concentration.)

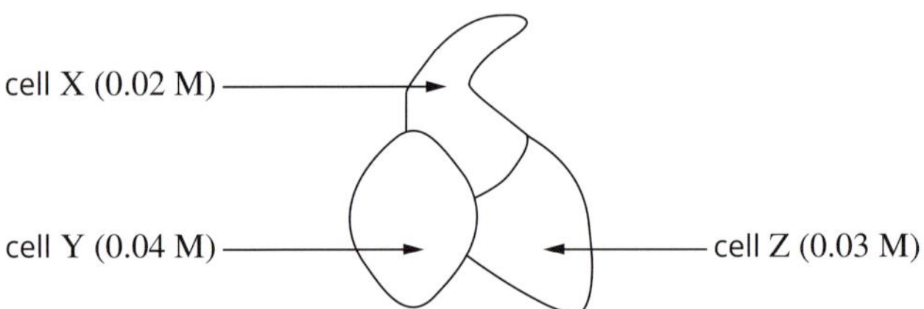

In which direction will osmosis occur?
- A from X to Y only
- B from Z to Y only
- C from X to Y, X to Z and Z to Y
- D from Y to Z, Z to X and Y to X

9 Potassium cyanide blocks the production of ATP in cellular respiration. Select the statement that describes how the potassium (K^+) and cyanide (CN^-) ions from potassium cyanide would enter a cell.
- A Ions will enter the cell by active transport.
- B Ions will enter the cell by facilitated diffusion.
- C Ions will enter the cell by osmosis.
- D Ions will not enter the cell.

10 Which of the following is required for facilitated diffusion but not simple diffusion to occur?
- A channel proteins and a concentration gradient
- B a concentration gradient only
- C channel proteins and ATP
- D channel proteins and membrane-bound enzymes

11 Which of the following is a feature of cancer cells that makes them different from normal cells?
- A Cancer cells are unable to synthesise DNA.
- B Cancer cells are arrested at the S phase of the cell cycle.
- C Cancer cells continue to divide even when they are tightly packed together.
- D Cancer cells are always in the M phase of the cell cycle.

12 The following figure represents the stages of the cell cycle, but they are not in the order in which they occur. Select the option with the correct order for the cell cycle.

1
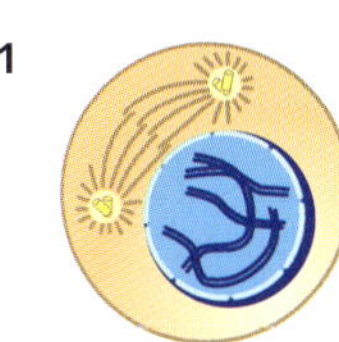

5
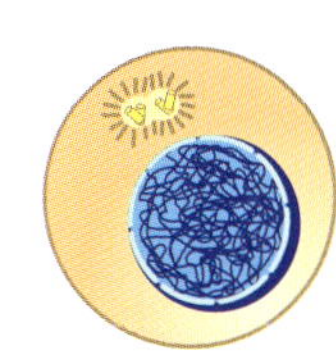

2

6
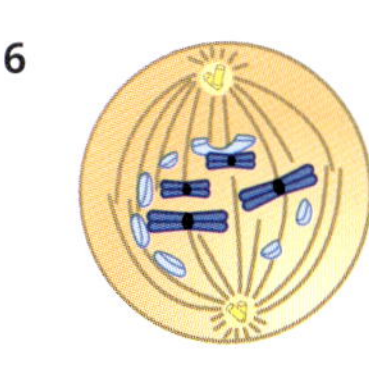

7
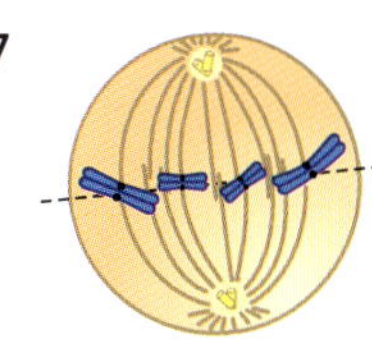

3
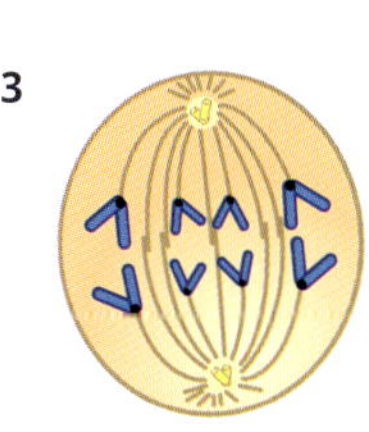

4
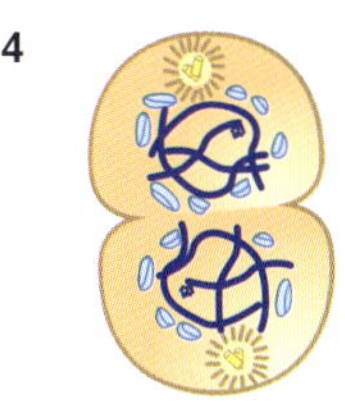

A 1, 6, 7, 3, 4, 2, 5
B 2, 5, 1, 3, 7, 6, 4
C 5, 1, 6, 7, 3, 4, 2
D 5, 1, 7, 6, 3, 4, 2

Short-answer questions

13 The following images, taken using transmission electron microscopy, show two cells, X and Y.

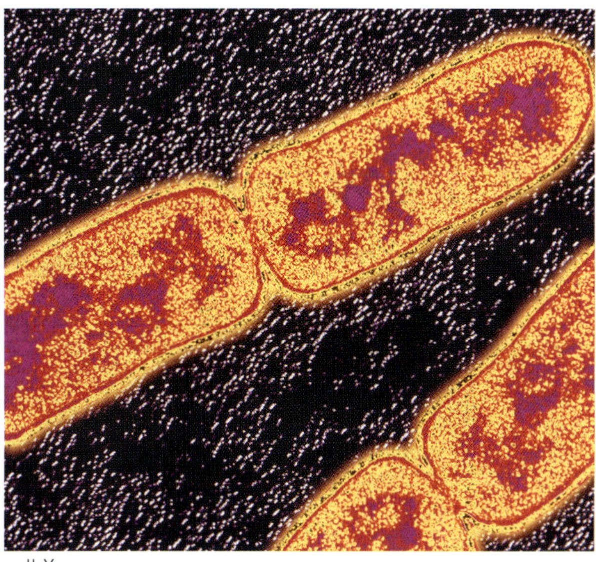

cell X

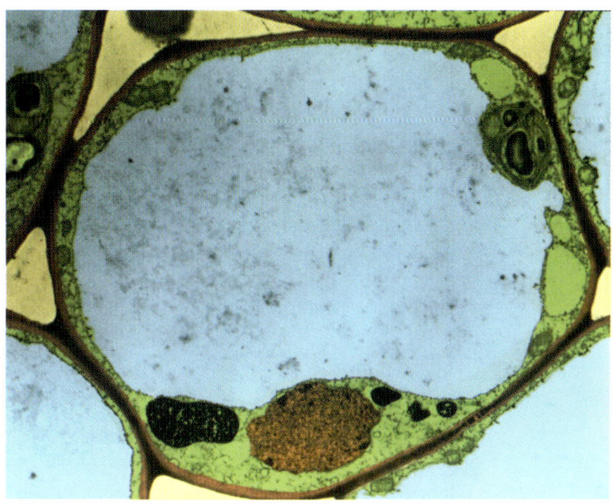

cell Y

a Which of the images shows a prokaryote? Give reasons for your answer.
b On each figure, label the structures where DNA can be found.
c Name two features in cell Y that are not visible in cell X.

The following image is a three-dimensional model of an animal cell. Use it to answer questions 14 and 15.

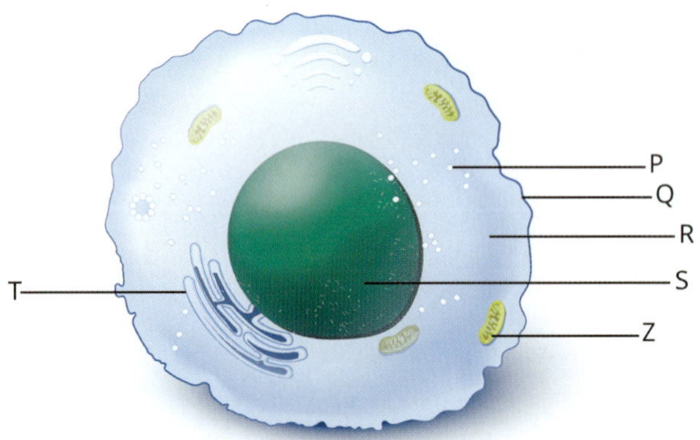

14 **a** Identify structures P, Q, R, S and T.

b What is a key feature of structure Q? How does this feature relate to its function?

15 Structure Z, from the above diagram, is shown enlarged in a transmission electron micrograph below.

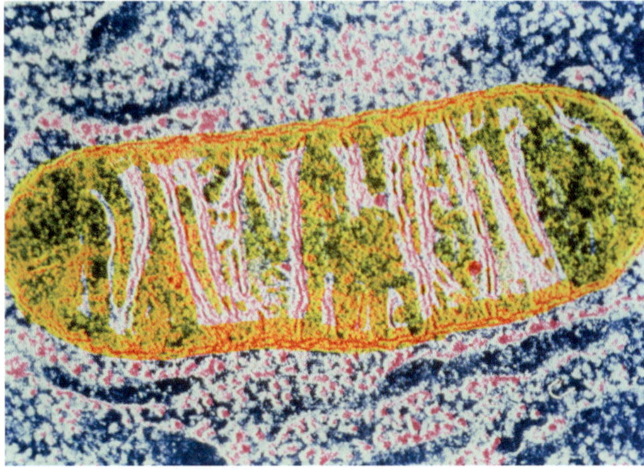

a Identify structure Z.

b Describe the internal structure of this organelle that facilitates its function.

16 The following diagram illustrates the structure of the plasma membrane.

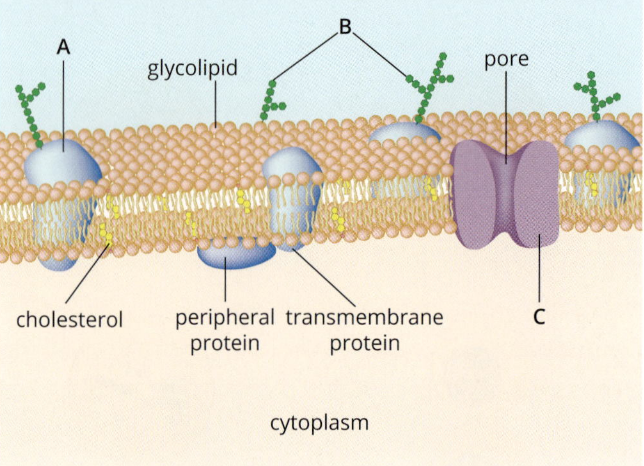

a Identify molecules A and B.

b **i** Name structure C.

ii Describe the function of structure C.

c Explain the function of cholesterol in the plasma membrane.

17 Many of the ribosomes found in a cell are attached to the endoplasmic reticulum, forming rough endoplasmic reticulum. Explain how this arrangement benefits the functioning of a cell. In your answer, you should refer to the way organelles form compartments.

18 Four tubes contain solutions with different sugar concentrations. One end of each tube is covered by a semi-permeable membrane. The tubes are placed in a tank containing a 2% sugar solution and are left until the fluid levels are stable, as shown in the diagram labelled final.

Estimate the original concentrations of solutions A, B, C and D. Justify your reasoning for each.

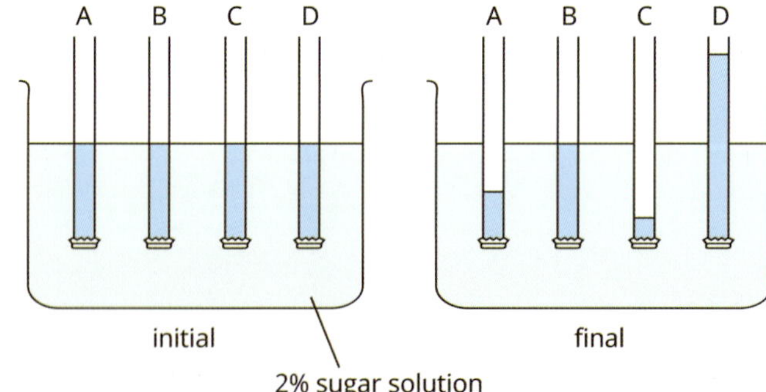

19 A student performed an investigation to model the selectively permeable nature of the plasma membrane.

- She placed distilled water in two bags made from dialysis tubing (pores of dialysis tubing are smaller than starch molecules).
- She weighed these bags and recorded their masses
- One of the bags (bag A) was placed in a beaker of distilled water
- The second bag (bag B) was placed in a 10% w/v solution of starch (w/v is weight/volume)
- After one hour, the student weighed the bags again.

a Draw a labelled diagram of the experimental set-up.

b Name the dependent variable in this experiment.

c Predict the student's results and explain your prediction.

20 Biology students prepared leaf specimens to observe under a light microscope in the school laboratory. They were then required to identify the components of the leaf cells they observed. Sketch a cell from the image below and add at least three labels.

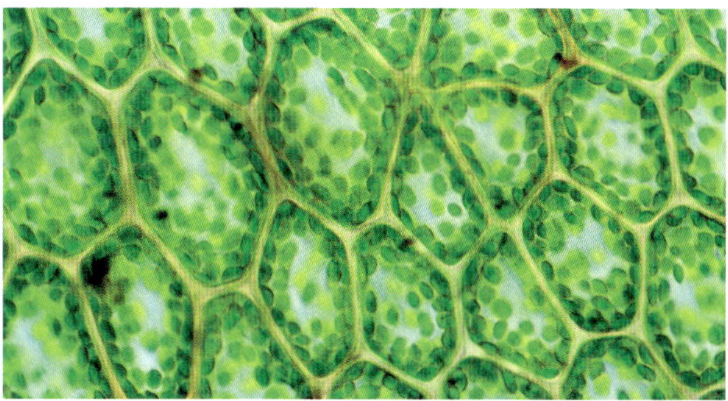

21 Complete the following table of cell organelles and their functions.

Organelle	Found in	Function
	all eukaryotic cells	contains DNA with genetic instructions for cell
chloroplast	eukaryotic plant cells and photosynthetic protists/algae	
		cellular respiration
ribosome		protein synthesis
vacuole	eukaryotic cells—many small vacuoles in animal cell; single large vacuole in plant cell	
plasma membrane		regulates transport of substances in and out of cell

22 To investigate the effect of surface area to volume ratio on the rate of diffusion, a student prepared different sizes of agar cubes containing phenolphthalein. The agar cubes were then suspended in a 4% sodium hydroxide solution for 10 minutes. When sodium hydroxide diffuses into the agar, the agar turns pink. After 10 minutes, the agar cubes were cut in half and the length of the colourless area (L) was measured. The illustration shows a cross-section of an agar cube.

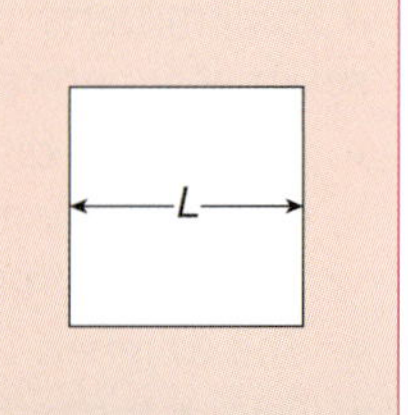

The following table shows the results of the experiment.

Length of cube side (cm)	Surface area of cube (cm^2)	Volume of cube (cm^3)	Surface area to volume ratio of cube	Length of colourless area, L (cm)	Volume of colourless area ($L \times L \times L$) (cm^3)	Percentage diffusion (%)
1.0				0.0		
1.5				0.4		
2.0				0.8		
2.5				1.8		

surface area of cube = 6 × length of cube × height of cube

volume of cube = length of cube × height of cube × width of cube

$$\text{percentage diffusion} = \frac{(\text{volume of cube}) - (\text{volume of colourless area})}{\text{volume of cube}} \times 100$$

- **a** Calculate the volume, surface area, surface area to volume ratio, and percentage diffusion of each cube, and complete the table.
- **b** Use graph paper to plot a graph of percentage diffusion against surface area to volume ratio.
- **c** Using the graph you plotted in part b, describe the relationship between surface area to volume ratio and diffusion in an agar cube.
- **d** Which size cube was the most effective for maximising diffusion?
- **e** Provide a reason for your answer to part d.
- **f** A large surface area is essential for quicker diffusion into the cell. Using the results from the experiment, explain why there is a limit to the size that individual cells can grow.
- **g** Suggest how the reliability of this experiment can be improved and what conclusion can be drawn.

23 The image below was taken by a transmission electron microscope (TEM) at a late stage of cell division.

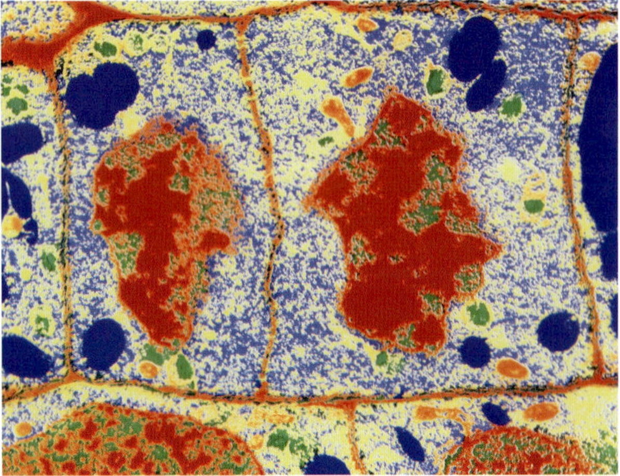

- **a** Identify the type of cell shown, giving at least one reason for your choice.
- **b** Name and describe a process of cellular division that these cells are undergoing.
- **c** In eukaryotic cells the final stage of cell division occurs in two different ways, depending on the type of cell.
  - **i** Draw and label a diagram to show the alternative process to that in the TEM.
  - **ii** Explain why the process has to be different for each type of cell.

24 The following image shows a light micrograph of a group of cells undergoing mitosis.

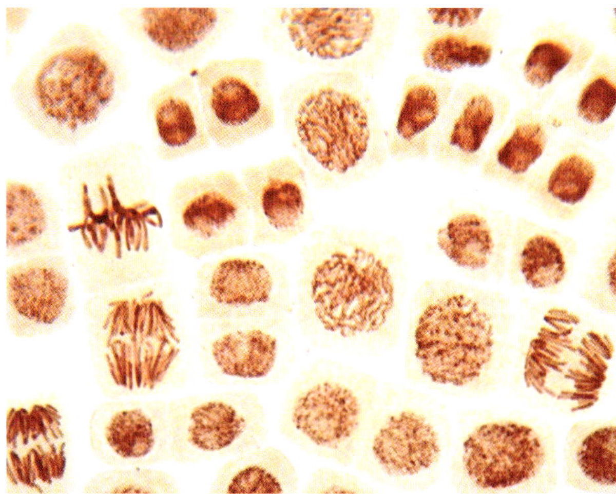

a Label one cell in each of the following stages: prophase, metaphase and anaphase.
b Describe the major events that occur during mitosis.
c Colchicine is a chemical substance that is used to prevent the formation of spindle fibres. What stage of mitosis will be prevented if dividing cells are treated with colchicine?

25 According to the cell theory, all cells arise from pre-existing cells. This figure shows the cell cycle for a eukaryotic cell of a diploid organism.

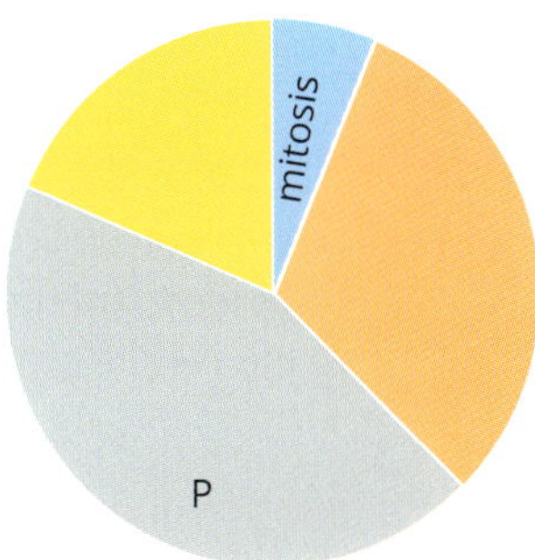

a Explain what is meant by the term 'cell cycle'.
b i Identify the phase of the cell cycle labelled P.
ii Describe the process that is occurring during part P of the cell cycle.
c i For some cells, there is a fourth phase in the cell cycle known as G_0. Copy the cell cycle diagram shown, and mark on it where the G_0 phase occurs.
ii What are some of the reasons why cells enter the G_0 phase?
d There are checkpoints in the cell cycle to ensure that two genetically identical daughter cells are produced at the end of the cell cycle.
i On your diagram, label the three checkpoints in the cell cycle.
ii Outline what happens during each of the three checkpoints.

26 Consider the following cross-section of a human embryo at an early stage of development when cells have differentiated into three separate germ layers.

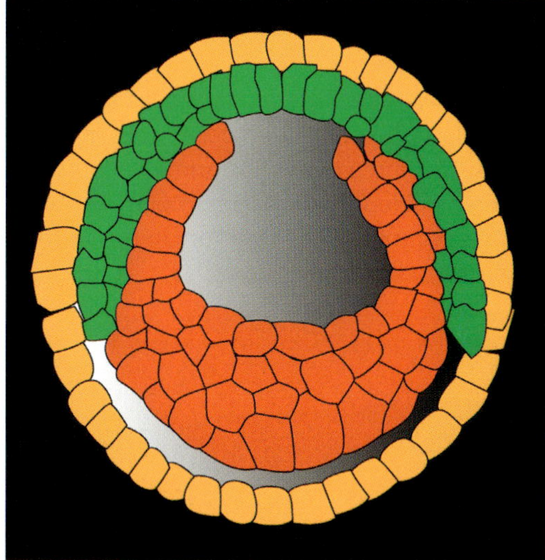

a Can the cells from this embryo be used as embryonic stem cells? Explain your answer.
b Describe the different types of stem cells.
c With the consent of the parents, blood can be collected from the umbilical cord of a newborn baby shortly after birth and stored in a cord blood bank. The umbilical cord blood contains haematopoietic stem cells similar to those found in the bone marrow, and also mesenchymal stromal cells, which can be grown into bone, cartilage and some other types of tissues.
i Discuss the advantage of using umbilical cord blood over adult blood in research studies.
ii Propose why parents might be using private cord banks for storage of their baby's umbilical cord blood.
d Compare the advantages and disadvantages of using adult stem cells and induced pluripotent stem cells. In your answer evaluate both ethical and scientific considerations.

27 Lectin is a type of glycoprotein produced by plants that can have toxic effects on cells. A group of students conducted an experiment to determine whether lectin has an effect on the rate of mitosis in onion root tips.

The roots of one onion bulb were immersed in distilled water for 48 hours, while the roots of another onion bulb were immersed in distilled water containing lectin for 48 hours. Three onion root tips from each onion were then harvested, stained and viewed under a microscope (see below). For each root tip, the number of cells in the field of view was counted. These cells were in interphase or undergoing mitosis in the apical meristem.

The results were recorded in tables, as shown on the right. Table 1 shows the number of cells in interphase or mitosis for three onion tips immersed in distilled water only. Table 2 shows the number of cells in interphase or mitosis for three onion tips immersed in distilled water containing lectin.

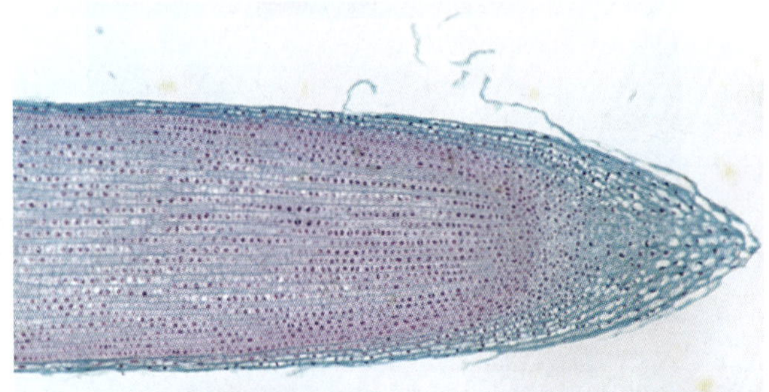

TABLE 1 Results for onion tips immersed in distilled water only

	Number of cells at:		
Onion root tip	**Interphase**	**Undergoing mitosis**	**Total**
1	47	34	81
2	36	29	65
3	37	30	67
Total	120	93	213

TABLE 2 Results for onion tips immersed in distilled water containing lectin

	Number of cells at:		
Onion root tip	**Interphase**	**Undergoing mitosis**	**Total**
1	52	44	96
2	83	25	108
3	90	54	144
Total	225	123	348

a Write a hypothesis for this experiment.

b Calculate the percentage of cells in interphase or undergoing mitosis for onion root tips treated with lectin and immersed in distilled water. Write down your answers in the following table.

Type of onion root tip	**Percentage of cells (%)**	
	Interphase	**Undergoing mitosis**
treated with lectin		
immersed in distilled water		

c Based on the results, what conclusion can you make about the effect of lectin on the rate of mitosis in onion root tips? Do the results support your hypothesis?

EQ U101

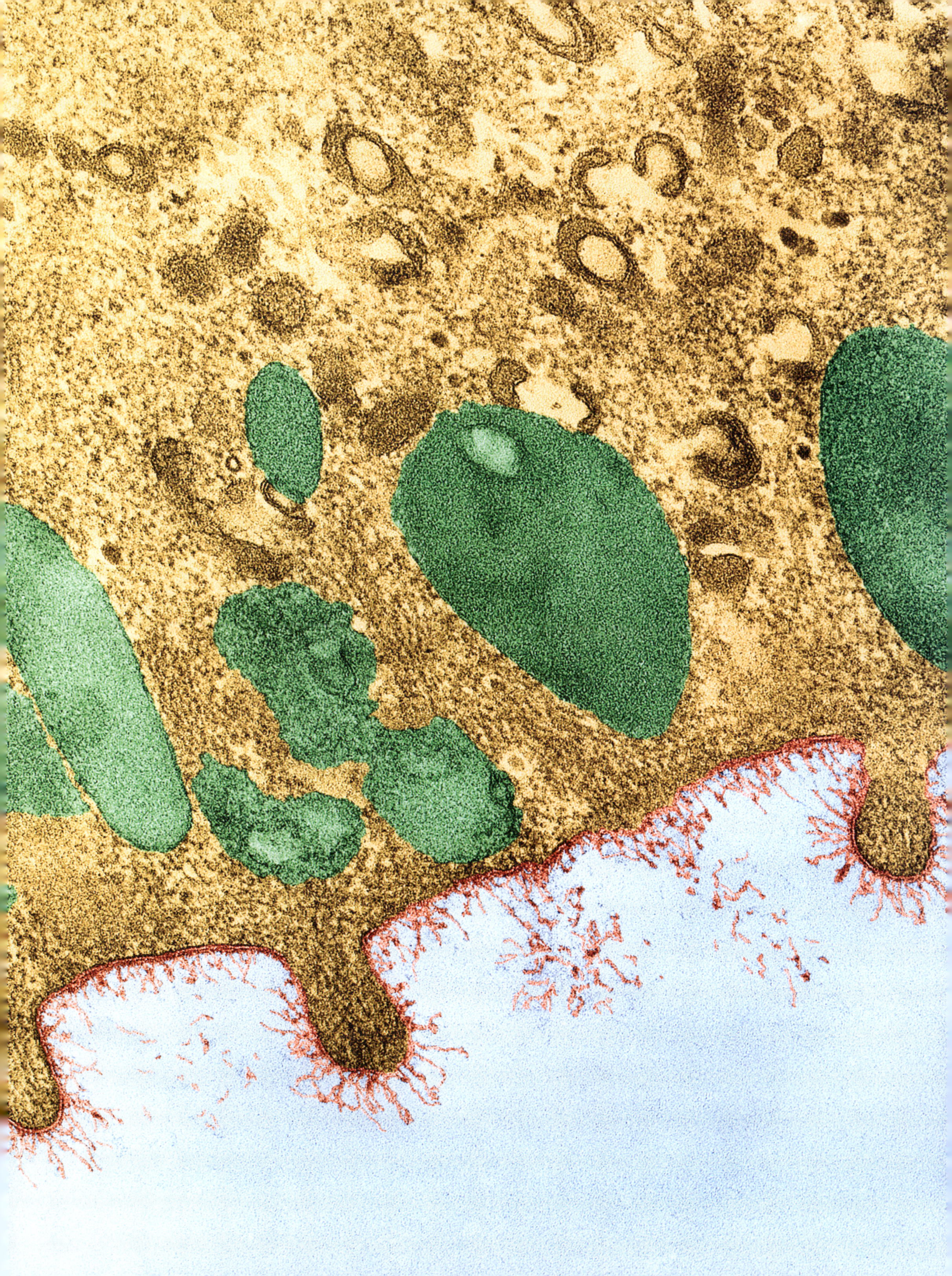

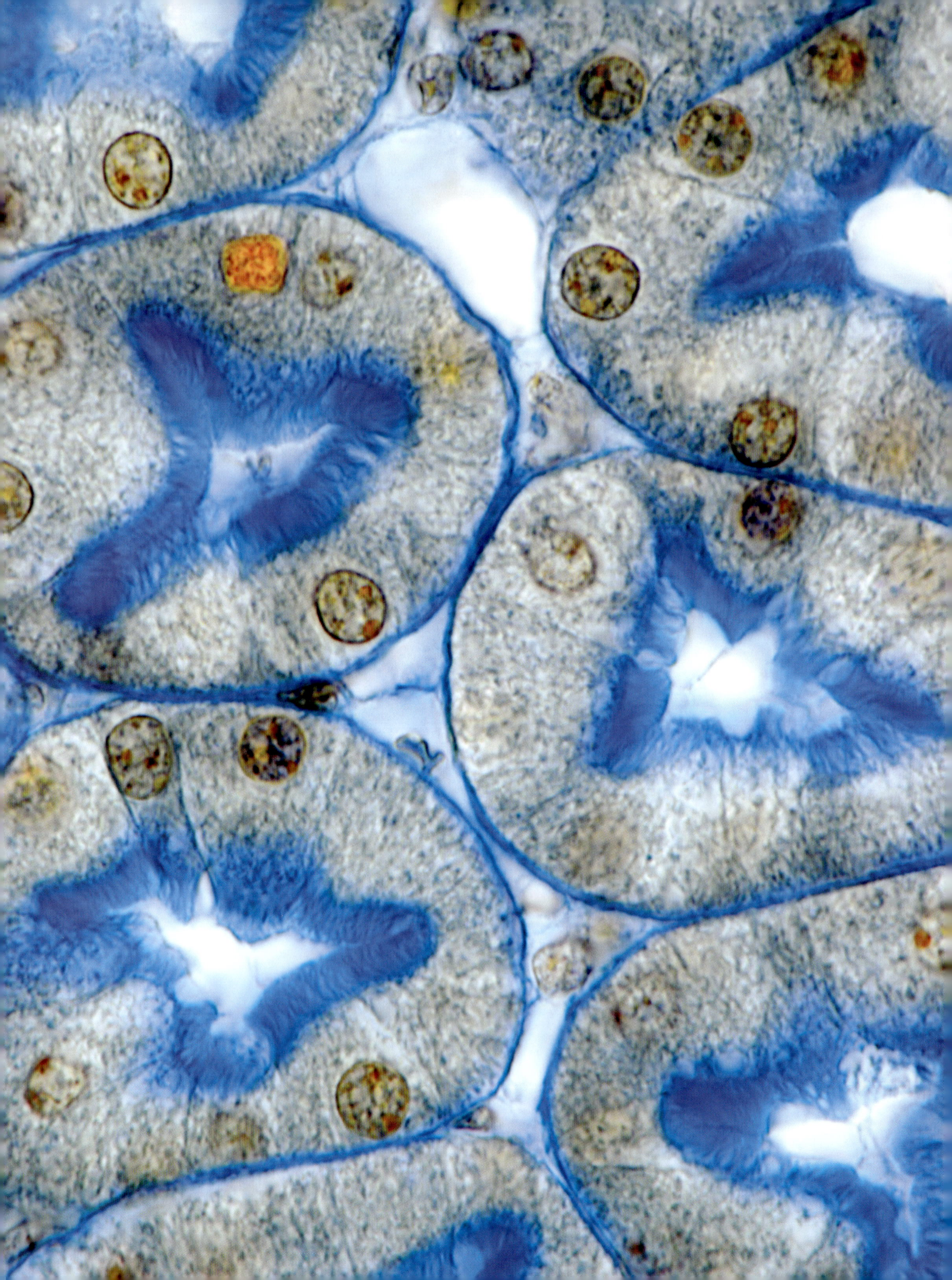

CHAPTER

04 Functioning systems

In this chapter you will learn about how the cells of multicellular organisms are organised to fulfil the needs of each cell and enable the whole organism to survive, grow, reproduce and take full advantage of multicellularity.

As multicellular organisms increase in complexity, their cells work together to achieve higher levels of organisation. The levels of organisation in multicellular organisms are: specialised cells, tissues, organs and systems. You will look at each of these levels of organisation and the specialised structures and functions that have evolved to meet the needs of complex organisms.

This chapter will examine the cellular organisation from tissues to systems in vascular plants to enable the intake, movement and loss of water, and the complex structure and function of the digestive, endocrine and excretory systems of animals.

Key knowledge

- specialisation and organisation of plant cells into tissues for specific functions in vascular plants, including intake, movement and loss of water **4.1, 4.2**
- specialisation and organisation of animal cells into tissues, organs and systems with specific functions: digestive, endocrine and excretory. **4.1, 4.3**

WS 10

OA

4.1 Multicellularity, cell specialisation and organisation

FIGURE 4.1.1 *Euglena* is an example of a eukaryotic protist.

Cells carry out all the functions necessary to sustain life, including obtaining nutrients and water, exchanging gases, sourcing energy, removing waste products and reproducing.

In **unicellular organisms** (single-celled organisms), such as the prokaryote *Escherichia coli* and the eukaryote *Euglena* (Figure 4.1.1), a single cell must carry out all of these functions. But in **multicellular organisms** (many-celled organisms) these functions are shared between different types of **specialised cells**—cells with features that allow them to perform specific roles.

> Specialised cells are cells in multicellular organisms that are adapted to carry out a specific function.

A multicellular organism is like a community of cells that work cooperatively for the survival and reproduction of the organism. All multicellular organisms consist of eukaryotic cells. There is an enormous diversity, from simple mosses, sponges and corals (Figure 4.1.2) to complex flowering plants, birds and mammals.

FIGURE 4.1.2 Corals look like single multicellular organisms, but they are actually groups of multicellular organisms (known as polyps) living in colonies.

Prokaryotes are not multicellular organisms. However, some bacteria such as cyanobacteria grow in chains of cells, and others form aggregates or colonies of cells that behave in a coordinated fashion, such as species that form biofilms. It is thought that multicellular organisms evolved from unicellular organisms to achieve a higher level of coordination. The colonial theory suggests that when a single cell divides, the new cells do not fully separate, forming a colony of individual cells that are then able carry out specialised functions (Figure 4.1.3).

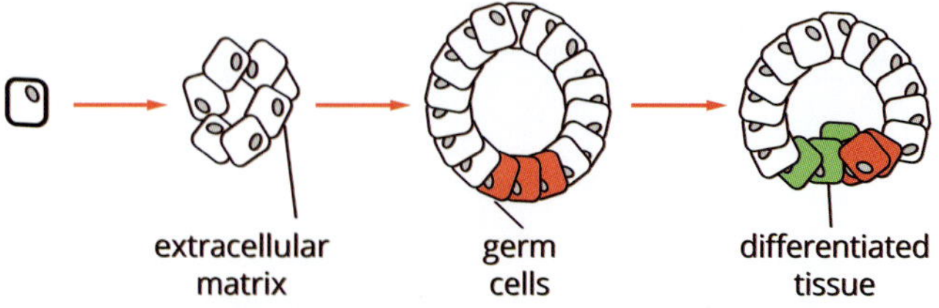

FIGURE 4.1.3 The colonial theory suggests that, during cell division, cells do not separate properly and eventually form a colony of specialised cells.

In this section you will explore the origin of multicellularity, the role of cell specialisation in multicellular organisms, and the advantages of being multicellular.

MULTICELLULARITY

For an organism to be considered truly multicellular:

- its cells (except the reproductive cells) must have the same DNA
- its cells must be connected and must communicate and cooperate to function as a single organism
- it must have different cells that are specialised and responsible for specific functions (one of which must be reproduction)
- its cells must be dependent on each other for survival.

The features of unicellular and multicellular organisms are listed in Table 4.1.1.

TABLE 4.1.1 Features of multicellular and unicellular organisms

Feature	Unicellular	Multicellular
number of cells	single cell	many cells
prokaryotes or eukaryotes	prokaryotes and some eukaryotes	eukaryotes only
functions	• one cell carries out all the functions to sustain life • functions are carried out by different organelles within the cell	• cells are specialised to perform specific functions required by the organism • functions are carried out at cellular, tissue, organ and system levels
size	microscopic size—surface area to volume ratio limits size	macroscopic size—increasing the number of cells allows increased body size
lifespan	short lifespan due to energetically expensive workload	long lifespan as work is efficiently divided between specialised cells
reproduction	• mostly asexual, clonal reproduction • whole organism is involved in reproduction	• mostly sexual reproduction • only cells specialised for reproduction—gametes—reproduce

Advantages and disadvantages of multicellularity

Almost all the life forms you can see around you are multicellular, eukaryotic organisms. The microscopic prokaryotes that surround you are invisible to the naked eye, yet they are incredibly abundant and diverse, survive in a remarkable range of environments and have been around for billions of years. Despite the success of prokaryotes, multicellular organisms continue to evolve and thrive, suggesting that being multicellular with specialised cells must have its advantages.

BIOFILE

The earliest multicellular animals

The earliest known multicellular organisms are the Ediacaran biota. Fossils of these organisms were first discovered in the Ediacara Hills in the Flinders Ranges of South Australia and are thought to have evolved 600 million years ago. These fossils represent a diverse range of the earliest known multicellular organisms on Earth. The Ediacaran biota had a wide variety of morphological features and some resembled modern-day sea jellies or worms.

a

b

FIGURE 4.1.4 (a) A representation of Ediacaran organisms, the earliest known multicellular organisms. (b) A fossil of one of the earliest known Ediacaran biota, the sea pen

BIOFILE

How many cells do you have?

A team of European scientists recently calculated that the human body is made up of about 37 000 billion cells. Most of these are red blood cells (26 000 billion), but you also have about 1.8 billion bone cells, 3 billion pancreas cells, 6 billion heart cells, 10 billion kidney cells, 15 billion muscle cells, 17 billion brain cells, 50 billion fat cells, 360 billion liver cells and 3100 billion nerve cells.

But the most common cells in your body are not part of you. In your digestive tract alone there are about 100 trillion bacteria and other unicellular microorganisms; that's almost three times the number of cells that make up your body!

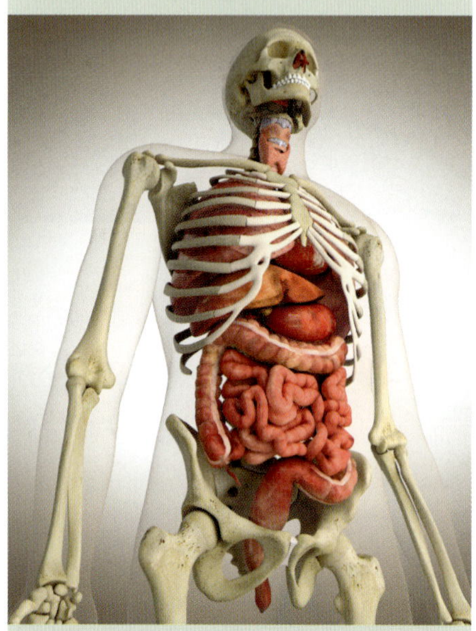

The digestive tract, where there are about 100 trillion bacteria and other unicellular microorganisms

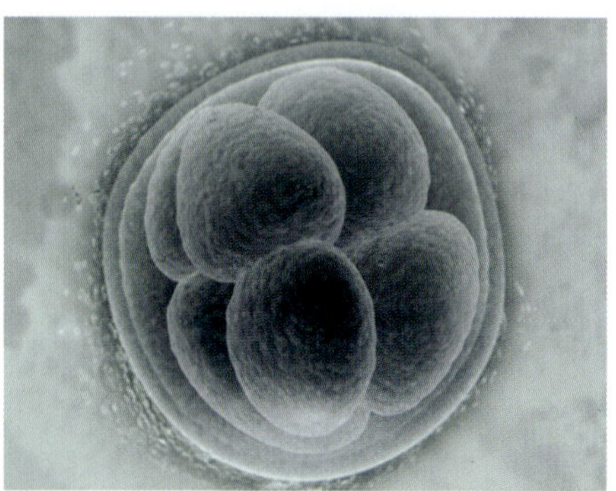

FIGURE 4.1.5 A zygote undergoing the first mitotic divisions leading to the development of a multicellular embryo

Table 4.1.2 lists some advantages and disadvantages of multicellularity and cell specialisation.

TABLE 4.1.2 Advantages and disadvantages of multicellularity and cell specialisation

Advantages	Disadvantages
1 Multicellularity is energy-efficient because specialised cells do not waste energy trying to complete all the functions necessary for life.	1 Having more cells means more energy is required for survival.
2 Multicellular organisms have longer lifespans than unicellular organisms because they are more energetically efficient.	2 The cells cannot function independently; they are dependent on the whole organism for survival.
3 Sexual reproduction and genetic recombination promote increasing complexity and specialisation over generations, compared to asexual, clonal reproduction in unicellular organisms.	3 More energy is required for reproduction; most animals need to find a mate to reproduce, and most plants need another plant in order to reproduce.
4 Multicellular organisms are less vulnerable to short-term changes in their environment. There are more systems to cope with change, and cell death does not necessarily affect the survival of the organism.	4 Populations of multicellular organisms take much longer to evolve and adapt to long-term changes in their environment because they have much longer generation times than unicellular organisms.
5 Multicellular organisms can grow significantly larger than unicellular organisms. Unicellular organisms must be small to obtain nutrients and remove waste efficiently by diffusion.	
6 Increased size and specialisation of limbs means multicellular organisms are more mobile and therefore more efficient at locating resources and avoiding predators and other negative stimuli.	
7 Multicellular organisms can have more complex responses to external stimuli.	

LEVELS OF ORGANISATION

Multicellular organisms, depending on their complexity, can be organised into the following levels to provide the needs of the entire organism:

- specialised cell—a cell with features that allow it to perform a specific function
- **tissue**—a group of similar specialised cells working together to carry out a specific function
- **organ**—two or more tissues that work together to perform one or more specialised tasks
- **system**—a group of organs that work together to perform vital functions.

CELL SPECIALISATION

All multicellular organisms begin life as a single cell that resulted from the fusion of two highly specialised cells called gametes. These gametes are called the egg (or ovum) and sperm. Gametes are unique in being able to fuse together to form a single cell, called a zygote. This one cell contains all the genetic information required to develop into a fully functional multicellular organism. The zygote develops by cell division into an embryo (Figure 4.1.5). It is through cell replication (see Chapter 3) and cell differentiation that one single cell can become the trillions of highly specialised cells that make up an organism.

Cell differentiation

Cell differentiation (also known as cell specialisation) is the process by which unspecialised cells, called **stem cells**, become specialised cells. You learnt about the development of specialised cells and stem cells in Chapter 3. Cell differentiation takes place in all multicellular organisms. Stem cells are present in the embryo and some adult tissues of animals, and in **meristem** tissue in plants. Stem cells retain the ability to divide while specialised cells usually can no longer divide.

In plants, specialised cells arise from cells in the meristem tissue. The unspecialised embryonic cells are found at the tips of shoots and roots, in a region known as the apical meristem (Figure 4.1.6). Organs such as leaves and flowers develop from cells in the shoot apical meristem, while root growth comes from the cells of the root apical meristem.

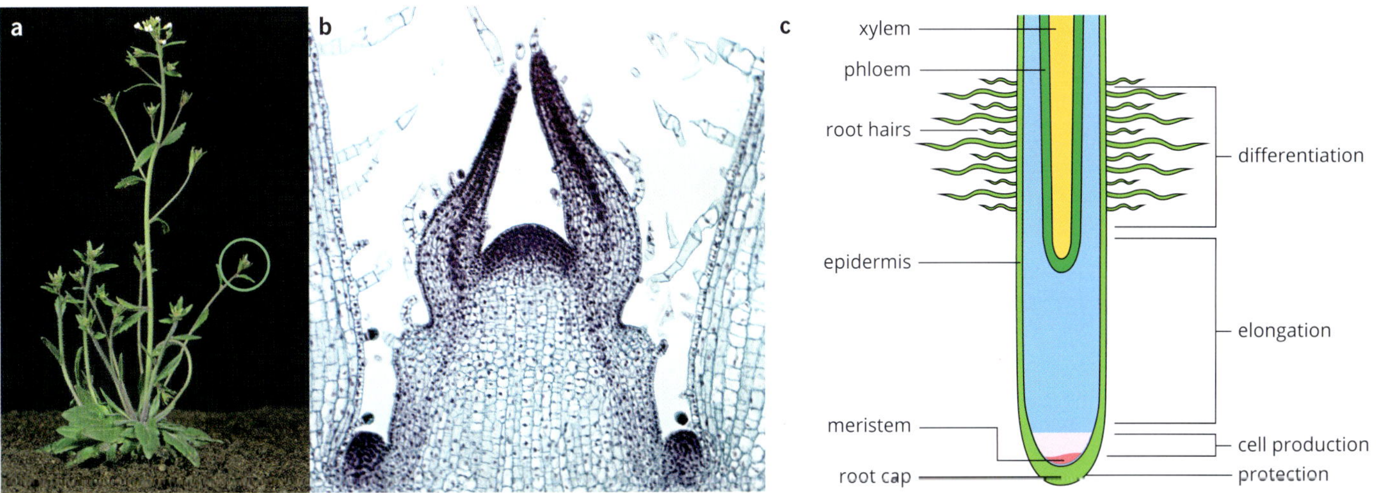

FIGURE 4.1.6 Plant cell production, growth and differentiation derives from unspecialised embryonic cells in the meristem. These cells are in the tips of shoots and roots in plants. (a) A thale cress (*Arabidopsis thaliana*) plant with the growing shoot tip circled. (b) Magnified image of stem cells in the shoot apical meristem. (c) The structures and regions of cell differentiation in the root apical meristem

Gene expression

All the genes required to produce every type of cell needed by an organism are present after fertilisation. Gene expression is the process in which the information stored in genes is used to build the different structures in a cell. Gene expression determines how a cell will differentiate and function. In specialised cells only some genes are active or expressed. For example, in developing red blood cells, the genes for haemoglobin are expressed, while in gland cells the genes that code for different hormones, such as insulin, are expressed.

The internal and external structure of a cell is the basis for the functions it performs and is also the result of regulated and controlled gene expression. Cell specialisation is an advantage because cells are more efficient when they have only one function rather than many. This makes multicellular organisms much more energy efficient than unicellular organisms.

Examples of the structures and functions of specialised cells in plants and animals are shown in Table 4.1.3.

TABLE 4.1.3 The structure and function of some specialised cells in plants and animals.

Cell function	Cell specialisation	
	Plant cells	Animal cells
exchange	root hair	gut epithelium cells
transport	companion cell sieve cell	red blood cells
strength/support	plant fibres (xylem, phloem)	bone cells cartilage cells
protection/defence	epidermal cells cuticle	ciliated epithelium cells white blood cells
photosynthesis	mesophyll cells	

CASE STUDY

Organisation in simple multicellular organisms

Some multicellular organisms are organised only at the cellular level. This includes simple multicellular organisms such as sponges and sea jellies. These animals are considered to be tissueless multicellular organisms, as their cells are not organised into discrete, functioning systems within the organism.

Although simple multicellular organisms are more complex than unicellular organisms like *Euglena*, they can survive without organising their cells into true tissues and organs because they are often only a few cells thick (Figure 4.1.7). This means that materials can diffuse easily into and between cells. This lack of organisational complexity also means that many simple multicellular organisms, such as sponges and sea stars, can regenerate, building new limbs or even an entirely new organism from just a tiny piece of their body or a single cell.

In sponges, the body is hollow and consists of two layers of eukaryotic cells separated by a jelly-like substance. The outer layer protects the sponge and also contains tiny pores through which water and food can enter. Sponges are filter feeders, filtering plankton, bacteria, dinoflagellates and many other microscopic organisms from the water around them. Digestion is carried out within food vacuoles inside the cells of the sponge. The inner layer consists of a number of cell types, including collar cells and amoebocytes (Figure 4.1.7).

Despite the simple organisation of these organisms, each of the different cell types found within them has a specialised function that contributes to their survival and reproduction (Table 4.1.4).

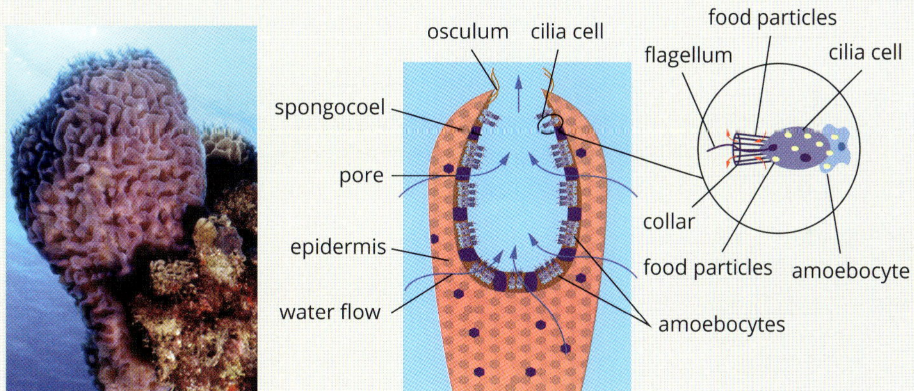

FIGURE 4.1.7 Sponges such as the azure vase sponge (*Callyspongia plicifera*) are organised at the cellular level, with different types of cells performing different functions. Although the cells work together, they do not form true tissues or organs. Sponges are often referred to as tissueless multicellular organisms.

TABLE 4.1.4 Structure and function of specialised cell types in sponges

Cell type	Function	Structure
epidermal cells	• protect the inner layer of cells	• thin, leathery • closely packed together
collar cells	• move water through the sponge's pores and into the central cavity (spongocoel) using the motion of their flagella • absorb nutrients	• flagella • hollow 'collar'
amoebocytes	• ingest and digest food caught by the collar cells • transport nutrients to the other cells of the sponge	• mobile and flexible

ORGANISATION IN VASCULAR PLANTS

A cellular level of organisation cannot meet the needs of larger and more complex organisms such as vascular plants. Consequently, cells in complex plants such as angiosperms (flowering plants) and conifers (cone-bearing plants) are organised into higher levels of organisation: tissues, organs and systems. **Vascular plants** are a group of land plants that have specialised tissues, known as **vascular tissues**, for conducting water, mineral ions and sugars.

In comparison, non-vascular plants, such as algae and mosses, do not have vascular tissue or true organs. Instead they have simplified tissues and absorb water directly through their cell walls, transporting it between cells via osmosis. The absence of vascular tissue in non-vascular plants also limits their size due to the lack of structural support and limited area over which they can transport water and nutrients.

FIGURE 4.1.8 Two major organs of vascular plants are the leaves and roots, both of which are visible in this Amazonian tree with exposed buttress roots.

Specialised cells in vascular plants

Some of the most important functions in vascular plants are involved in the transport of nutrients and water and acquiring energy via photosynthesis. There are many specialised cells within the vascular tissue of plants for these functions. You will learn more about these specialised cells in Section 4.2.

Tissues in vascular plants

The characteristic tissues in vascular plants (and the basis of this type of plant's name) are the vascular tissues, which are involved in the transport of water and nutrients throughout the plant. There are two types of vascular tissue: xylem and phloem. You will learn more about vascular tissue in Section 4.2.

Organs in vascular plants

The major organs of vascular plants are as follows:

- Roots—responsible for absorbing and storing water and nutrients (mineral ions) required by the plant from the soil. Roots also function to support and anchor the plant to the ground. Root systems are often extremely complex and can be much larger than the above ground structures of the plant. The large root systems of many trees in nutrient-poor rainforest soils do not penetrate deep into the soil layers and instead grow above ground (Figure 4.1.8).
- Leaves—the primary organ of photosynthesis. Photosynthesis is carried out to convert light energy into the chemical energy that fuels the organism's cells. The overall shape and organisation of leaves makes them well suited for their purpose. The major tissues making up a leaf are the epidermis, photosynthetic tissue and vascular tissue. The vascular tissue (xylem and phloem) is visible as veins in the leaf structure (Figure 4.1.9).
- Stems—primary functions of the stems are to support the plant's leaves, flowers and fruit; to store nutrients; to transport water and nutrients between the roots and the shoots; and to grow new plant tissue. The stem is made up of three tissue types: dermal tissue, ground tissue and vascular tissue. The structure of stems varies widely between different species. The stems of strawberry runners are flexible and fleshy, while the stem or trunk of an oak tree is thick and woody. Some stems are even edible, such as asparagus and celery stalks.

FIGURE 4.1.9 Several of the major organs of vascular plants can be seen on this orange tree, including leaves, flowers, fruits and stems.

- Flowers—the reproductive structures found in angiosperms. The function of a flower is to facilitate the fertilisation of the ovules (contained within the ovary) by the sperm (contained within pollen). The structures of many flowers are highly specialised to attract pollinators, such as bees, moths and fruit bats, to disperse the pollen from one flower to another. Other flowers produce pollen that is specialised for wind dispersal. Following fertilisation, the seeds develop and the surrounding ovary grows into a fruit.
- Fruits—protect the developing seeds of the plant and help seeds to disperse from the parent plant. Fruits develop from the mature ovaries of flowers and often have a fleshy outer layer that surrounds the seeds. The outer structure of the fruit is often specialised to attract animals that aid in the dispersal of the seeds. Some animals, such as birds, eat the fruit and later excrete the seeds, while other animals disperse seeds that have attached to their fur. Examples of fruits are berries, peaches, tomatoes, nuts and legumes.

Systems in vascular plants

Vascular plants have two systems: the root system and the shoot system. The root system is usually underground and functions to support the structure of the plant and absorb water and nutrients from the soil. The shoot system is made up of two parts: the non-reproductive (vegetative) parts of the plant, such as leaves and stems, and the reproductive parts, such as flowers and fruits. Figure 4.1.10 summarises the levels of organisation in vascular plants. You will learn more about vascular plants in Section 4.2.

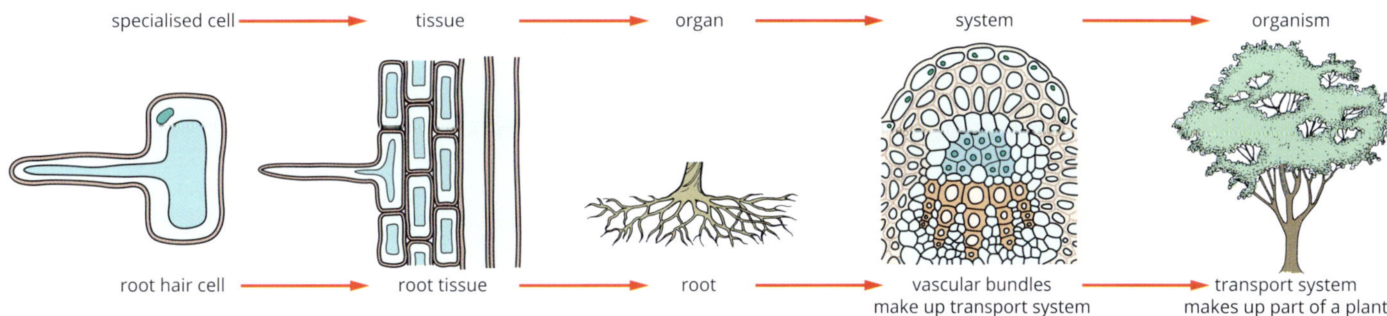

FIGURE 4.1.10 The levels of organisation in a vascular plant: cell, tissue, organ, system and organism

ORGANISATION IN COMPLEX ANIMALS

The animal kingdom includes the most complex types of multicellular organisms. An organ level of organisation is not enough to meet the needs of the most complex animals. For this reason, specialised cells of complex animals are organised into tissues, organs and systems.

Specialised cells in complex animals

Most complex animals are made up of hundreds of different cell types that are specialised to perform different functions. The roles of these cells are critical to the healthy functioning of the tissues, organs and systems of animals.

The human body consists of about 210 different types of cells. Many of these cell types differentiate from unspecialised stem cells during embryonic development. Examples of specialised cells in complex animals include smooth muscle cells, red blood cells and neurons.

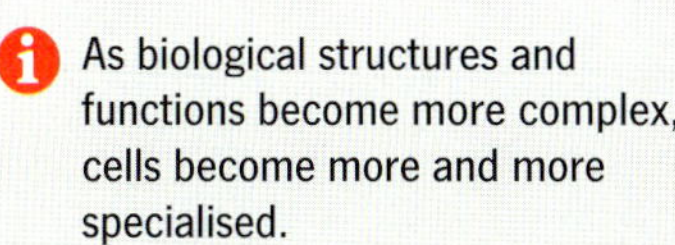
As biological structures and functions become more complex, cells become more and more specialised.

Tissues in complex animals

Specialised cells working together to complete a specific function are called a tissue. For example, a human red blood cell is perfectly adapted to absorbing and releasing oxygen as it travels around the body. However, one red blood cell cannot possibly carry all the oxygen that a human body needs. Billions of red blood cells need to work together to meet the needs of a human.

Cells do not need to be identical to be considered a tissue; they just need to be working together to carry out a particular function. Blood, for example, is a tissue that consists of red blood cells, white blood cells and platelets all working together.

A vertebrate is an animal that has a spine. Vertebrates are complex animals and include mammals, birds, amphibians, reptiles and fish.

Tissues in vertebrates are grouped into four basic types:

- Muscle tissue—formed by cells that can contract (for example, smooth muscle cells that can be found in the oesophagus, allowing peristaltic contractions).
- Nerve tissue—consists of highly specialised cells called neurons that sense stimuli and transmit signals (Figure 4.1.11). This is essential for communication and coordination in complex multicellular animals.
- Connective tissue—forms the supporting and connecting structures of the body (for example, bone and blood).
- Epithelial tissue—formed by one or more layers of cells that cover most internal and external surfaces of the organism (for example, skin and intestinal lining).

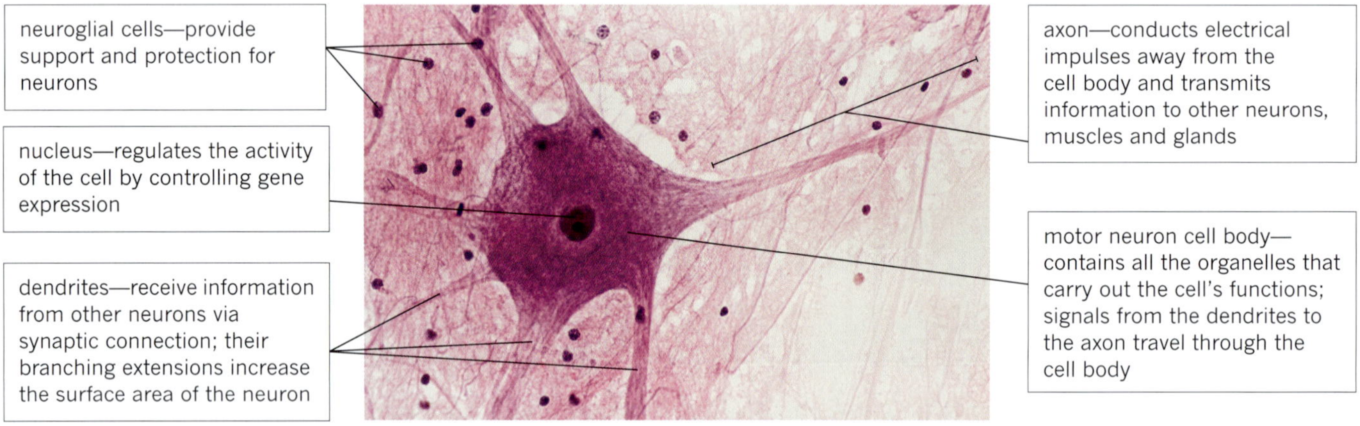

FIGURE 4.1.11 Motor neuron cell and surrounding neuroglial cells from the spinal cord. These specialised structures are part of the tissue of the nervous system that is responsible for communicating signals that regulate and control bodily functions and activity in animals.

Figure 4.1.12 shows examples of each basic tissue type found in vertebrates.

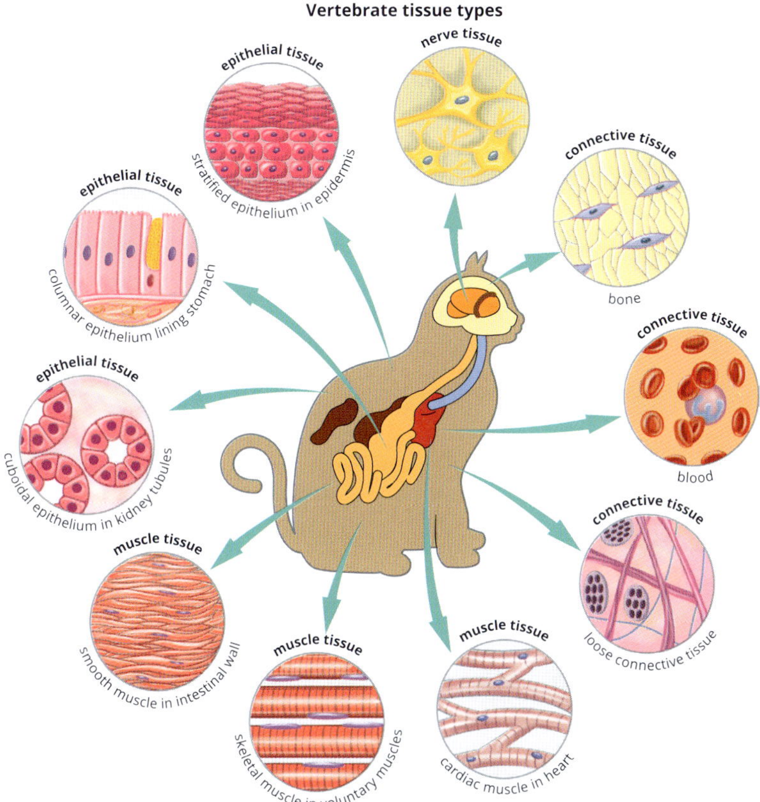

FIGURE 4.1.12 Complex multicellular organisms are made up of a diverse array of tissue types, specialised for many different functions.

Organs in complex animals

An organ is a structure made up of two or more tissues that perform a specific function. Some of the many organs in complex animals include the eye, skin, kidney and heart.

The kidney

The function of the **kidney** is filtration. Filtration is an important function of the body to remove wastes and prevent them from building up in the body, which may lead to toxicity. The kidney is made up of many tissues and specialised cells that allow it to remove wastes efficiently (Figure 4.1.13). One complex structure in the kidney that is associated with filtration is the nephron. The nephron allows the filtration of water, salts, proteins and saccharides in the blood. You will learn more about the kidney's role in the excretory system in Section 4.3.

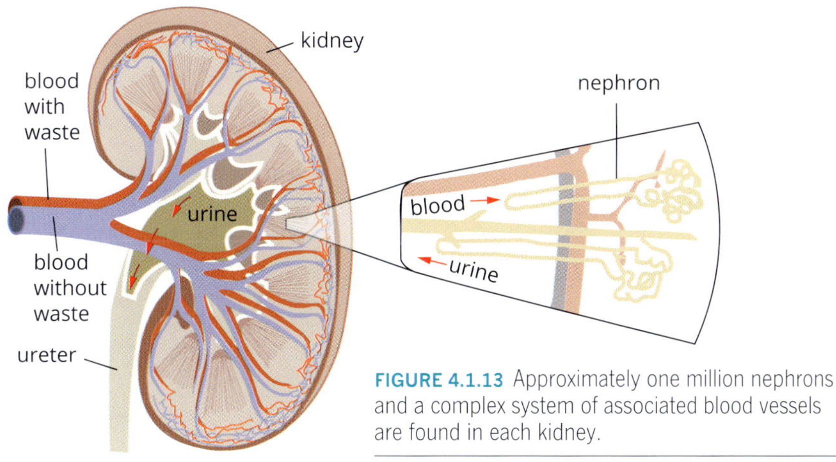

FIGURE 4.1.13 Approximately one million nephrons and a complex system of associated blood vessels are found in each kidney.

CASE STUDY

Organs-on-chips could end animal testing

The development of micro-devices that simulate organ-level functions could revolutionise the research and development of drugs and potentially end the need for animal testing. The organs-on-chips are being developed by researchers at Harvard University to mimic the mechanical and chemical functions of organs on a micro-scale. The technology will enable much faster, cheaper and accurate research and testing of the safety and effectiveness of drugs, cosmetics, cleaning products and environmental pollutants, while reducing the use of animals in the laboratory.

The memory-stick-sized chips are made from a clear, flexible polymer membrane that is lined with living human cells. The membrane sits at the centre of the chip, surrounded by fluid and air-conducting micro-channels. Because the chips are transparent, scientists are able to observe the internal physiological processes of the organs in real time (Figure 4.1.14). The organs-on-chips give scientists insight into the functioning of cells, tissues and organs in a way that has never before been possible.

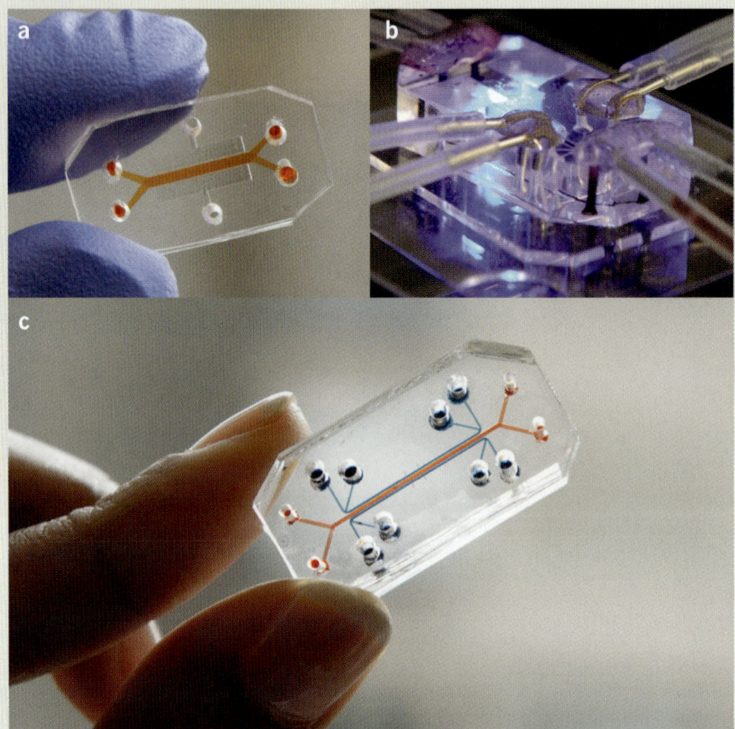

FIGURE 4.1.14 Organ-on-a-chip technology, including samples of (a) intestine, (b) lung and (c) both

Lung, kidney, liver, bone marrow and peristaltic gut-on-a-chip have already been completed, and skin-on-a-chip is being further refined to simulate a complete layer of skin.

The lung-on-a-chip consists of a porous membrane coated with human capillary cells on one side and human lung cells on the other side. Air flows through a channel on the lung cell side, while blood-like fluid containing red and white blood cells flows on the capillary side. A vacuum simulates the motion of breathing, stretching and relaxing the cells on the chip as if they were in a lung in a living organism (Figure 4.1.15). Using the lung-on-a-chip, researchers have mimicked the effects of a lung infection. After introducing bacteria to the air channel, scientists observed white blood cells crossing the membrane to engulf bacteria that were attaching to the lung cells.

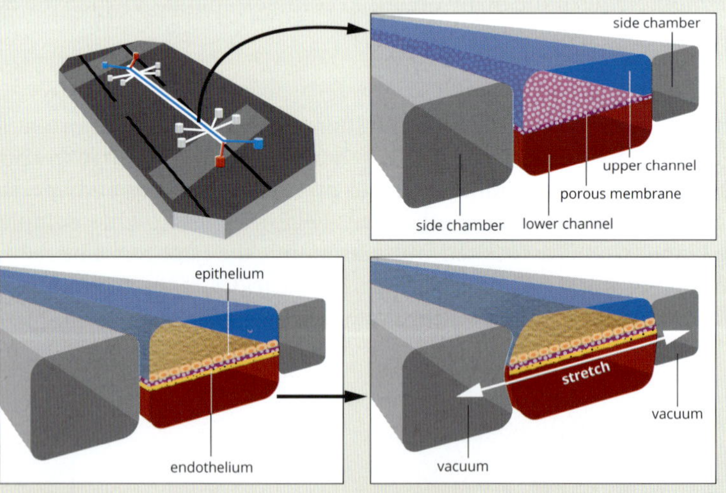

FIGURE 4.1.15 Internal view of a lung-on-a-chip lined with living human lung cells (epithelium) and capillary cells (endothelium)

Once enough micro-organs have been developed, scientists can connect them to simulate a whole human system. This research paves the way for truly personalised medicine—your cells can be used to test drug compatibility and effectiveness.

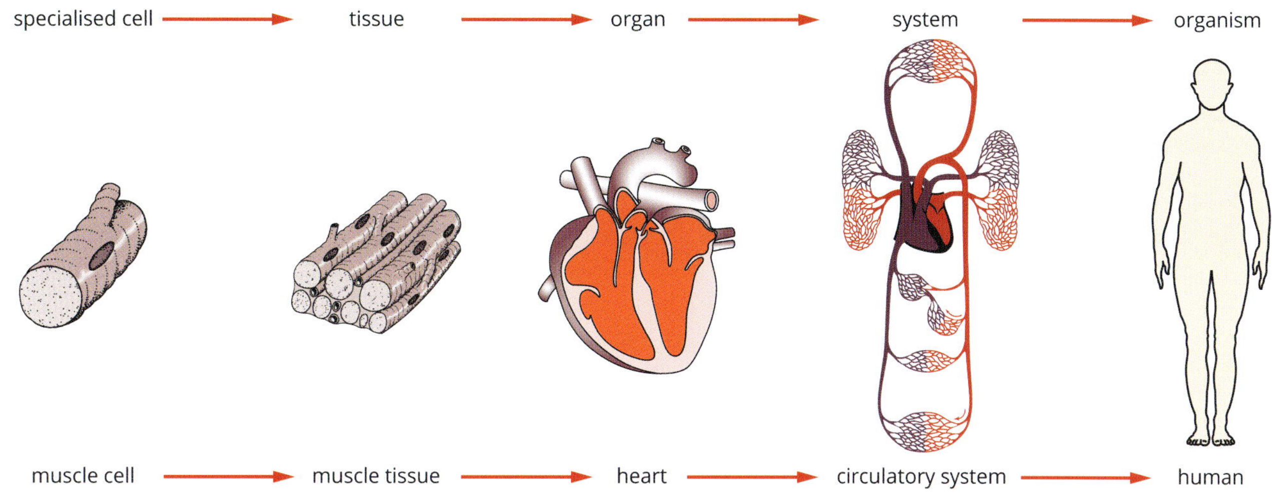

FIGURE 4.1.16 The levels of organisation in a complex multicellular animal: cell, tissue, organ, system and organism

Systems in complex animals

The organisation of cells into tissues may be enough to fulfil the biological requirements of simple animals, but more complex animals require further organisation of their organs. Systems are groups of functionally similar organs working together as a unit.

The major systems in complex animals are as follows:

- respiratory system
- circulatory system
- digestive system
- excretory system
- immune system
- nervous system
- endocrine system (glands and hormone secretion)
- reproductive system
- muscular system
- skeletal system
- integumentary system (skin, hair, nails and sweat glands).

The mammalian body is a complex structure that relies on the interaction of these systems in order to adequately function. Most mammals will use these systems in similar ways; however, some mammals have evolved independent characteristics that make their body systems unique.

Organisms

The final level of organisation is the organism itself. In complex plants and animals, systems work together and contribute to the successful functioning and reproduction of the whole organism (Figure 4.1.16).

BIOFILE

Dolphins are mammals too

Despite looking somewhat similar to aquatic gill breathers (fish), dolphins (see figure below) are mammals with a similar respiratory system to humans. Dolphins have evolved a range of characteristics that allow them to live in an aquatic environment. For example, dolphins have a blow hole, which replaces both the mouth and nostril for breathing, and can collapse the alveoli in their lungs, giving them a greater capacity to inhale and exhale quickly. These adaptations allow dolphins to remain primarily underwater when they breathe and then dive deep below the water's surface for long periods of time.

While dolphins have poorer vision than other mammals, they have developed a unique nervous system that allows them to use echolocation to visualise three-dimensional structures and detect movement in their underwater environment.

A bottlenose dolphin

4.1 Review

OA ✓✓

SUMMARY

- For an organism to be considered truly multicellular:
  - its non-reproductive cells must have identical DNA
  - its cells must be connected and must communicate and cooperate to function as a single organism
  - it must have different cells that are specialised to carry out specific functions, one of which must be reproduction
  - its cells must be dependent on each other for survival.
- Multicellularity gives organisms several advantages, such as increased efficiency, lifespan, size and mobility. However, more energy is required for survival and reproduction, and it takes longer for populations to adapt to changing environments because such organisms have generation times.
- In multicellular organisms, greater cooperation and coordination is required between cells because:
  - not all cells have direct access to the external environment, so they need some means of receiving nutrients and removing their accumulated wastes
  - specialised cells fulfil their own needs, but alone they cannot maintain a whole multicellular organism.
- Specialised cell function results from expression of particular sets of genes.
- Specialised cells perform a specific function, and this is reflected in their structure.
- Tissues are groups of specialised cells working together to carry out a particular function.
- An organ consists of two or more tissues that work together to perform a specialised task. It is often recognisable as a distinct structure.
- A system is a group of organs that work together to perform a vital task.
- The highest level of organisation is the organism itself.

KEY QUESTIONS

Knowledge and understanding

1 Put the following levels of organisation in the correct order, from simplest to most complex.
organ, specialised cell, system, tissue

2 Why is it important for multicellular organisms to be organised into cell groups, tissues, organs and systems?

3 Define tissue (as used in biology), and give an example of a tissue in vascular plants and animals.

4 Give three examples of organs in vascular plants and describe their functions.

Analysis

5 Describe a system in vertebrates and how it supports the survival of the organism.

6 Research a multicellular organism found in your local area and identify three specialised cell types. For each cell type:
  - **a** explain how its structure relates to its function
  - **b** list the tissue, organ and system the cell belongs to.

4.2 Transport in vascular plants

In unicellular and simple multicellular organisms, such as algae and sponges, all cells exchange substances directly with the environment to obtain required substances and remove wastes. This direct exchange is sufficient to meet their needs.

Direct exchange of substances in unicellular and simple multicellular organisms is possible because they have a relatively large surface area to volume ratio. This means that the distance substances need to travel within the organism to get to the surface area for exchange is quite short. You learnt about the importance of surface area to volume ratio in Chapter 2.

In more complex multicellular organisms such as mammals and vascular plants, specialised cells organised into tissues and organs have evolved to move substances, such as nutrients and wastes, around the organism. For example, most terrestrial plants have specialised tissues for fluid transport. In vascular plants, transport occurs inside closed vessels that move water, mineral ions and sugars around the plant (Figure 4.2.1).

In this section you will learn about the tissues and organs that transport water and mineral ions in vascular plants.

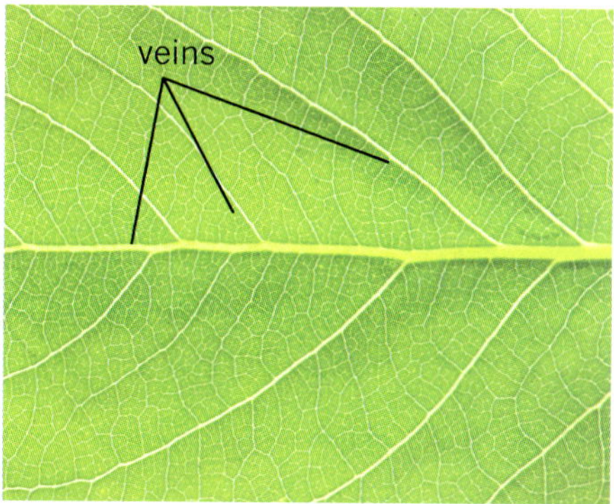

FIGURE 4.2.1 Veins are vascular tissue that transport materials both to and from a leaf.

VASCULAR PLANTS

Vascular plants include ferns, cycads, conifers and flowering plants, and usually grow in terrestrial environments. They are characterised by the presence of vascular tissue, which is tissue that is specialised for transporting fluids. Vascular tissue can be strengthened by a sheath of lignin to support the tissue, enabling greater flow rates of substances around the plant.

Plant tissue is also organised into organs. Two of the major organs in plants are leaves and roots. In vascular plants, vascular tissue is found within both of these organs.

Transport of water, mineral ions and sugars

Like all plants, vascular plants are **autotrophs** (producers), manufacturing their food from light energy by photosynthesis.

For photosynthesis, plants need water, carbon dioxide and sunlight for energy. Water is absorbed through the roots, and carbon dioxide is absorbed via the leaves. Photosynthesis occurs in the leaves and produces the sugars that are needed by all active cells of the plant.

In large vascular plants, the leaves can be a long way from where water is absorbed via the roots. The active cells in the roots that require sugar for energy can also be a long way from where photosynthesis occurs in the leaves.

Transport of these substances to the locations where they are needed is made possible by the presence of vascular tissue.

Vascular tissue transports:

- water and mineral ions obtained from the soil by the roots throughout the plant
- sugars made in the leaves to other parts of the plant.

Vascular tissue is easily visible in leaves as parallel veins in grasses, branching veins in many other leaves and the stringy parts of celery.

The tallest trees in the world are the giant sequoias (*Sequoiadendron giganteum*) of California.

BIOFILE

Water transport in giant sequoias

In tall trees like the giant sequoias of California, water and nutrients need to be transported over 100 m to the top of the plant. Their root system is relatively shallow, but water still needs to travel a great distance from the root hairs, where it is absorbed, to the leaves, where it is used in photosynthesis. It is thought that the ultimate limit to tree height is set by the pathway water has to take from root to leaf.

STRUCTURE AND FUNCTION OF VASCULAR TISSUE

In vascular plants, there are two types of vascular tissue:

- **Xylem** transports water and inorganic nutrients (mineral ions) absorbed from the soil up the plant.
- **Phloem** transports dissolved sugars, which are produced in the leaves by photosynthesis, throughout the plant. As substances transported in the phloem can be moved in any direction, sugars made by the photosynthetic tissue can be stored by the plant and moved around again for later use. Other organic substances, such as amino acids, are also transported in the phloem.

Xylem and phloem contain continuous, closed tubular pathways through roots, stems and leaves (Figure 4.2.2). Fluids flow through these tubules to all parts of the plant. All cells are close to vascular tissue.

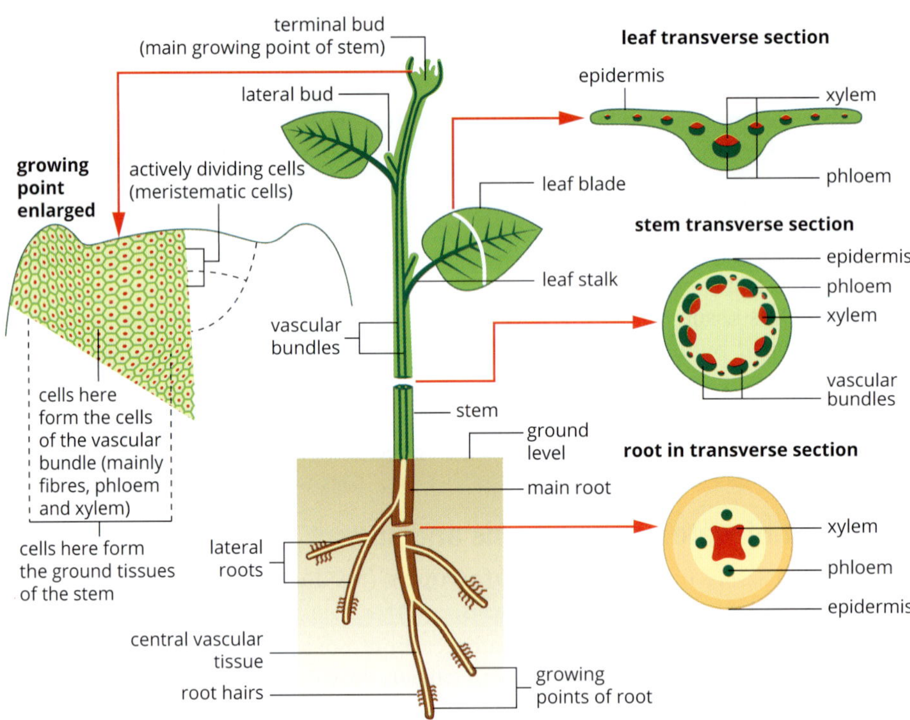

FIGURE 4.2.2 The transport system of a vascular plant has a range of structures to support the movement of substances.

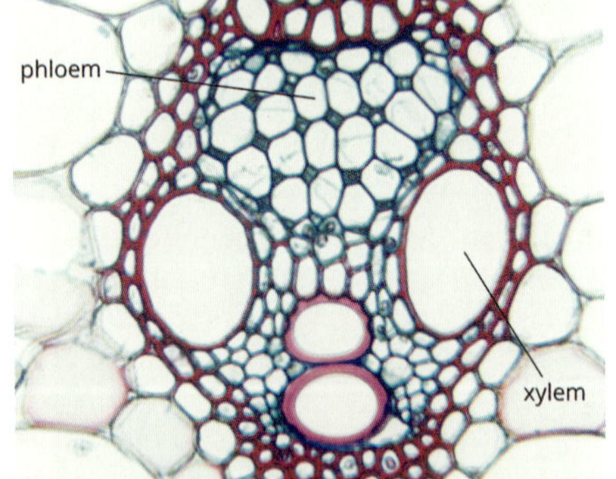

FIGURE 4.2.3 A cross-section through a stem, showing a vascular bundle containing xylem and phloem

The arrangement of xylem and phloem tissues in roots, stems and leaves is distinctive. Roots have either a central core of xylem in a star or cross shape, with phloem between the arms of the xylem, or a ring of phloem with a central xylem core. In stems and leaves the xylem and phloem are grouped into **vascular bundles**, as shown in Figure 4.2.3. These vascular bundles extend into the leaves.

Xylem

Xylem is the vascular tissue that transports water and mineral ions obtained from the soil throughout the plant. It is composed mainly of xylem vessels and tracheids.

Xylem vessels

A mature **xylem vessel** (also called a vessel element) is a long, water-filled tube consisting of specialised elongated cells joined end to end (Figure 4.2.4a). As these cells mature, the cell wall is strengthened with **lignin**, becoming stronger and more rigid, and the cytoplasm and nucleus disintegrate. Mature xylem vessels have:

- cylindrical skeletons of dead cells joined end to end to form continuous tubes
- perforated or complete openings at each end, like a straw, so that fluid can flow directly through them
- pits (non-thickened areas) and perforations in the sidewalls that allow sideways movement of substances between neighbouring vessels in the vascular bundle
- no nucleus or cytoplasm.

Tracheids are single, large, tapering water-filled specialised cells that form part of the xylem tissue in all vascular plants (Figure 4.2.4b). When mature, tracheids lose their nucleus and cytoplasm. This leads to cell death but creates an open structure for water to flow through. Mature tracheids have:

- cylindrical skeletons of dead cells joined to form continuous tubes, like xylem vessels
- pits and perforations in their lignified cell walls
- no nucleus or cytoplasm.

Unlike xylem vessels, tracheids are not connected end to end. Instead their ends overlap and water is transferred horizontally through the adjoining pits.

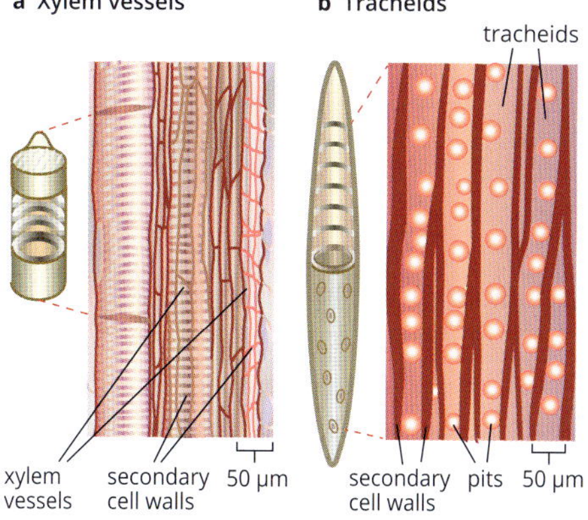

FIGURE 4.2.4 Specialised cells for transport in xylem tissue: (a) xylem vessels, (b) tracheids

Phloem

Phloem transports organic solutes, especially sugars such as sucrose, from the site of synthesis (leaves) to the site of use or storage (stems and roots) and back again when necessary. Phloem tissue, unlike xylem tissue, is alive and is composed of the following:

- **sieve tubes**, which form linear rows of elongated cells lacking a nucleus and lignin from their cell walls that transport substances directly from one cell to another through small pores. They are usually found with adjoining companion cells (Figures 4.2.5 and 4.2.6)
- **parenchyma cells**, which make up the soft tissue of a plant and contain chloroplasts to enable photosynthesis to occur
- **companion cells**, which are a type of parenchyma cell that give metabolic support to the sieve tube cells. Without the metabolic products from companion cells, sieve tubes would not survive
- **sclerenchyma cells**, which have a very thick cell wall to provide structural support for the plant. These are most commonly found in the more mature stems of a plant.

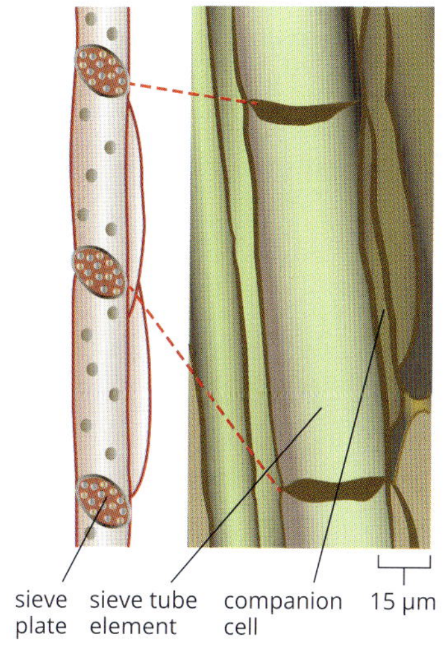

FIGURE 4.2.5 Phloem sieve tube elements

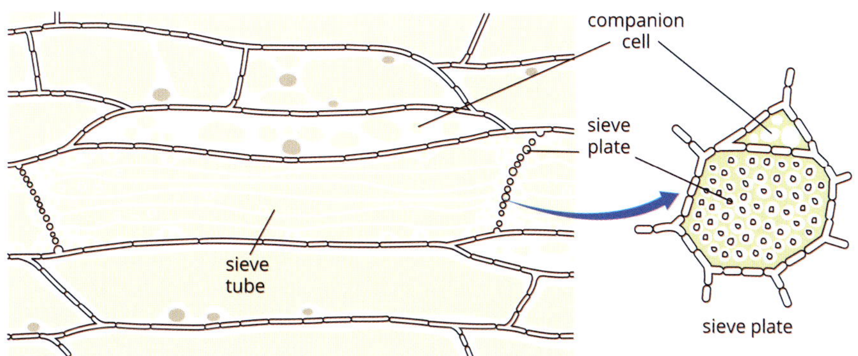

FIGURE 4.2.6 The cytoplasm of sieve tubes in phloem is continuous from cell to cell through the sieve plates.

CASE STUDY **ANALYSIS**

The Separation Tree

On 15 November 1850, citizens gathered under the canopy of a majestic river red gum (*Eucalyptus camaldulensis*) in the newly established Royal Botanic Gardens in Melbourne. Superintendent Charles La Trobe announced that Victoria would separate from New South Wales and become a new colony of the United Kingdom. The tree was commemorated with a plaque and named the Separation Tree. The tree is now heritage-listed and is believed to be over 400 years old.

Sadly, the Separation Tree was attacked in 2010 and 2013 by vandals, who removed deep strips of bark from the trunk (Figure 4.2.7).

This is known as ring-barking or girdling, and involves removing a strip of bark from the whole circumference of the trunk. The bark that is removed contains the cork cambium, phloem tissue, vascular cambium and sometimes the xylem tissue (Figure 4.2.8).

Although the leaves of the Separation Tree were still able to photosynthesise and produce sugars, the sugars could no longer be transported to the roots, and the roots starved. Despite efforts to save the tree, it was confirmed in February 2015 that it was dying. For safety reasons, the branches and upper trunk of the Separation Tree have now been removed.

FIGURE 4.2.7 The Separation Tree following the first vandalism attack in 2010

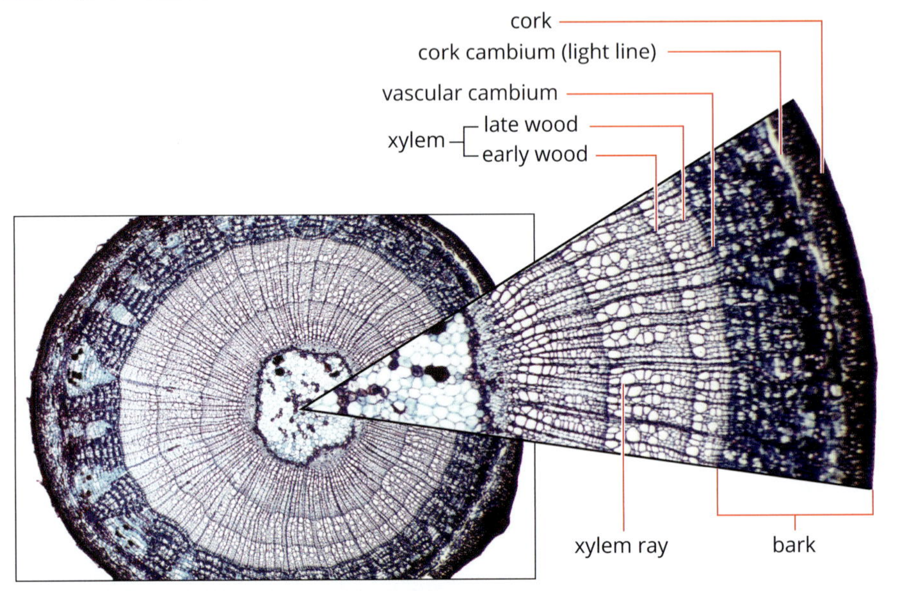

FIGURE 4.2.8 Cross-section of the woody stem of a three-year-old *Tilia* tree, showing the different tissues that contribute to the healthy functioning of a tree. The pith is at the centre, surrounded by three annual rings and the bark, which is the outer portion of the stem.

Analysis

1 Given that ring-barking is the removal of bark from the full circumference of the trunk, suggest how this may occur in nature.

2 Use Figure 4.2.8 to help you identify the structures affected by the deep ring barking of the Separation Tree.

3 The tree bark was damaged but the leaves could still function and photosynthesise to produce 'food' in the form of sugars. If the tree was still able to produce food, why was it confirmed as dying?

4 Research online to find the ways that organisations such as the Royal Botanic Gardens Victoria and the local council have conserved the seeds of the Separation Tree to commemorate its loss.

Leaves

In vascular plants a leaf is an organ composed of three distinct layers of specialised cells, or tissues (Figure 4.2.9):

- upper epidermis
- mesophyll
- lower epidermis.

The epidermis is a layer of cells covering the entire leaf. It secretes a waterproof waxy layer called the **cuticle**. Together the epidermis and cuticle provide a barrier that protects the cells and tissues inside the leaf and prevents excessive water loss. The epidermal cells lack chloroplasts but are transparent, allowing sunlight to reach the photosynthetic cells below.

Within the **lower epidermis** are **stomata** (singular stoma). Each stoma consists of two highly specialised epidermal cells called **guard cells**. The guard cells surround a pore, creating an opening through the epidermis and cuticle. They regulate gas exchange and water loss by changing shape, which causes the pore to open or close.

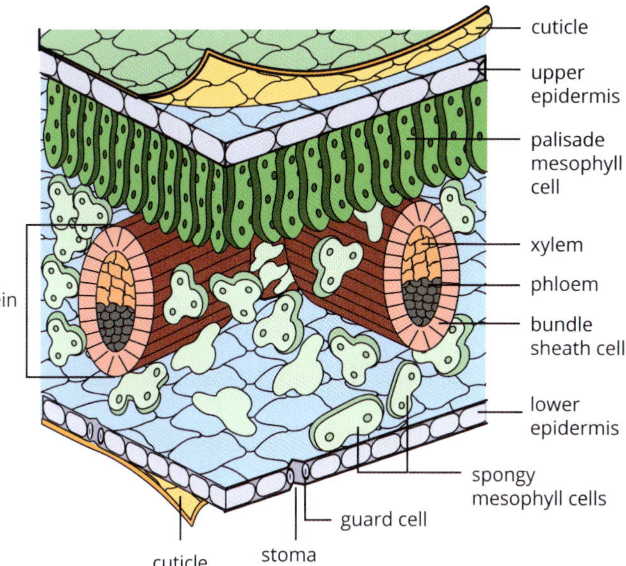

FIGURE 4.2.9 The three distinct layers of cells in leaves are the upper epidermis, the mesophyll and the lower epidermis.

Between the epidermal layers are the **mesophyll cells**, where photosynthesis takes place. The cells closest to the **upper epidermis** are the palisade mesophyll cells. These cells contain many chloroplasts and are tightly packed together. The spongy mesophyll cells below the palisade mesophyll cells are loosely packed together, with air spaces between them to allow gas exchange. These cells contain fewer chloroplasts.

The vascular tissue (xylem and phloem) is also located between the two layers of epidermal cells. Vascular tissue is often visible in leaves as veins.

Table 4.2.1 summarises the structure and function of the parts of leaves.

TABLE 4.2.1 The structure and function of the specialised cells and tissues of leaves.

Leaf tissue	Structure	Function
upper epidermis	a thin and transparent tissue covered by a waxy cuticle	protects the inner cells, prevents water loss and allows sunlight to penetrate
lower epidermis	a thin and transparent tissue covered by a waxy cuticle; contains stomata	protects the inner cells and allows the stomata to open and close depending on the needs of the plant
mesophyll	palisade mesophyll; tightly packed column-shaped cells with many chloroplasts, close to upper epidermis	photosynthesis
	spongy mesophyll; loosely packed, with air spaces around the cells	allows gas exchange, including the diffusion of carbon dioxide throughout the leaf
vascular tissue (veins)	tubular vessels	transports fluids

Roots

The major function of roots is to take in water and mineral ions (e.g. nitrogen, phosphorus and potassium) from the soil. Water is essential for photosynthesis and nutrient transfer, and mineral ions are needed to manufacture a range of organic compounds, including amino acids, proteins and lipids.

Roots have branched structures, known as **root hairs**, that increase the surface area of the roots and their capacity to absorb water and mineral ions (Figures 4.2.10 and 4.2.11).

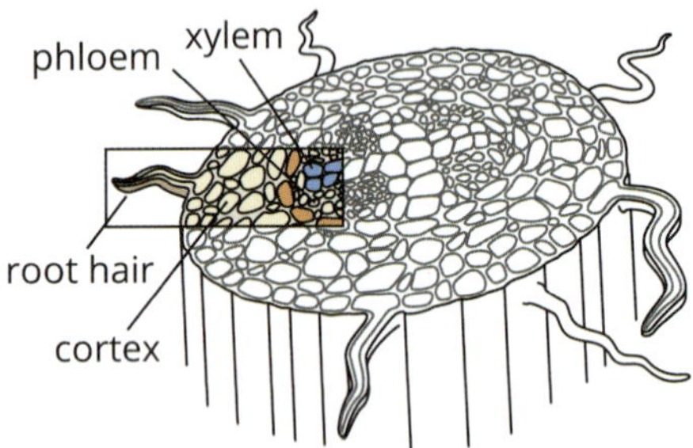

FIGURE 4.2.10 Cross-section of a root showing root hairs and vascular tissue

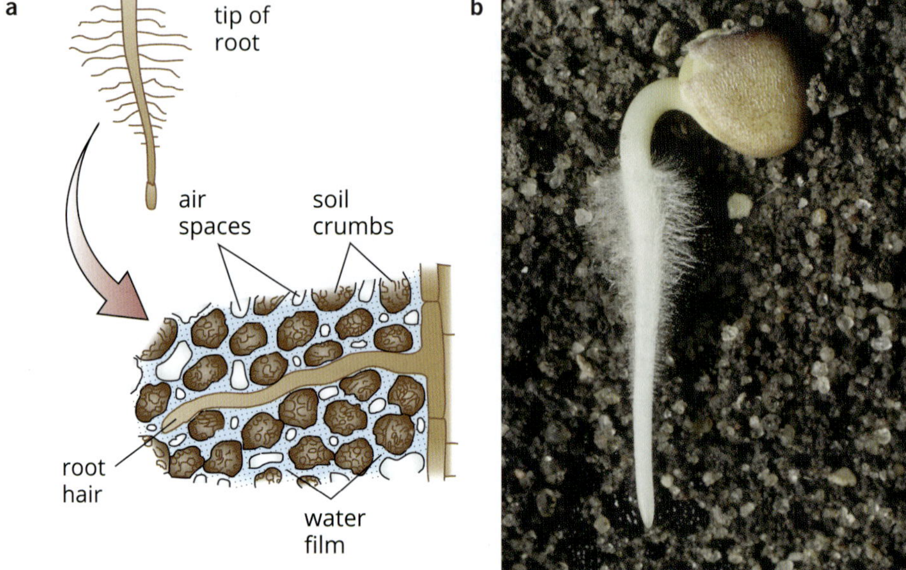

FIGURE 4.2.11 (a) Water and inorganic nutrients are absorbed by roots from soil water through many fine root hairs. (b) Root hairs on a radish seedling. The branched structure of the fine root hairs provides a greater surface area for the radish seedling to absorb water.

BIOFILE

Transport in 'non-vascular' plants

Bryophytes (mosses, liverworts and hornworts) are often called non-vascular plants because they do not have a system like that in vascular plants for transporting fluids. Most species can take nutrients and water directly into their cells by diffusion.

However, a large number of bryophyte species have an internal transport system for conducting fluids, so 'non-vascular' is not really correct. The transport system in bryophytes consists of long cells that are bundled together to form a tissue called the hydrome. Whilst bryophytes do not have lignin present to strengthen their cell walls, which means that they usually sprawl horizontally, other species can have a lignin-like substance that allows them to grow much taller, like vascular plants.

One of these is the largest non-vascular plant in the world, the giant dawsonia (*Dawsonia superba*). It is a moss that grows in wet forests and rainforests in Australia and New Zealand, and reaches 65 cm in height—taller than many vascular plants.

The giant dawsonia (*Dawsonia superba*) has an internal transport system and is taller than many vascular plants.

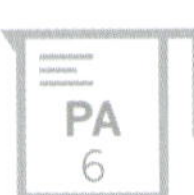

4.2 Review

OA ✓✓

SUMMARY

- Most terrestrial plants, including ferns, conifers and flowering plants, have vascular tissues (xylem and phloem) that are specialised for transporting fluid.
- The vascular tissues are:
  - xylem, which carries water and mineral ions from roots to leaves
  - phloem, which carries sugars and other organic molecules from leaves to roots.
- Xylem vessels:
  - are the skeletons of dead elongated cells
  - have perforations at each end
  - are joined end to end to form continuous tubes and allow the flow of fluid
  - have pits (thinner areas) in the side walls that enable the movement of substances into and out of the adjacent companion cells.
- Xylem tracheids:
  - like xylem vessels, are dead and have pits in their lignified cell walls, and have no nucleus or cytoplasm
  - unlike xylem vessels, are not connected end to end; their ends overlap and water is transferred horizontally through the adjoining pits.
- Phloem tissue is composed of sieve tubes made from adjoining elongated cells to transport substances.
- Sclerenchyma and parenchyma cells, including companion cells, assist the sieve tubes in transporting substances by providing energy and structural support.
- Water and inorganic nutrients (mineral ions) are absorbed by the root hairs from the soil. Roots have specialised vascular tissue to support this.

KEY QUESTIONS

Knowledge and understanding

1 What are the two types of vascular tissue in plants? State the function of each.

2 Why is there a limit to the size non-vascular plants can grow?

3 List the key features of mature xylem vessels.

4 What is the difference between mature xylem vessels and tracheids?

5 List the specialised cells found in phloem tissue.

6 What is the function of a palisade mesophyll cell and where would it be found?

Analysis

7 The following image is a transverse section through a buttercup stem.

a Label the phloem and xylem tissue on this image.

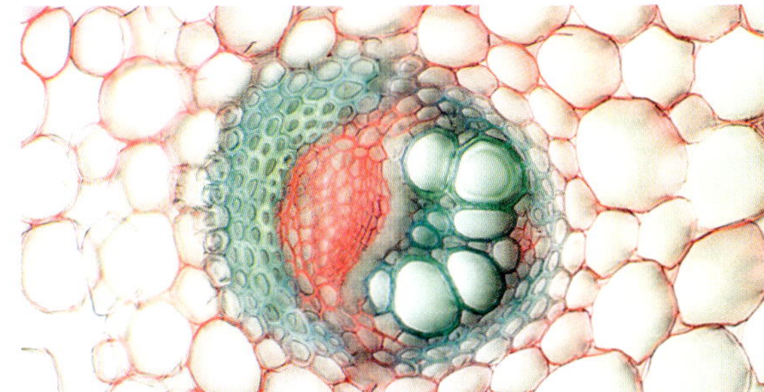

b Outline the features shown in the diagram that helped you identify these tissues.

c Outline the structural differences between these vascular tissues.

8 An experiment was conducted to determine how applying a sticky gel onto a leaf affects the rate of water loss from the leaf tissue. The graph below shows the results of the experiment.

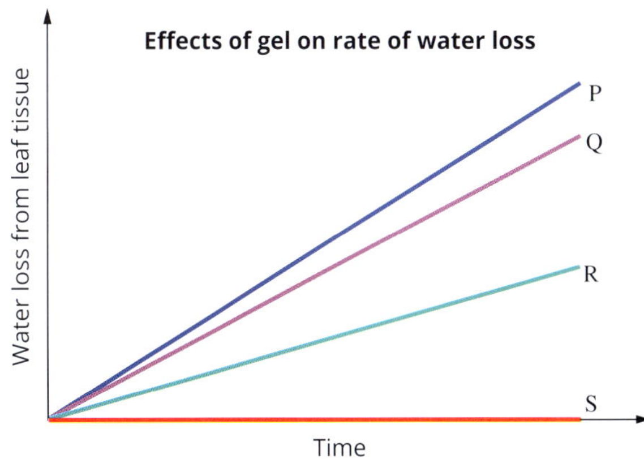

a Differentiate each line labelled on the graph by matching it to the correct experimental condition from the list below.

i no gel applied

ii gel applied to the lower side of the leaves

iii gel applied to the upper side of the leaves

iv gel applied to the lower and upper sides of the leaves

b Explain how the cellular structures of the leaf tissue support the chosen answers to part **a**.

4.3 Animal systems

Animals, including humans, are composed of billions or even trillions of specialised cells organised into tissues, organs and systems. There are many advantages to this level of complexity but, as with vascular plants, there are also challenges. For example, specialised cells cannot survive independently and must rely on other cells and the survival of the organism as a whole.

The grouping of organs into systems is the highest level of biological complexity. There are eleven systems in mammals, each with specialised roles that are essential for the correct functioning of the organism (Figure 4.3.1).

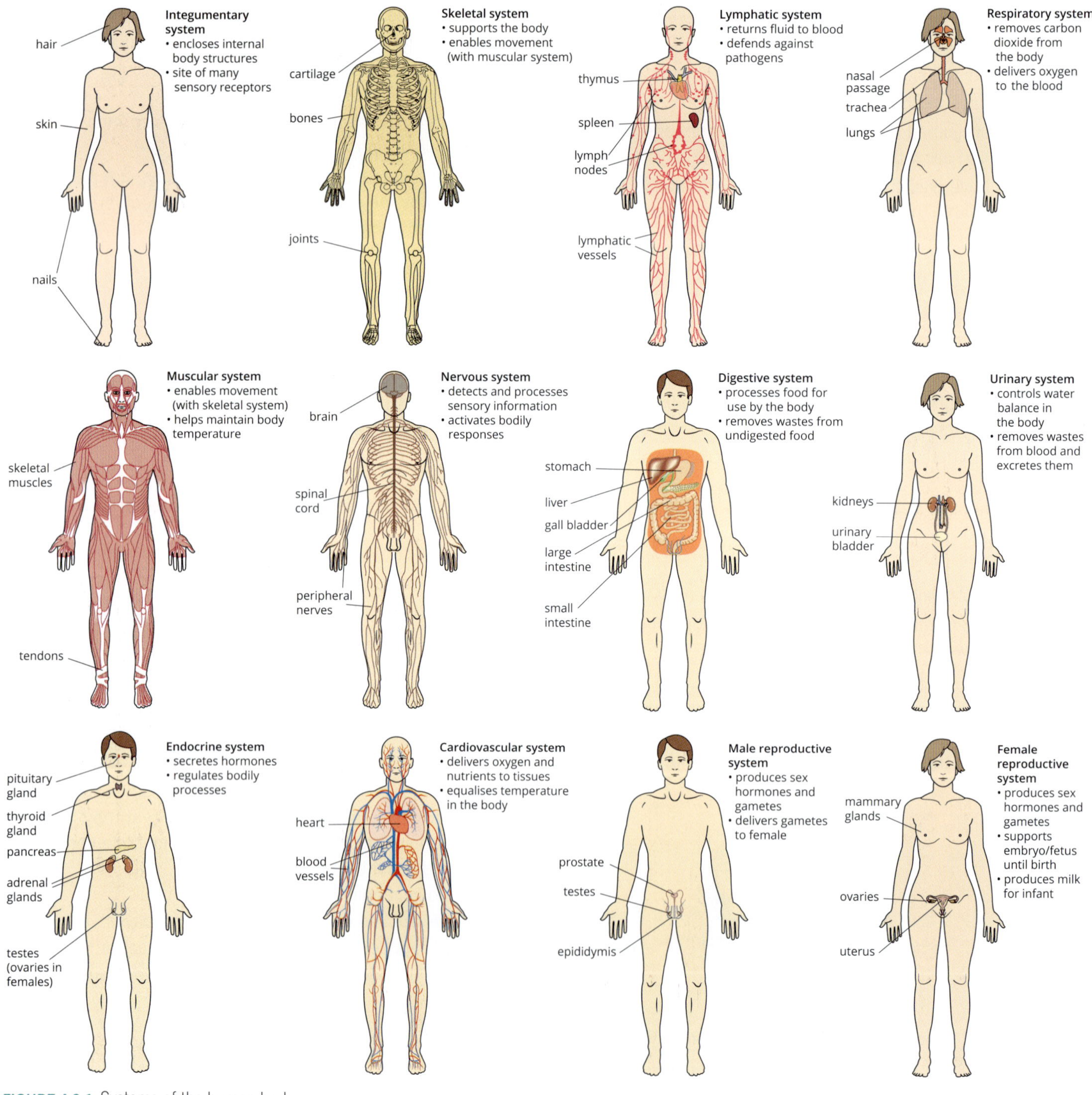

FIGURE 4.3.1 Systems of the human body

These systems do not work in isolation; they have vital connections to one another, and many of their functions overlap. Each of the systems ultimately functions to maintain homeostasis and ensure the survival and reproduction of the organism.

This section will examine each of the following systems in more detail:

- the **digestive system** and its ability to break down and take up nutrients
- the **endocrine system** and its ability to produce hormones to assist in metabolism and growth
- the **excretory system** and its ability to remove waste products from the body.

OBTAINING NUTRIENTS: THE DIGESTIVE SYSTEM

Mammals are **heterotrophs**; unlike plants, they cannot make organic molecules from inorganic materials. So they must consume other organisms or their products to obtain organic molecules. As well as needing organic molecules to provide chemical energy, heterotrophs also require other organic molecules such as vitamins, amino acids and fatty acids. Their diet must also contain minerals and water.

Nutritional requirements

Carbohydrates are an important source of immediate energy for all living organisms. The monosaccharide glucose is broken down to produce adenosine triphosphate (ATP) during cellular respiration. Animals store carbohydrates in the form of the polysaccharide glycogen.

Lipids include fats and oils and are an important energy store in animals. They are required for plasma membranes, hormones and vitamins.

Amino acids are required for **protein** synthesis. Animals cannot make all the amino acids they need, but they can change some amino acids into others. However, there are nine amino acids that cannot be made in this way. These are called the essential amino acids because they must be included in the diet. Because amino acids are not stored, all required amino acids must be present in the blood for protein synthesis to proceed smoothly. This means that all essential amino acids should be eaten regularly to maintain their levels in the blood.

Vitamins are a diverse group of organic compounds made by plants and by some simple animals and microorganisms. They are not used to supply energy, but are required in very small amounts for cellular processes. Many vitamins are important because they are needed to make certain enzymes.

Minerals are also essential for cellular processes. Dietary minerals are chemical elements that are required as essential nutrients by an organism.

Food digestion

Organisms are composed of many different types of complex organic molecules. When eaten as food, these molecules are too large to be simply absorbed into an animal's body. Regardless of the type of animal, food molecules must be small enough to pass across plasma membranes into the cells lining the gut. This is the purpose of **digestion**—to rapidly break down organic food into molecules small enough to be able to pass through plasma membranes and into cells.

The food you eat does not become part of your body until it has been absorbed by the cells lining the walls of your intestine. The digested food then passes into the bloodstream and is carried throughout the body. If food is not absorbed, it continues through the intestine and is passed out again as faeces (**egestion**).

Do not confuse egestion with excretion. **Excretion** refers to the removal of substances that were once part of the body, and occurs largely in the kidneys.

BIOFILE

Naming vitamins

Vitamins were named alphabetically (i.e. A, B, C) before their chemical structure was understood. This way of naming them is still used, although we now know their chemical formulae and dietary sources. We also know a great deal about their functions in the body.

Vitamins A, D, E and K are fat soluble, and are therefore obtained from food containing fats and oils. The remaining nine vitamins (the B group and vitamin C) are water soluble. Fat-soluble vitamins are stored in the liver, whereas excess water-soluble vitamins are excreted. This is why we need to regularly eat food containing vitamins B and C.

More than 20 minerals are also required in our diet. The main minerals required are calcium, phosphorus, magnesium, iron, sodium, potassium and iodine. Others are needed only in trace (small) amounts. Mineral ions occur in the cytoplasm of cells, in structural components (e.g. bone) and in the molecules of many enzymes and vitamins.

Digestion is the breakdown of food into a form that can be used by an organism for metabolism. This involves physical and chemical breakdown.

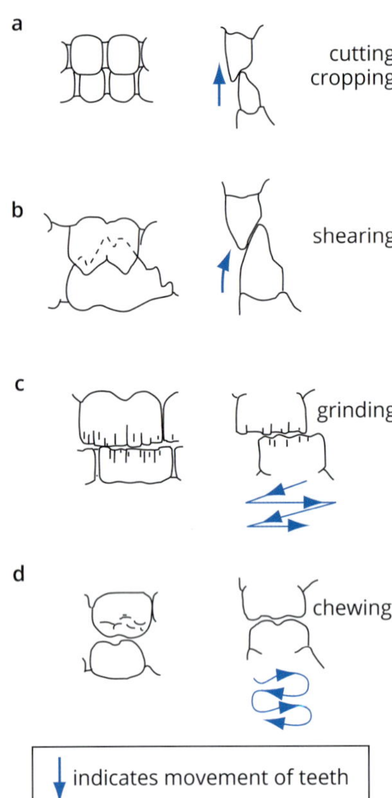

FIGURE 4.3.2 In mammals, tooth structure is adapted for the mechanical breakdown of different types of foods. (a) Incisors are typically used for cutting and tearing. (b) Carnivores have large powerful cheek teeth that shear through tough sinews and bones. (c) Herbivores have molars that grind fibrous plant foods. (d) Omnivores, such as humans, have molars that roll and crush a variety of foods.

Enzymes used in digestion are often named according to the substance on which they act, with the common ending –ase. For example, protease digests proteins and lipase digests lipids.

Physical breakdown

Digestive enzymes can only act on the outside surface of food. If food is swallowed in large pieces, the enzymes have a relatively small surface area to work on. Unless the digestive system is extraordinarily long, most of the food would remain undigested. Given the relationship between surface area and volume (see Section 2.2), digestion is much faster if food is in small pieces and the enzymes have a proportionally larger area to act upon.

So it is important to have a mechanism for breaking down large food into pieces to increase its surface area. Animals have developed a variety of structures to break down food physically—for example, the teeth of vertebrates, which break food into pieces small enough to be swallowed (Figure 4.3.2).

To improve the efficiency of digestion, this physical breakdown should take place before chemical digestion is completed. In contrast to chemical digestion, physical breakdown does not chemically change food molecules.

Bile is important in the physical breakdown of fats (lipids), but it is not an enzyme. Bile is produced by specialised hepatocyte cells in the liver and released into the small intestine where it acts like a detergent to emulsify fats—breaking up large fatty masses into small droplets. This increases the surface area of fats available for chemical digestion by lipases.

Chemical digestion

The process of breaking apart complex molecules into simple molecules is called **chemical digestion**, and is carried out by the action of **digestive enzymes**. Enzymes are important in digestion because they greatly increase the rate of breakdown of food molecules.

Most digestive enzymes split food molecules by the process of hydrolysis (from Greek *hydro* water and *lysis* split). This means they split the food molecule at a particular point by adding a water molecule. There are three main kinds of digestive enzymes (Figure 4.3.3):

- **amylases**, which act on carbohydrates
- **proteases**, which act on proteins
- **lipases**, which act on lipids.

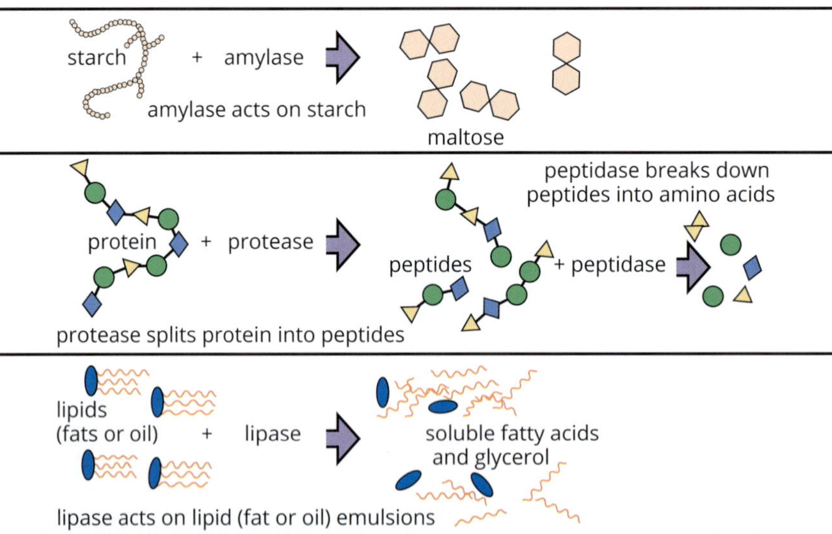

FIGURE 4.3.3 Digestion involves enzymes that split food molecules into components small enough to pass across plasma membranes and into cells.

Enzymes used in digestion are often named according to the substance on which they act, with the common ending –ase. For example, protease digests proteins and lipase digests lipids. Digestive enzymes are manufactured by specific cells. The salivary glands contain specialised cells known as serous cells that secrete amylase. Gastric chief cells found in the stomach are able to secrete proteases. The pancreas contains a range of specialised cells that can secrete amylase, protease and lipase. Many very large food molecules can be broken down only by several enzymes acting one after the other. In this case, the different enzymes are produced at appropriate sites along the digestive system.

As enzymes are proteins, they are sensitive to changes in the pH of a solution. Altering the pH changes the shape of protein molecules, which in turn alters their chemical properties. The change in shape alters the way that an enzyme binds with the molecule upon which it acts. Enzymes, therefore, have certain pH ranges over which they operate best. Different regions of the gut have different pH values to suit the enzymes found in that region. For example, pepsin and trypsin are both enzymes that digest proteins, but they have very different pH requirements. For example, pepsin is released in the stomach and is most active in the stomach's acidic environment. Trypsin is most active in the slightly alkaline small intestine (Figure 4.3.4).

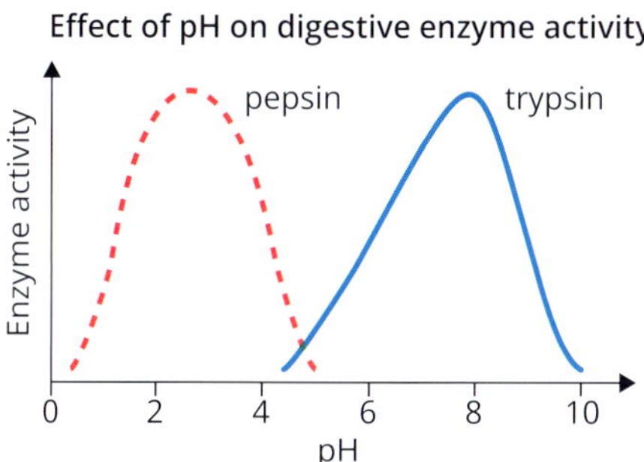

FIGURE 4.3.4 Enzyme activity of pepsin and trypsin at the different pH levels of the digestive system

Extracellular digestion

Chemical digestion can be extracellular or intracellular. **Extracellular digestion** is chemical digestion that occurs outside cells. For example, cells release enzymes into the lumen (central cavity) of the small intestine, where enzymes split the food molecules, and the resulting smaller molecules are absorbed. Sometimes, digestive enzymes are located on the actual surface of cells. As the food is digested into smaller molecules, the molecules pass immediately into the cells. Mammals and most other animals rely on some form of extracellular digestion.

Features of efficient digestive systems

Every animal's digestive system must support their nutritional requirements. Large animals, including vertebrates, require higher levels of energy and nutrients for their normal activities. Because mammals are endothermic (meaning they maintain a stable body temperature, usually higher than their environment), they require a lot of energy to maintain their body temperature. They therefore need digestive systems that can efficiently extract large amounts of energy and nutrients from food resources, and these systems are found in more active animals. Characteristics of these highly efficient digestive systems include:

- effective mechanisms for capture and preliminary handling of food
- appropriate physical breakdown of food
- a one-way gut with separation of tasks along its length
- efficient transport and storage of ingested food
- efficient sequential release of digestive enzymes
- an adequate surface area for maximal absorption of nutrients and water
- efficient egestion of unwanted materials.

Digestive systems in mammals

All mammals need food and water, but different species have different food requirements, feeding behaviours and digestive systems. Cows are slow-moving and spend much of the day eating grass and chewing. In contrast, dogs are energetic and active, and may spend only 5 or 10 minutes each day gulping down food. Dogs and cows have many other differences that relate to their eating habits. Their teeth are very different, and cows have much larger and more complex intestines than dogs (Figure 4.3.5a, b). Humans are different again. Our teeth are unlike those of dogs or cows—we are not very good at chewing bones or grass. Our preferred foods include both meat and plant material, and we often cook our food first. Humans spend about 30 to 90 minutes each day eating, although the social aspects of eating may extend this time. The human digestive system is proportionally longer than that of a dog, but shorter than that of a cow (Figure 4.3.5c).

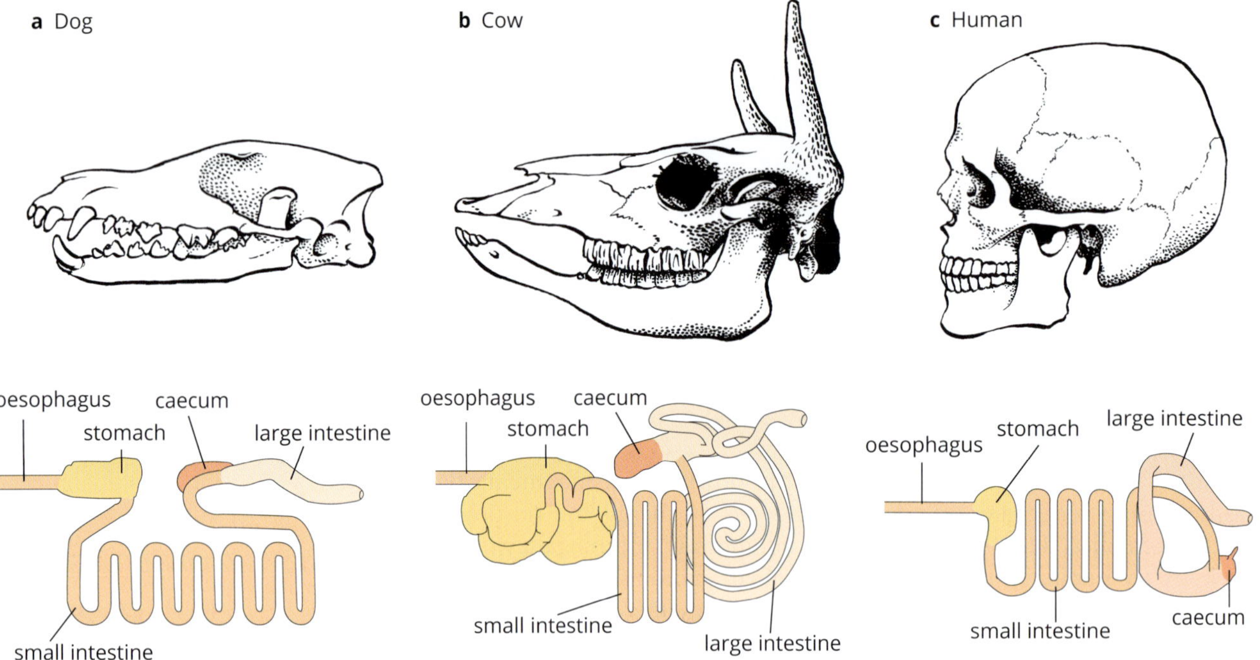

FIGURE 4.3.5 Skulls and digestive systems of (a) the dog (a carnivore), (b) the cow (a herbivore) and (c) the human (an omnivore). The teeth and digestive systems of different species are adapted to the type of food that they eat.

Cows, dogs and humans are examples of animals with three different dietary patterns. Animals that eat only plants, such as cows, rabbits, kangaroos and koalas, are herbivores. Herbivores typically spend much of the day eating. Carnivores, including dogs and cats, consume animals. They spend much less time eating; sometimes animals in the wild, such as lions, may not eat for days between meals. Humans, on the other hand, are omnivores (from Latin *omnivorus*, which means eating everything), because they eat both plant and animal foods.

Animal matter has a much higher proportion of extractable energy per gram than plant matter. The carnivore gut produces all the enzymes needed for the complete digestion of meat so digestion is quicker and more efficient. Digestive systems are shorter and simpler in carnivores than in herbivores.

The reason for the difference in feeding behaviour between herbivores and carnivores is clear. Plant material must be repeatedly ground by the teeth to expose as much surface area as possible for enzyme action and to release the contents from broken cells. As a food, it provides much less energy than meat and it takes a long time to extract that energy.

Humans are omnivores

The digestive system in mammals has the principal function of digesting and absorbing food. In other words, the digestive system breaks down food, making it simple enough to pass across plasma membranes and be useful to cells.

Before food passes into the digestive system of a mammal, it is physically broken into pieces by the teeth. Mucus is secreted to protect the lining of the gut and to lubricate food for easier passage. The food then moves along the gut past a series of digestive enzymes that sequentially break down the various compounds for absorption. Proteins are broken down to amino acids, fats and lipids to fatty acids and glycerol, and complex carbohydrates such as starch to simple sugars. Useful substances, such as water, are absorbed, leaving unwanted and undigested substances to be eliminated in the faeces.

The main regions of the human digestive system are the mouth and mouth cavity, oesophagus, stomach, small intestine, large intestine, rectum and anus (Figure 4.3.6). The salivary glands, pancreas and liver are digestive glands that develop as outgrowths of the digestive system.

Key steps in the process of digestion in humans occur at the following sites:

- **mouth**—Teeth mechanically break food into small pieces. Saliva lubricates food and the enzyme amylase, produced by serous cells, digests starch into maltose.
- **epiglottis**—This flap, at the entrance to the larynx, prevents food from entering the trachea and respiratory system, directing it down the oesophagus. The epiglottis is also associated with the gag and cough reflex.
- **oesophagus**—Food travels down this tube to the stomach, aided by muscular contractions (**peristalsis**).
- **stomach**—Protein-digesting enzymes (proteases) and gastric juices are secreted by **gastric chief cells** and other specialised **epithelial** cells in the stomach to aid in food digestion (Figure 4.3.7). Peristalsis of the stomach muscles further breaks the food down and pushes it through the digestive system.
- **liver**—The liver has important roles in regulating metabolism, removing toxins and processing nutrients. It stores excess glucose as glycogen (a polysaccharide or carbohydrate) for later conversion back to glucose when needed for energy. The liver is also the site where bile is produced by specialised hepatocyte cells, for the breakdown of fats.

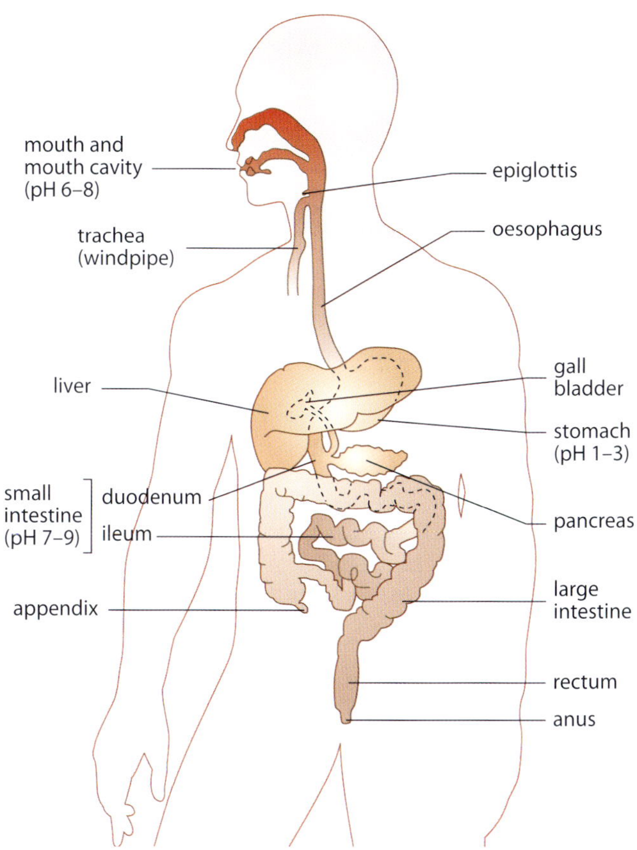

FIGURE 4.3.6 Components of the human digestive system

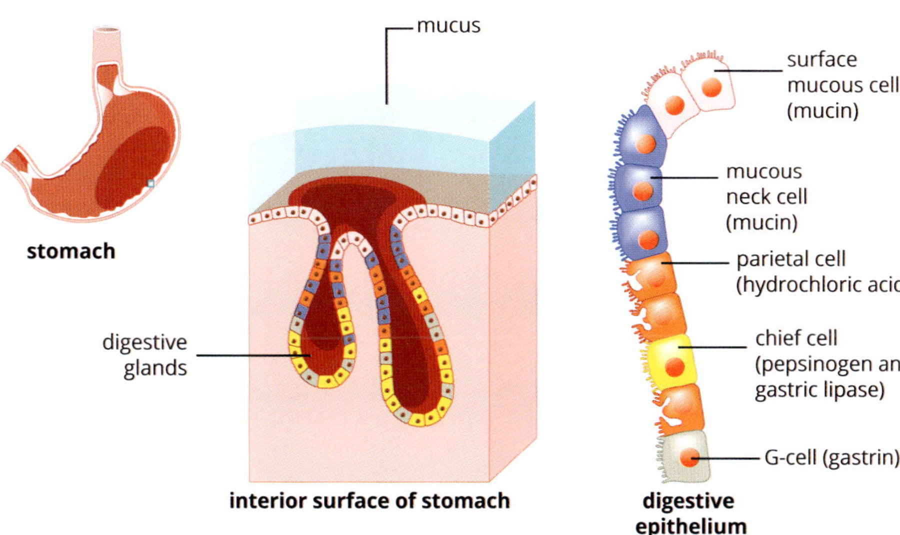

FIGURE 4.3.7 Specialised cells are found within the epithelial tissue of the stomach lining. These cells have highly developed endoplasmic reticulum, Golgi apparatus and secretory vesicles to synthesise and secrete a range of substances. Mucin-secreting cells protect the stomach lining from being damaged by gastric acids.

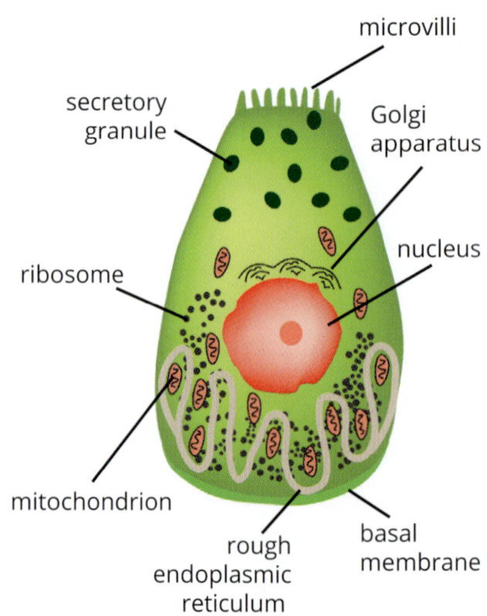

FIGURE 4.3.8 Diagram of the acinar cell of the pancreas, showing highly developed endoplasmic reticulum involved in the production of enzymes to be released into the digestive system

- **gall bladder**—Stores and concentrates bile before releasing it to the small intestine.
- **pancreas**—Digestive enzymes are produced in the **pancreatic acinar cells** (Figure 4.3.8) and activated when the food reaches the duodenum (first part of the small intestine). The pancreas also produces the hormones insulin and glucagon from specialised alpha and beta cells, which regulate sugar levels in the blood, and sodium bicarbonate, which neutralises stomach acids in the food.
- **small intestine**—The primary function of the small intestine is to absorb nutrients and minerals from food. Enzymes produced in the pancreas and the small intestine and bile from the liver and gall bladder further break down food products to facilitate nutrient and water absorption. The small intestine's many blood vessels absorb the nutrients and waste products of digestion and deliver them to the circulatory system.
- **large intestine**—Water is absorbed with soluble compounds like vitamins and minerals; undigested food leaves the body as faeces.

The small intestine

The principal organ of absorption is the small intestine. 'Small' refers to the diameter of this part of the intestine. The small intestine consists of many specialised cells, including epithelial cells, known as **enterocytes**, that line the internal surface of the small intestine and absorb nutrients, and **goblet cells** that secrete mucin to protect the lining of the lumen. The small intestine is long and has a large surface area, making it well-suited for absorption. The internal surface area of the small intestine is increased by millions of tiny folds called **villi** (singular, villus), which are small finger-like projections (Figure 4.3.9). The specialised epithelial cells (enterocytes) on the surface of the villi are covered with **microvilli**, further increasing the surface area for absorption (Figure 4.3.9).

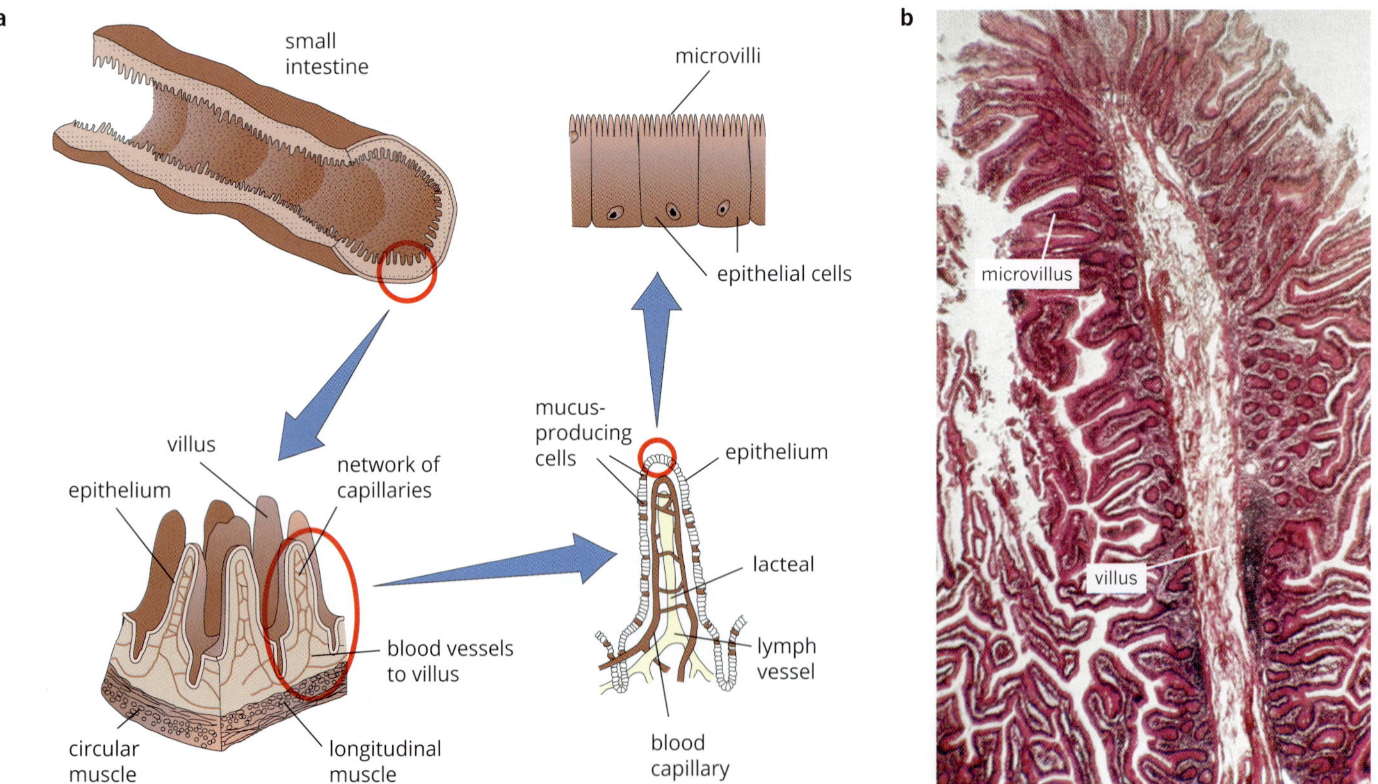

FIGURE 4.3.9 (a) The internal surface of the human small intestine, showing the villi and microvilli (b) Cross-section of a villus (plural villi) from the small intestine

The epithelial lining in the small intestine is only one cell thick, allowing a rapid transfer of nutrients to the many blood and lymphatic vessels beneath the surface, which transport nutrients away to the body tissues. Nutrients pass through the lining of the small intestine by facilitated diffusion or active transport, along or against the concentration gradient.

Lipid-soluble molecules, which are the products of fat digestion (fatty acids and glycerol), diffuse easily through the membranes of the epithelial cells along a concentration gradient. They then reassemble into fats before passing into the lacteals. Lacteals are capillaries of the lymphatic system near the intestine and have a milky appearance because of their high fat content after a fatty meal. Lipid-soluble vitamins also pass through the intestinal epithelium by passive diffusion.

Water-soluble molecules, including amino acids, simple sugars (monosaccharides such as glucose), and water-soluble vitamins and minerals pass through the membranes of the epithelial cells by active transport and facilitated diffusion. This can occur down or against a concentration gradient, ensuring that these essential nutrients are absorbed quickly.

Most of the water (90–95%) that enters the small intestine is also absorbed. This absorption is passive. Water diffuses across the lining of the intestine osmotically as the products of digestion are absorbed.

Blood leaving the intestine passes first into the liver through the hepatic portal vein, where absorbed nutrients can be removed and stored in the liver, before passing into the general venous circulation.

Herbivores utilise cellulose

Cellulose is the main component of plant cell walls, but its molecules are too large to be absorbed without digestion. Although many species of animals are herbivores, only a few can make the enzyme cellulase that is needed to digest cellulose. To get around this problem, herbivores have a symbiotic partnership, called mutualism, with bacteria that can produce cellulase. The bacteria live in the gut of the animal. They receive shelter and free nutrients for themselves, and in return convert cellulose into simpler molecules that can be absorbed by the gut. The bacteria also supply important vitamins such as the B group and vitamin K.

The environment inside the gut is warm and wet but there is little or no oxygen, so the breakdown of cellulose must occur anaerobically by **fermentation**. Because of this, the part of the intestine in which the breakdown of cellulose occurs is sometimes called a fermentation chamber.

In herbivorous mammals, fermentation takes place in different parts of the intestine in different species, with varying degrees of efficiency. Generally, herbivorous mammals belong to either of two groups—hindgut or foregut fermenters.

BIOFILE

Good bacteria in the gut

Bacteria play a vital role in our digestive system. Without them, our bodies would not function correctly. The key to optimal digestive function is maintaining digestive balance. This means that beneficial bacteria outnumber potentially harmful bacteria.

Probiotics are live, beneficial bacteria that are added in high numbers to food and drink products to improve the balance of good bacteria within the digestive system. Probiotics are available in different forms, including fermented milk drinks, yoghurts, capsules and powders. Most bacteria used in food production are lactic acid bacteria, such as species of *Lactobacillus* and *Streptococcus*.

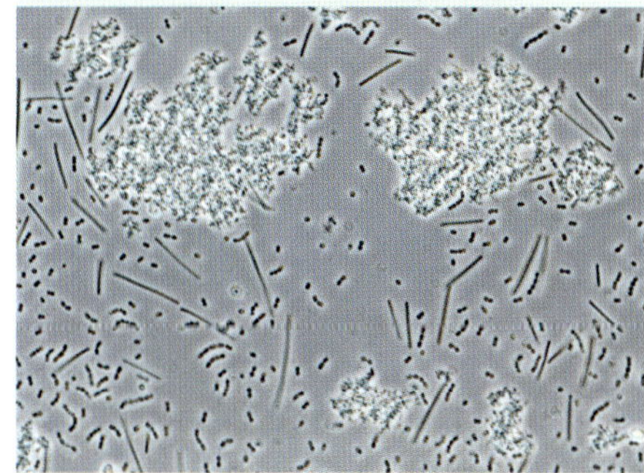

Phase contrast light micrograph (LM) of lactic acid bacteria—*Lactobacillus* and *Streptococcus*

Hindgut fermenters

In hindgut fermenters, fermentation occurs in the **caecum** (an enlarged pouch where the small and large intestines join) (e.g. the rabbit), or the first part of the large intestine (e.g. the wombat, Figure 4.3.10a), or both (e.g. the koala, Figure 4.3.10b). Both of these are located after the small intestine, which is the region where most absorption takes place. This arrangement limits the advantage obtained from the symbiotic relationship, because the products of their digestion are not completely absorbed.

Horses are hindgut fermenters, and the relative inefficiency of their system can be seen by the fact that horse faeces contain large amounts of undigested plant material. Some hindgut fermenters, such as possums and rabbits, overcome this problem by producing two types of faeces. One of these comes directly from the caecum at night and is reingested so that it can go through the intestine again. This means that the vitamins and products of cellulose digestion from the bacteria are available for absorption from the small intestine.

Foregut fermenters

In the foregut fermenters (Figure 4.3.10c), the fermentation chamber is located before the stomach. In ruminants such as cattle and sheep, it is called the rumen. Food can be regurgitated back into the mouth for further physical breakdown (rumination), then returned to the rumen for continued chemical breakdown by bacteria. This regurgitated food is called cud.

Foregut fermentation has the obvious advantage that the products of digestion by microorganisms are available for absorption along the entire length of the small intestine. Kangaroos and wallabies are the only marsupial foregut fermenters.

Digestion in the rumen has some drawbacks. The complete digestion of plant material in the rumen by microorganisms can take a long time—hours or even days, with constant regurgitation and chewing of the cud. If the quality of food is very low (that is, mostly cellulose and not much fresh, young plant growth) an animal may be starved of food that is digested enough for absorption, even though the animal has a very full rumen.

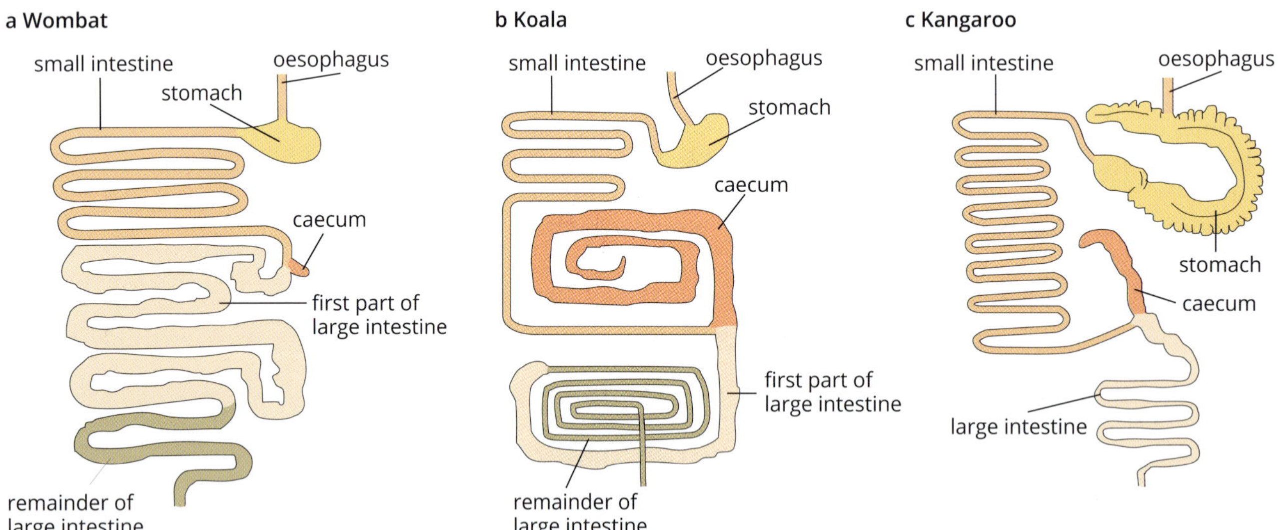

FIGURE 4.3.10 (a) Wombats, (b) koalas and (c) kangaroos are herbivores and use symbiotic bacteria for the digestion of cellulose. Wombats and koalas are hindgut fermenters whereas kangaroos are foregut fermenters.

Virtual dissections

Over the course of history, animal dissection has been an important way for humans to investigate the complexity of organisms. Early investigations that used animal dissections advanced our understanding of the anatomy and physiology of living things and made progress in medicine possible.

Animals are still an important part of scientific research today. However, the use of animals in research is not always necessary. With today's technology, virtual dissections can be used to investigate the internal workings of organisms such as frogs, pigeons and rats, without using a dead specimen (see figure).

The 3Rs of animal welfare—replacement, reduction and refinement—should be used as a guiding principle when using animals in research. Virtual dissections can reduce the number of animals used for dissections and could potentially replace real animal dissections in school and university laboratories.

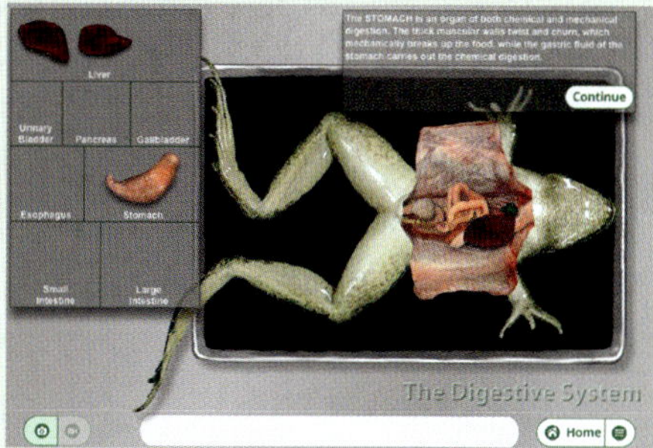

Virtual dissection of frogs can reduce the need for real animals in educational laboratories.

HORMONE SIGNALLING: THE ENDOCRINE SYSTEM

Multicellular organisms are complex in structure and require communication systems to ensure that their tissues, organs and systems are regulated and coordinated. Many of these responses involve negative feedback mechanisms through which a stable internal environment (homeostasis) is maintained. You will learn more about homeostasis in Chapter 5. Growth, metabolism and cellular regulation in animals rely on chemical messenger systems to ensure survival.

Hormones are chemical messages that are used as **signalling molecules** to have relatively long-lasting effects on cells within tissues. In vertebrates, the endocrine system is made up of many **glands** and organs within the body that, along with some specialised tissues, synthesise and secrete hormones into the bloodstream (or in some cases the lymphatic system). The main glands and organs of the human endocrine system are shown in Figure 4.3.11.

Once hormones are secreted into the bloodstream, they are transported to the site where they are needed. Hormones generally affect specific organs, and often only one type of cell within the tissue of that organ, known as the **target cell**. However, there are a few hormones that are able to affect most cells within the body.

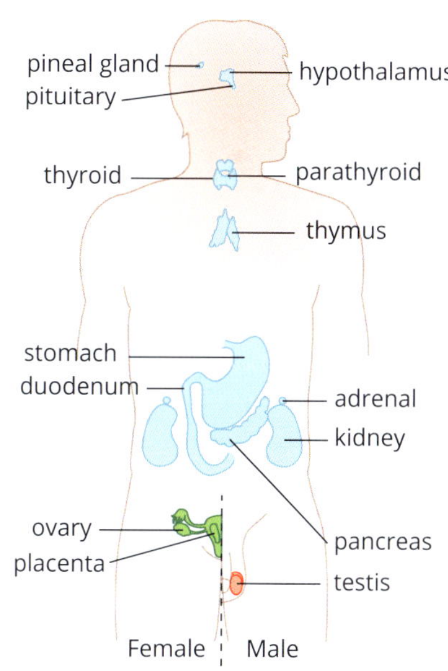

FIGURE 4.3.11 The organs and glands of the human endocrine system

Animal hormones

Hormones help regulate the rates of chemical reactions within the cell (metabolism), the transport of substances in and out of the cell, the production and secretion of other hormones, and the growth and reproduction of cells. Hormones are highly specific in their action. In order for a hormone to be able to have an effect on a target cell, the cell must have a specific **receptor** to that hormone. The hormone binds with the receptor to result in cellular change (Figure 4.3.12). For example, a sudden shock causes the release of the hormone known as adrenaline from the adrenal gland. Only those cells that have adrenaline receptors in their plasma membranes, such as muscle cells of the heart and blood vessels, can respond to the adrenaline circulating in the blood.

Hormones can exert their effects on a target cell by either directly passing through the plasma membrane to reach an intracellular receptor or by interacting with a receptor found on the surface of the plasma membrane.

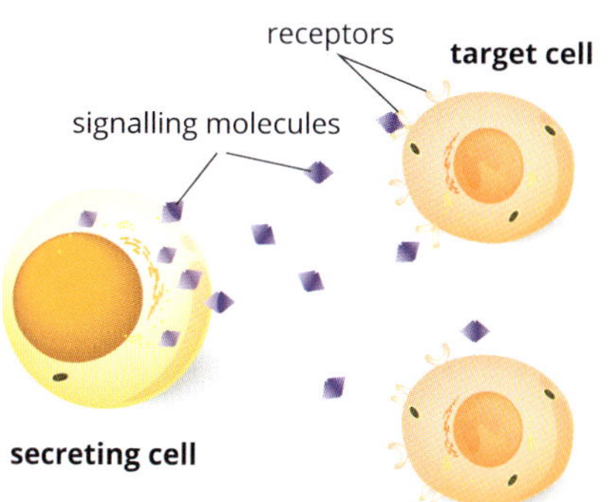

FIGURE 4.3.12 Each type of signalling molecule (hormone) is designated for certain cells (target cells). Receptors are specific and will only bind to a particular signalling molecule (hormone).

Types of hormones

Hormones can be broadly grouped into three main classes:

- Lipid hormones are a class of hydrophobic signalling molecules derived from fatty acids (eicosanoids) or cholesterol (steroids). Eicosanoids include prostaglandins, which are involved in cell growth, fever and inflammation. Steroid hormones help to regulate metabolism, salt and water balance, inflammation and sexual function. Examples of steroid hormones include testosterone, oestrogen and cortisol.
- Peptide and protein hormones are a class of hydrophilic signalling molecules. An example of a peptide hormone is insulin and an example of a protein hormone is growth hormone.
- Amino acid–derived hormones are a class of small signalling molecules derived from the amino acids tyrosine and tryptophan. They can be further divided into catecholamines and thyroid hormones. Thyroid hormones such as thyroxine are hydrophobic. Catecholamines are hydrophilic; examples include adrenaline and dopamine. (Dopamine acts as both a neurotransmitter and a hormone and is secreted by dopaminergic cells within the hypothalamus.)

A single hormone can trigger different responses in multiple target cells at the same time. Adrenaline (produced by specialised zona fasciculata cells in the adrenal cortex) targets cardiac muscle cells, vascular smooth muscle cells, and the various glands and organs of the digestive system. An increase of adrenaline in the bloodstream will result in an increase in heart rate and blood pressure and will simultaneously decrease digestive functions, preparing for a 'fight or flight' response.

Some common mammalian hormones, their sources, target tissues and functions are listed in Table 4.3.1.

TABLE 4.3.1 Common mammalian hormones, their sources, target tissues and functions

Gland (source)	Hormone(s)	Hormone class	Hydrophobic or hydrophilic	Target cell location	Function
adrenal cortex	glucocorticoids	steroid	hydrophobic	many cell types	regulate glucose metabolism and stimulate fat breakdown
	mineralocorticoids	steroid	hydrophobic	kidney tubule cells	regulate reabsorption of salts
hypothalamus	dopamine	amino acid–derived	hydrophilic	anterior pituitary	inhibits release of prolactin
	growth hormone releasing hormone (GHRH)	peptide	hydrophilic	somatotroph cells in the pituitary gland	stimulates release of growth hormone
anterior pituitary	adrenocorticotropic hormone (ACTH)	peptide	hydrophilic	adrenal cortex	promotes release of adrenal cortex hormones
	growth hormone (GH)	protein	hydrophilic	bone muscle	promotes protein synthesis and growth
	follicle stimulating hormone (FSH)	protein	hydrophilic	ovaries	promotes development of follicle and secretion of oestrogen
	luteinising hormone (LH)	protein	hydrophilic	ovaries	promotes ovulation, development of corpus luteum and secretion of progesterone
	prolactin	protein	hydrophilic	mammary glands	stimulates milk synthesis and secretion
	thyroid stimulating hormone (TSH)	protein	hydrophilic	thyroid	promotes production and release of thyroxine
pancreas	insulin	peptide	hydrophilic	many cell types	regulates blood glucose levels
thyroid	thyroxine	amino acid–derived	hydrophobic	many cell types	regulates cellular metabolic rate

Cell sensitivity and response

The sensitivity of a cell to a specific hormone depends directly upon the number of receptors that cell has for that particular hormone. The more receptors for a particular hormone that a cell has, the greater degree of sensitivity that cell has to the hormone.

The processes involved in a cell detecting and responding to a signalling molecule (like a hormone) are together known as **signal transduction**. The general characteristics of signal transduction in hormones depend on whether the hormone is hydrophobic or hydrophilic.

Signal transduction can be considered in terms of a stimulus–response model (Figure 4.3.13).

The **stimulus–response model** is a three-step process:

1 reception—the detection of the hormone by a receptor
2 transduction—the relay of the hormones signal into the cell
3 cellular response—the activation of a cellular activity or process.

Reception involves the detection of a hormone by a cell. The receptor that detects a hormone can be located on the surface of the plasma membrane or in the cytoplasm or nucleus of the cell. The position of the receptor depends on whether the hormone is hydrophobic (Figure 4.3.14) or hydrophilic (Figure 4.3.15). Receptors are specific and will only bind to a particular hormone.

Transduction involves converting the signal into a form that can be relayed to reach its final destination within the cell and bring about a cellular response.

Following transduction, a response is initiated. **Cellular responses** include any cellular activity, such as the activation of genes, the activation of enzymes or the secretion of other molecules by the cell. Responses can occur in the:

- nucleus
- cytoplasm
- plasma membrane.

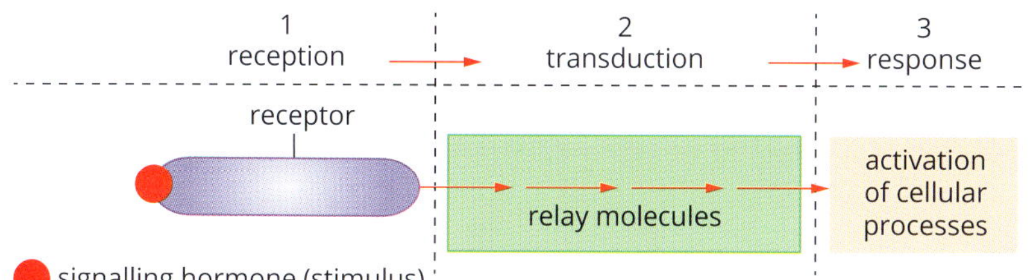

FIGURE 4.3.13 The stimulus–response model applied to the cell in terms of signal transduction

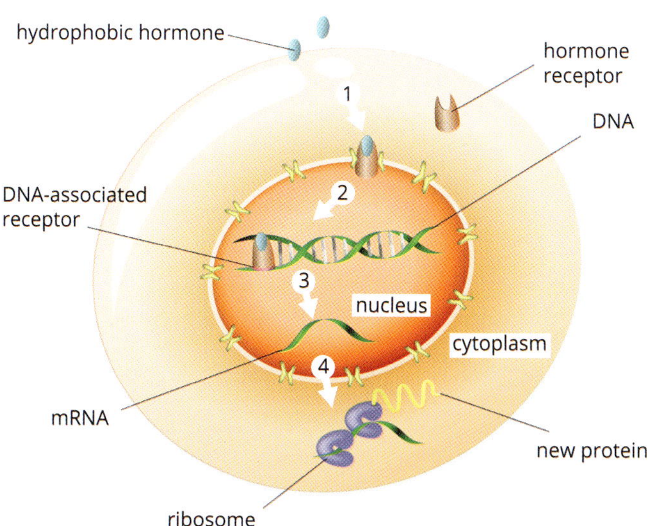

FIGURE 4.3.14 An example of signal transduction events for a hydrophobic hormone

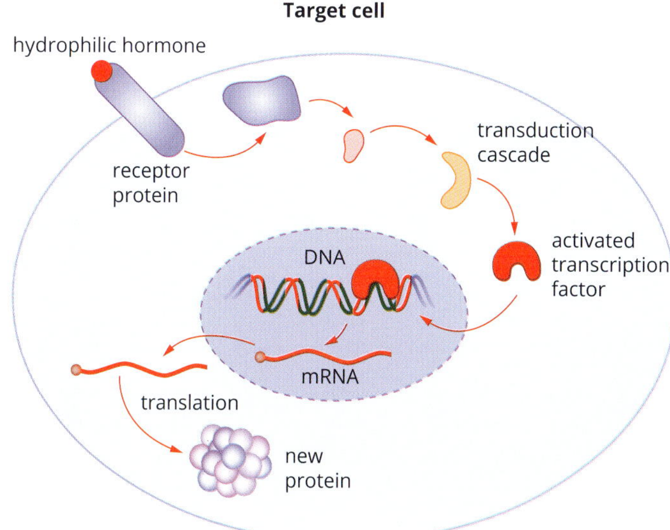

FIGURE 4.3.15 An example of signal transduction events for a hydrophilic hormone

CASE STUDY **ANALYSIS**

The role of hormones in bone repair

You are probably aware that if you break your arm, the broken bone becomes thinner and weaker during the period in which it is in plaster and out of action (Figure 4.3.16). For people suffering from osteoporosis, weight-bearing exercises are recommended to build up bone strength. Both of these situations relate to the ability of bone cells to detect and respond to physical stress. Physical stress on bones causes them to become thicker. Removal of stress causes bone material to be resorbed into the circulation.

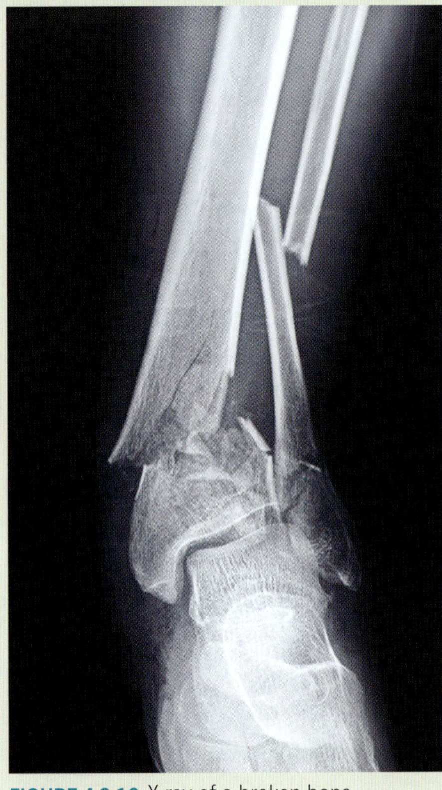

FIGURE 4.3.16 X-ray of a broken bone

Bones are not permanent structures—they are dynamic. They are a reservoir of calcium used to maintain blood calcium levels of approximately 2.1–2.5 mmol/L. Parathyroid hormone and calcitonin are hormones involved in the deposition and resorption of calcium salts in bone.

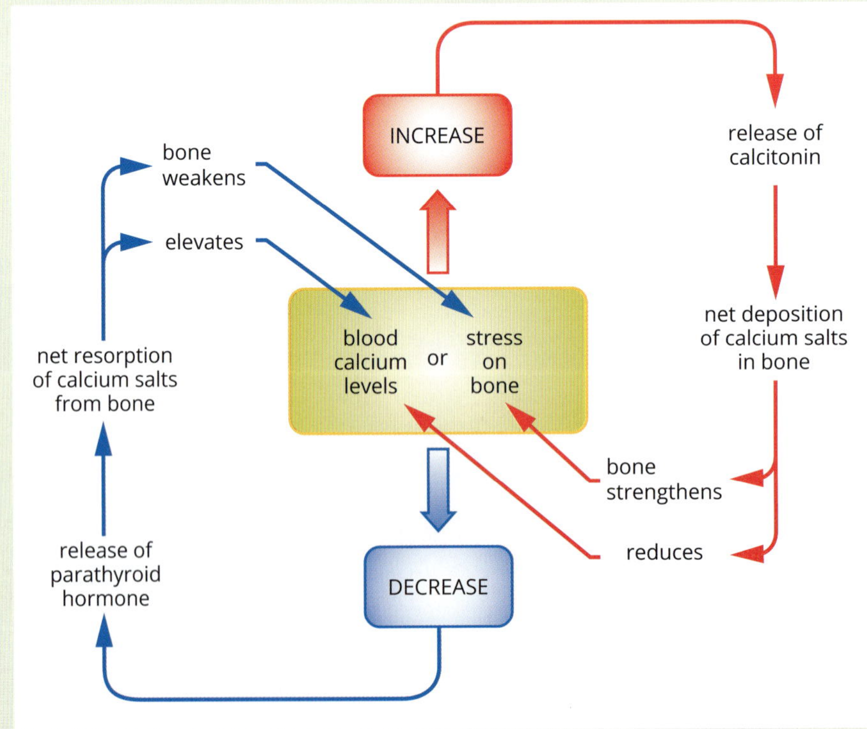

FIGURE 4.3.17 Bones respond to increased physical stress by becoming stronger. If the stress on bones decreases they become weaker.

The hormone calcitonin is produced by parafollicular cells of the thyroid gland. Calcitonin undergoes the stimulus–response model sequence in order to exert its effects. It binds with a receptor found on the membrane of the target cell. This activates a transduction cascade to occur, and the final response is an increase in reabsorption of calcium salts from the blood into the bone during new bone formation.

When bone grafts are necessary to repair broken or fractured bone tissue, a balance between parathyroid hormone and calcitonin is able to assist in the regeneration of the new bone (Figure 4.3.17). A piece of bone from the fibula in the leg can be grafted into the spinal column. It will soon be reshaped by physical stress to suit its new location and the work it has to do.

Analysis

1 Is calcitonin hormone a hydrophilic or hydrophobic hormone? Explain your choice.

2 The graph demonstrates the blood calcium range of an individual over 12 months.

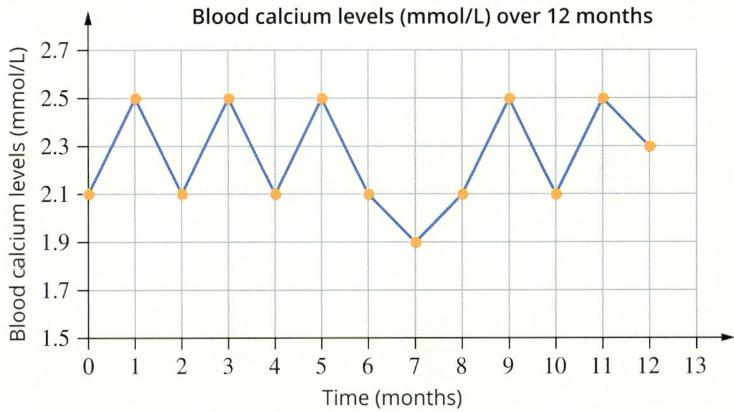

a At one point in the testing the individual had a calcium deficiency. When did this occur?

b Based on this individual's data, determine the normal range of blood calcium levels in a human.

3 What is the cell's response to an increase in calcitonin hormone?

4 Many people in Victoria become vitamin D deficient through the winter months because of lack of sun exposure. In the absence of vitamin D, dietary calcium is not absorbed efficiently from the digestive tract.

a At what blood calcium levels would this deficiency be expected to be seen?

b Infer the likely effect that this will have on the relative concentrations of parathyroid hormone in the blood.

Endocrine glands

Animals usually have specialised cells for producing hormones, such as the alpha and beta cells in the pancreas that are responsible for the production of the hormones glucagon and insulin. These cells are often clustered into tissues and then into organs known as endocrine glands. Initially, the term 'hormone' was restricted to the products of endocrine glands. However, it is now evident that hormones are secreted by a wide variety of tissues and that they are able to reach their site of action by simple diffusion. Mammals have many major hormones and many more molecules that act as minor hormones.

BIOFILE

Biorhythms and the pineal gland

All vertebrates have a pineal gland that detects and responds to light. In mammals the gland is deep within the brain and does not sense light directly, but instead receives messages from the eyes about the brightness of light.

The pineal gland produces the hormone melatonin. When light levels are high, the gland stops producing the hormone, decreasing levels of melatonin during the day and increasing levels at night. This allows animals to sense the time and seasons internally. The pineal gland therefore functions like an internal 'biological clock'.

Travel from one time zone to another disrupts the internal rhythms, producing the symptoms of jet lag, which include disorientation, sleepiness and wakefulness. Similar problems are experienced by shift-workers. Melatonin or bright light can synchronise the internal clock, which has led to the use of melatonin as a 'jet-lag pill'. Alternatively, spending a day out in the bright sunlight will also help to reset your biological clock.

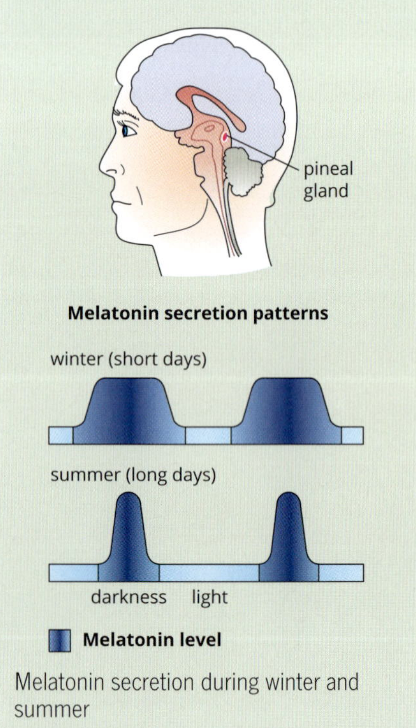

Melatonin secretion during winter and summer

Pituitary gland

The **pituitary gland** is located at the base of the brain, just above the roof of the mouth. In humans, it is about 1 cm in diameter and weighs about 0.5 g. Despite its relatively small size, the pituitary gland is often called the 'master gland' of the endocrine system because it produces many of the body's hormones, a number of which help regulate the production of other hormones around the body. Hormones secreted by the pituitary gland are also involved in a range of cellular processes including growth, reproduction, lactation, kidney function, skin pigmentation and regulation of the activity of the thyroid and the adrenal glands.

Research has shown that the pituitary gland consists of two distinct parts, called the anterior and posterior pituitary gland. There is a range of specialised cells found within the pituitary gland that are capable of producing hormones. Some of these cells include thyrotropes that secrete thyroid stimulating hormone, gonadotropes that secrete follicle stimulating hormone and luteinising hormone, and somatotropes that secrete growth hormone. All of these specialised cells contain specific cellular structures and organelles to support the production and export of hormones, such as rough endoplasmic reticulum and Golgi apparatus. As these cells secrete hormones via exocytosis, a process that requires ATP energy, it would be expected that there would be increased demands of the cells' mitochondria to undergo cellular respiration.

Another endocrine gland, the **hypothalamus**, is located above the pituitary gland, and connects directly to it (Figure 4.3.18). The hypothalamus is responsible for detecting internal stimuli from all over the body and determining whether or not optimal conditions are being maintained. These internal stimuli trigger the production of releasing hormones from the hypothalamus. Releasing hormones are those that control and regulate specific hormone production in the pituitary gland. The combined functions of the hypothalamus and pituitary gland are vital for homeostasis.

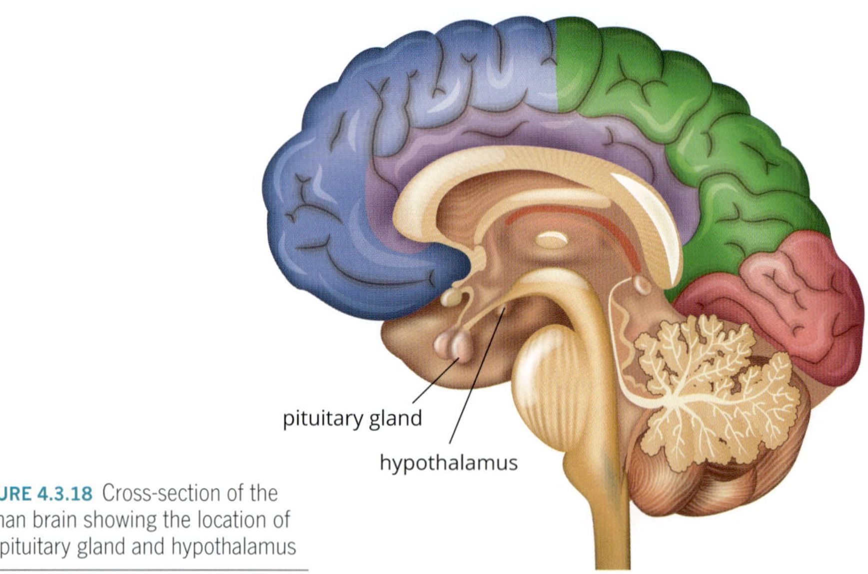

FIGURE 4.3.18 Cross-section of the human brain showing the location of the pituitary gland and hypothalamus

REMOVING WASTES: THE EXCRETORY SYSTEM

As cells function, they produce substances that are no longer useful to them. The accumulation of these waste substances, such as carbon dioxide from cellular respiration and **nitrogenous wastes** from protein breakdown, can prevent cells from functioning properly.

In mammals, the function of the kidneys is to excrete nitrogenous wastes. Excretion usually involves the loss of water and is therefore closely linked to water balance in terrestrial animals. This system works closely with the circulatory system, filtering waste products from the bloodstream and collecting them in **urine**.

For heterotrophs, it is sometimes inevitable that toxic substances are absorbed from the food they eat; these toxic substances must also be excreted. Excretion is the removal of substances that once formed part of the body of the organism. (This is different from egestion, which is the removal of undigested food from the gut in faeces.)

The internal environment of animals is extracellular fluid, which is separate from the external environment and has a highly regulated composition. Salts form ions in solution. The concentrations of certain ions in cells are held within narrow limits. Some of these ions are also important for regulating the pH of body fluids, which must be at a suitable pH for enzymes and other molecules to function efficiently.

In animals the removal of wastes and toxic substances, and the control of pH, ion concentrations and water balance, are carried out mainly by excretory organs. These processes vary with the activity of the animal and the external conditions.

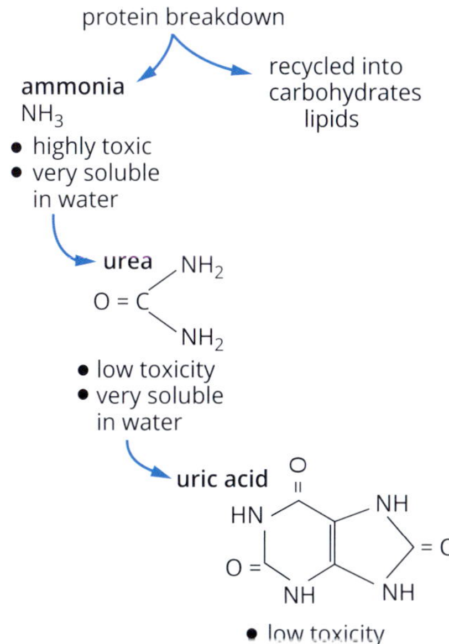

FIGURE 4.3.19 Three important nitrogenous wastes produced from the breakdown of proteins in animals: ammonia, urea and uric acid. Mammals excrete their nitrogenous wastes in the form of urea.

The nature of wastes

During normal activity, animal cells break down and replace carbohydrates, lipids and proteins, producing waste products that usually cannot be used by the body.

Carbon dioxide

When carbohydrates or lipids are broken down during cellular respiration to release energy, carbon dioxide and water are produced. These are released into the surrounding environment across membranes in the respiratory system, which in mammals are in the lungs. Water produced during cellular respiration is incorporated into body fluids, and excess water is expelled from the body.

Nitrogenous wastes

Protein consists of amino acids, which contain nitrogen. When amino acids are broken down, the nitrogenous parts are split off and the remainder of the molecule is converted into carbohydrates or lipids, which can be used for energy. The remaining nitrogenous wastes must be removed from the cell, because they can become toxic.

The first nitrogenous waste to be formed from the breakdown of protein is ammonia. Ammonia can be converted into **urea** or uric acid (Figure 4.3.19), but this process requires energy. Neither urea nor uric acid are of any further use to most animals. A few animals, such as sharks, are adapted to maintain high levels of nitrogenous wastes, particularly urea, within their body to aid in water balance.

Excretory mechanisms in mammals

Kidneys

The kidneys of all vertebrates, from fishes to mammals, function by filtering blood, then reabsorbing useful substances and secreting unwanted ones. Blood is filtered through blood vessel walls to form a primary filtrate that has the same composition as plasma, except that the large proteins have been filtered out. Most of the useful substances in the primary filtrate are reabsorbed as it passes through the kidney tubule. Some unwanted substances may be secreted into the fluid in the tubule before it passes out of the kidney to the bladder. These processes regulate the concentration of different salts in the blood, including those salts that are responsible for maintaining the pH of body fluids within closely controlled limits.

Mammals are able to conserve water by producing urine that is more concentrated than body fluids. The ability to produce concentrated urine is related in some mammals to the degree of water stress experienced in their normal environments. Desert-adapted mammals are able to excrete highly concentrated urine (Figure 4.3.20).

FIGURE 4.3.20 Desert-dwelling animals such as the bilby (*Macrotis lagurus*) can produce highly concentrated urine, which minimises water loss.

> **BIOFILE**
>
> **Urea**
>
> Urea is a larger molecule than ammonia and also contains carbon and oxygen. It is much less toxic than ammonia and is highly soluble in water, but converting ammonia into urea requires energy. So although urea is less toxic, an animal spends more energy excreting urea instead of ammonia.
>
> Land vertebrates need to conserve water. Mammals excrete their nitrogenous waste largely in the form of urea, but their kidneys are capable of regulating and minimising water loss. This strategy is a successful adaptation to life on land.

Liver

The liver performs many different functions and has a central role in the maintenance of a stable internal environment. In addition, it is responsible for preparing various substances for excretion. It detoxifies a variety of harmful chemicals such as alcohol and some drugs. It is also responsible for breaking down amino acids to release ammonia, which it then converts largely into urea. The waste products from these processes travel in the bloodstream to the kidneys for excretion.

The liver also destroys worn-out red blood cells, producing bile pigments from the breakdown of haemoglobin. Bile pigments, along with bile salts, which emulsify fats as part of digestion, are stored in the gall bladder before they are released into the lumen of the intestine. Bile pigments are one of the few substances excreted into the gut.

The mammalian kidney

Mammals have two kidneys at the back of the abdominal cavity. Blood flow to the kidneys is always kept high, because they are so important in maintaining the stability of the internal environment. Although kidneys are only about 1% of body tissue, they receive approximately 25% of the body's blood flow.

Blood enters the kidney from the aorta through the renal artery, and leaves through the renal vein. Blood vessels branch throughout the kidney in a complex fashion (Figure 4.3.21). Urine, formed in the kidneys, drains via the **ureters** into the **bladder**, for storage, until an appropriate time for release through the **urethra** (Figure 4.3.21).

The functions of the mammalian kidney are carried out by **nephrons**, which are the functional units of the kidney. There are approximately one million nephrons in a human kidney, and their combined function carries out the work of the kidney. A nephron is composed of a **Bowman's capsule** surrounding a glomerulus, and a tubular region consisting of the proximal convoluted tubule, **loop of Henle** and distal convoluted tubule, which leads into a collecting tubule (Figure 4.3.21). The formation of urine involves passive **filtration**, selective **reabsorption** and **secretion**, and the passive removal of water.

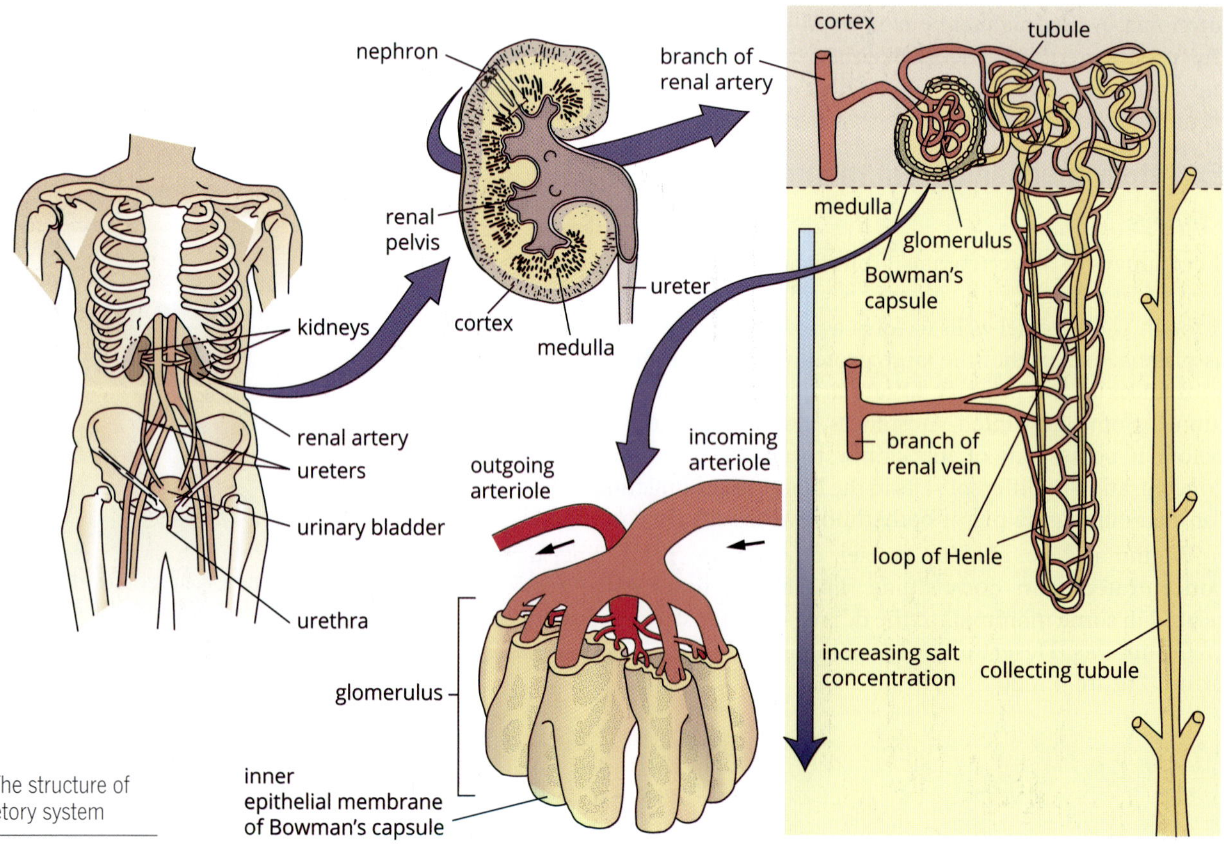

FIGURE 4.3.21 The structure of the human excretory system

The nephron is very closely associated with blood vessels, particularly the **glomerulus** (plural glomeruli), which is a clump of looping capillaries that have a thin layer of flattened endothelial cells, embedded in the Bowman's capsule, and networks of capillaries wrapped around the remainder of the tubule.

There are two distinct regions in the kidney: the outer cortex and the inner medulla. Glomeruli are located in the cortex (Figure 4.3.21). There can be as many as 50 capillaries with very thin walls within each glomerulus. Their narrow diameter, large surface area and close contact with the membranes of the Bowman's capsule ensure a fast and voluminous filtering process. The blood vessel walls contain numerous pores (called fenestrae) 50–100 nm in diameter. These pores allow for the free filtration of fluid, plasma solutes and small protein molecules. However, they are not large enough that red blood cells can pass through. The outside surface of the glomerular capillaries is lined with **podocyte cells**, a group of specialised cells that form tissue that control the filtration of protein molecules. Podocyte cells form a tissue layer and have specialised branching structures that allow the cells to wrap around capillaries to increase the surface area available for exchange of waste materials. These very complex structures and processes are vital to maintaining homeostasis and healthy functioning of the body. If the walls of the glomerular capillaries are damaged in a diabetic patient, this can allow larger protein molecules to pass into the urine, a condition called nephropathy.

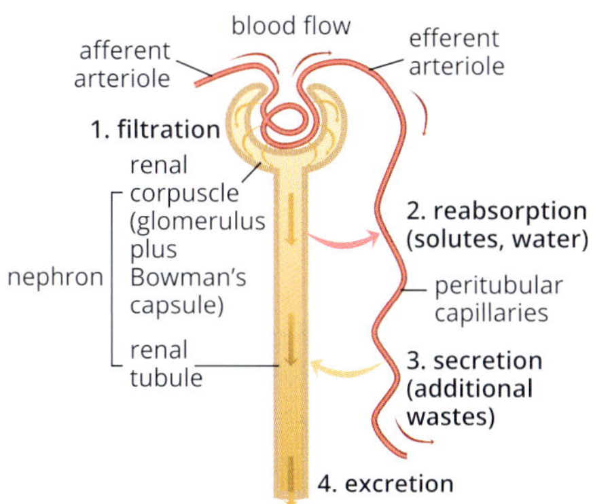

FIGURE 4.3.22 The basic steps in urine formation and filtration

Filtration

Filtration occurs across the glomerulus into the Bowman's capsule. The high pressure of blood in the glomerular blood vessels forces fluid through the walls of glomerular capillaries and into the Bowman's capsule (Figure 4.3.21).

Only small molecules and water can pass through the wall membranes; blood cells and large blood proteins remain behind in the glomerular capillaries. This primary filtrate has the same composition as blood plasma, without large proteins.

If red blood cells or large proteins are found in urine, this indicates that the normal filtration mechanism has broken down and blood is leaking from the glomerulus into the Bowman's capsule. This may occur as a result of damaged glomerular blood vessels, or very high blood pressure.

Reabsorption

Approximately 99% of the primary filtrate—including salts, glucose, amino acids and water, but only half or less of the urea—undergoes reabsorption along the length of the nephron (Figure 4.3.22). Virtually all amino acids and glucose are reabsorbed in the convoluted tubules by active transport against a concentration gradient. The presence of glucose or amino acids in urine therefore indicates a possible kidney malfunction. Specific salts, particularly sodium chloride, are also actively reabsorbed. These active processes consume a lot of energy.

Water is reabsorbed from the urine passively, along an osmotic gradient. The mechanism by which the kidney is able to produce concentrated urine involves the loop of Henle. A large amount of sodium chloride pumped out of the loop of Henle is retained in the medullary region of the kidney, producing a very high salt concentration. The osmotic concentration within the kidney therefore increases considerably from the outer cortex to the medulla. When the urine finally passes down the collecting tubules towards the ureter, it passes through this region of high salt concentration (Figure 4.3.23). Because the collecting tubule is permeable to water, but not to salt, water passes from the collecting tubule back into the kidney and into blood vessels. As a result, the urine becomes concentrated. **Antidiuretic hormone (ADH)** from the pituitary gland increases the permeability of the collecting tubule to water, increasing reabsorption of water and causing urine to become concentrated.

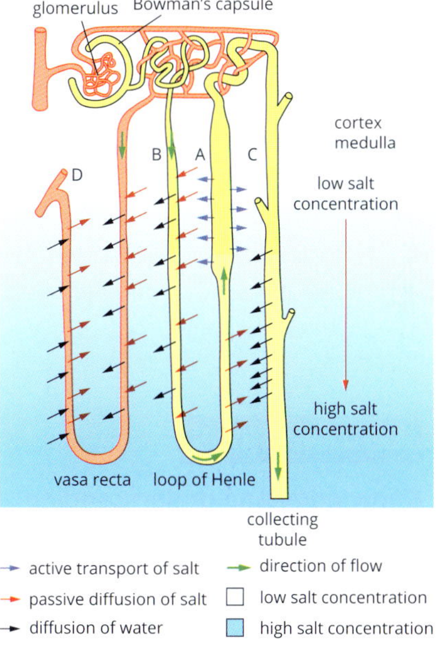

FIGURE 4.3.23 Regulation of water and salts in the medulla through active transport and diffusion

Secretion

Secretion is the active removal (excretion) of particular substances by the cells of the tubule wall. Ammonium, potassium and hydrogen ions are actively secreted into the convoluted parts of the tubules. Various dyes and drugs such as penicillin and aspirin are also eliminated by tubular secretion. These substances are added to the filtrate as it passes through the nephron.

4.3 Review

SUMMARY

The digestive system

- Animals are heterotrophs. They must consume other organisms or their products to obtain organic molecules.
- The purpose of digestion is to rapidly break down organic food into molecules small enough to be able to pass through membranes and into cells.
- The internal surface of the small intestine is lined with villi that have specialised epithelial cells that are covered in microvilli, increasing the surface area available for absorption of nutrients.
- Chemical digestion involves breaking apart complex molecules into simple molecules by the action of enzymes (amylase, protease and lipase) that are secreted by specialised cells in the salivary glands, stomach and pancreas.
- Physical breakdown of large food into smaller pieces increases the surface area available for enzyme action and increases the efficiency of digestion.

The endocrine system

- Hormones are signalling molecules that regulate the growth or activity of specific target cells, as a result of interaction with specific receptors.
- Animal hormones:
  - are produced by specialised cells found in organs and glands of animals
  - can be hydrophilic or hydrophobic.
- The stimulus–response model involves three steps:
  - reception—the detection of a signalling molecule (the stimulus) by its specific receptor (including the physical binding of the signalling molecule to the receptor)
  - transduction—the transformation of the signal in terms of form, type of signalling molecule and passage into and out of a cell
  - response—the change in cellular activity as a result of the initial stimulus.
- Complex animals have endocrine glands that typically release hormones directly into the circulatory system.
- In vertebrates the pituitary gland has a pivotal role in overall endocrine regulation.

The excretory system

- Excretion is the removal of substances that once formed part of the body of the organism.
- In animals, removal of waste and toxic substances, and control of pH, ion concentrations and water balance, are carried out largely by excretory organs, such as the kidney.
- Proteins are broken down into carbohydrates or lipids, which can be used for energy, and nitrogenous wastes, which must be removed from the cell, because they can become toxic.
- The nephron is the functional unit of the mammalian kidney. A nephron consists of a Bowman's capsule (surrounding a glomerulus with specialised cells involved in filtration) leading into a tubular region (proximal convoluted tubule, loop of Henle and distal convoluted tubule) and then into the collecting tubule.
- The three main stages of urine formation are filtration, reabsorption and secretion.

KEY QUESTIONS

Knowledge and understanding

1 Name three types of specialised cells found in the digestive, endocrine and excretory systems, and outline their function.

2 Explain chemical digestion and give an example of chemical digestion occurring within the body.

3 Describe the feature of a target cell that makes it receptive to a particular hormone.

4 Explain how hormones that circulate throughout the blood can act only on a specific type of target cell.

5 Name the structural and functional unit of the kidney.

6 Explain the process of filtration.

7 The small intestine is a site of absorption.

a Describe the features of the small intestine that make it well suited to its absorptive role. Use diagrams to illustrate your answer.

b What is different about the absorption of the products of fat digestion compared with the absorption of other products?

Analysis

8 Evaluate (using one paragraph) the statement, 'The pituitary gland is the master gland of the body'.

9 Explain why it is important that the permeability to water of the collecting tubule of the mammalian kidney can be regulated.

Chapter review

KEY TERMS

amino acid
amylase
antidiuretic hormone (ADH)
autotroph
bile
bladder
Bowman's capsule
caecum
carbohydrate
cell differentiation
cellular response
cellulose
chemical digestion
companion cell
cuticle
digestion
digestive enzyme
digestive system
egestion
endocrine system
enterocyte
epiglottis
epithelium (adj. epithelial)
excretion
excretory system
extracellular digestion
fermentation
filtration
gall bladder
gastric chief cell
gland
glomerulus
goblet cell
guard cell
heterotroph
hormone
hypothalamus
kidney
large intestine
lignin
lipase
lipid
liver
loop of Henle
lower epidermis
meristem
mesophyll cell
microvillus (pl. microvilli)
mineral
mouth
multicellular organism
nephron
nitrogenous waste
oesophagus
organ
pancreas
pancreatic acinar cell
parenchyma cell
peristalsis
phloem
pituitary gland
podocyte cell
protease
protein
reabsorption
reception
receptor
root hair
sclerenchyma cell
secretion
sieve tube
signal transduction
signalling molecule
small intestine
specialised cell
stem cell
stimulus–response model
stoma (pl. stomata)
stomach
system
target cell
tissue
tracheid
transduction
unicellular organism
upper epidermis
urea
ureter
urethra
urine
vascular bundle
vascular plant
vascular tissue
villus (pl. villi)
vitamin
xylem
xylem vessel

REVIEW QUESTIONS

Knowledge and understanding

1 Unicellular organisms are often unable to perform multiple cellular functions simultaneously in order to survive. How do multicellular organisms overcome this?

2 Match each of the following specialised cells with its function.

epidermal cell	produces and secretes calcitonin hormone
guard cell	produces and secretes growth hormone
meristem cell	produces and secretes mucin to protect epithelial lining
gastric chief cell	prevents water loss and regulates gas exchange
goblet cell	secretes enzymes into the digestive system
parafollicular cell	barrier against environmental stressors
somatotrope cell	gives rise to specialised cells

3 Describe the process of cell differentiation and specialisation, mentioning at least three of the key terms from the list above.

4 a List five advantages of being multicellular.
b Referring to your answers to part **a**, explain how being multicellular provides each of these advantages compared to unicellular organisms.

5 Explain how the structure of specialised podocyte cells facilitates their function in the kidney.

6 What is the function of the pit structures in xylem tissue?

7 In autumn, the leaves of deciduous trees change colour and eventually fall. The change in colour is due to the movement of nutrients out of the leaves for storage. This involves:
A xylem and phloem
B only the xylem
C only the phloem
D diffusion

8 Describe each of the levels of organisation in multicellular organisms.

9 State the level of organisation that each of the following illustrations shows: tissue, organ, system or organism.

a 

b 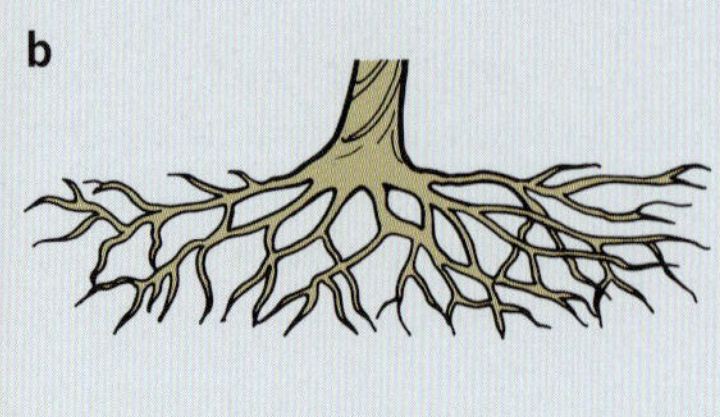

c 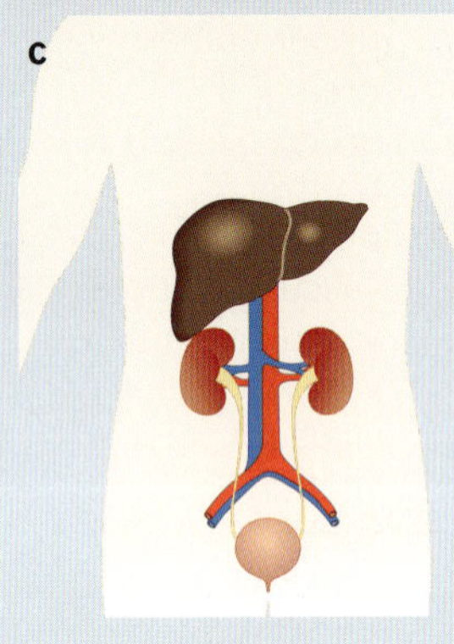

d 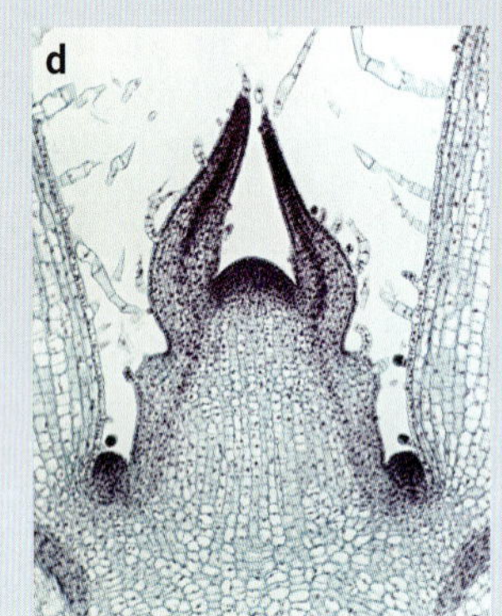

10 When the blood concentration of thyroid hormones increases above a certain threshold, neurons that secrete thyrotropin-releasing hormone (TRH) in the hypothalamus are inhibited and stop the secretion of TRH.

- **a** Which endocrine glands are affected by the secretion of TRH?
- **b** Name two other hormones that are secreted from the hypothalamus.

11 Which one of the following statements relating to fermentation in herbivores is true?

- **A** In foregut fermenters, cellulose is digested in the caecum.
- **B** Fermentation in the gut requires oxygen.
- **C** The rumen is located between the oesophagus and the stomach.
- **D** All native Australian mammals are hindgut fermenters.

12 What is the difference between foregut fermentation and hindgut fermentation?

13 Indicate which statements are true or false.

- **a** One hormone can affect every cell.
- **b** Hormones affect target cells.
- **c** Target cells contain receptors.
- **d** Receptors recognise hormones specific for them.
- **e** Receptors recognise groups of hormones that are specific for them (e.g. peptide hormones).
- **f** Receptors for steroid hormones are located in the cytoplasm and receptors for peptide hormones are located on the surface of the plasma membrane.

14 Describe the functions of the glomerulus, Bowman's capsule, proximal and distal convoluted tubule, loop of Henle and collecting tubule.

Application and analysis

15 Coeliac disease causes the destruction of the villi cells in the small intestine. Which one of the following is most likely to happen to people with coeliac disease?

- **A** damage in the oesophagus caused by increase in acid reflux
- **B** incomplete digestion of proteins
- **C** increased levels of glucose in blood
- **D** poor absorption of calcium

16 Specialised gonadotrope cells and specialised thyrotrope cells are both involved in the production of hormones. However, these cells are only able to produce specific types of hormones. How is this type of cell specialisation beneficial to humans?

17 Both gastric chief cells and enterocytes are found in the digestive system. Provide a comparison that explains how their cellular specialisation assists their function.

18 Figure A below shows a cross-section through the small intestine. Figure B shows a longitudinal section through a villus.

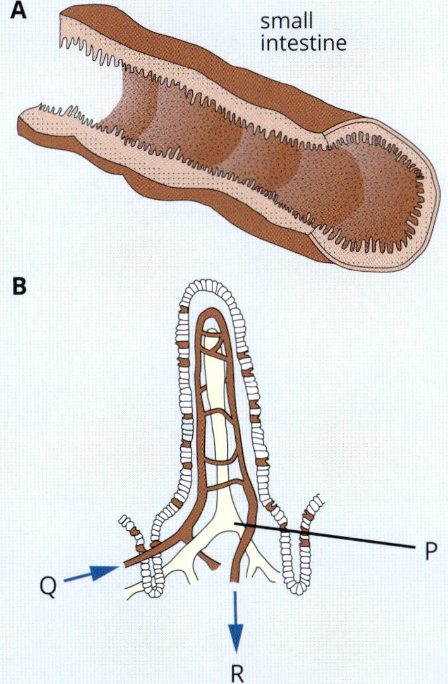

a Using Figures A and B, outline three ways in which the structure of the small intestine is related to its function of absorbing products of digestion.

b Referring to Figure B, identify structure P and state its function.

c The arrows in Figure B indicate the direction of blood flow. State how the composition of blood entering from Q would be different from blood leaving at R.

19 Cortisol is an important human hormone. It has a role in glucose regulation, immune system regulation and regulation of metabolic rate.

a Despite its role in many aspects of human physiology, not all cells respond to cortisol. Explain why.

b Receptors for cortisol are found in the cytoplasm of the cell. What does this indicate about the chemical nature of this hormone?

c Insulin is the hormone that stimulates the uptake of glucose by cells. Fat and muscle cells are generally particularly sensitive to insulin, but cortisol is known to limit their response to this hormone. Propose a reasonable intracellular mechanism of cortisol that could reduce the normal response by fat and muscle cells.

20 The table below shows the relative concentrations of urea, glucose, amino acids, salts and proteins in the primary filtrate and urine of mammals as a percentage of the concentration in blood plasma.

Substance	Primary filtrate (%)	Urine (%)
urea	100	700
glucose	100	0
amino acids	100	0
salts	100	200
proteins	0	0

a What is the explanation for each value, for both primary filtrate and urine?

b People with diabetes have difficulty removing glucose from their blood, and their urine can contain higher than normal concentrations of glucose. Propose a reason for this.

OA ✓✓

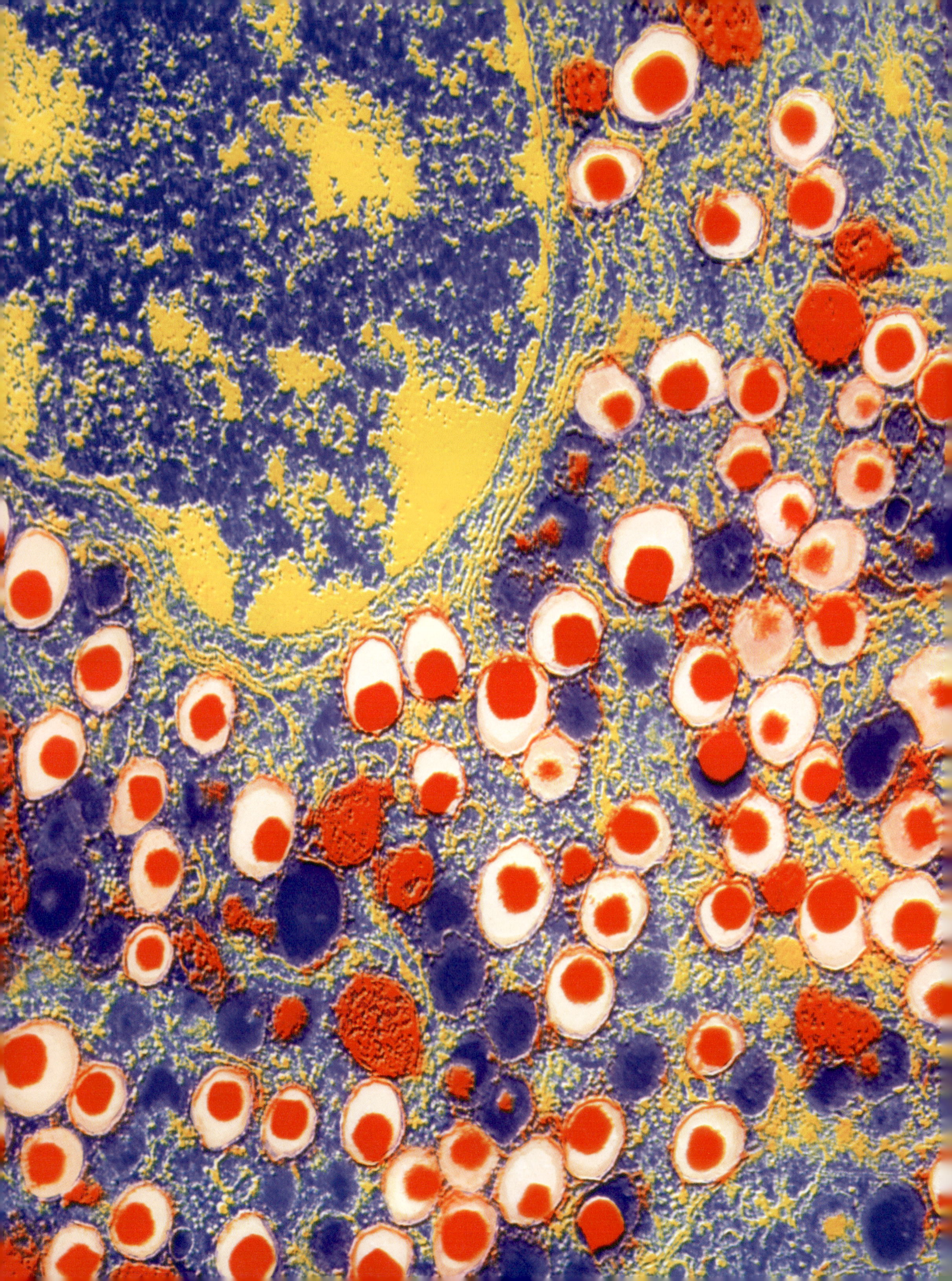

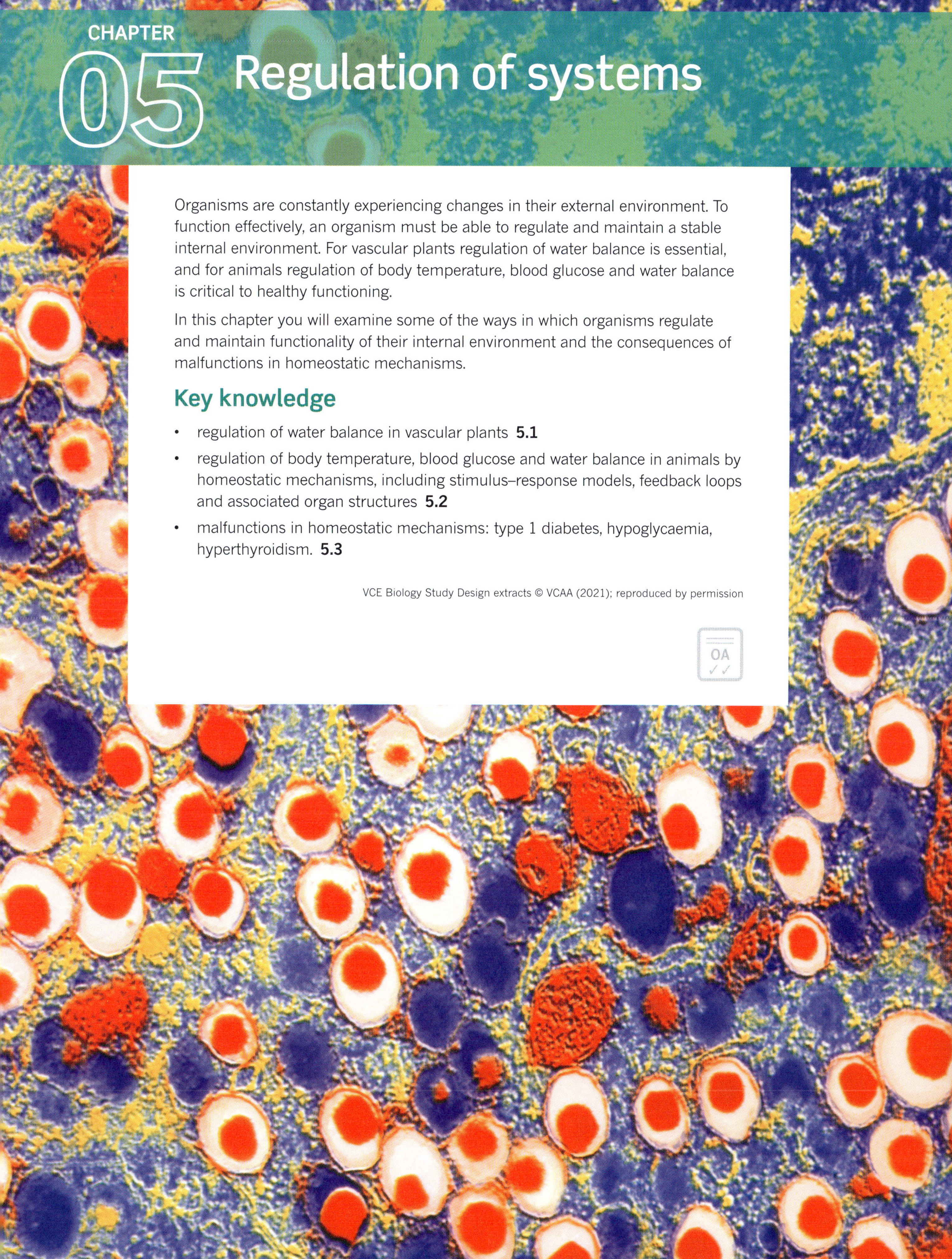

CHAPTER

05 Regulation of systems

Organisms are constantly experiencing changes in their external environment. To function effectively, an organism must be able to regulate and maintain a stable internal environment. For vascular plants regulation of water balance is essential, and for animals regulation of body temperature, blood glucose and water balance is critical to healthy functioning.

In this chapter you will examine some of the ways in which organisms regulate and maintain functionality of their internal environment and the consequences of malfunctions in homeostatic mechanisms.

Key knowledge

- regulation of water balance in vascular plants **5.1**
- regulation of body temperature, blood glucose and water balance in animals by homeostatic mechanisms, including stimulus–response models, feedback loops and associated organ structures **5.2**
- malfunctions in homeostatic mechanisms: type 1 diabetes, hypoglycaemia, hyperthyroidism. **5.3**

OA

5.1 Regulatory mechanisms in plants

Plants carry out two energy-transforming processes. Cellular respiration occurs throughout all of the cells in vascular plants, requiring oxygen and producing carbon dioxide. However, photosynthesis occurs primarily in cells in the leaves, where chloroplasts in the cells convert water and carbon dioxide, in the presence of sunlight, into glucose, water and oxygen. To undergo both of those processes there are specialised tissues and structures in place to exchange gases and move substances. You learnt about the specialised cells and tissues for water transport in vascular plants in Chapter 4. In this section, you will learn about how plants use these specialised structures to regulate water balance.

GAS EXCHANGE IN VASCULAR PLANTS

Most plants do not have specialised organs for gas exchange. Simple plants, such as mosses, have leaves that are small and extremely thin—only one cell thick—so each cell is in direct contact with the surrounding environment. Gases such as oxygen and carbon dioxide can easily diffuse directly between the air and the contents of each cell. In **vascular plants** (plants with conducting tissue), the exchange of oxygen and carbon dioxide in the leaves, stems and roots occurs by diffusion through special openings in the epidermis called **stomata** (singular stoma).

> Stomata in the leaves regulate the exchange of gases between a plant and its external environment.

Recall from Chapter 4 that the rate of movement of gases between plants and the atmosphere is regulated by the stomata, the main route through which gas exchange occurs. When the stomata are closed, the exchange of oxygen, carbon dioxide and water vapour between the plant and its environment virtually stops. Only small quantities of gases are able to pass directly through the epidermis and the overlying cuticle (the outer waxy layer).

Plant cells are loosely packed, allowing rapid diffusion of gases through intercellular spaces, which are filled with air (Figure 5.1.1). During gas exchange, oxygen and carbon dioxide diffuse from these air spaces through the water film covering the cells and into the cells along concentration gradients. Diffusion also occurs against the concentration gradient.

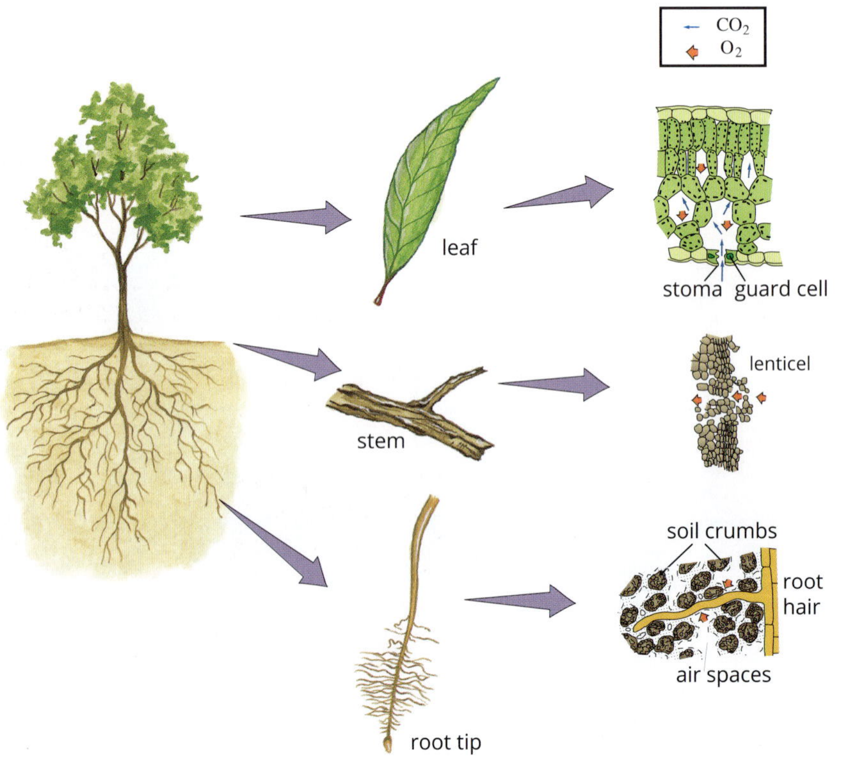

FIGURE 5.1.1 Routes of gas exchange with cells of leaves, stems and roots

Stomata

Stomata are tiny pores in the epidermis, bordered by two highly specialised epidermal cells called **guard cells** (Figure 5.1.2). Unlike other epidermal cells, guard cells contain chloroplasts. Stomata can occur on any part of a plant except the roots, but in most species they are most abundant on the lower epidermal surface of the leaves.

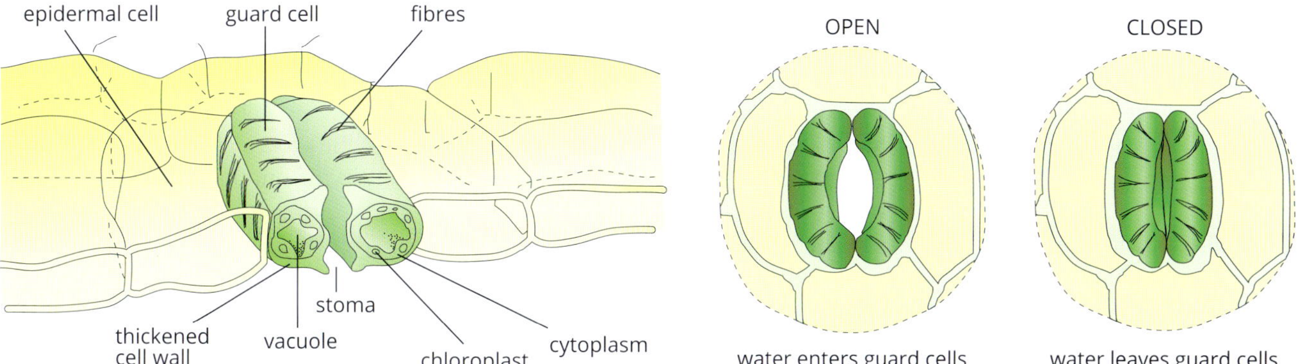

FIGURE 5.1.2 Gas exchange in leaves occurs through stomata. When water enters guard cells, they expand, opening the stoma. Guard cells expand lengthwise because they have a thickened inner cell wall with cellulose fibres that prevent the cells expanding in width.

The number and size of stomata on a leaf vary according to the plant species and the environmental conditions under which it has grown. In a typical plant, most stomata are on the underside of the leaves (Figure 5.1.3), away from the drying effect of the Sun's rays. In contrast, stomata in floating aquatic plants, such as water lilies, are confined to the upper epidermis. In plants such as eucalypts, which are adapted to dry conditions, stomata are often in sunken pits in the surface of the leaves. This reduces the direct flow of air across the leaves and so reduces water loss.

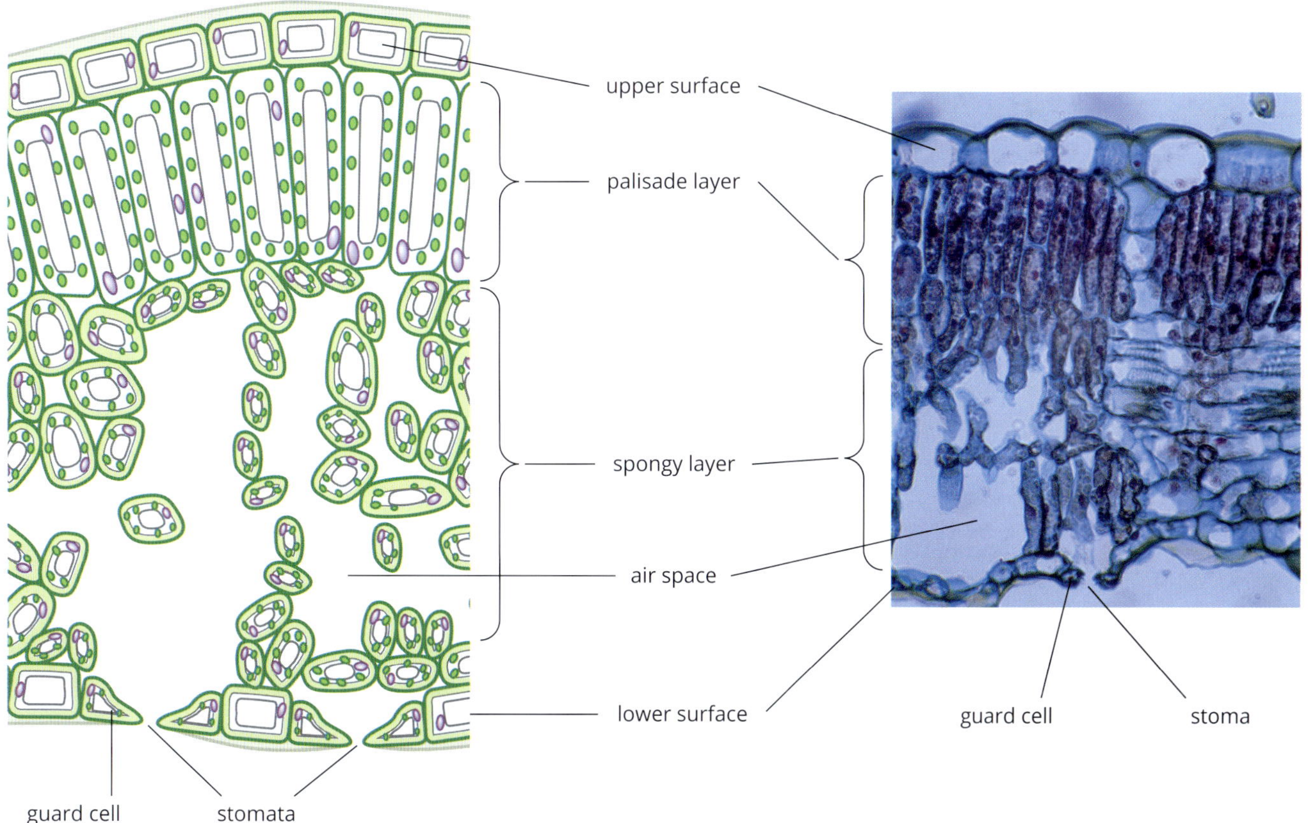

FIGURE 5.1.3 Cross-section of a leaf. Most stomata are on the underside of the leaves, away from the drying effect of the Sun's rays.

Controlling guard cells

Guard cells have the following structural features relating to their function.

- They are joined at their ends in pairs.
- Their cell walls are thicker on the side adjacent to the stoma.
- Bands of inelastic fibres run around each cell wall.

When water passes into the guard cells, their internal fluid pressure, or **turgor**, increases. This causes them to expand in the only direction possible: lengthways. The guard cells buckle and open the stoma.

Terrestrial plants, like terrestrial animals, must reduce loss of water resulting from evaporation. The moist surfaces that they use for gas exchange are major sites of water loss. Therefore, the stomata act to balance the plant's need to obtain carbon dioxide for photosynthesis against the dangers of drying out due to the loss of water from the leaves.

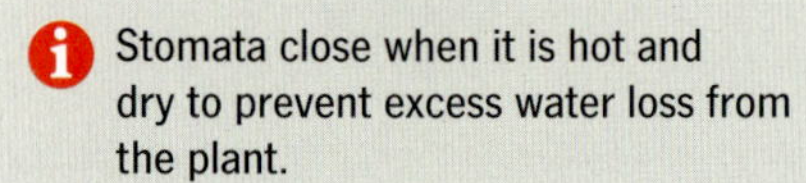

Stomata close when it is hot and dry to prevent excess water loss from the plant.

During daylight, when plants undergo photosynthesis, large volumes of carbon dioxide and oxygen are exchanged with the environment through open stomata. At night, when photosynthesis is not occurring, stomata are usually closed. Stomata also close during the day if it is very hot and dry. This prevents excessive water loss, but also drastically reduces the rate of photosynthesis.

Thus, conditions that usually favour the opening of stomata are abundant water, bright light and low internal carbon dioxide concentrations.

Stems and roots

In the epidermis of green stems, as in the leaf epidermis, there are stomata through which gas exchange takes place. In woody stems and mature roots, the epidermis is replaced by a layer of cork cells that are waterproof and airproof. Air passes freely through groups of these loosely packed cells to the cells beneath. Each group of loosely packed cells is called a **lenticel**.

Roots exchange gases with the air in spaces in the soil. Oxygen readily diffuses into the film of moisture surrounding **root hairs**, and then into the roots themselves. When soil is waterlogged due to excessive rain or poor drainage, the spaces in the soil are filled with water instead of air. Because the amount of oxygen dissolved in water is so much less than the amount of oxygen in air, the roots may not be able to get enough oxygen for their needs. The root cells may die, killing the plant. Many indoor plants are killed by over-watering.

MOVEMENT OF WATER AND SOLUTES

Vascular plants use **vascular tissue** to transport water and mineral ions absorbed from the soil, and sugars produced in the leaves, to cells throughout the plant. Vascular tissue forms continuous, closed tubular pathways through roots, stems and leaves and is visible as veins in leaves and the stringy parts of celery. There are two types of vascular tissue: xylem and phloem. You learnt about the structure of xylem and phloem in Chapter 4 and you will learn more about the function of these vascular tissues in this section.

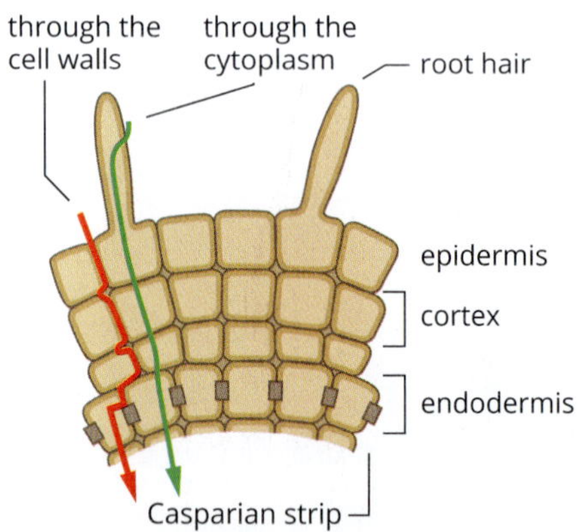

FIGURE 5.1.4 Water and mineral ions move through the roots via the extracellular pathway (red arrow) and the cytoplasmic pathway (green arrow). From the Casparian strip, water can no longer travel along the extracellular pathway and is forced into the cytoplasm before moving into the xylem.

Root absorption

There are two possible pathways for movement of water and mineral ions absorbed from the soil through the roots. These are the extracellular pathway and the cytoplasmic pathway. In the **extracellular pathway**, most water and some mineral ions pass through or between cell walls (Figure 5.1.4). In the **cytoplasmic pathway**, most mineral ions and some water pass through the cytoplasm of living root cells (Figure 5.1.4).

The cytoplasmic pathway involves substances entering a root hair cell by crossing the cell's plasma membrane, and then passing from cell to cell through microscopic channels called **plasmodesmata** (singular plasmodesma). The three types of transport that move substances across plasma membranes along the cytoplasmic pathway are as follows:

- **Active transport**—Most dissolved mineral ions are selectively taken into roots by active transport. Proteins in the plasma membrane of root cells, specific for each ion, are used for this purpose. As a result, the concentration of ions in the vascular tissue of roots can be more than 100 times their concentration in the water of the surrounding soil.
- **Osmosis**—The high concentration of ions in the vascular tissues of terrestrial plants creates a very large osmotic gradient. The **osmotic gradient** is the difference between the concentration of solutions on either side of a semi-permeable membrane. Large amounts of water move into root cells along the osmotic gradient.
- **Diffusion**—Some mineral ions such as potassium and phosphate enter the roots by diffusion. The uptake of these nutrients therefore depends on the rate of water uptake.

You learnt about active transport, osmosis and diffusion in Chapter 2.

Water and mineral ions are absorbed from the soil through the roots via the extracellular pathway or the cytoplasmic pathway.

Entering the xylem

From either of the two pathways through the roots, water and mineral ions must then reach the xylem tissue. You learnt in Chapter 4 that the **xylem** consists of hollow chains of dead cells that transport water and mineral ions upwards from the roots. Between the roots and the xylem is a waterproof layer of cells that form a barrier known as the **Casparian strip** (Figure 5.1.4). At this barrier, water travelling through the extracellular pathway is forced into the cytoplasm. In this way, the Casparian strip ensures the regulation of the substances entering the xylem.

Root pressure

In some plants the osmotic gradient draws in so much water from the roots that it can travel up to 10 metres up the stem. This is known as **root pressure** (Figure 5.1.5). Root pressure causes the rising of sap (water and mineral ions) in spring in deciduous plants such as birch trees, but does not occur in all plants.

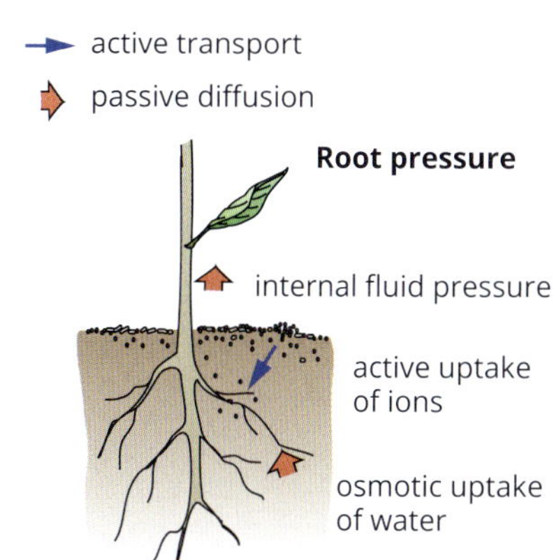

FIGURE 5.1.5 Internal fluid pressure (root pressure) in the roots of some plants causes fluid to rise through the xylem vessels.

Transpiration

Transpiration is the passive movement of water through a plant from the roots, including its evaporation through the stomatal pores in leaves. The plant uses a small amount of water for metabolic processes, but 99% of the water absorbed by the roots is lost via transpiration.

Transpiration is the movement of water through a plant—from absorption by the roots to its evaporation from the leaves.

Transpiration is a passive process: it does not require energy expenditure by the plant. It is driven by the heat energy in sunlight, which breaks the **cohesive bonds** between water molecules, allowing evaporation through the stomata.

Water molecules are very cohesive; that is, they have a strong tendency to stick together. When water evaporates from the cell walls of the leaf, cohesion between the water molecules remaining in the leaf draws water from nearby xylem vessels to replace the lost water.

In this way thousands of leaf cells, each drawing water from xylem, create a differential pressure that pulls water up xylem vessels from the roots. This continuous one-way flow of water from roots to leaves is called the **transpiration stream**. The pull of transpiration can be strong enough to draw water to the top of the tallest tree, more than 100 metres high.

Although transpiration is the cause of 99% of a plant's water loss, it is vital because it enables plants to:

- absorb the water necessary for the process of photosynthesis
- transport mineral ions to leaf cells and fruits
- cool down and not become overheated.

Factors that affect transpiration rates

Water vapour is lost from leaves mainly by transpiration through open stomata. The total surface area across which transpiration takes place is related to the degree of opening of all stomata. This is by far the most important factor affecting the rate of transpiration. The greater the number of stomata and the more open they are, the more surface area there is from which water can be lost.

Other factors that affect the rate of transpiration (Figure 5.1.6) include:

- humidity—transpiration rates decrease when there is a lot of water vapour in the air (i.e. a high level of humidity), because this reduces the water concentration gradient between leaf spaces and air, so fewer water molecules evaporate into the air.
- temperature—transpiration rates increase as temperature increases because heat energy increases the rate of evaporation of water.
- wind—air currents increase the rate of transpiration by moving water vapour away from the leaf and therefore increasing the rate of evaporation of water.

Environmental factors such as sunlight and humidity affect the rate of transpiration and therefore the rate of water uptake by the roots of the plant. The rate of transpiration is low at night because it is cooler and more humid, and because stomata are usually closed. The leaves of some plants that live in exposed conditions have developed structural features that reduce the rate of transpiration. For example, some plants have hairs on the leaf surface, which create a layer of relatively undisturbed, humid air.

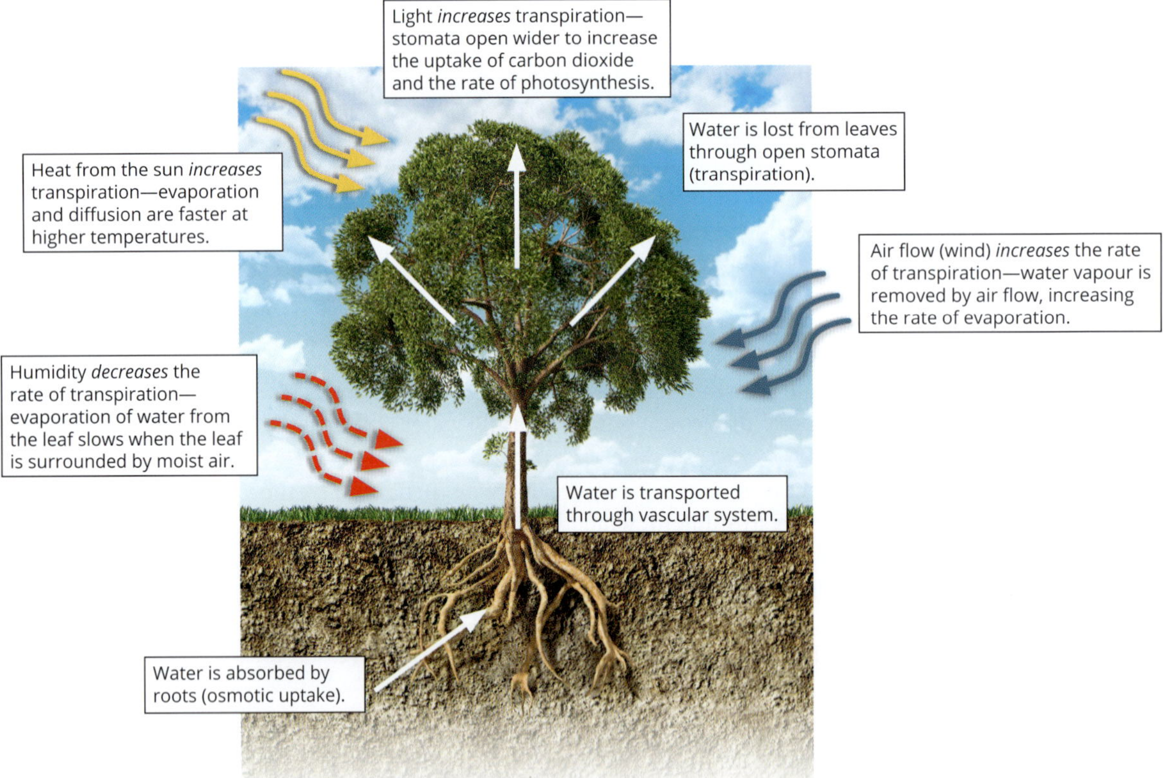

FIGURE 5.1.6 Transpiration is the movement of water through the xylem vessels of vascular plants and into the atmosphere through leaf stomata, in the form of water vapour.

CASE STUDY **ANALYSIS**

Water transport adaptations in desert plants

Plants that live in deserts need specialised strategies to survive the hot, dry conditions (Figure 5.1.7). In an environment where water is scarce, plants have developed special structures that enable extremely efficient uptake and storage of this precious resource. Cactus plants are specialised to hold large volumes of water in their fleshy leaves, stems and roots. When water does come along, they need to be able to absorb as much as possible, as fast as possible. Their roots are shallow and cover a large area, enabling them to efficiently absorb water from the soil.

Because cacti need to hold onto water once they have it, most cacti are spiny, bitter tasting or toxic, which deters thirsty animals. A thick, waxy cuticle also protects the leaves from damage and reduces water evaporation. While most plants open their stomata during the day, in a hot, dry environment this would lead to substantial water loss through transpiration (Figure 5.1.8). To overcome this problem, cacti open their stomata at night and use a type of photosynthesis called crassulacean acid metabolism (CAM). At night when stomata are open, carbon dioxide is taken in and converted to malic acid, which is stored in the vacuoles of mesophyll cells. In daylight, when the stomata stay closed to reduce water loss, the stored malic acid is broken down, releasing carbon dioxide which diffuses into chloroplasts for conversion into glucose and carbohydrates, completing the photosynthetic process. CAM photosynthesis is excellent for conserving water, but the rate of photosynthesis is slow. This is why many cacti grow very slowly.

Analysis

1 Plants live in a wide variety of environments, including the dry deserts where cacti live, warm and wet tropical rainforests, and slightly drier and cooler temperate savannas with grasses and scattered trees. Determine which plant in Figure 5.1.8 is the:
 a tropical plant
 b desert plant
 c temperate plant.

2 At what time are the stomata of the desert plant most likely to be open? Explain why this would be the case.

3 At what time are the stomata of the temperate plant most likely to be open? Explain why this would be the case.

4 Describe the trend in transpiration rate for the tropical plant in comparison to the other plants, and explain why this trend exists.

FIGURE 5.1.7 Cacti have many special adaptations that enable them to absorb and store water in harsh, dry environments.

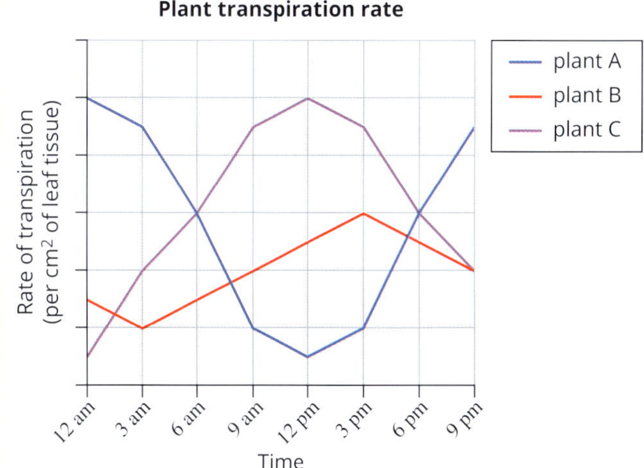

FIGURE 5.1.8 Rate of transpiration in three plants in different environments over 24 hours

Translocation: sources and sinks

Translocation is the transport of organic solutes (e.g. sugars) from the leaves (sources) to other plant tissues (sinks).

The transport of organic solutes from the leaves to other tissues in the plant is known as **translocation**. Leaves produce carbohydrates in the form of sugars during photosynthesis. The non-photosynthetic tissues of the plant also need these carbohydrates and other organic compounds, such as amino acids, hormones and proteins, so these nutrients are transported from the **sources** (the leaves) to the **sinks** (regions where the nutrients are needed, such as roots, stems, flowers and fruits).

The vascular tissue through which these organic solutes move is the **phloem**, and the material that flows through it is known as phloem sap. This sap is composed of around 90% sucrose. Sucrose is a disaccharide that dissolves easily in water, making it a good transport material. It is produced in the chloroplasts of the chlorenchyma (parenchyma cells with chloroplasts) and pumped into the companion cells. From the companion cells the sucrose flows into the sieve tube cells (Figure 5.1.9). Transport in individual sieve tube cells is in one direction only, but bundles of sieve tube cells can transport sap in both directions: upwards to leaves and fruit, or downwards to the roots.

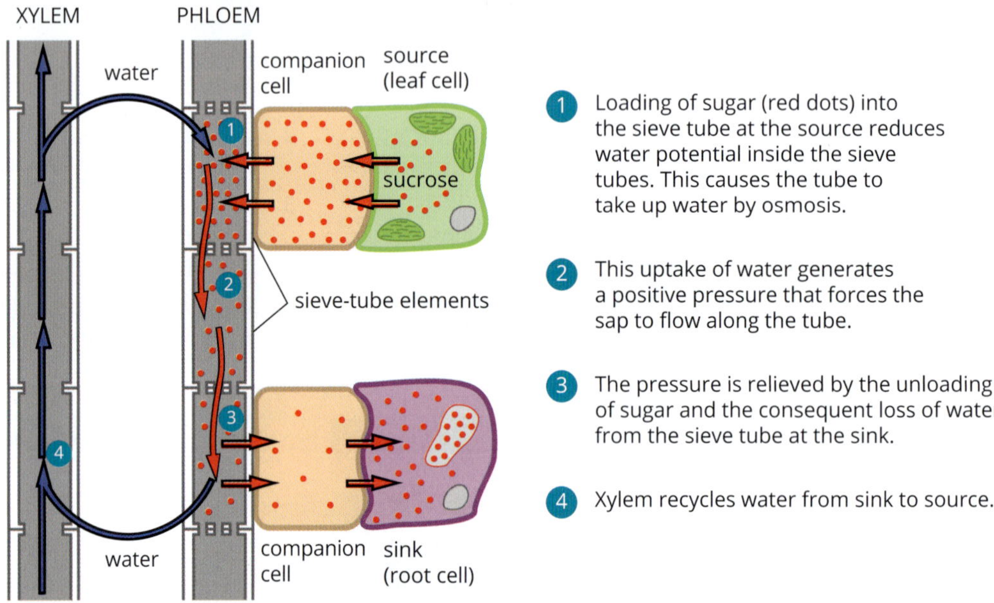

FIGURE 5.1.9 The movement of fluid through the phloem is the result of active pumping of sugars, with water flowing along an osmotic gradient. Sugars and water enter the phloem sieve tubes in leaves in this way and are translocated throughout the plant. Sugars are actively unloaded from sieve tubes where they are required.

Translocation is an active process. It involves the flow of cytoplasm in sieve tubes driven by a pressure gradient, and requires the expenditure of energy by the plant. This pressure gradient begins in the leaves, where sucrose is actively pumped into phloem sieve tube cells. This creates an osmotic gradient that draws water passively into the sieve cells. As water enters, it increases the fluid pressure (turgor) in sieve cells, which pushes fluid from these cells into adjacent sieve cells.

While this is happening in the leaves, sucrose is being actively removed from sieve cells in roots, growing shoots and developing fruit. This causes an osmotic gradient that draws water out of sieve cells and lowers their turgor pressure.

Fluid pressure is therefore high in sieve tube cells in leaves and low in sieve tube cells in roots and growing shoots. A bulk flow of the contents of sieve tubes occurs along this fluid pressure gradient, from sources to sinks. Translocation stops if the cells in the stem die.

5.1 Review

OA ✓✓

SUMMARY

- Stomata are found in the epidermis of leaves and some stems. They are the main route through which gas exchange occurs in plants.
- Conditions favouring the opening of stomata are abundant water, bright light and low internal carbon dioxide concentrations.
- Roots exchange gases with air in well-aerated soil spaces.
- Most terrestrial plants, including ferns, conifers and flowering plants, have vascular tissues (xylem and phloem) that are specialised for transporting fluid.
- The vascular tissues that support the regulation of water in vascular plants are:
  - xylem, which carries water and mineral ions from roots to leaves
  - phloem, which carries sugars and other organic molecules from leaves to roots and other tissues as required.
- Water and inorganic nutrients (mineral ions) are absorbed by the root hairs from the soil by one of two pathways:
  - the extracellular pathway
  - the cytoplasmic pathway.
- Water and mineral ions are transported through xylem vessels as sap. This transportation occurs in one direction only: from roots to leaves.
- Transpiration is the evaporation of water from stomata in leaves. It is a passive process (driven by energy from sunlight) that also draws water up from the roots, through the xylem, following what is known as the transpiration stream.
- The rate of transpiration is affected by:
  - the number of stomata and their degree of opening
  - temperature
  - humidity
  - wind.
- Translocation is the transport of organic materials from the leaves to the roots, stems, flowers and fruits of the plant, through the sieve tube cells and companion cells of the phloem tissue.
- Translocation is an active process and requires an expenditure of energy by the plant.
- Translocation is driven by a pressure gradient that begins in leaves (sources), where sucrose is actively pumped into phloem sieve cells while being actively removed from sieve cells in roots, growing shoots and developing fruit (sinks).

KEY QUESTIONS

Knowledge and understanding

1. Explain how the structural features of guard cells relate to their function in opening and closing the stomata.
2. Xylem and phloem have different transport functions in vascular plants. Outline the functions of these vascular tissues.
3. Root cells have no chlorophyll, and being underground they are not exposed to sunlight, yet they continue to grow. How do roots obtain their nutrients?
4. Explain why transpiration is vital to plants.
5. Determine whether each of the following environmental factors increases or decreases transpiration rates in plants.
   - **a** high temperature
   - **b** high humidity
   - **c** darkness
   - **d** strong wind

Analysis

6. Predict what will happen to gas exchange in a pot plant if you add excessive amounts of water to it.
7. A student set up an experiment at home to measure and compare the amount of water two different plants release through transpiration and evaporation. The student used the leaves of roses from their garden and the leaves of sunflowers that were in a vase away from sunlight. The student securely taped a clear bag around the leaf of the sunflowers but was not able to completely seal the bag around the leaves of the roses due to the thorns. The bags were left on both of the plants overnight. As water transpired and evaporated from the leaves, the vapour condensed and was collected in the bag in the form of liquid water. The sunflower leaves released 35 mL of water whereas the rose leaves released 10 mL. The student concluded that the leaves of the sunflower have a higher transpiration rate than the leaves of roses. Evaluate the design of the experiment and the student's conclusions.

5.2 Homeostatic mechanisms in animals

FIGURE 5.2.1 Sweating results in evaporative cooling and is one of the human body's homeostatic mechanisms to regulate body temperature.

Organisms and cells are constantly experiencing changes in their environment. These changes to the internal and external conditions can adversely affect the survival, growth and functioning of the organism. The internal environment of an organism must always remain within tolerable limits, even when conditions in the external environment fluctuate widely. When a change occurs in the external environment, an adjustment must be made to the internal environment.

Living organisms rely on their external environments to provide adequate levels of nutrients, water and oxygen and suitable physical conditions, such as light and temperature. Organisms have a range of mechanisms that allow them to adapt to changing conditions while maintaining a stable internal environment (Figure 5.2.1). If an organism is not able to adapt to its external environment, it will suffer cellular damage and possibly death when conditions change.

In this section you will look at how animals regulate their internal environment to maintain stability at the organ, tissue, cellular and intracellular level to sustain life.

Body temperature, blood glucose and water balance are some of the most important factors that are regulated by homeostatic mechanisms. These mechanisms will be explored in detail in this section.

MAINTAINING EQUILIBRIUM

An animal is able to bring about balance or equilibrium in its internal environment by coordinating a number of systems, such as circulatory, respiratory, immune, digestive and excretory, and by changing its behaviour.

Homeostasis

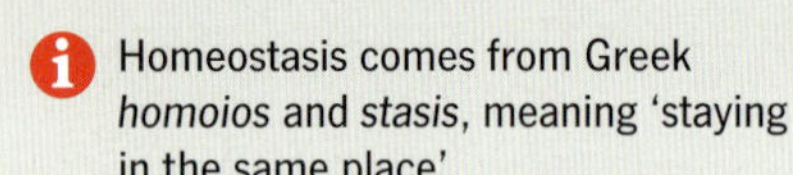

Homeostasis comes from Greek *homoios* and *stasis*, meaning 'staying in the same place'.

Homeostasis is the maintenance of a stable internal environment within an organism. When an organism is healthy and functioning well, its systems are in homeostasis. Homeostasis is achieved by a variety of mechanisms that respond to keep internal environments within certain limits. This maintains conditions at an optimum level when the internal or external environment changes.

Animals coordinate the activities of their cells, tissues, organs and systems so that **responses** occur in an integrated and controlled manner. Detecting and responding to a stimulus requires an effective internal communication system. Communication in animals is achieved by hormonal and nervous system mechanisms, which transmit information between different parts of the organism and translate environmental disturbances into signals that can be interpreted and responded to. For example, the iris of the human eye detects light and responds by dilating or constricting to regulate the amount of light that enters the eye (Figure 5.2.2).

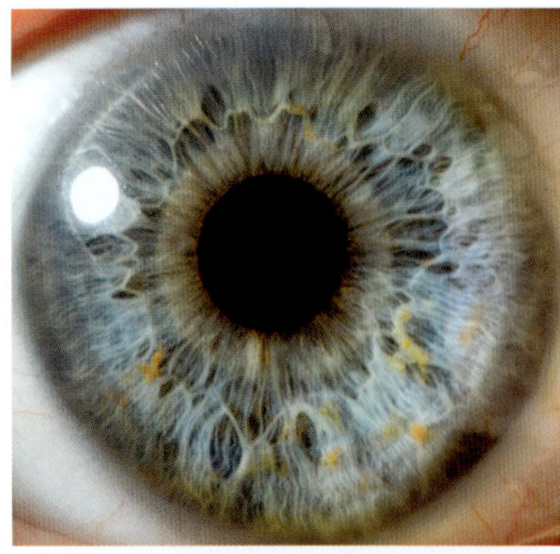

FIGURE 5.2.2 The iris is a pigmented muscle that responds to the stimulus of light by dilating and constricting to regulate the amount of light entering the eye.

The two most important systems in maintaining homeostasis in animals are the **endocrine system** (which produces hormones) and the **nervous system**. Physiological and behavioural responses to environmental change that are carried out by both the endocrine and nervous systems include:

- short-term and long-term regulation of growth
- maturation and reproduction
- homeostatic regulation of the internal environment.

The nervous system also provides rapid responses to:

- produce efficient coordinated movement to detect and avoid predators
- find and capture prey.

These regulatory systems are the most developed in mammals, which are able to maintain a relatively stable internal environment in the face of changing conditions.

Although in many ways hormonal systems and nervous systems appear distinctly different, they share one common feature: they both involve chemical communication. Chemical communication involves signals being passed from one cell to the next by the release of specific **signalling molecules**, known as **hormones** and **neurotransmitters**. Hormones are released from glands or other tissues, and neurotransmitters are released from nerve endings. These molecules exert their effects by highly specific interactions with a receptor on, or within, the responding or target cell. **Receptors** are specialised structures that can detect a specific stimulus and initiate a response.

A response is a physiological or behavioural change in an organism as a result of receiving a stimulus.

Feedback loops

Feedback loops may involve the endocrine and nervous systems working together to regulate the internal environment.

Negative feedback loops promote stability in the internal environment and maintain homeostasis by responding to changes in the body and adjusting the variables to their original or optimal state. They are **stimulus–response mechanisms** in which the response produced reduces the effect of the original stimulus by reversing its direction. For example, if the concentration of a substance in the blood is too high, a negative feedback loop will lower the concentration. If the concentration is too low, a negative feedback loop will increase the concentration. Most feedback loops in biological systems are negative.

A stimulus is an environmental factor that an organism can detect and respond to.

Negative feedback loops are called negative because the information produced by the feedback causes a reversal of the size or effect of the stimulus. Negative feedback loops maintain stability through the action of the nervous or hormonal systems, or both acting together.

An example of a negative feedback loop is the regulation of blood glucose levels by the hormone **insulin**. When blood glucose levels are high, receptors detect the change and the pancreas secretes insulin. This lowers the blood glucose levels until homeostasis is reached, at which point the pancreas stops releasing insulin.

A negative feedback loop acts as follows (see Figure 5.2.3):

1 The system is in a stable state. A change (**stimulus**) occurs.
2 The change is detected by an appropriate receptor.
3 The receptor sends a signal to a **control centre** (**hypothalamus** or transmission molecules) via the efferent pathway (carries signals to the central nervous system).
4 The control centre sends a signal to an appropriate **effector** or a specific effector cell, tissue or organ via the efferent pathway (carries signals away from the central nervous system).
5 The effector responds to the signal, and the original state is restored.

The hypothalamus is known as the control centre of the brain—it receives information from all parts of the body and regulates the internal environment through the secretion of hormones.

An effector is a cell or tissue that responds to a stimulus.

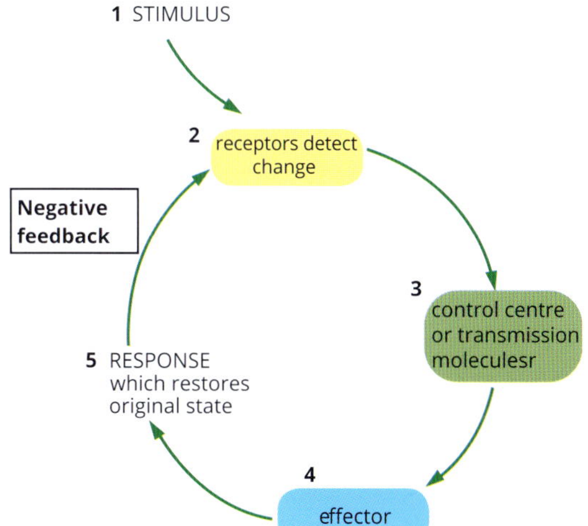

FIGURE 5.2.3 When the response reduces the initial stimulus or disturbance, it is operating as a negative feedback mechanism.

In the control centre, information from sensory receptors is received and compared with a set-point (the optimal value for the functioning of that organism). This information is processed with other information about the state of the organism, and an appropriate response is initiated.

Regulation therefore involves fluctuations around the set-point. The size of the fluctuations depends on:

- the sensitivity of the receptor
- the tolerance of the control centre to variation from the set-point
- the efficiency of the effector.

Some features of the internal environment, such as blood glucose levels, can vary considerably; others, such as body temperature in mammals, are tightly controlled.

In contrast to negative feedback loops, **positive feedback loops** force an organism out of homeostasis by maintaining the direction of the stimulus, and sometimes increasing the stimulus. An example of a positive feedback loop is uterine contractions during childbirth. The hormone oxytocin stimulates the uterus to contract, causing pain. Rather than the nervous system signalling the endocrine system to lower the oxytocin and reduce the pain, more oxytocin is produced to stimulate stronger contractions. The contractions work to push the baby into the birth canal and continue until the baby is born.

Table 5.2.1 summarises the differences between negative and positive feedback loops.

TABLE 5.2.1 Comparison between negative and positive feedback loops

	Negative feedback loop	Positive feedback loop
Result	The response is opposite to the stimulus. Hence if a decrease was detected, the response would be to increase the signal, and vice versa.	The response is the same as the stimulus. Hence if an increase was detected, the response would be to further increase the signal.
Homeostasis maintenance	supports homeostasis by bringing back a balance internally	breaks down homeostasis by causing an imbalance internally
Frequency	occurs often	is less common
Examples	• regulation of body temperature; for example, if there was a decrease in body temperature (stimulus), the response would be to increase the body temperature, and vice versa • regulation of blood pressure and blood glucose levels in humans	• release of oxytocin during childbirth to stimulate uterine contractions • release of prolactin during breastfeeding to promote breastmilk production

Receptors—detecting external and internal stimuli

Animals have sensory receptors to detect aspects of their environment that may affect their ability to survive and reproduce. The types of sensory receptors present and their sensitivity differ substantially between animals, and are related to the way animals have adapted to their environments. For example: a wombat has less visual acuity for distinguishing small objects than an eagle; dogs use chemical scents much more than humans; some moths have chemoreceptors that can detect a single molecule of pheromone; and platypuses can detect weak electric currents. Some animals respond to different parts of the electromagnetic spectrum—for example, snakes can detect infrared radiation and bees see ultraviolet light.

In humans, the five senses (vision, hearing, taste, smell and touch) are perceived through sense organs (eyes, ears, tongue, nose and skin) that collect and process sensory information. Receptors that detect external stimuli are known as **exteroreceptors**. These are usually located close to the surface of the body and detect stimuli such as pain and pressure. Some receptors detect internal states, such as blood pressure and blood chemistry (e.g. oxygen and carbon dioxide levels), and are known as **interoreceptors** or visceral receptors. From a functional point of view, the types of sensory receptors can be classified as **photoreceptors** (vision), **chemoreceptors** (taste, smell, communication), **mechanoreceptors** (hearing, balance, pressure, touch) and **thermoreceptors** (temperature) (Table 5.2.2).

TABLE 5.2.2 Types of receptors and examples found in complex animals, and the stimuli each of the receptors responds to

Receptor type	Examples	Received stimuli
mechanoreceptor	pressure (touch) receptors blood vessels: • baroreceptors	 stretching of blood vessel wall
	skin: • Meissner corpuscle • Pacinian corpuscle	 • light touch • heavy touch
	proprioreceptors: • muscle spindles • Golgi cells • joint receptors	 • movement, position of body • gravity • movement with ligaments
	labyrinth in the vertebrate ear: • sacculus and utriculus • semicircular canals • ciliated cells in the cochlear duct	 • gravity and linear acceleration • angular acceleration • sound waves
chemoreceptor	taste buds, olfactory epithelium	specific chemical compounds
thermoreceptor	• thermoreceptors in blood-sucking insects and ticks • pit organs in pit vipers • nerve endings and receptors in the skin and tongue of many animals	heat
electroreceptor	organs in the skin of some fish	electric currents in water
photoreceptor	• eyespots • ommatidia in arthropods • rods and cones in the retina of the vertebrate eye	light energy

CASE STUDY ANALYSIS

Dietary calcium deficiency

Many people in Victoria become vitamin D deficient through the winter months because of the lack of sun exposure. Vitamin D is necessary at all times as it is required to help your body reabsorb an important mineral, calcium. Therefore vitamin D deficiency may lead to a decrease in levels of blood calcium.

Calcium is stored in the bones, and is essential for keeping them healthy and strong. Many of the cells in the body need calcium for proper cell functioning. Any imbalance could have serious effects on the human body. Hence, for homeostasis to be maintained specific responses are put into place (Figure 5.2.4).

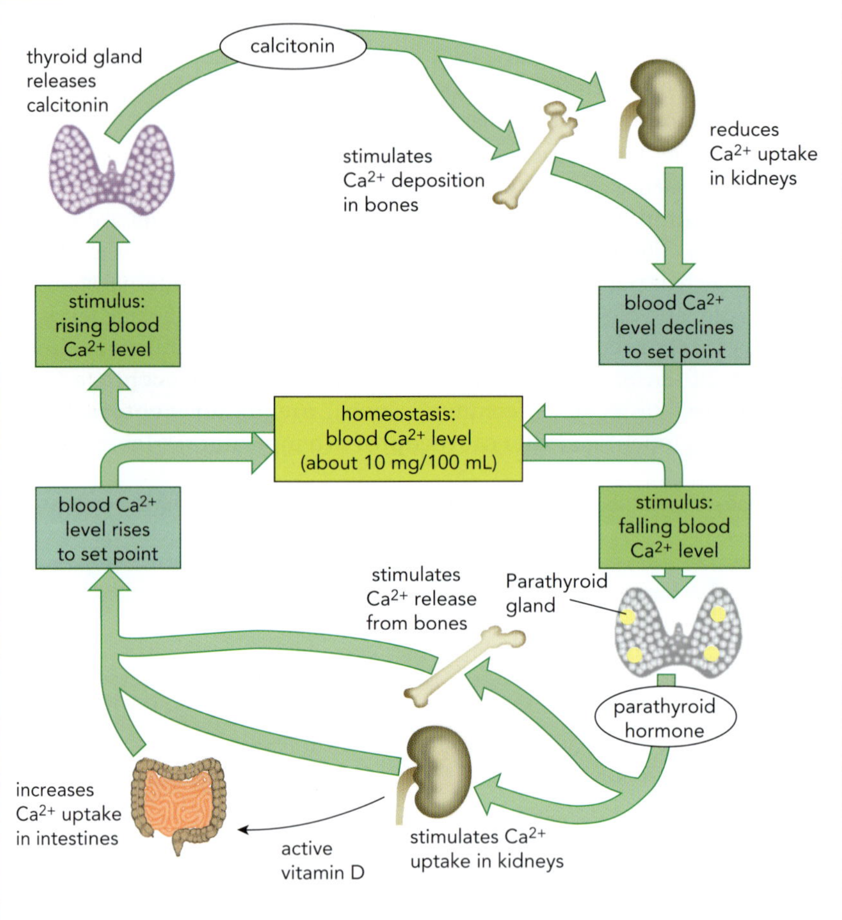

FIGURE 5.2.4 Feedback loop showing the response to the changes in blood calcium levels in humans

Analysis

1. What type of feedback loop is observed in Figure 5.2.4? Explain your answer.
2. A chemoreceptor receives a signal that there has been an increase in the levels of calcium in the blood. A message is sent to the hypothalamus (the control centre) which then activates the appropriate signalling mechanisms. Identify the effector and response using the diagram, and define each term.
3. 'Low blood calcium levels may cause the bones to weaken.' Do you agree with this statement or not? Use Figure 5.2.4 to explain your reasoning.
4. The parathyroid gland releases parathyroid hormone into the blood after it receives a signal from the hypothalamus (the control centre). How does this reinforce the fact that, to reach equilibrium, the endocrine and nervous systems need to work together?

REGULATION OF BODY TEMPERATURE

The maintenance of core body temperature within a specific range is called **thermoregulation**. Regulation of body temperature in humans involves a complex negative feedback pathway with several sensory inputs and many effector responses that act together to maintain a stable body temperature. The control centre for measuring the body temperature set-point (37 °C) is in the hypothalamus. A change in the temperature of the hypothalamus initiates regulatory responses that can reduce heat loss or initiate heat production or heat exchange (Figure 5.2.5).

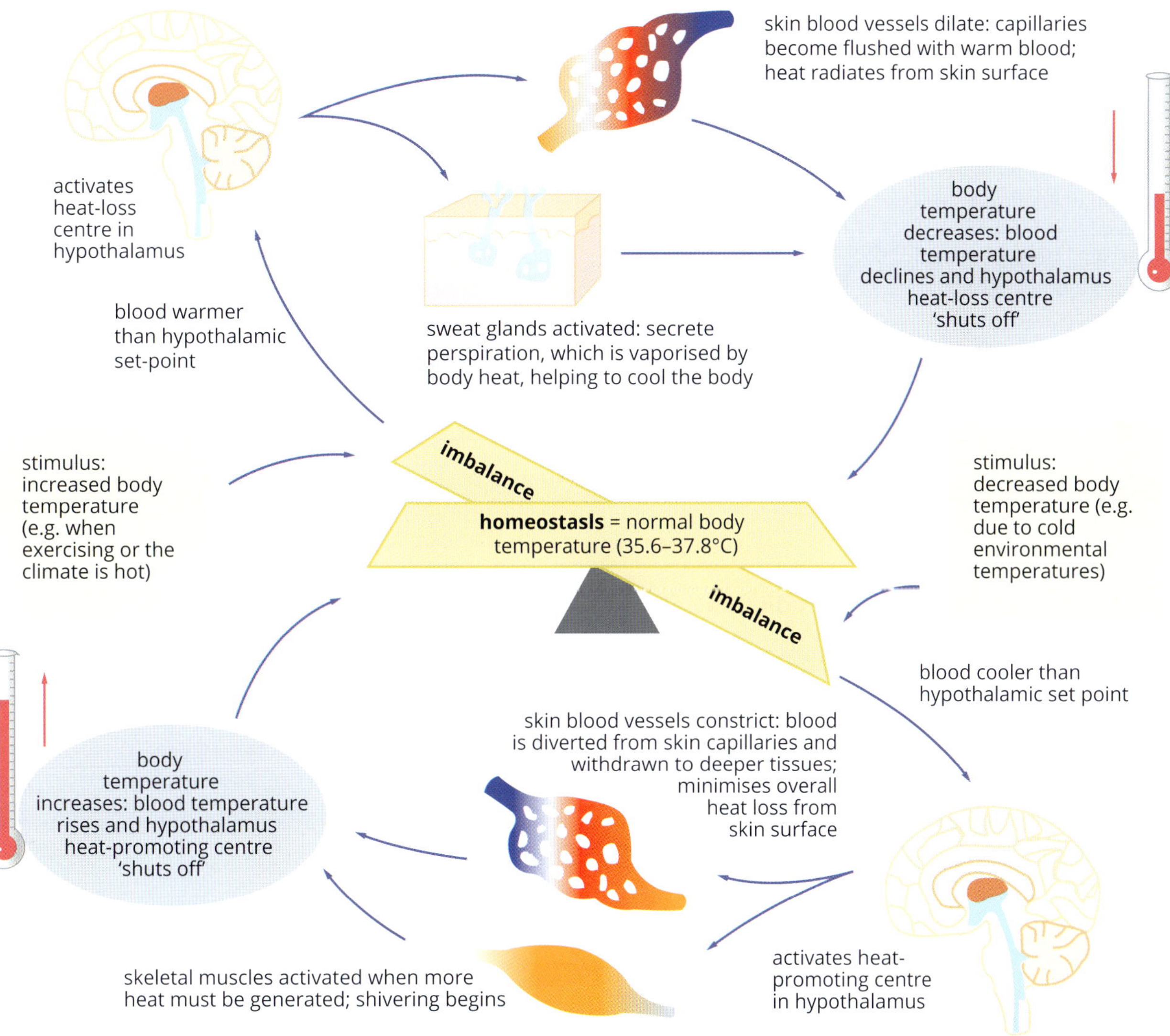

FIGURE 5.2.5 Thermoregulation in humans involves a range of regulatory mechanisms that function to maintain thermal homeostasis (normal body temperature) for optimal functioning of the organism.

Detecting temperature change

Regulation of temperature in humans is an example of the way different sensory receptors work together to produce an integrated response. Arterial blood has the most constant temperature. The relatively constant temperature of many other parts of the body indicates that they are well-supplied with arterial blood.

In **endotherms** (e.g. mammals and birds), a group of temperature-sensitive cells in the hypothalamus act as misalignment detectors, triggering homeostatic responses if blood temperature deviates from the optimal temperature range, or set-point. Lowering or raising the temperature of the hypothalamus initiates regulatory changes in heat production or heat exchange.

Temperature receptors are also found in the skin. A decrease in environmental temperature detected by these receptors will initiate regulatory responses such as decreased blood flow to the skin to reduce heat loss, and behavioural changes such as moving into a warmer or more sheltered environment. These responses take place long before there has been any change in the internal temperature of the body. Skin temperature receptors act as disturbance detectors, detecting changes in the external environment and triggering responses before there is a change in core body temperature.

As the environmental temperature falls, disturbance detectors stimulate responses that reduce heat loss and increase heat production. The reverse occurs as environmental temperature increases. If the arterial temperature falls despite the regulatory responses that have been initiated, or if it rises because the responses made have been too effective, these changes will be detected by the misalignment receptors in the brain, which will fine-tune the temperature-regulating mechanisms.

The value of the disturbance detectors in the skin is to reduce fluctuations in arterial blood temperature, providing a more precise control around the set-point level than there would be if misalignment detectors (the brain's temperature receptors) alone were involved.

BIOFILE

Sweating: an efficient cooling mechanism

Humans are one of the few animals that produce sweat to cool down. Adults can sweat up to 4 litres per hour during vigorous exercise. Even when you are not exercising, sweating plays an important role in thermoregulation through evaporative cooling. When warm sweat (which is about 99% water, with sodium chloride and some other substances) comes into contact with cooler air, it evaporates, carrying heat away and lowering your body temperature. It does this through a process of energy (and therefore heat) transfer.

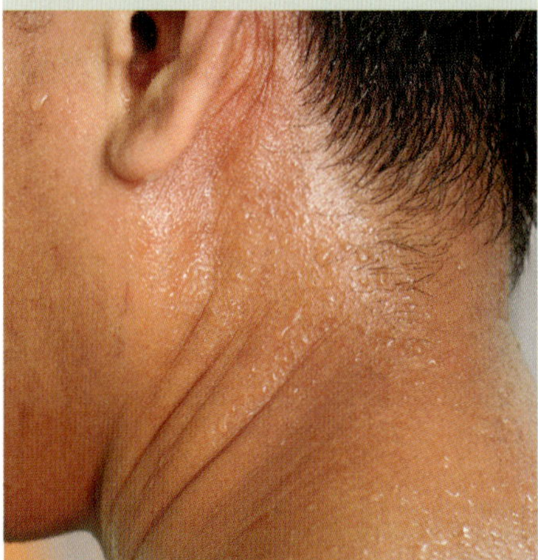

Sweating is an important physiological adaptation in humans. It uses evaporative cooling to regulate the body's temperature and prevent overheating.

Heat loss

Organisms are constantly exchanging heat with their environment. This heat exchange occurs through four mechanisms (Figure 5.2.6):

- **conduction**—Occurs when the temperature of the organism and the environment are different. Heat exchange is a result of direct contact (e.g. a lizard basking on a warm rock).
- **convection**—Transmission of heat from a warmer region to a colder region, resulting from the movement of liquid or gas (e.g. heat is lost from the body surface by convection, which transfers heat through the movement of liquid or air).
- **radiation**—Occurs all the time, without direct contact, regardless of temperature differences between the organism and their environment (e.g. heat radiating from dark coloured surfaces).
- **evaporation**—Heat loss by water evaporation. This occurs most rapidly when the air is hot and dry (e.g. sweating).

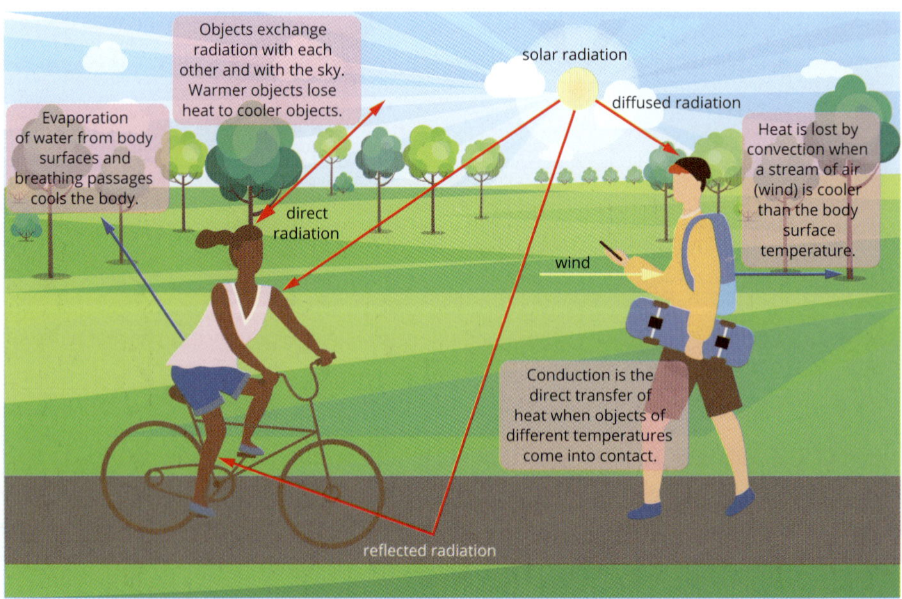

FIGURE 5.2.6 Methods of heat exchange between a human and the environment

CASE STUDY

Fever

A fever is an increase in your body temperature above the normal range or set-point and is a common symptom of an infection (Figure 5.2.7). Fever is one of your body's defences against infection, working to raise your body's internal temperature above the tolerable limits of the invading pathogens.

Pathogenic viruses and bacteria contain substances called pyrogens. An example of a pyrogen is the lipopolysaccharides found in some bacterial cell walls. Pyrogens induce fever by triggering your body's immune response, signalling the hypothalamus to increase your body's temperature.

The body responds to these signals from the hypothalamus by initiating a variety of warming activities in order to generate and retain heat. Peripheral blood vessels are constricted to reduce blood flow and heat loss through the skin. The reduction in blood flow to your skin makes you look pale and feel cold, even though your body is working to retain heat. If your body temperature is still too low, you might start shivering to generate more heat. Your body temperature will continue to increase until it reaches the new higher set-point of the hypothalamus. The fever is maintained until the invading organisms are eliminated and their effect on the hypothalamus ceases. The fever then makes you feel hot and flushed, and the mechanisms that were used to warm your body are reversed; blood vessels dilate, shivering stops and sweating works to cool your body back to the normal temperature range.

A body temperature above 41.5°C is considered extremely high and requires immediate medical attention. Extreme fevers are known as hyperpyrexia and are most commonly caused by a haemorrhage inside the skull. Some infections and sepsis (blood infection) may also lead to hyperpyrexia.

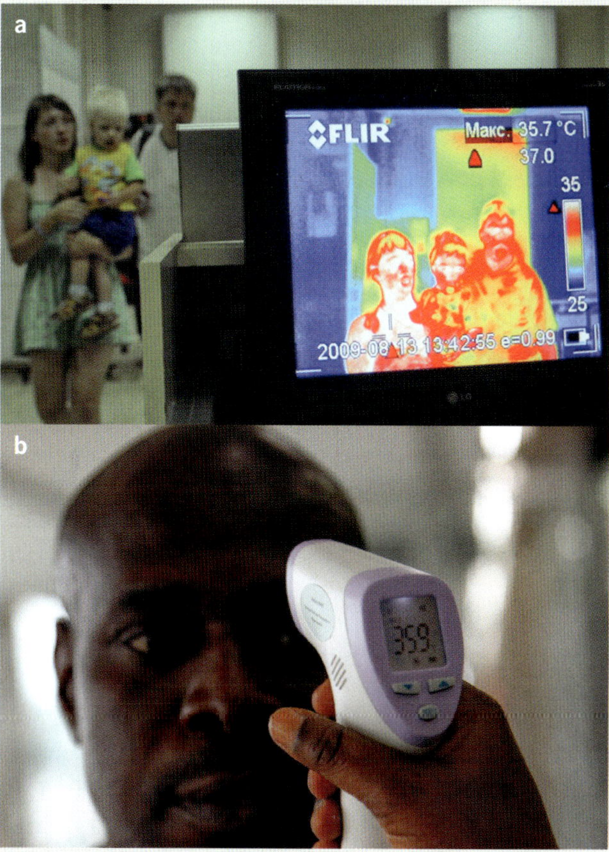

FIGURE 5.2.7 (a) Thermal scanning cameras are used to monitor the skin temperature of passengers arriving at airports in an effort to prevent the global spread of infectious diseases. (b) Teams in airports around the world were equipped with thermal guns to measure the temperatures of travellers from Liberia, Sierra Leone and Guinea in an effort to prevent an Ebola pandemic in 2014.

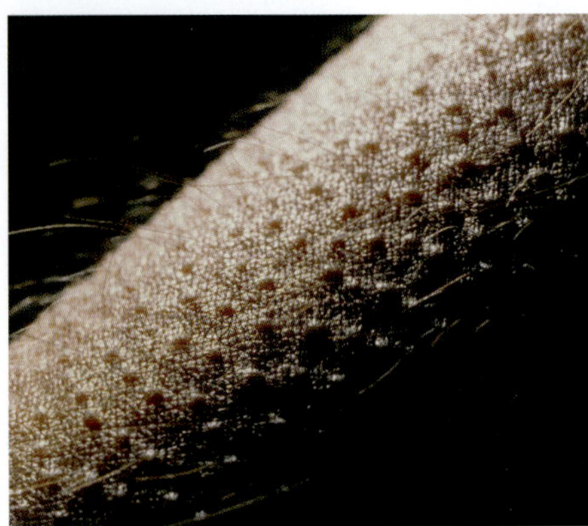

FIGURE 5.2.8 A close-up of a human forearm with goose bumps. The contraction of blood vessels and small muscles (arrectores pilorum) that are attached to the base of each hair follicle pull the hair into an upright position. In this position the skin resembles plucked goose skin.

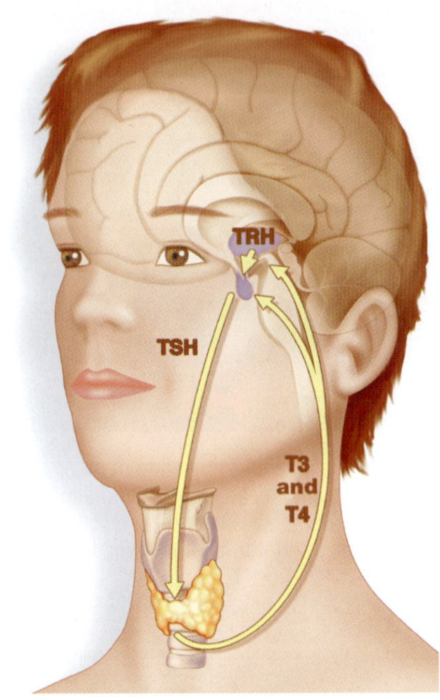

FIGURE 5.2.9 The thyroid is a gland that produces hormones that stimulate cellular respiration. The hypothalamus releases TRH, which stimulates the secretion of TSH by the pituitary gland, which in turn stimulates production of T3 and T4 hormones by the thyroid. The increase in cellular respiration creates thermal energy.

Responding to cold

A number of nervous and endocrine responses occur rapidly to reduce heat loss from the body and increase heat production when the body becomes too cold.

The following involuntary responses reduce heat loss from the body:

- **vasoconstriction** (constriction of the blood vessels in the skin)—This reduces heat loss from the skin, as the amount of blood moving close to the exposed surface is reduced.
- **piloerection**—The constriction of the piloerector muscles around hair follicles ('goose bumps'), which increases the insulating effect of the hairs (Figure 5.2.8). This response has a minimal effect in humans but in animals with thick fur, the layer of trapped air increases significantly and reduces heat loss from the body.
- **shivering thermogenesis**—The production of metabolic heat is increased through shivering. This involuntary movement of the muscles generates large amounts of heat. Shivering thermogenesis is stimulated by adrenaline.
- **non-shivering thermogenesis** in brown fat (brown adipose tissue or BAT)—Increased cellular activity in BAT, a tissue specialised for heat production, causes the tissues to warm up. The heat produced is carried to other parts of the body in the blood. Brown fat contains many mitochondria (which give it its brown colour), fat-metabolising enzymes and an extensive vascular network. Brown fat is capable of high rates of aerobic metabolism using a pathway that breaks down fats to produce large amounts of heat, but little ATP (energy). This mechanism is crucial in many baby animals and hibernators.
- increasing **metabolism** (the rate of cellular respiration)—Metabolic processes in the internal organs are the main source of heat when the organism is at rest (Table 5.2.3). In humans, around 60% of the energy released during cellular respiration is transformed into thermal energy. In humans, the overall metabolic rate, and therefore the rate of heat production, is controlled by hormones.
- **thyrotropin releasing hormone (TRH)** secretion by the hypothalamus—TRH acts on the anterior pituitary to secrete **thyroid stimulating hormone (TSH)**. As the name suggests, TSH acts on the thyroid gland to release thyroid hormones, tri-iodothyronine (T3) and thyroxine (T4) (Figure 5.2.9). T3 and T4 hormones regulate metabolic processes, increasing heat production and body temperature. The amount of T3 and T4 in the bloodstream is regulated by the pituitary gland via a negative feedback loop; if there is too much or too little T3 or T4, the pituitary gland reduces or increases the amount of TSH it secretes. This mechanism allows a very delicate regulation of the level of thyroid hormones in the blood.

TABLE 5.2.3 Major sources of heat production in humans. Metabolic processes in the internal organs are the main source of heat when a person is at rest. During physical activity, the heat generated in the muscles increases.

Organs	Participation in heat production at rest (%)	Participation in heat production during physical effort (%)
brain	16	2
internal organs	56	22
skeletal muscles	15	73
other organs	13	3

Animals, including humans, may also change their behaviour to reduce heat loss and increase heat production in the body. Some examples include changing body shape or decreasing surface area (e.g. curling up to make yourself small or huddling together), seeking shelter to reduce exposure to cold temperatures and increasing physical activity. You will learn more about behavioural adaptations of animals in Chapter 9.

Responding to heat

As well as responding to cooler temperatures, animals can adjust their internal environment in response to high external temperatures. Many of the responses to heat work in the opposite way to the responses to cold temperatures. Some of the processes that take place in animals when it is hot include:

- **evaporative cooling**—This is a very effective way of losing heat energy from the body as water (sweat) changes to water vapour. Sweat glands are distributed over much of the human body and release sweat onto the skin surface via pores when your body temperature rises. Most animals do not sweat in the same way as humans, but many use panting, where their breathing is fast but shallow, to increase evaporative cooling. Spraying water on your skin (Figure 5.2.10) produces the same evaporative cooling effect as sweating.
- **vasodilation** (dilation of the blood vessels in the skin)—This means more blood is sent to the extremities. Heat is lost to the environment by radiation and convection, especially if it is windy. Furry animals often have areas of bare skin that are rich in blood vessels in order to allow vasodilation to take place.

Animals may also change their behaviour to release excess heat or to avoid increasing their body temperature. Some examples include changing body shape or increasing surface area (e.g. standing with your arms outstretched), swimming or bathing in cool water, seeking shade during the hottest parts of the day and decreasing physical activity.

FIGURE 5.2.10 Long-distance runners spray water on their skin to take advantage of evaporative cooling. This enables their bodies to work at maximum efficiency, and helps to prevent them from becoming dangerously hot.

BIOFILE

Rosy cheeks

Children often have rosy cheeks in cold weather. This is a result of increased blood flow to the cold tissues following vasodilation (opening of the blood vessels). Cold-induced vasodilation functions to warm parts of the body that have been exposed to the cold. By increasing the blood to the exposed area, the risk of injury from extreme cold is reduced. As vasodilation allows heat to escape the body, it can be maintained only for short periods in cold conditions. If the body does not warm up after a while, vasoconstriction occurs to minimise heat loss.

Rosy cheeks from exposure to cold are caused by the dilation of blood vessels (vasodilation). Vasodilation increases blood flow to exposed skin to counteract the cold.

BIOFILE

Apocrine glands

The apocrine glands are associated with hair follicles and occur mainly in the armpits and groin. These glands extend deep into the dermis and secrete proteins and fats into canals of the hair follicles. The sweat secreted by apocrine glands is usually thicker than that secreted by the eccrine glands, and is the source of 'body odour'.

REGULATION OF BLOOD GLUCOSE

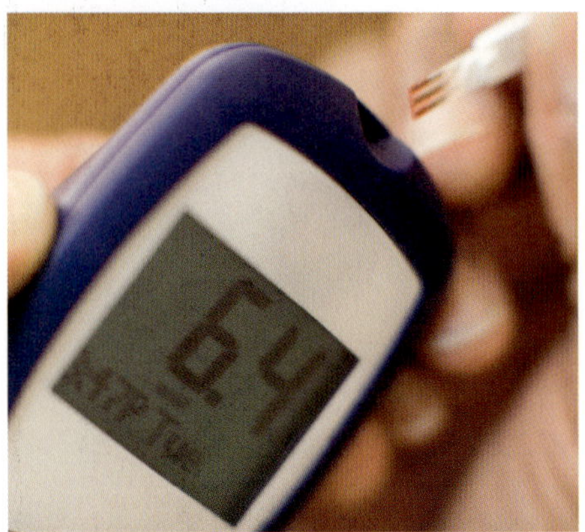

FIGURE 5.2.11 Blood glucose levels can be tested by pricking the finger and placing a small drop of blood on a piece of absorbent material, which is tested by an electronic device. The person pictured here has a BGL of 6.4 mmol/L.

Blood glucose level (BGL), sometimes called blood sugar level, is the concentration of glucose in the blood of mammals, including humans. This level is constantly changing in your body and is tightly regulated by homeostatic mechanisms. **Glucose** is the main source of energy for your body's cells. Eating carbohydrates and doing physical activity will change blood glucose levels throughout the day. The more carbohydrates you eat, the more insulin you will produce. Glucose is stored in the body in the form of a polymer called **glycogen**. When the body needs glucose, the glycogen is broken back down into usable glucose for cellular respiration, which is the energy-producing reaction in cells. If BGL is not maintained within its optimal range, **hyperglycaemia** (BGL too high) or **hypoglycaemia** (BGL too low) can develop, leading to a range of long-term health problems, such as diabetes (Figure 5.2.11).

Detecting blood glucose level change

Cells in the pancreas detect changes in BGL. Blood glucose concentration is regulated so it remains within a range of about 3.5–8 mmol/L. A deviation from these levels in either direction will result in a response by clusters of specialised cells in the pancreas, called the islets of Langerhans (Figure 5.2.12). These cells detect blood glucose levels and release insulin and glucagon to maintain blood glucose levels within the normal range. **Glucagon** is a hormone that stimulates the conversion of glycogen to glucose, which raises BGL.

There are also glucose-sensing neurons in the hypothalamus in the brain. Scientists believe that these glucose-sensing neurons play a key role in food intake, thus helping to regulate blood glucose concentrations.

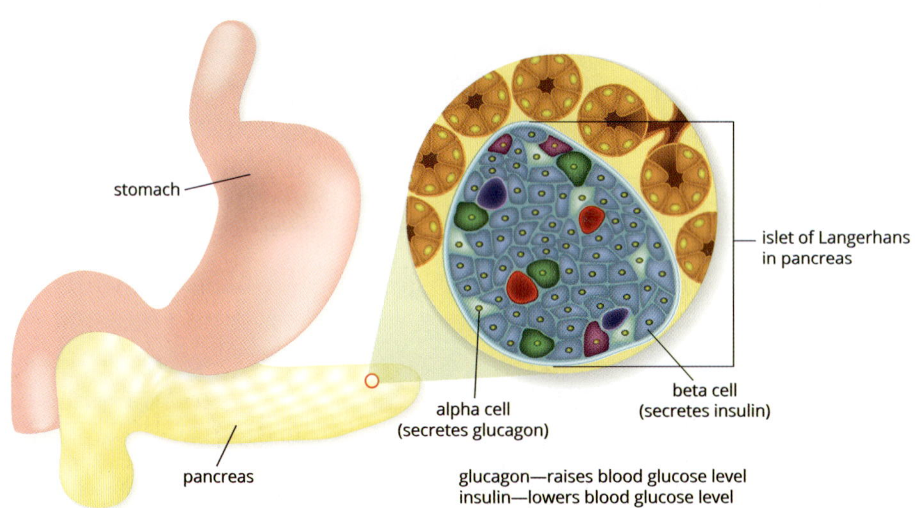

FIGURE 5.2.12 Groups of specialised cells in the pancreas, called the islets of Langerhans, detect blood glucose levels and secrete either insulin (from beta cells) or glucagon (from alpha cells) to maintain blood glucose levels within the normal range.

BIOFILE

Carbohydrates and blood glucose levels

Simple carbohydrates consist of only one or two sugar molecules (monosaccharides and disaccharides). Because of this they can be rapidly digested and absorbed, raising blood glucose levels quickly. Some sources of simple carbohydrates are table sugar, soft drinks and lollies. These foods provide a short burst of energy but contain very little nutrition.

Complex carbohydrates (polysaccharides) are starches made up of longer chains of sugars, and take longer to break down and absorb. Foods such as white bread and cakes contain refined starches, while unrefined starches are found in whole grain foods, such as brown rice and wholemeal bread, and starchy vegetables. Foods that contain unrefined complex carbohydrates provide a steady energy supply, minimising spikes in blood glucose levels, and also provide beneficial nutrients and fibre.

Responding to high blood glucose levels

When glucose levels rise above about 5 mmol/L, the islets of Langerhans in the pancreas release insulin. Insulin increases the conversion of glucose to glycogen, fats or fatty acids for storage in the liver and skeletal muscles. The overall effect of insulin is to lower BGL (Figure 5.2.13, Table 5.2.4).

Responding to low blood glucose levels

When glucose levels fall below about 5 mmol/L, the islets of Langerhans in the pancreas release glucagon (Figure 5.2.13, Table 5.2.4). Adrenaline also raises BGL by its actions on fat cells and the liver.

BIOFILE

Glycogen storage

It is often said that glycogen is synthesised and stored in the liver, but this is only partly true. Most glycogen is actually synthesised and stored in skeletal muscle cells. The human liver can store about 100 g of glycogen, whereas skeletal muscle can store about 500 g. When the glycogen-storing capacity of the body is full, excess glucose is stored mainly as triglycerides (fats).

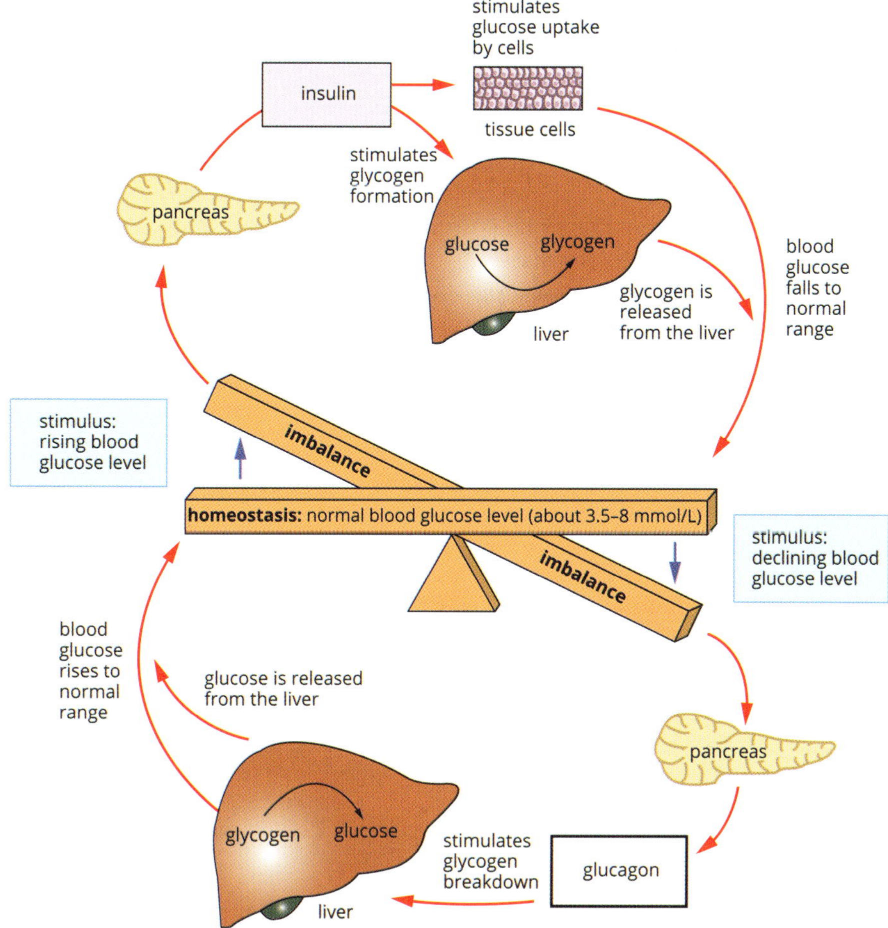

FIGURE 5.2.13 The regulation of blood glucose level (BGL) by insulin and glucagon. When BGL is too high, insulin is secreted by the beta cells in the pancreas. This stimulates glucose uptake and storage as glycogen in the liver, reducing BGL. When BGL is too low, glucagon is secreted by the alpha cells in the pancreas. This stimulates glycogen breakdown and the release of glucose from the liver, increasing BGL to within the normal range.

TABLE 5.2.4 Changes observed in blood glucose levels—hormones secreted and effects on the body

Blood glucose level	Hormone secreted	Effect on blood glucose concentration in the body
higher than normal	Insulin is produced and secreted by the beta cells of the islets of Langerhans.	Insulin stimulates glucose uptake into muscle and liver cells. Glucose is then converted into glycogen, and blood glucose levels fall.
lower than normal	Glucagon is produced and secreted by the alpha cells of the islets of Langerhans.	Glucagon stimulates the breakdown of glycogen into glucose in the liver. This releases glucose into the blood leading to raised blood glucose levels.

BIOFILE

Effects of dehydration

The effects of dehydration can range from mildly uncomfortable to life-threatening. The following table shows the typical symptoms of dehydration in humans. The progressively worsening symptoms are typical of those experienced by someone without sufficient water in a hot, dry environment, such as a person lost in a desert.

Progressive symptoms caused by dehydration

Water volume lost (%)	Symptoms
0	no symptoms
1	thirsty but not uncomfortable
2	uncomfortably strong thirst
3	dry mouth, fatigue, stumbling, headache, irritability
4	strong fatigue, nausea, lack of concentration
5	decision-making and coordination strongly impaired, sleepiness
6	reduced ability to sweat, rapid, pounding heartbeat, tingling and numbness in fingers and toes
7	fever, confusion, dry skin, inability to stand
8	dizziness, laboured breathing, extreme fatigue
9	muscle spasms, delirium
10	blood circulation impaired, kidney failure
> 10	unconsciousness, likely general organ failure and death

REGULATION OF WATER BALANCE

Some animals maintain water balance simply by living in environments where fresh water is freely available. Others can regulate the composition of their internal environment, allowing them to live in drier or saltier environments. The maintenance of an internal balance between water and solute concentrations is called **osmoregulation**.

Animals that control their cells' solute concentrations and maintain a stable internal environment are known as **osmoregulators**. Most animals are osmoregulators, including land animals, freshwater animals and some marine (saltwater) animals. Animals that conform to changes in their external environment, matching the solute concentrations in their cells to their surroundings, are known as **osmoconformers**. Osmoconformers are mostly marine invertebrates, such as sea jellies and crabs.

Maintaining water balance is necessary to control salt concentrations. Salts form ions in solution, and cells require the concentrations of ions to be held within narrow limits for biochemical processes to occur efficiently (Figure 5.2.14). Some ions (such as the hydrogen carbonate ion) are also important for regulating the pH of body fluids, which must be at a suitable pH for enzymes and other molecules to function efficiently. Maintaining the correct concentrations of ions is achieved by regulating both water and salt balance.

Water balance involves regulating the intake and loss of both water and salts. In organisms, net movement of water occurs as a result of osmosis, which is regulated by solute concentrations. Water moves across a semi-permeable membrane from regions of lower solute concentration (higher concentration of free water molecules) to regions of higher solute concentration (lower concentration of free water molecules).

The amount of water lost or gained throughout the day differs between individuals and depends on the amount of exercise, temperature, humidity, and food and fluid intake. Urination rather than water intake is a better indicator of whether an individual has good water balance. A healthy person with adequate hydration would usually urinate four to eight times per day, and the urine would be pale yellow.

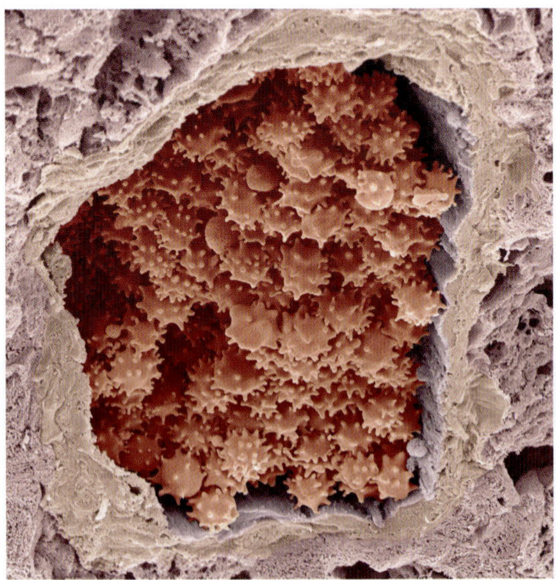

FIGURE 5.2.14 The maintenance of constant osmotic pressure in the blood is important because it prevents red cells from dehydrating or bursting. This scanning electron micrograph (SEM) of a section through an arteriole shows crenated (wrinkled) red blood cells, a condition caused by dehydration.

Water gain and loss

The total volume of fluid taken into the body depends on diet and activity levels, and typically varies from about 2 to 16 L. The minimum water requirement for fluid replacement in a 70 kg person in a cool climate is about 3000 mL per day. Of this, about 400–600 mL is obtained by eating, and about 400 mL is produced by cellular respiration. (This is called metabolic water because it is produced in cellular respiration.) The remainder of about 2000–2200 mL must be obtained by drinking.

For a person of this weight, water will be lost mainly in urine (500–1500 mL per day), evaporation from the respiratory system (400–800 mL per day), sweat (100–800 mL per day), and faeces (100–200 mL per day).

Salt gain and loss

Salt intake varies greatly depending on diet. The three major salt groups in the human diet are sodium salts, potassium salts and calcium salts.

Daily sodium salt intake (mostly as sodium chloride) ranges from about 1 to 10 g per day, mainly in bread, meat and processed cereal products. Highly processed foods usually contain more sodium salts than unprocessed foods. The recommended daily intake for Australians is 1.6 g.

Daily potassium salt intake (mainly potassium chloride and potassium citrate) ranges from about 2.0 to 4.0 g per day. The recommended daily intake for Australians is 4.7 g. Highly processed foods usually have a much lower potassium salt content that unprocessed foods.

Daily calcium salt intake (mainly in dairy foods and green vegetables) is up to about 2.4 g per day. The recommended daily intake for Australians is 1.0–1.3 g, depending on age.

Salts are lost mainly in urine but also in sweat and faeces. The kidney filters excess salts from the blood and excretes them into the urinary system. However, most salts are reabsorbed into the blood plasma for recirculation to tissues.

Hormonal control of water balance

Water and solute concentration are monitored by **osmoreceptors** in the hypothalamus and **baroreceptors** in the atria of the heart. Osmoreceptors are sensitive to blood solute concentrations, while baroreceptors detect changes in blood pressure, which is an indication of the volume of blood. Collectively, these receptors detect the solute concentration in blood and extracellular fluid. The unit of measurement used for these blood solute concentrations is **osmolality**, because they contribute to osmotic effects on cells.

Osmolality is a measure of the concentration of particles (such as sodium and chloride ions) that affect osmosis.

Because plasma membranes are permeable to water, the osmolality in the extracellular fluid is about the same as it is in the intracellular fluid (cytosol). Changes in the osmolality of the extracellular fluid will therefore affect cytosol concentrations, which can cause problems with cellular metabolic reactions. Compared to extracellular fluid, the cytosol of cells is high in potassium and magnesium and low in sodium and chloride ions.

Antidiuretic hormone (ADH), also called vasopressin, regulates water reabsorption. It is synthesised in the hypothalamus and transported to the posterior pituitary gland, where it is stored. When osmoreceptors in the hypothalamus detect an increase in the osmolality of the blood, a signal is sent to the posterior pituitary gland, and ADH is released.

ADH acts on the kidneys to increase the permeability to water of the distal tubules and collecting ducts. The collecting ducts run through the medulla of the kidney, which has high salt levels (and therefore a higher osmotic potential). This causes the absorption of water from the tubules back into the blood by osmosis, decreasing urine output; the urine becomes more concentrated and has a darker yellow colour. As the blood returns to a normal concentration, negative feedback stops the production of ADH.

Conversely, if the osmoreceptors detect a decrease in osmolality (e.g. if too much water has been taken into the body), ADH release will be stopped. This reduces the reabsorption of water and consequently increases urine volume; the urine becomes more dilute and has a paler yellow colour (Figure 5.2.15).

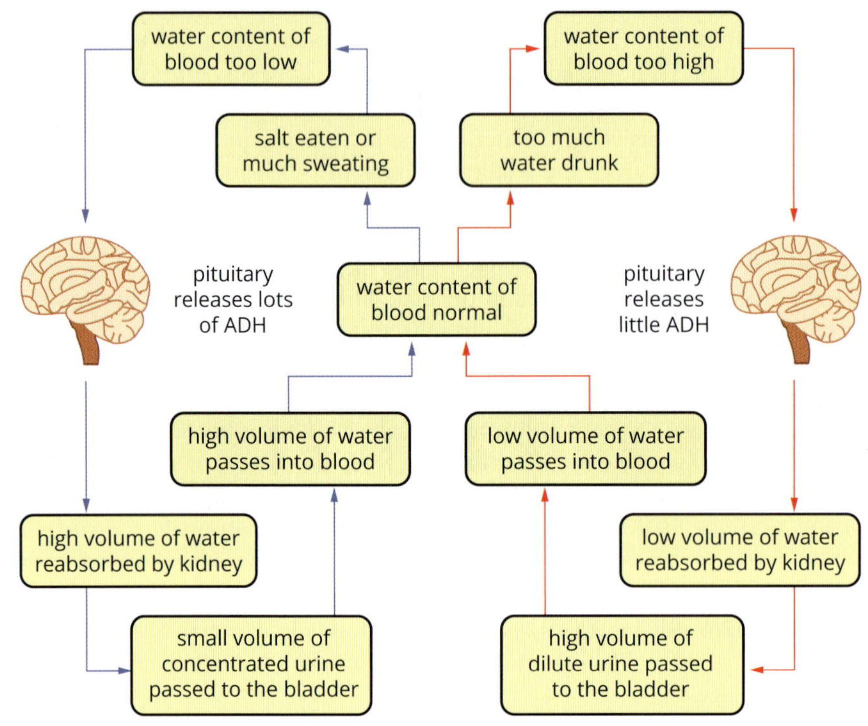

FIGURE 5.2.15 Hormonal control of water balance by antidiuretic hormone (ADH)

A number of substances such as nicotine, alcohol and narcotics can interrupt the feedback control of water balance in the body. This can also occur because of pain, stress or hypothermia (lowered body temperature).

Changes in blood osmolality or blood pressure also stimulate counteracting response. Initially an enzyme called **renin** is secreted from the kidneys in response to these changes (Figure 5.2.16). Renin then triggers a series of reactions involving other hormones that results in the release of **aldosterone** from the adrenal glands situated above the kidney. Aldosterone simultaneously regulates sodium and potassium levels by increasing potassium excretion into the urine and causing sodium reabsorption into the blood. This causes more water to be drawn into the blood by osmosis, thus increasing blood volume and pressure.

A lack of aldosterone can result in low sodium levels, high potassium levels and high acid levels in the blood. These are potentially dangerous conditions. People with an aldosterone deficiency suffer from Addison's disease and must take a synthesised hormone called fludrocortisone acetate.

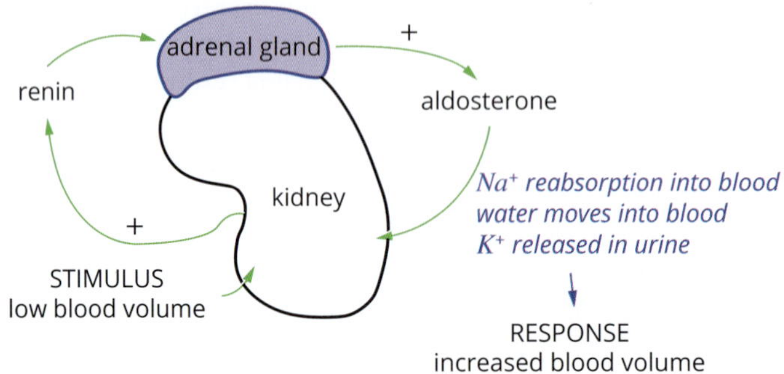

FIGURE 5.2.16 Hormonal control of sodium and potassium levels by renin and aldosterone

CASE STUDY

Salmon: osmoregulation in fresh and salt water

Some animals, like salmon, are able to survive in a relatively wide range of salt concentrations, making them different to other animals, like humans, who can only function efficiently in a narrow range of salt concentrations. The gills of salmon are adapted to transport salt ions through specialised transport mechanisms. This unique adaptation allows salmon to spend part of their lives in fresh water and part in salt water (Figure 5.2.17).

Salmon in salt water

When salmon are in salt water there will be more water molecules inside the body cells than outside the cells, so water will move out of the cells via osmosis. This leads to a high concentration of salt in the blood, which is recognised by receptors in the hypothalamus. Those receptors stimulate the pituitary gland to release into the blood a hormone that initiates the transport of salt ions (Na^+ and Cl^-) out of the body via the gills. Salmon also rely on their kidneys to remove excess salts. Since living in salt water leads to dehydration, salmon have to drink many litres of water a day. In addition, another hormone released by the pituitary gland, antidiuretic hormone (ADH), signals to the kidneys to reabsorb more water before excretion. This leads the kidneys to remove excess salts through the production of very concentrated urine.

Salmon in fresh water

When salmon are living in fresh water there will be more water molecules outside the body cells than inside the cells, so water molecules will move into the body cells via osmosis. This increased movement of water into the body cells leads to salt loss and high water levels. The hypothalamus detects the high water level and stimulates the pituitary gland to release into the blood a hormone that stimulates the gills to actively absorb salts from the water. Due to their high water levels while living in fresh water, salmon do not actively drink water and they excrete large volumes of dilute urine from their kidneys.

FIGURE 5.2.17 Red salmon (*Oncorhynchus nerka*) migrate from a freshwater river to a saltwater ocean as juveniles and return to the same river as adults to spawn (reproduce).

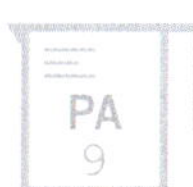

5.2 Review

Homeostasis

- Homeostasis is the maintenance of a stable internal environment within an organism.
- Regulation in animals involves internal communication by the endocrine (hormone) and nervous systems to integrate and coordinate the activities of cells, tissues, organs and systems.
- In both endocrine and nervous systems, signals are passed from one cell to the next by chemical communication—the release of signalling molecules (hormones and neurotransmitters) and their detection by matching receptors on the target cells.
- The nervous system provides rapid responses to produce efficient coordinated movement.
- Hormones are specific, effective in low concentrations and they produce responses that are generally slower and more indirect than nervous responses.

Feedback systems

- Negative feedback loops are stimulus–response mechanisms that respond to changes in the body by adjusting variables back to their original or optimal state, reversing the direction of the stimulus.
- Positive feedback loops are the opposite of negative feedback loops. They promote a process rather than reversing the effect of the stimulus.

Regulation of body temperature

- A change in the temperature of the hypothalamus initiates regulatory responses that can involve heat production or heat exchange.
- Temperature receptors are found in the skin and the hypothalamus.
- Heat is lost from the body by conduction, convection, radiation and evaporation.
- Responses to cold environmental temperatures include:
  - vasoconstriction
  - piloerection
  - shivering thermogenesis
  - non-shivering thermogenesis
  - increasing metabolism
  - TRH secretion by the hypothalamus.
- Responses to hot environmental temperatures include evaporative cooling and vasodilation.

Regulation of blood glucose

- Blood glucose levels are detected by receptor cells in the pancreas and neurons in the hypothalamus.
- When glucose levels rise, insulin is released from the beta cells in the islets of Langerhans in the pancreas. Insulin causes a decrease in BGL by acting on a number of tissues to:
  - increase conversion of glucose to fat in fat cells
  - increase uptake of glucose in muscle and fat cells
  - increase conversion of glucose to the storage compound glycogen for storage in the liver.
- When glucose levels decrease, glucagon is released from alpha cells in the islets of Langerhans and stimulates the conversion of glycogen to glucose.
- Adrenaline acts on:
  - skeletal muscle and the liver to increase breakdown of glycogen to glucose
  - fat cells to increase fat breakdown for energy.

Regulation of water balance

- Water enters body cells throughout the day from drinking, eating and cellular respiration.
- Water is mainly lost from the body in urine, faeces, sweat and from the lungs.
- Osmoreceptors in the hypothalamus and baroreceptors in the atria of the heart detect the osmolality of the blood.
- An increase in blood osmolality causes:
  - release of ADH from the pituitary; ADH acts on the kidney to increase water absorption back into the blood
  - increase in urine concentration and decrease in urine output.
- A decrease in blood osmolality causes:
  - a decrease in ADH levels
  - an increase in urine volume.
- Low blood volume stimulates the secretion of aldosterone, involving the following steps:
  - renin is secreted from the kidneys
  - renin causes release of aldosterone
  - aldosterone causes absorption of sodium into the blood
  - aldosterone causes potassium excretion into the urine
  - blood volume and blood pressure increase.

KEY QUESTIONS

Knowledge and understanding

1 What is homeostasis and why is it important?

2 List the five parts of the negative feedback loop.

3 a What are three mechanisms animals use to produce heat?
  b What are three mechanisms animals use to lose heat?

4 a What does osmolality measure?
  b Which two receptors detect changes in osmolality in the blood?

5 What is ADH and what is its role?

6 How do negative feedback loops function? Explain, using an example.

7 Compare and contrast the roles of insulin and glucagon in the human body.

Analysis

8 The urine of a healthy person with adequate hydration is a pale yellow colour, whereas the urine of a less-hydrated person is darker yellow. Explain this colour difference.

5.3 Malfunctions in homeostatic mechanisms

> A disease is any condition that impairs, or has the potential to impair, the normal functioning of the body.

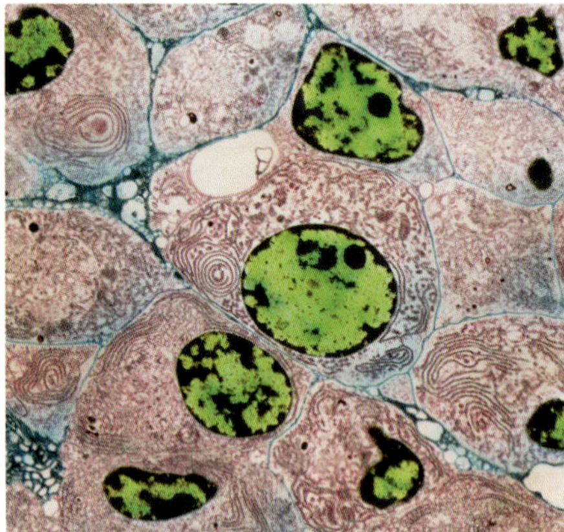
FIGURE 5.3.1 Coloured transmission electron micrograph (TEM) of thyroid cancer cells

> *'Hyper'* is a prefix that indicates an excess of a certain kind, whereas *'hypo'* is a prefix that indicates a deficiency.

> An autoimmune disease occurs when a person's immune system mistakenly targets the body's own cells. Examples of autoimmune diseases are type 1 diabetes, Graves' disease and multiple sclerosis.

The regulation of the internal environment, and thus the functionality and health of your body, is maintained by homeostatic mechanisms. These mechanisms are extremely sensitive to changes in internal conditions. However, homeostatic mechanisms can malfunction because of factors such as genetic disorders, ageing, poor nutrition, insufficient physical activity or exposure to harmful substances.

A malfunction in a homeostatic mechanism causes an imbalance and a subsequent oversupply or undersupply of substances to cells. Many diseases are associated with malfunctions in homeostatic mechanisms.

Hormonal balance plays a key role in regulating the internal environment. The endocrine system makes, stores and releases hormones, which act as chemical messengers that trigger and direct functions in the body. The endocrine system controls vital functions in growth and development, metabolism, reproduction and tissue repair, among many others. Malfunctions in the endocrine system often lead to a disruption in a homeostatic mechanism, which can have an adverse effect on the body (Figure 5.3.1).

Diseases of the endocrine system fall into three groups:

- **hypersecretion** (oversupply) of hormones
- **hyposecretion** (undersupply) of hormones
- cancers of endocrine glands.

Examples of conditions and diseases involving the endocrine glands include hypoglycaemia, diabetes, Graves' disease, Cushing's disease, acromegaly, congenital hypothyroidism, Addison's disease and hyperthyroidism. In this section you will explore type 1 diabetes, hypoglycaemia and hyperthyroidism in detail.

TYPE 1 DIABETES

Type 1 diabetes is caused by a malfunction of the pancreas, which leads to a deficiency in insulin secretion. It is an **autoimmune disease** in which the body's immune system destroys the insulin-producing beta cells in the islets of Langerhans in the pancreas (Figures 5.3.2). Insulin allows the muscle, liver and fat cells to absorb glucose from the blood and store it until the body needs energy, for example in between meals. Treatment of type 1 diabetes involves artificially increasing the insulin supply by injections or an insulin pump.

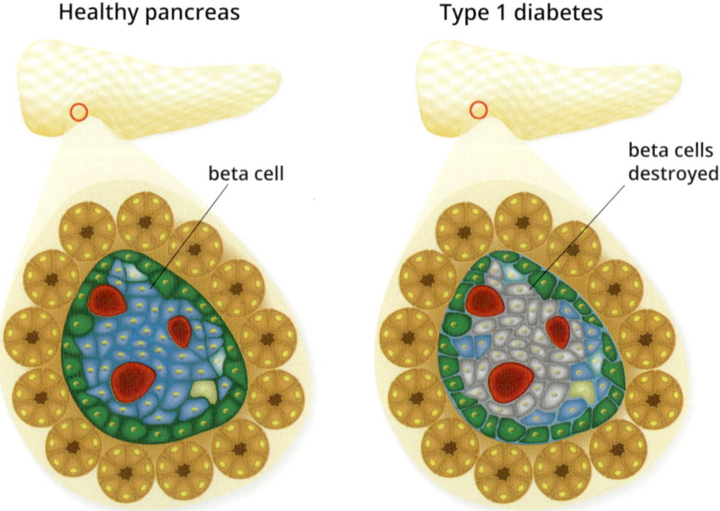

FIGURE 5.3.2 In type 1 diabetes, beta cells in the islets of Langerhans are destroyed. This means that the pancreas cannot secrete enough insulin to convert glucose to glycogen, and blood glucose levels can increase to dangerous levels.

The onset of type 1 diabetes symptoms often occurs in childhood to early adulthood, and there is currently no cure or way of preventing the disease. People with type 1 diabetes must monitor their blood glucose levels (BGL) and inject or pump insulin into their bodies every day. Type 2 diabetes is also a disturbance of glucose homeostasis. In this disorder the pancreas still makes insulin, but the body's cells do not respond. This form of diabetes often appears later in life, but is increasingly common in younger people, and is more common in those who are overweight or obese and have a sedentary lifestyle. It can often be managed by diet, exercise and medication.

Causes of type 1 diabetes

Scientists are unsure what causes the beta cells in the islets of Langerhans in the pancreas to be destroyed in type 1 diabetes. There is some evidence for a link between the Coxsackie A and B4 viruses (which are common in children) and the onset of the autoimmune disease. Other childhood viruses, including enterovirus, mumps, polio and rubella, have also been suggested as triggers for type 1 diabetes. Without functioning beta cells, the body cannot secrete the insulin required to convert glucose to glycogen in the liver and to stimulate glucose uptake into muscle and fat. This causes blood glucose levels to increase to dangerously high levels (Figure 5.3.3).

Symptoms of type 1 diabetes

Insulin deficiency results in hyperglycaemia (high blood glucose levels) and accelerates the breakdown of fat for the body to use as energy. Symptoms of the disease include:

- glucose in the urine
- increased urine production
- excessive thirst
- excessive hunger
- ketosis
- weight loss
- fatigue
- blurred vision
- irritability
- muscle cramps
- skin infections
- delayed wound healing
- tingling or numbness in the feet.

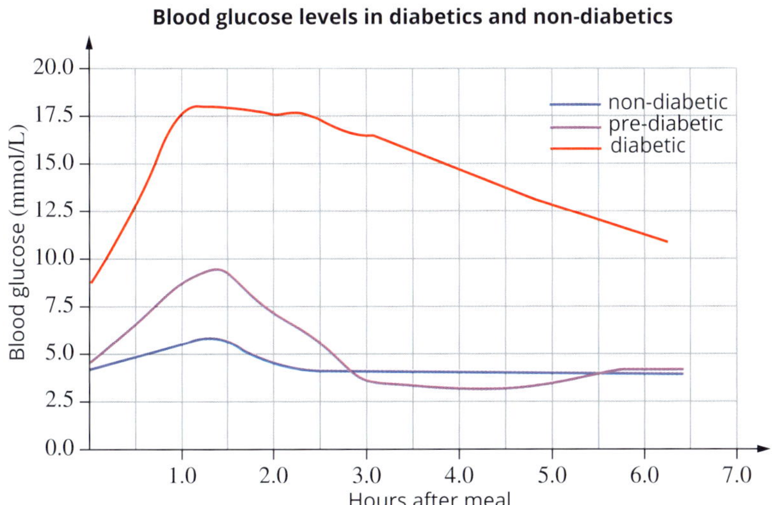

FIGURE 5.3.3 Graph indicating blood glucose levels after a meal in pre-diabetic and diabetic patients compared with a non-diabetic control group

BIOFILE

Diabetic complications

People with type 1 diabetes are at risk of several serious and even fatal conditions. In the short term, ketoacidosis (the release of ketones into the blood for energy production when insulin is not available) can lead to severe dehydration, vomiting, blurred vision and fainting. In extreme cases the person may become comatose.

Long-term risks include heart failure, vision impairment and possibly blindness, obesity, slow wound healing, skin infections, nerve damage causing a loss of sensation in the limbs, and insufficient blood supply to the hands and feet.

Injecting too much insulin, or over-exercising after injecting insulin, can result in dangerously low blood glucose levels (hypoglycaemia). This can result in a rapid loss of consciousness if not treated promptly. The usual first aid treatment is drinking fruit juice or sucking on a lolly such as barley sugar to boost blood glucose levels quickly (see figure below). This can be followed by more complex carbohydrates once the patient is more alert.

First aid for hypoglycaemia usually involves drinking fruit juice or sucking on a lolly to boost blood glucose levels quickly.

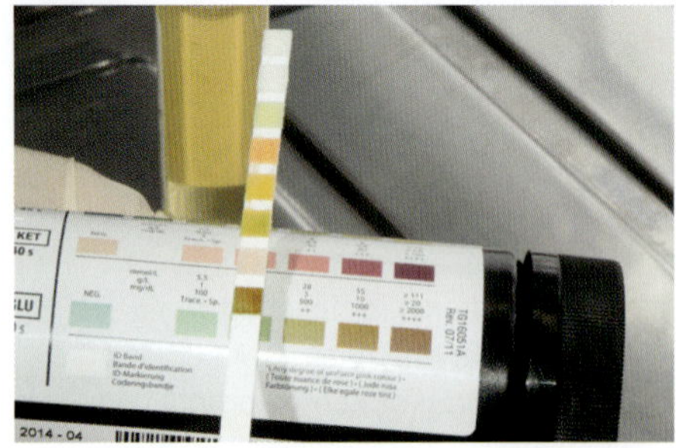

FIGURE 5.3.4 A dipstick test for glucose in urine. The pad is dipped in the urine, and the colour of the pad is checked against the chart. This gives an estimate of the glucose level in the urine. A high level is usually an indication of diabetes, although other conditions or the use of certain medications can cause high glucose levels.

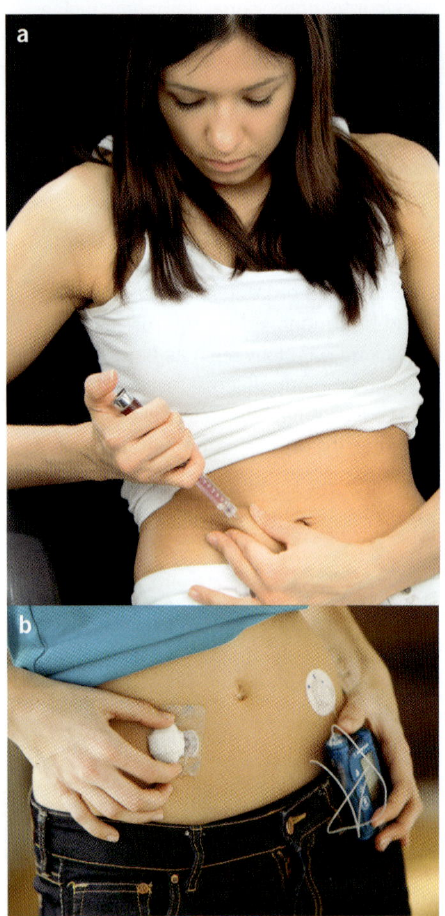

FIGURE 5.3.5 (a) A diabetic woman injecting insulin with an insulin pen, which delivers the correct dose of insulin. (b) This person is wearing a continuous glucose monitoring device and insulin pump. The monitor measures blood glucose levels and sends the levels to the pump to help the pump's internal calculator recommend an insulin dose.

Longer-term consequences are kidney and eye disease. All of these symptoms occur because of the elevated levels of glucose in the blood. Glucose is excreted in the urine because the raised blood glucose levels exceed the filtration capacity of the kidneys (normally the kidneys prevent glucose from entering the urine). Glucose escaping into the nephron tubules draws in more water, by osmosis, increasing the volume of urine produced. As a result, more frequent urination leaves the body dehydrated and feeling thirsty. The presence of glucose in urine is a simple test for diabetes (Figure 5.3.4).

Dehydration can lead to blurred vision as the lens loses moisture and the blood vessels are damaged. This can result in blindness if left untreated. The raised glucose levels in blood cause chemical reactions with molecules on the surface of neurons and cells lining the small blood vessels. The resulting damage to the body's nerves can result in loss of sensation in limbs, while damage to capillaries contributes to kidney malfunction and eye disease (diabetic retinopathy).

Management of type 1 diabetes

Artificial insulin

As well as managing their diet, people with type 1 diabetes must receive insulin artificially. This is usually by injection (Figure 5.3.5a). Blood glucose levels are monitored by pricking a finger and testing a small drop of blood with a blood glucose meter or a chemical strip.

Alternatively, an electrode placed under the skin and connected to a continuous glucose monitoring device warns a person when their glucose level is reaching a high (or low) level (Figure 5.3.5b). The monitor can be coupled to an electronic pump that delivers insulin when blood glucose levels reach a predetermined level. An improved system called an 'artificial pancreas' is currently being tested in several countries, including Australia. This system uses a monitoring and feedback system to deliver insulin as the body requires it, in the same way that the pancreas produces insulin.

Transplants

Pancreas transplants from deceased donors are usually given to patients with serious complications from diabetes. Human pancreas cells can also be transplanted into a patient's liver, where they begin to produce insulin. This process is called pancreatic islet transplantation. Although it is still in the experimental stage, it may become widely available in the next few years. Recipients of pancreas or pancreatic islet transplants must take **immuno-suppressant drugs** for the rest of their lives to prevent their bodies from rejecting the transplanted organ. These drugs can have side effects such as high blood pressure, fatigue, and increased risk of bacterial and viral infections.

Gene therapy

Gene therapy, in which the gene that codes for insulin is inserted into the patient's cells, is a potential future treatment for diabetes. Trials in the USA have been successful in diabetic rats, targeting the liver because of the organ's regenerating ability. A major benefit of gene therapy is that patients would not require immunosuppressant drugs.

PA 7

HYPOGLYCAEMIA

As discussed in Section 5.2, hypoglycaemia is a lower than normal blood glucose level (BGL), i.e. less than 4 mmol/L. A lot of people mistakenly think that hypoglycaemia only happens to people with diabetes, but it may develop in non-diabetic people. Hypoglycaemia may be caused by an overproduction of insulin or an underlying reason such as anorexia, excessive alcohol consumption or even pregnancy.

Causes of hypoglycaemia

Hypoglycaemia is divided into two main categories: reactive hypoglycaemia and fasting (non-reactive) hypoglycaemia.

Reactive hypoglyceamia is the most common form, and occurs 3–5 hours after a meal. The cause of reactive hypoglyceamia is an overproduction of insulin (Figure 5.3.6). Insulin released from the pancreas stimulates glucose uptake by tissue cells and glycogen formation in the liver, which in turn leads a decrease in blood glucose levels.

Fasting hypoglycaemia is rare and does not occur after a meal like reactive hypoglycaemia. Fasting hypoglycaemia appears in people suffering from severe diseases such as pancreatic tumours, extensive liver damage (e.g. from severe alcoholism), prolonged starvation (e.g. anorexia) and various cancers. It can also occur during pregnancy.

FIGURE 5.3.6 Excessive insulin production causes low blood glucose levels due to an increased uptake of glucose in muscle and fat cells and a reduction in glucose production.

Symptoms of hypoglycaemia

Glucose is the primary fuel for the brain, so when glucose levels are low the first effects observed are brain related. Symptoms of hypoglycaemia may differ depending on the cause, but common symptoms include:

- shaking
- confusion
- headaches
- dizziness
- moodiness
- drowsiness
- seizures or convulsions
- sweating
- feeling anxious and nervous
- nausea
- hunger
- feeling sleepy
- faster heartbeat
- pale skin.

Management of hypoglycaemia

To be able to manage and treat hypoglycaemia it is important to identify the cause. In some cases simply eating sugary foods may help raise the blood glucose levels fast. Other, more severe cases of hypoglycaemia may be managed by an injection of glucagon. Glucagon is a hormone that stimulates the conversion of glycogen to glucose, which raises blood glucose levels.

BIOFILE

Insulin from microbes

Insulin for human use was once extracted from animals, mostly pigs, but it is now produced mainly by genetically engineered microbes. The gene that codes for insulin is inserted into yeast or bacteria (e.g. *Escherichia coli*), which are cloned so they can produce large quantities of insulin (see figure below). These organisms are grown in culture and produce insulin on

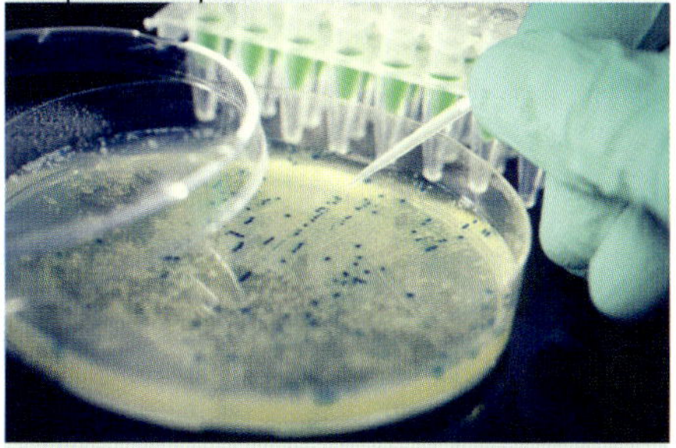

Bacteria with the insulin gene are cultured then grown to produce large quantities of insulin.

CASE STUDY

Brown fat and diabetes management

A study led by endocrinologist Dr Paul Lee at the Garvan Institute of Medical Research in Sydney is investigating the link between brown fat and diabetes. Brown fat is rich in mitochondria and plays an important role in generating heat in hibernating animals and babies. Cool environments promote the growth of brown fat, while warm environments suppress it. Dr Lee wanted to investigate how brown fat is regulated in humans, its role in metabolism, and how this influences blood glucose levels and diabetes.

In this research, study participants slept in temperature-controlled rooms for four months: 24°C for the first month (the body does not have to work to produce or lose heat at this temperature), 19°C for the second month, 24°C for the third month and then 27°C for the last month. Participants completed a 'thermal metabolic evaluation' at the end of each month.

As expected, brown fat increased during the cooler month (19°C) and decreased during the warmer month (27°C).

The researchers also found that increased brown fat led to heightened insulin sensitivity. This means that people with more brown fat need less insulin to bring their blood glucose levels down. With increased efficiency in household heating over the last few decades, the average home temperature in the UK and USA has risen from 19°C to 22°C. This temperature change is enough to reduce brown fat production.

The researchers speculate that this shift in household temperature, along with unhealthy diets and lack of exercise, may have contributed to the rise in obesity in these populations. The findings from this research indicate that people with diabetes may be able to regulate their brown fat deposits, making themselves more sensitive to insulin and therefore less reliant on large doses of insulin. This research could also open new avenues for diabetes management.

HYPERTHYROIDISM

Hormones secreted by the thyroid gland interact with cells throughout the body. They are responsible for regulating growth, development and metabolic rate, along with many other vital functions. Malfunction of the thyroid can therefore have widespread and serious effects on a range of organs and bodily functions (Table 5.3.1).

Hyperthyroidism is a condition in which excess amounts of the hormones triiodothyronine (T3) and thyroxine (T4) are secreted by the thyroid gland. T3 and T4 are made from an amino acid (tyrosine) and contain iodine. Blood tests for these hormones and thyroid stimulating hormone (TSH) are used to diagnose the condition. When T3 and T4 are oversupplied by the thyroid, a negative feedback message is sent to the hypothalamus to decrease the release of thyrotropin releasing hormone (TRH). This in turn decreases the synthesis of TSH from the pituitary gland. Increased metabolic activity is only one of the effects of hyperthyroidism listed in Table 5.3.1, which causes the body to work harder and faster.

A positive blood test for hyperthyroidism shows elevated levels of T3 and T4 and decreased TSH levels. Thyroid malfunctions affect about 6–7% of the population. Hyperthyroidism occurs in about 2% of the Australian population. The disease most commonly affects people over the age of 60 and women are 5 to 8 times more likely to develop hyperthyroidism.

In **hypothyroidism** the thyroid produces less T3 and T4 than the body needs. Hypothyroidism also can have serious effects on health, including decreased glucose metabolism, low heart rate and blood pressure, sluggish muscle action and depressed ovarian function.

TABLE 5.3.1 The major effects of thyroid hormones T3 and T4 in the human body. Hyperthyroidism results from an excess of T3 and T4, secreted from an overactive thyroid gland (hypersecretion).

Process or system affected	Normal physiological effects (T3 and T4 hormones within normal range)	Effects of hyperthyroidism (excess T3 and T4 hormones)
basal metabolic rate (BMR)/ temperature regulation	promotes normal oxygen use and BMR; heat production via the digestion of food and thyroid hormones; enhances effects of sympathetic nervous system	BMR above normal; increased body temperature, heat intolerance; increased appetite; weight loss
carbohydrate/lipid/ protein metabolism	promotes glucose metabolism; mobilises fats; essential for protein synthesis; enhances liver's synthesis of cholesterol	enhanced breakdown of glucose, proteins and fats; weight loss; loss of muscle mass
nervous system	promotes normal development of nervous system in fetus and infant; promotes normal adult nervous system function	irritability, restlessness, insomnia, personality changes, bulging eyes (in Graves' disease)
cardiovascular system	promotes normal functioning of the heart	increased sensitivity to adrenal gland hormones (e.g. adrenaline and dopamine) leads to rapid heart rate and possible palpitations; high blood pressure; if prolonged, heart failure
muscular system	promotes normal muscular development and function	muscle atrophy and weakness
skeletal system	promotes normal growth and maturation of the skeleton	in children, excessive skeletal growth initially, followed by early epiphyseal closure and short stature; in adults, demineralisation of skeleton
gastrointestinal (GI) system	promotes normal GI motility and tone; increases secretion of digestive juices	excessive GI motility; diarrhoea; loss of appetite
reproductive system	promotes female reproductive ability and lactation	in females, depressed ovarian function; in males, impotence
integumentary system	promotes normal hydration and secretory activity of skin	skin flushed, thin and moist; hair fine and soft; nails soft and thin

Causes of hyperthyroidism

There are many causes of hyperthyroidism. The most common cause is an autoimmune disease called Graves' disease. In patients with Graves' disease, the immune system makes an antibody called thyroid stimulating immunoglobulin (TSI), which mimics TSH, stimulating the thyroid to make more T3 and T4 hormones than the body needs (Figures 5.3.7 and 5.3.8 on page 230). The trigger for the production of the antibody is unknown, but a combination of environmental and genetic factors are thought to contribute. Infection caused by some viruses and bacteria, stress, childbirth, excess iodine (through food or contrast dyes used for imaging) and some medications have been linked to the onset of Graves' disease. People with other autoimmune diseases such as type 1 diabetes, as well as smokers and those with tumours of the testes or ovaries, have a higher risk of developing the disease.

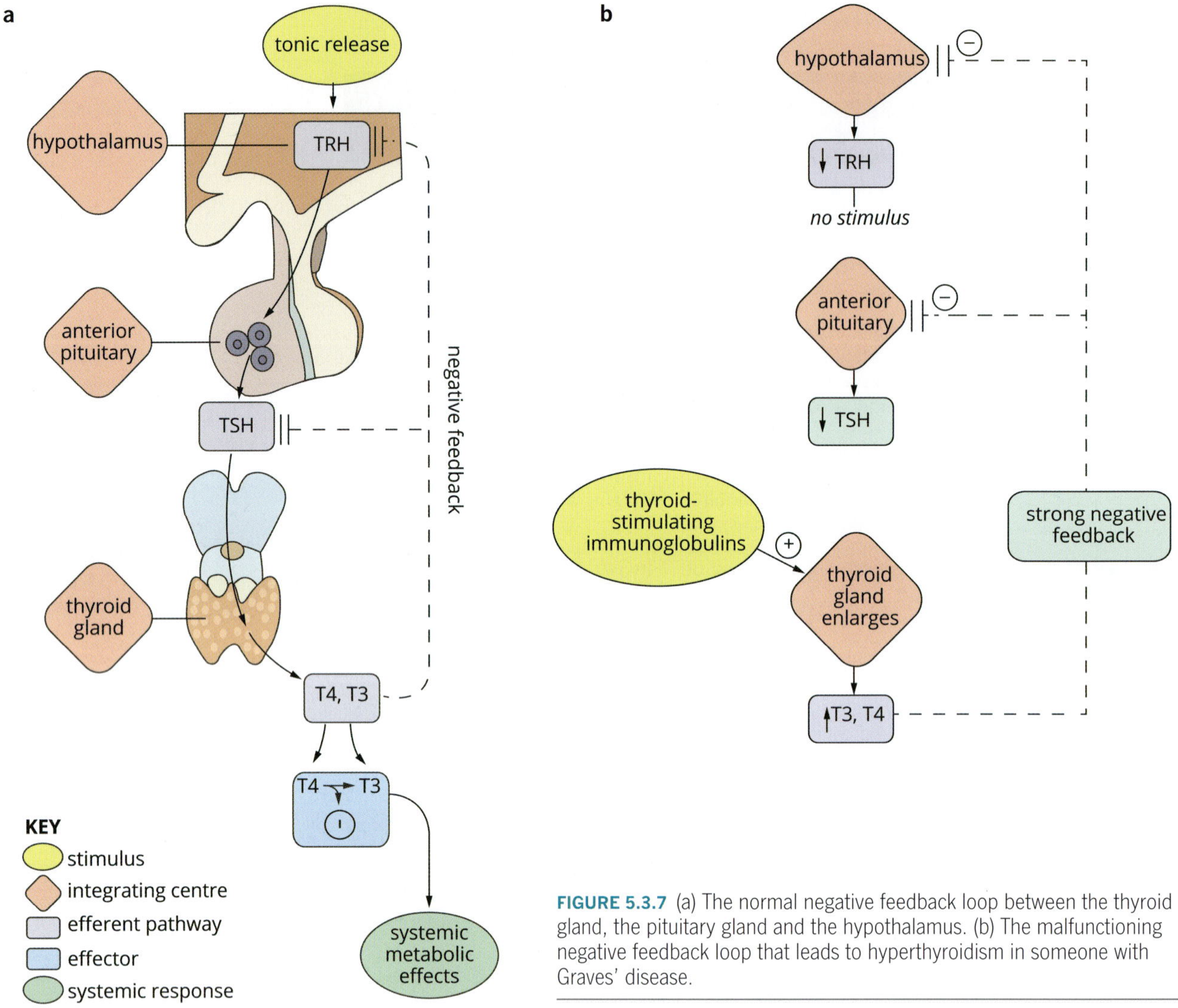

FIGURE 5.3.7 (a) The normal negative feedback loop between the thyroid gland, the pituitary gland and the hypothalamus. (b) The malfunctioning negative feedback loop that leads to hyperthyroidism in someone with Graves' disease.

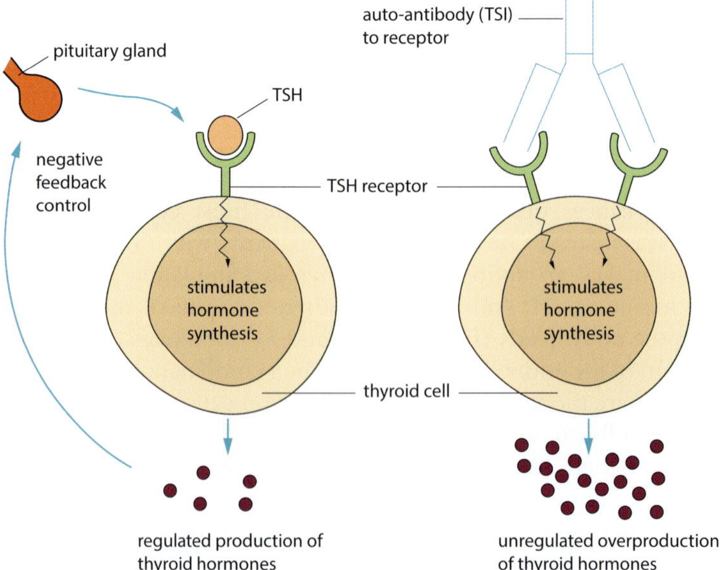

FIGURE 5.3.8 Thyroid stimulating immunoglobulin (TSI) acts on cells in the same way as thyroid stimulating hormone (TSH) to trigger the synthesis and release of T3 and T4. This results in an excess of these thyroid hormones and hyperthyroidism in people with Graves' disease.

Symptoms of hyperthyroidism

The symptoms of hyperthyroidism can include some or all of the following:

- goitre (a visibly enlarged thyroid) or thyroid nodules (Figure 5.3.9a)
- weight loss
- rapid heartbeat (tachycardia)
- irregular heartbeat (arrhythmia)
- pounding heart (palpitations)
- increased appetite
- nervousness, anxiety and irritability
- changes in bowel patterns (more frequent)
- fatigue, muscle weakness
- tremor—usually a fine trembling in hands and fingers
- breast development in men
- bulging eyes (exophthalmos) (Figure 5.3.9b)
- nausea and diarrhoea
- sweating and heat intolerance
- changes in menstrual patterns, possibly including light or absent menstrual periods
- trouble sleeping
- skin thinning, blushing, flushing or being itchy or clammy
- fine, brittle hair or hair loss.

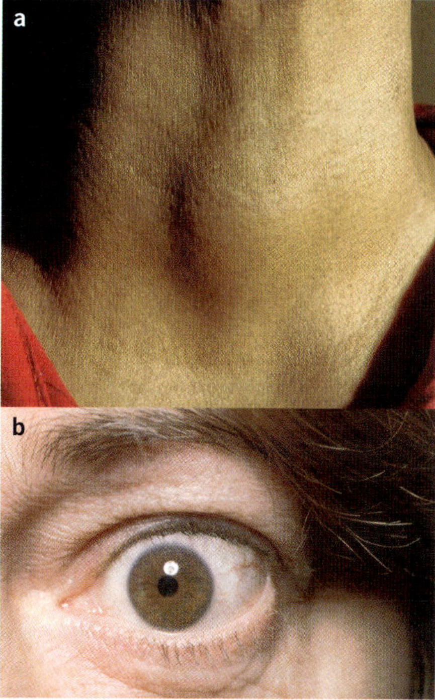

FIGURE 5.3.9 Two common symptoms of hyperthyroidism. (a) An enlarged thyroid gland, called a goitre. (b) In Graves' disease, bulging eyes result when thyroid stimulating immunoglobulin (TSI) produces inflammation and swelling of the soft tissues in the eye.

Management of hyperthyroidism

The symptoms of hyperthyroidism can be treated with drugs called beta-blockers, which act on the circulatory and nervous systems to slow down the increased heartbeats and tremors associated with the disease. However, these drugs do not have an effect on the thyroid itself.

Sometimes anti-thyroid drugs are prescribed to interfere with the thyroid's ability to make hormones. These drugs act on the thyroid gland to slow the production of hormones to normal levels and reduce or eliminate the symptoms. Only 20–30% of patients have long-term success in treating hyperthyroidism with anti-thyroid drugs.

Radioactive iodine treatment is the most widely used permanent treatment for hyperthyroidism. (Iodine is needed to make T3 and T4 hormones.) Thyroid cells are the only cells in the body that absorb iodine. Taken orally, the radioactive iodine (I-131) is absorbed by the thyroid cells. I-131 emits beta radiation, which kills thyroid cells.

Surgery is another permanent cure and involves the removal of all or part of the thyroid. Removal or destruction of the thyroid means the patient will suffer from hypothyroidism, and must take thyroid pills such as levothyroxine for the rest of their life.

BIOFILE

Hypothyroidism

In contrast to hyperthyroidism, underactivity of the thyroid gland can cause hypothyroidism as a result of insufficient production of the thyroid hormones. Symptoms of this disease in adults are weight gain, lethargy, slow heart rate, hair loss and sensitivity to cold. In children, hypothyroidism can cause delays in growth and mental development. Babies are screened for this disorder a few days after birth. Raised levels of TSH in the blood indicate possible hypothyroidism. The main cause of hypothyroidism is lack of iodine in the diet, which is essential for the production of thyroid hormones. This deficiency can be resolved by adding potassium iodide to table salt. In countries where iodine deficiency is not a widespread problem, the most common cause of hypothyroidism is an autoimmune disease called Hashimoto's thyroiditis.

5.3 Review

OA ✓✓

SUMMARY

- Malfunctions in homeostatic mechanisms lead to imbalances and a subsequent oversupply or undersupply of substances needed by cells.
- Many diseases are associated with malfunctions in homeostatic mechanisms.
- The endocrine system is particularly important for maintaining homeostasis, and malfunctions in this system can affect the whole body.

Diabetes

- There are two types of diabetes: type 1 (genetic and may be caused by a virus) and type 2 (often late onset and related to lifestyle).
- The cause of type 1 diabetes is the autoimmune destruction of the insulin-producing beta cells in the islets of Langerhans. People with type 1 diabetes do not produce enough insulin. Without treatment, blood glucose concentration can rise to dangerous levels.
- Symptoms of type 1 diabetes include glucose in the urine, increased urine production, excessive thirst, weight loss, fatigue, blurred vision, irritability, muscle cramps, skin infections and delayed wound healing.
- Treatment and management of type 1 diabetes includes daily insulin supplements by injection or insulin pump, and a strictly controlled diet. Pancreas transplants may be required for people with serious complications caused by type 1 diabetes. Potential treatments include pancreatic islets transplants, an artificial pancreas and gene therapy.

Hypoglycaemia

- Hypoglycaemia is the condition where a person experiences low blood glucose levels (BGL), less than 4 mmol/L.
- Hypoglycaemia can be caused by an overproduction of insulin or other underlying conditions, such as excess alcohol consumption.
- Symptoms of hypoglycaemia include shaking, dizziness, anxiety, hunger, nausea and fatigue.
- Treatment for hypoglycaemia includes eating sugary foods or an injection of glucagon to stimulate the conversion of glycogen to glucose to raise BGL.

Hyperthyroidism

- Hyperthyroidism is the over-secretion of the thyroid hormones triiodithyronine (T3) and thyroxine (T4).
- Symptoms of hyperthyroidism include goitre, weight loss, heartbeat irregularities, increased appetite, nervousness, changes in bowel movements, fatigue, tremor, bulging eyes, nausea, heat intolerance, changes in menstrual patterns, and hair loss.
- Treatment for hyperthyroidism involves the use of beta-blockers, anti-thyroid drugs or radioactive iodine I-131, or surgery.

KEY QUESTIONS

Knowledge and understanding

1 Type 1 diabetes is an autoimmune disease. What effect does type 1 diabetes have on the pancreas and blood glucose levels?

2 What is the difference between reactive hypoglycaemia and fasting hypoglycaemia?

3 Define hyperthyroidism.

4 **a** How is thyroid disease diagnosed?
  b How would blood test results differ between someone with hyperthyroidism and hypothyroidism?

5 Discuss how malfunctions in homeostatic mechanisms can lead to disease, using an example.

Analysis

6 Use Figure 5.3.3 on page 225 to answer the following questions.
  a What is the blood glucose concentration of each of the three patients an hour after the meal?
  b Describe two trends that can be seen in the graph of the diabetic patient.
  c Explain why there is such a big difference between the blood glucose levels of the diabetic patient and the non-diabetic patient an hour after the meal.

7 Why do people with diabetes sometimes need to have sweet drinks or food?

8 Infer how brown fat might be important in diabetes management.

Chapter review

OA ✓✓

05

KEY TERMS

active transport
aldosterone
antidiuretic hormone (ADH)
autoimmune disease
baroreceptor
blood glucose level (BGL)
Casparian strip
chemoreceptor
cohesive bond
conduction
control centre
convection
cytoplasmic pathway
diffusion
effector
endocrine system
endotherm
evaporation
evaporative cooling
exteroreceptor
extracellular pathway
glucagon
glucose
glycogen
guard cell
homeostasis
hormone
hyperglycaemia
hypersecretion
hyperthyroidism
hypoglycaemia
hyposecretion
hypothalamus
hypothyroidism
immuno-suppressant drug
insulin
interoreceptor
lenticel
mechanoreceptor
metabolism
negative feedback loop
nervous system
neurotransmitter
non-shivering thermogenesis
osmoconformer
osmolality
osmoreceptor
osmoregulation
osmoregulator
osmosis
osmotic gradient
phloem
photoreceptor
piloerection
plasmodesmata (sing. plasmodesma)
positive feedback loop
radiation
receptor
renin
response
root hair
root pressure
shivering thermogenesis
signalling molecule
sink
source
stimulus
stimulus–response mechanism
stomata (sing. stoma)
thermoreceptor
thermoregulation
thyroid stimulating hormone (TSH)
thyrotropin releasing hormone (TRH)
translocation
transpiration
transpiration stream
turgor
type 1 diabetes
vascular plant
vascular tissue
vasoconstriction
vasodilation
xylem

REVIEW QUESTIONS

Knowledge and understanding

1 What is meant by the terms 'sources' and 'sinks'?

2 Create a summary table comparing xylem and phloem transport. Outline what each tissue transports and the direction of transport.

3 List the key factors that influence the rate of transpiration in plants.

4 Describe the pathway of water absorption in the diagram of plant root tissue below.

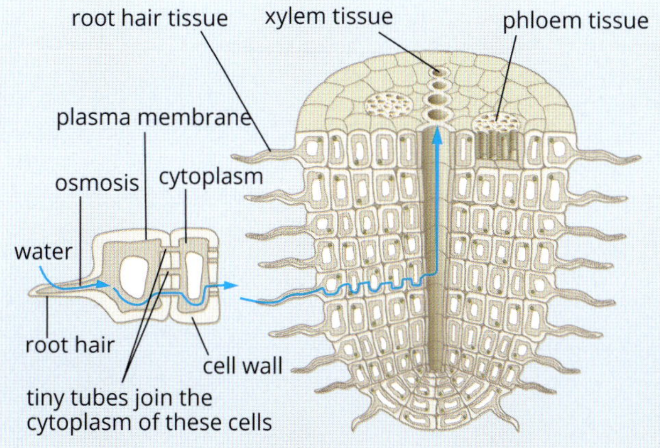

5 Why is it important for an organism to be able to regulate its internal environment?

6 Arrange the following terms from first to last in the order of their involvement in a physiological response: control centre, effector, receptor, response, stimulus.

7 What is the similarity between neurotransmitters and hormones?

8 How do organisms exchange heat with their environment? Explain each of the four methods of heat exchange, using examples.

9 Summarise the thermoregulatory mechanisms that occur during and immediately after a fever.

10 **a** Explain the principle of negative feedback in homeostasis.

b Using a diagram, explain how a low body temperature can be increased. In your diagram, draw and label an arrow to show where negative feedback occurs.

11 Draw a table to list ways a healthy person may gain or lose water in a day.

12 What change in the blood acts as a stimulus for ADH release?

13 Describe how negative feedback is involved in ADH action and water balance.

14 Diseases of the endocrine system fall into three groups. What are they? Provide an example of each.

15 List three ways of managing type 1 diabetes.

Application and analysis

16 Celery curls are an attractive way of serving celery. They are made by taking sections of celery stalk, making several lengthwise cuts in one end and submerging them in cold water. The cut parts of the celery start to curl. If the cuts are made too close together, long strings of celery will peel away. Explain what causes the celery ends to curl.

17 **a** What is the association of gas exchange with transpiration in plants?

b How does wind affect transpiration rates?

c Why is the rate of transpiration lower at night?

18 The rates of transpiration in plants were measured in three different environments over a period of 6 hours. The results are listed below.

Environment 1—23°C, night, with wind speeds of 15 km/h and an average humidity of 88%

Environment 2—24°C, day, with wind speeds of 15 km/h and an average humidity of 84%

Environment 3—33°C, day, with wind speeds of 20 km/h and an average humidity of 80%

a Order the environments from highest to lowest expected rates of transpiration and give reasons to support your answer.

b Identify the variables that would need to be kept constant when measuring rates of transpiration for your conclusion to be valid.

19 Occasionally a person is born without sweat glands, so they cannot lose heat through sweating. A person without sweat glands and a person with normal sweat glands were placed in cool, dry conditions and their skin and mouth temperatures were recorded. The two people were then placed in a moist, hot environment, and further recordings were made. The results of the experiment are recorded in the table. Deduce which person (A or B) was born without sweat glands. Explain your reasoning.

Response	Person A		Person B	
	Cool, dry	Hot, moist	Cool, dry	Hot, moist
skin temperature (°C)	33.8	40.5	32.7	37.4
oral temperature (°C)	36.9	38.6	36.8	37.2
water loss from skin and lungs (mL)	not recorded	20.0	not recorded	282.0
urine volume (mL)	not recorded	280.5	not recorded	12.6

20 **a** Copy and complete the following table to summarise the two hormones that play a vital role in blood glucose regulation.

Hormone	Site of production	Target organs	Main functions
insulin			
glucagon			

b Name the hormone that is produced in insufficient amounts in type 1 diabetes.

c Blood glucose levels in a non-diabetic and a diabetic person after eating similar meals are shown in the following graph. Which line on the graph represents the diabetic individual? Explain your reasoning, using the data presented in the graph.

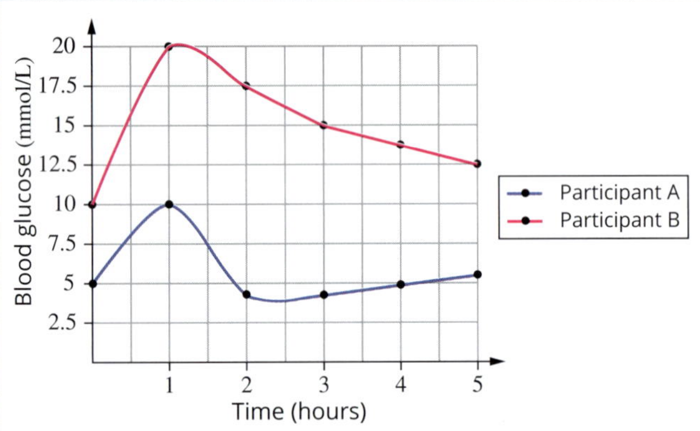

OA ✓✓

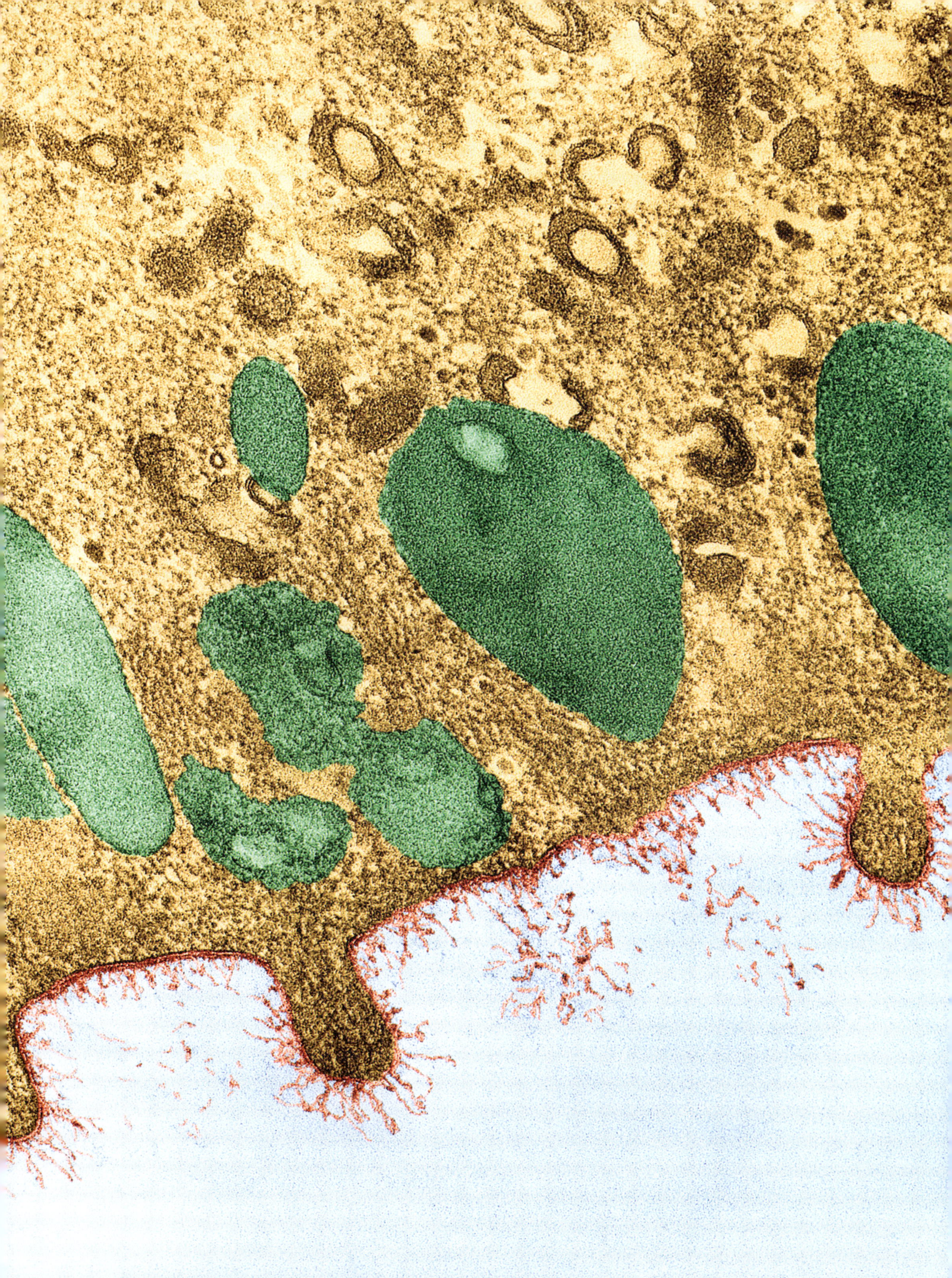

REVIEW QUESTIONS

How do plant and animal systems function?

Multiple-choice questions

1 Identify the list that contains only the names of organs.
  A root, leaf, stem, fruit
  B artery, eye, skin, liver
  C stoma, flower, fruit, chloroplast
  D kidney, nucleus, ovary, lung

2 Identify the list that contains only the names of tissues.
  A xylem, phloem, cuticle, vacuole
  B muscle, nerve, connective, blood
  C cardiac, digestive, epithelial, spleen
  D chloroplast, mitochondrion, nucleus, cytoplasm

3 Identify the list that contains only the names of systems.
  A excretory, immune, respiratory, epidermal
  B circulatory, muscular, nervous, nuclear
  C root, phloem, xylem, fruit
  D endocrine, skeletal, reproductive, digestive

4 Identify the microscopic structures labeled X and Y in this transverse section of a plant stem.

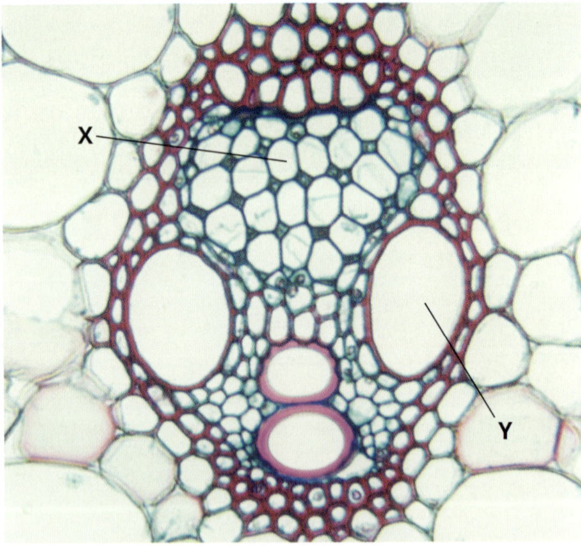

  A X is a tracheid and Y is a companion cell.
  B X is a vascular bundle and Y is pith.
  C X is phloem and Y is xylem.
  D X is xylem and Y is phloem.

5 Which group of structures lists the correct sequence for the digestive process in humans?
  A mouth, oesophagus, liver, small intestine, large intestine, anus
  B mouth, stomach, pancreas, large intestine, small intestine, anus
  C mouth, oesophagus, stomach, small intestine, large intestine, anus
  D mouth, oesophagus, stomach, gall bladder, small intestine, anus

6 Identify the list that contains only the names of human endocrine glands.
  A pineal, adrenal, thyroid, pituitary
  B ovary, placenta, cervix, oestrogen
  C oestrogen, testosterone, progesterone, prolactin
  D pancreas, hypothalamus, neurons, cerebellum

7 Which row correctly identifies hypoglycaemia and hyperthyroidism?

	Hypoglycaemia	Hyperthyroidism
A	underproduction of hormones T3 and T4	BGL is too high, more insulin required
B	overproduction of hormones T3 and T4	BGL is too low, more glucagon required
C	BGL is too high, more insulin required	underproduction of hormones T3 and T4
D	BGL is too low, more glucagon required	overproduction of hormones T3 and T4

8 Select the most accurate description of a negative feedback mechanism in an endotherm.
  A increase in temperature → thermoreceptor → vasodilation → hypothalamus → heat loss
  B decrease in temperature → hypothalamus → vasodilation → heat gain → thermoreceptor
  C increase in temperature → hypothalamus → thermoreceptor → vasodilation → hypothalamus → heat loss
  D increase in temperature → thermoreceptor → hypothalamus → vasodilation → heat loss

9 A person is swimming in water that has a temperature of 20°C. Which row correctly describes how that person's body temperature is regulated?

	Blood circulation to the skin	Sweat glands	Skeletal muscle
A	increased blood flow	increased secretion	decreased shivering
B	decreased blood flow	decreased secretion	increased shivering
C	decreased blood flow	increased secretion	increased shivering
D	increased blood flow	decreased secretion	increased shivering

10 Salmon lay their eggs in freshwater streams, where they are fertilised. Soon after hatching, the young salmon (smolts) swim downstream and out to sea. Once in the ocean, the smolts grow into fully developed adult salmon. These adults later make their way back to their original freshwater stream to lay eggs. Identify the homeostatic process used when the smolts move from a freshwater to a marine environment to maintain internal health and survival.

A The concentration of water in the blood plasma would decrease.

B A smaller volume of more concentrated urine would be produced.

C A larger volume of more dilute urine would be produced.

D Less solutes, including urea, would be filtered out of the blood.

11 A scientist was studying how water balance and environment interacted in maintaining a stable body temperature. The scientist hypothesised that environmental temperature and humidity would cause variations in how water was lost from the body. A subject was provided with the same amount of food and drink for four days and their water output was measured. The results of the experiment were graphed as shown below.

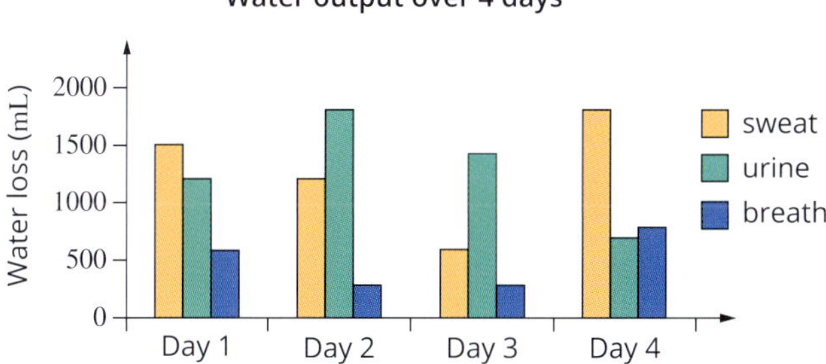

Environmental conditions varied over the four days. Analyse the data and select the situation that would not support the scientist's hypothesis.

A Day 4 was hot and dry.

B Day 2 was cold and wet.

C Day 1 was cool and dry.

D Day 3 was cool and humid.

Short-answer questions

12 The diagram below shows the organisation of tissues in a multicellular organism. Biologists describe a hierarchical organisation of functional levels for organelles, cells, tissues, organs and systems for a complex animal such as this.

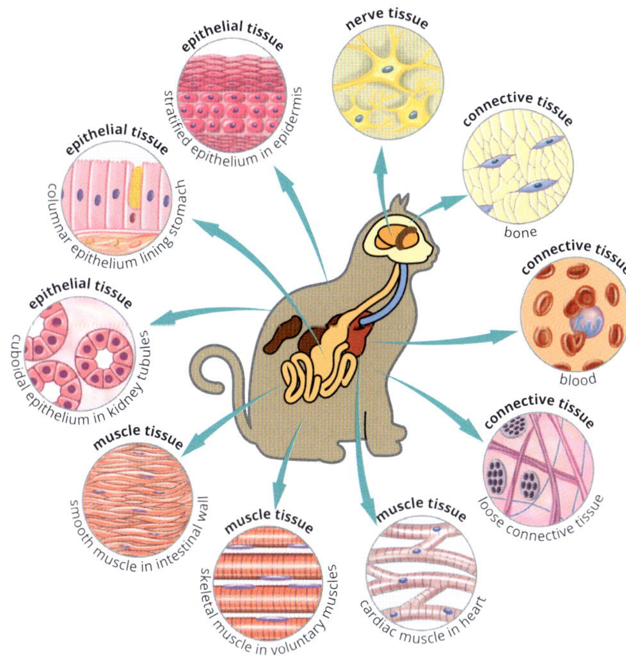

a Construct a flow chart diagram to depict the structural hierarchy that forms a system in a multicellular organism, adding a definition and an example for each structural level.

b Explain why it is useful to use a hierarchy like this, using an example.

13 a Identify each of the three structures A, B and C shown below.

b Describe what the structures have in common and relate these common features to their functions.

A

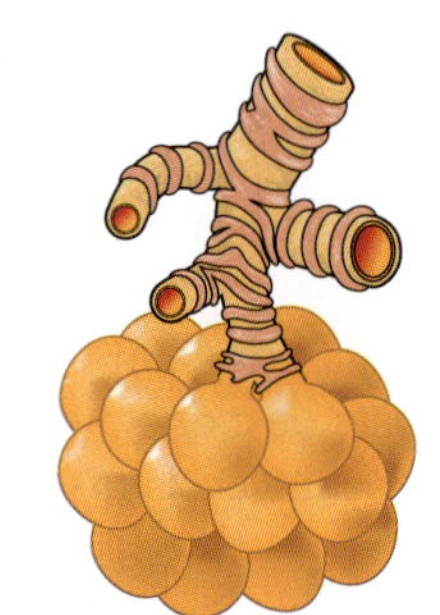

B

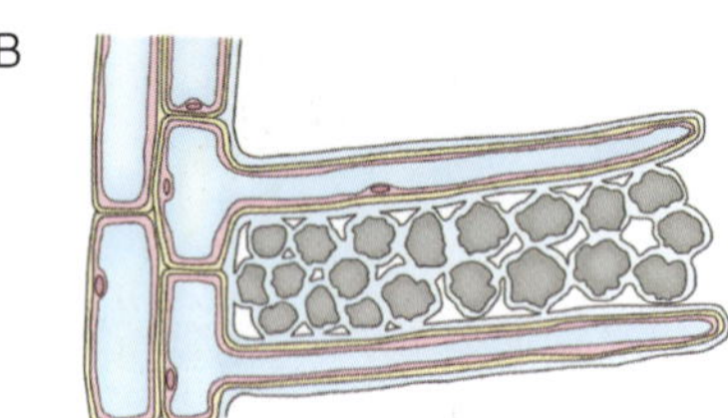

C

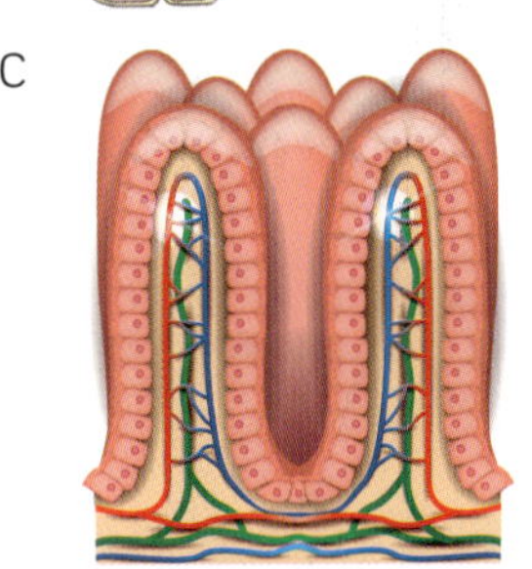

14 The following diagram represents a nephron from a mammalian kidney.

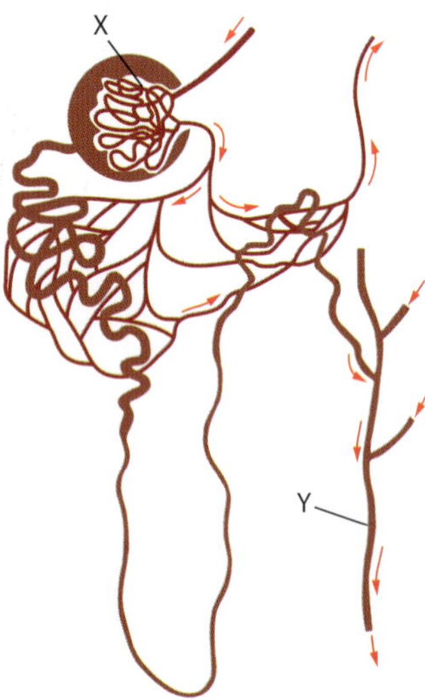

a State the function of a kidney in the mammalian excretory system.

b i Identify at least two differences that would be expected between the fluid at locations X and Y in a healthy person.

ii Explain these differences.

15 The diagram below shows an apparatus called a plant potometer. It is used to measure the rate of transpiration of a plant by timing the rate at which the bubble moves along the horizontal glass tube.

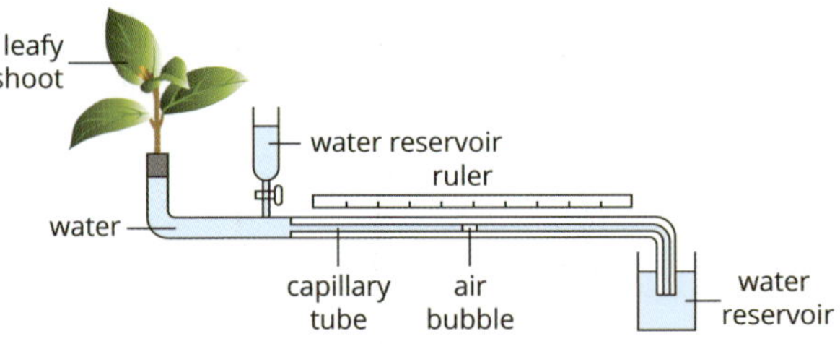

a Design a controlled experiment using a plant potometer to test the hypothesis that the leaves of Australian eucalypts do not lose water as fast as leaves of an English elm. In your answer identify the three types of variables.

b Propose the results you would expect if the hypothesis were to be supported.

16 The following image is a photomicrograph of a plant root taken using specialised microscopy.

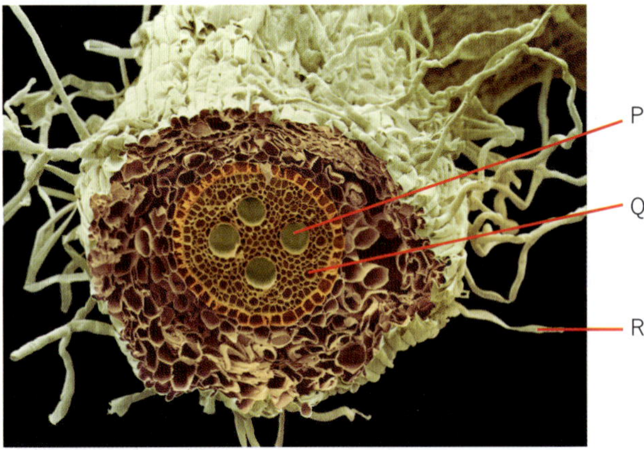

a Explain how cell R is specialised in its function.

b Describe how structure P is involved in the transport and eventual loss of water.

c Describe the role that structure Q plays in maintaining structure R.

d The root is an organ. Using the root as an example, outline how cells are organised to form an organ.

17 **a** Vascular is a word commonly used to refer to human blood vessels; for example, a vascular surgeon treats problems with arteries, veins and capillaries. Analyse why the word vascular can also be used to describe many plants.

b Clarify why translocation in vascular plants is sometimes referred to as involving 'sources and sinks'.

c Construct a table to compare the two transport mechanisms in vascular plants—transpiration and translocation.

18 **a** The image below is a micrograph of a leaf epidermis. Name the circled structure and outline its function.

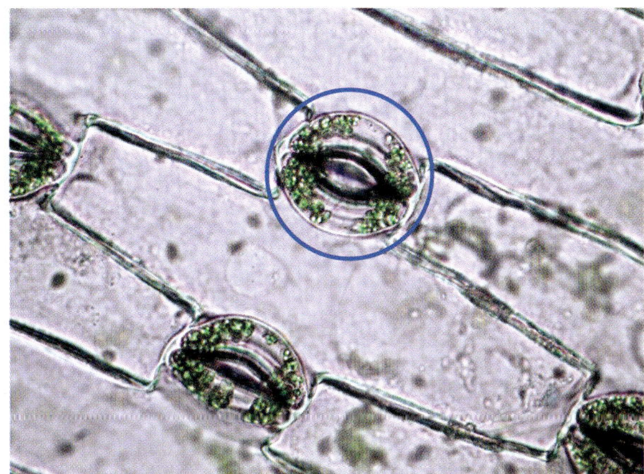

b Identify structures M–V in the diagram below, which depicts a transverse section of a leaf.

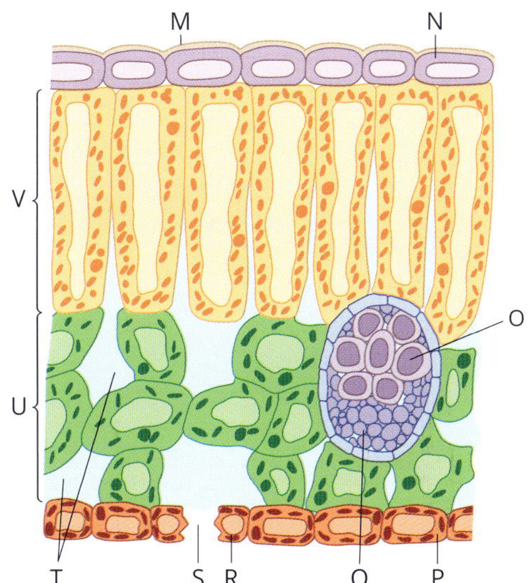

c Explain the advantage to the plant of structure S.

d Some studies have indicated that different carbon dioxide concentrations in a controlled environment during a plant's growth period can induce changes in the density of leaf stomata. Analyse the graph and answer the questions below.

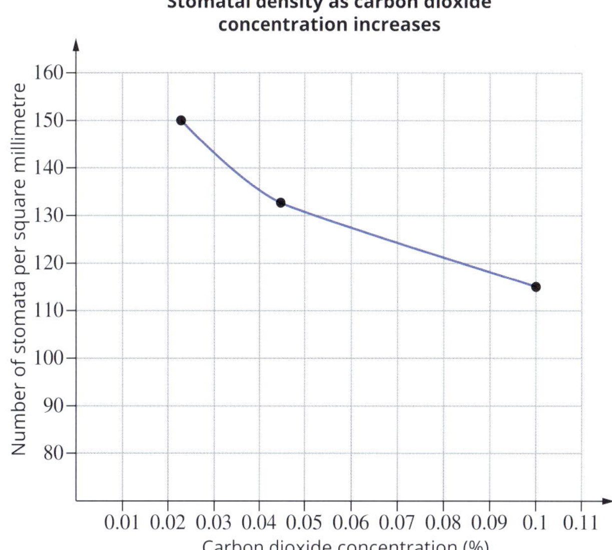

i Describe the trend in number of stomata as carbon dioxide concentration increases.

ii Propose what advantage exists for the plant that results in this trend.

19 **a** Design an experiment to determine the effects of air humidity on the rate of transpiration in geranium plants in a hot climate. Identify your controlled, independent and dependent variables.

b Explain why humidity can affect the rate of transpiration.

c Explain why temperature must be controlled for all of the plants.

d **i** Outline the basic safety precautions for conducting this investigation.

ii List two ways the experiment could be improved.

20 Identify and outline a stimulus–response mechanism used by endothermic animals for homeostatic regulation.

21 Scientists investigated the effect of ambient (surrounding) temperature on metabolic rate and core body temperature of Arctic ground squirrels (*Spermophilus parryii*). The scientists placed the squirrels in a cool environment and then slowly increased the temperature of the environment, without harming the squirrels. At the same time, they measured changes to core body temperature and respiratory quotient (the ratio of carbon dioxide produced to oxygen consumed) of the small mammal. Results of the experiment are shown in the graph below.

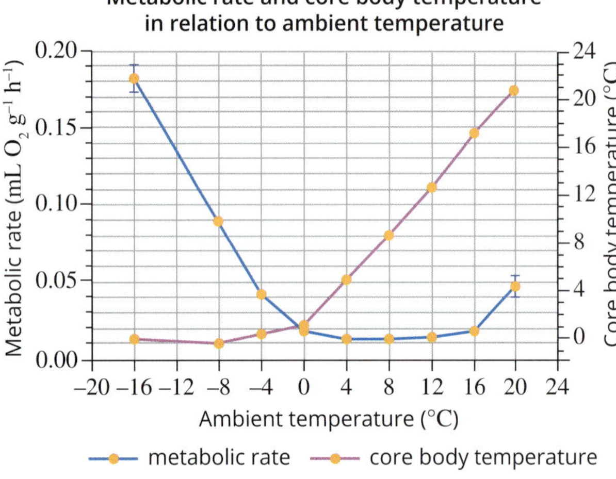

a To estimate the metabolic rate of the squirrels, the scientists measured their rate of oxygen consumption. Suggest why this is appropriate.

b Outline the effects of ambient temperature on the core body temperature and the metabolic rate.

c Predict why when the core body temperature is low, the metabolic rate of the squirrel is high.

22 The graph below shows changes in the core body temperature and skin surface temperature of a jogger. He jogged from $t = 0$ to $t = 60$ minutes. From $t = 60$ to $t = 90$ minutes, the jogger stopped and sat on a chair.

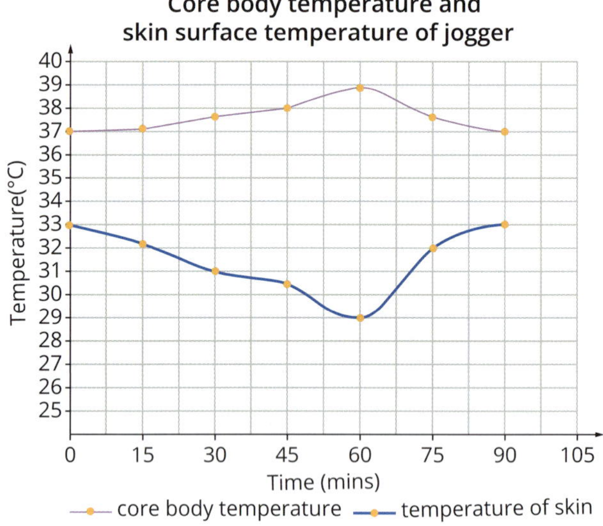

a i Describe the changes in the core body temperature and the temperature of the skin from 0 to 60 minutes.

ii Explain the patterns described in part **i**.

b Suggest why the temperature of the jogger's skin started to rise after 60 minutes.

23 The following diagram shows how the body regulates the glucose concentration in blood.

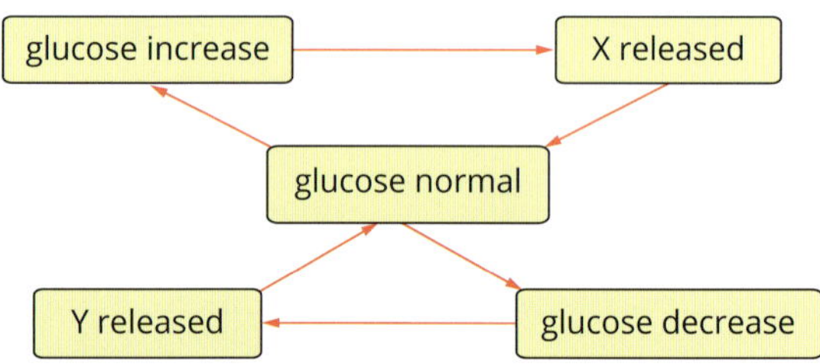

a Identify X and Y.

b Explain how the body regulates the glucose concentration in blood.

c i Draw a circle on the diagram above to show which process is stopped when a person suffers from type 1 diabetes.

ii Summarise the cause and symptoms of type 1 diabetes.

24 Antidiuretic hormone (ADH) regulates the amount of water lost via the kidneys. The following diagram outlines the stimulus–response pathway for ADH in a mammal.

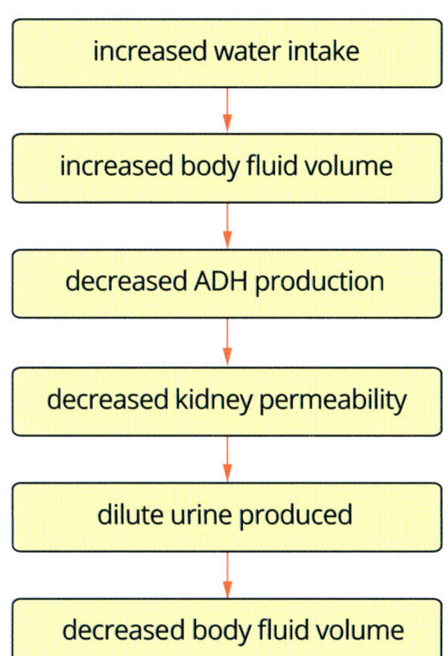

a Identify the stimulus that causes release of ADH.

b i Use the diagram to predict the response to a decrease in body fluid volume.

ii Is the feedback mechanism of ADH production considered negative or positive? Explain why.

c Alcohol inhibits ADH production. Suggest why alcohol consumption can cause dehydration.

d Diabetes insipidus (DI) is a rare malfunction that may cause ADH deficiency. It is not related to the more common form of diabetes which affects the level of insulin hormone. Predict the symptoms that would be caused by DI.

25 By filtration in the kidneys, 100 litres of fluid is removed from the blood in 24 hours. In this same time, only 1.5 litres of urine is formed. The blood plasma contains on average 0.03% urea in solution, but urine contains 2% urea in solution. The table shows the ratio of nephron concentration to plasma concentration measured at various points along the nephron tubule, and the diagram shows the reference points used for these measurements.

Chemical	A	Mid B	Start E	Mid E	End E	Mid C	D
glucose	0.8	0.08	0	0	0	0	0
amino acids	0.4	0.5	0	0	0	0	0
urea	1.0	1.2	1.6	7.0	15.0	20.0	50.0
Cl^-	1.0	1.0	1.0	2.0	0.35	0.4	0.8
Na^+	1.0	1.0	1.0	2.0	0.2	0.25	0.7
K^+	1.0	1.0	1.0	2.0	0.3	0.5	3.0
creatinine	1.0	2.0	5.0	6.5	16.0	20.1	50.0

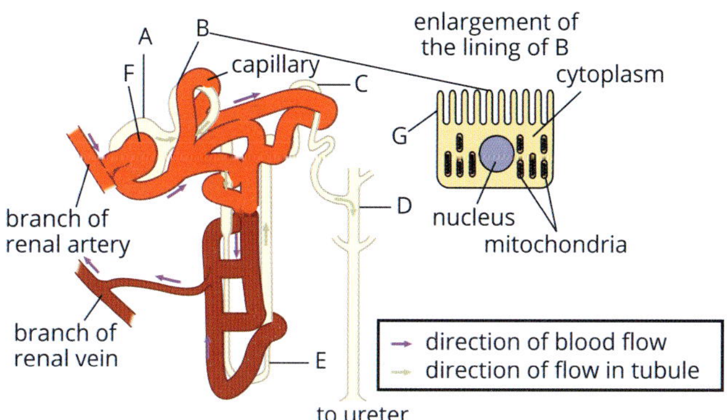

a Calculate how many times (to the nearest whole number) urea is more concentrated in urine than it is in blood.

b i Explain what a ratio of 1 for the nephron : plasma concentration indicates.

ii Suggest why less glucose and amino acids are found in the filtrate compared to ions such as Na^+ and K^+.

c Deduce which letter codes in the table represent points closest to the start, middle and end of the nephron tubule. Support your answers with evidence from the table.

d Account for why glucose and amino acid concentrations are both zero in the urine excreted from kidney tubules.

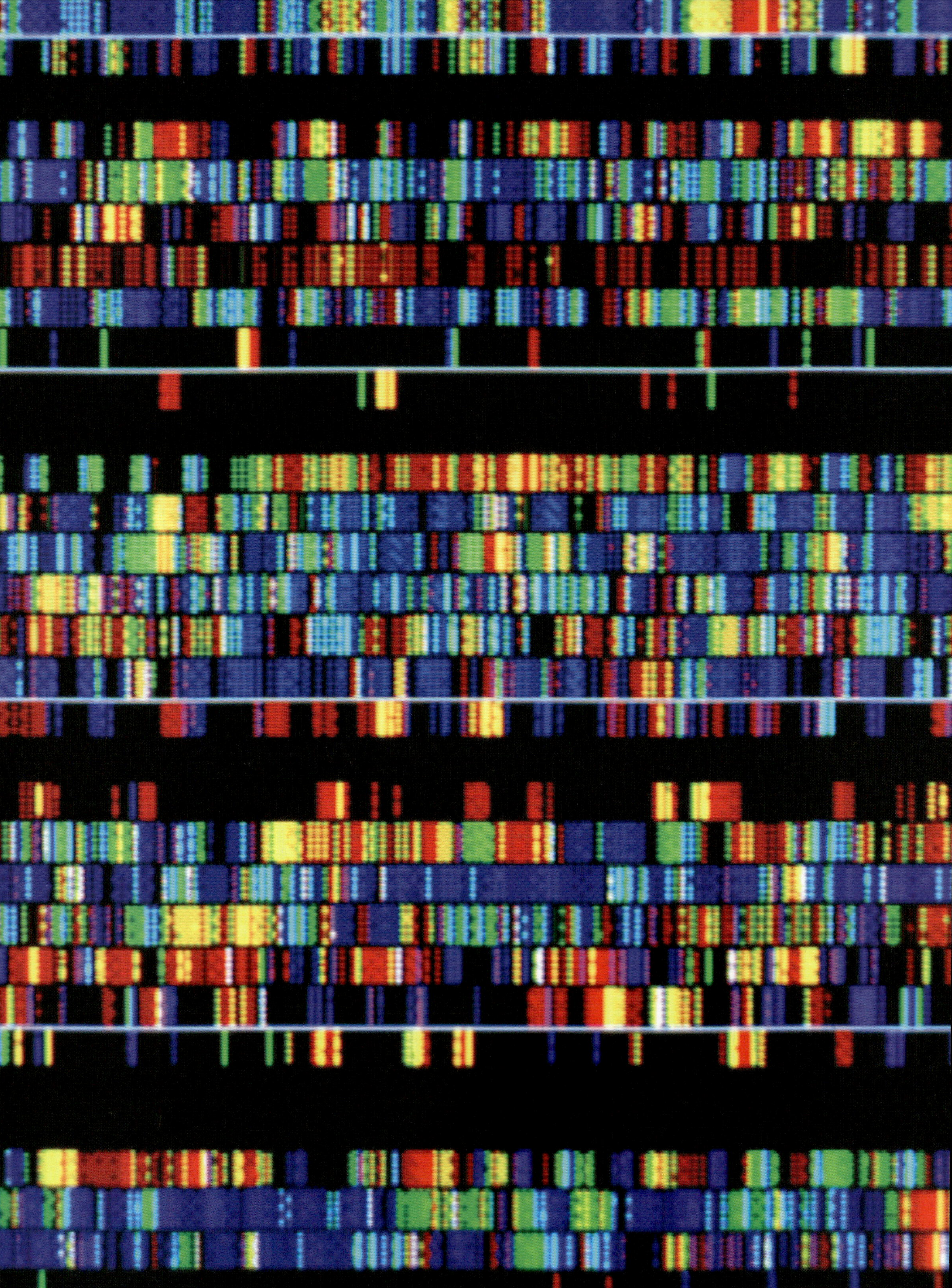

UNIT 2 How does inheritance impact on diversity?

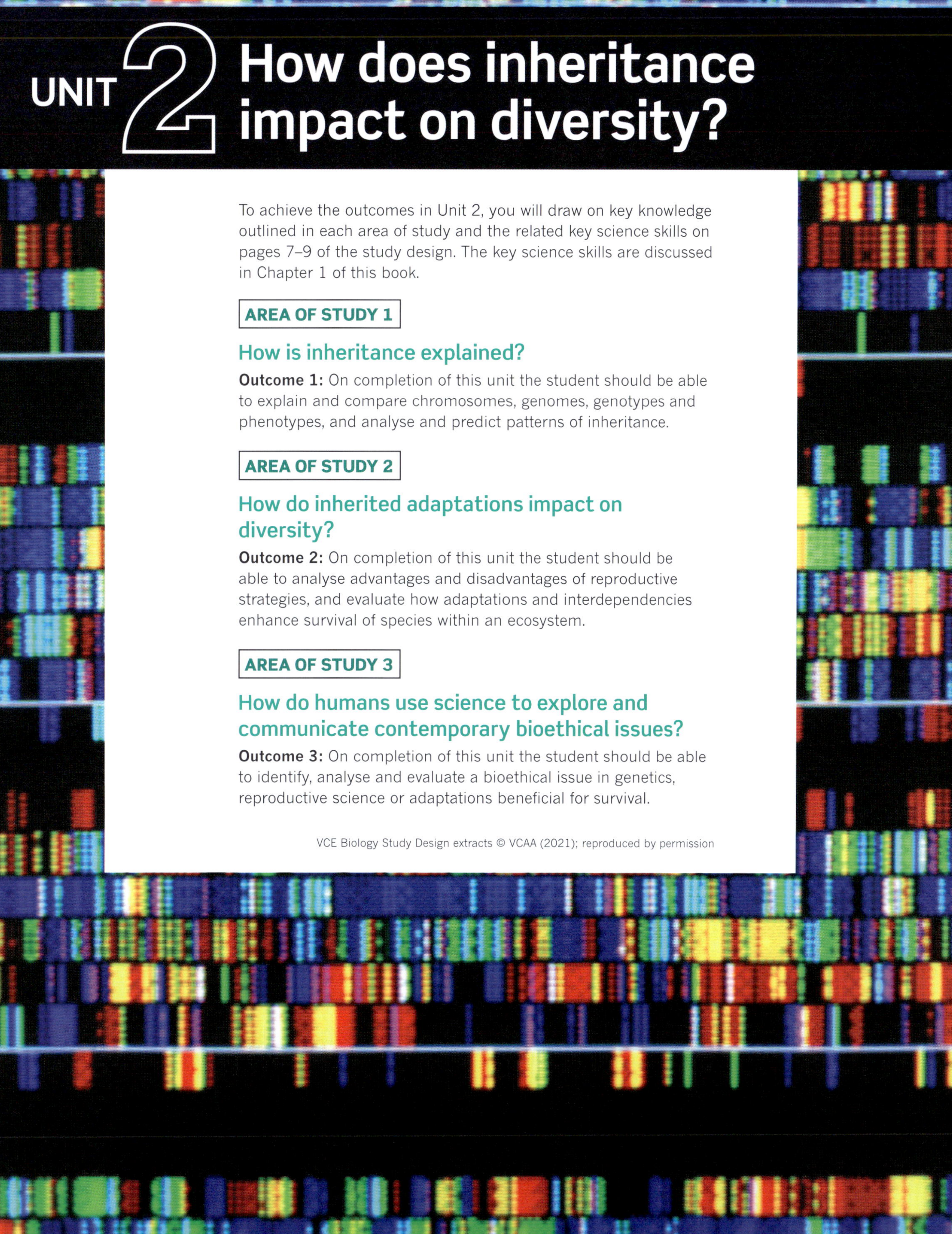

To achieve the outcomes in Unit 2, you will draw on key knowledge outlined in each area of study and the related key science skills on pages 7–9 of the study design. The key science skills are discussed in Chapter 1 of this book.

AREA OF STUDY 1

How is inheritance explained?

Outcome 1: On completion of this unit the student should be able to explain and compare chromosomes, genomes, genotypes and phenotypes, and analyse and predict patterns of inheritance.

AREA OF STUDY 2

How do inherited adaptations impact on diversity?

Outcome 2: On completion of this unit the student should be able to analyse advantages and disadvantages of reproductive strategies, and evaluate how adaptations and interdependencies enhance survival of species within an ecosystem.

AREA OF STUDY 3

How do humans use science to explore and communicate contemporary bioethical issues?

Outcome 3: On completion of this unit the student should be able to identify, analyse and evaluate a bioethical issue in genetics, reproductive science or adaptations beneficial for survival.

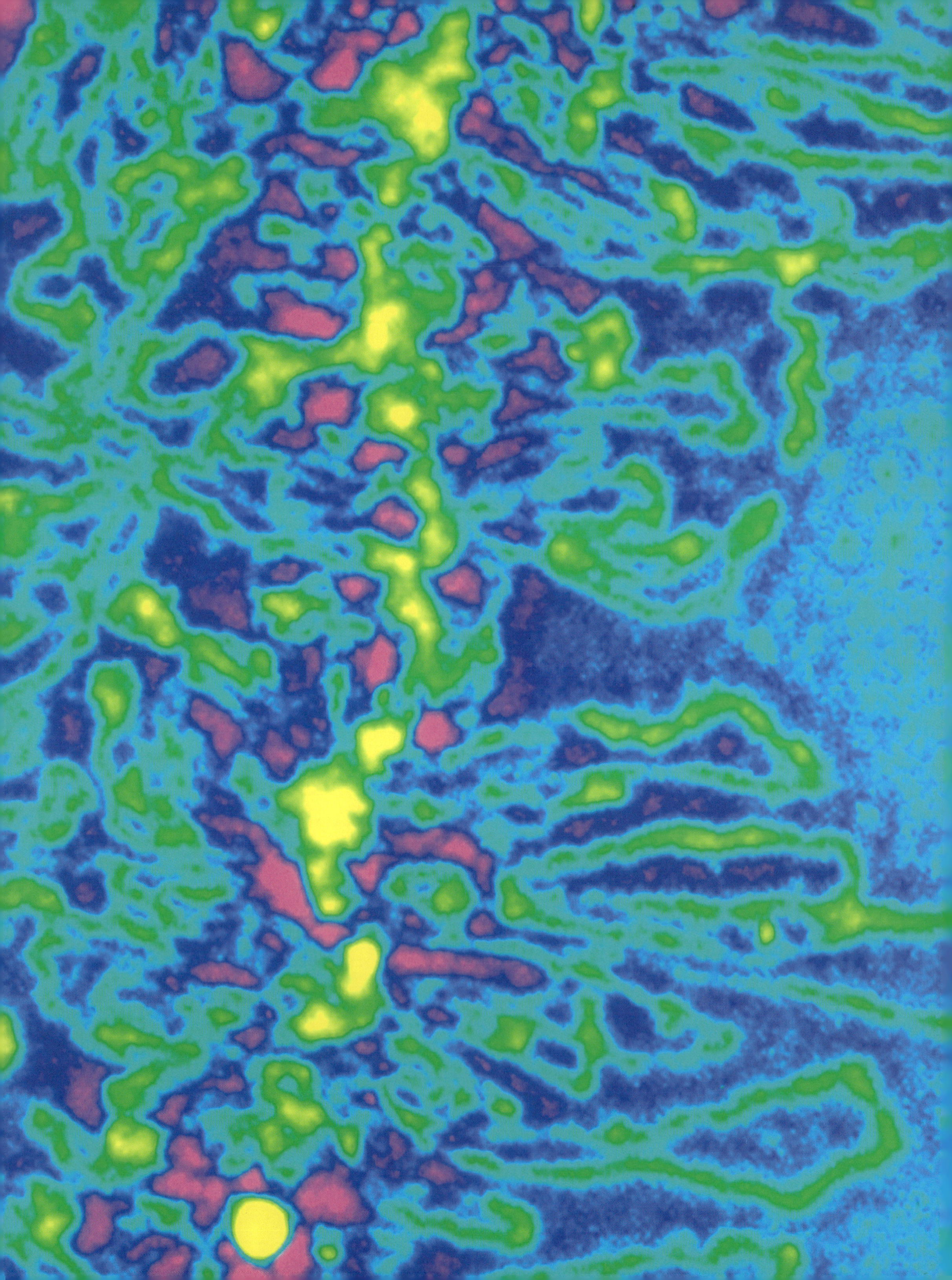

CHAPTER 06

The nature of genes

Sexual reproduction results in offspring with a set of unique characteristics that are inherited from their parents. In this chapter, you will learn about the nature of homologous chromosomes and their role in carrying and passing on genetic information from parents to offspring. You will also learn how the inheritance and expression of this genetic information determines the characteristics of an organism, and the influence of environmental and epigenetic factors.

Key knowledge

From chromosomes to genomes

- the distinction between genes, alleles and a genome **6.2**
- the nature of a pair of homologous chromosomes carrying the same gene loci and the distinction between autosomes and sex chromosomes **6.1**
- variability of chromosomes in terms of size and number in different organisms **6.1**
- karyotypes as a visual representation that can be used to identify chromosome abnormalities **6.1**
- the production of haploid gametes from diploid cells by meiosis, including the significance of crossing over of chromatids and independent assortment for genetic diversity **6.1**

Genotypes and phenotypes

- the use of symbols in the writing of genotypes for the alleles present at a particular gene locus **6.3**
- the expression of dominant and recessive phenotypes, including codominance and incomplete dominance **6.3**
- proportionate influences of genetic material, and environmental and epigenetic factors, on phenotypes. **6.3**

6.1 Chromosomes

Throughout history, people have wondered why children resemble their parents more than they resemble unrelated individuals. Today we know that many characteristics are inherited, such as the colour of our hair, eyes and skin, and also that many conditions such as cystic fibrosis are inherited. In all organisms, inherited characteristics are determined by genes located on DNA in chromosomes (Figure 6.1.1).

In this section you will learn about the variability of chromosomes in terms of size and the number of genes they carry in different organisms. You will also learn about the distinction between an autosome and a sex chromosome and the nature of a homologous pair of chromosomes.

Using karyotypes to identify chromosome abnormalities in humans will also be explored, along with the significance of crossing over and independent assortment of chromosomes during gamete production.

FIGURE 6.1.1 Human chromosomes photographed just before alignment along the metaphase plate (metaphase) in mitosis

THE ROLE AND STRUCTURE OF CHROMOSOMES

A **chromosome** is a structure containing a single DNA molecule and its associated proteins. Chromosomes therefore carry genes. Chromosomes can have various shapes and sizes, and their appearance changes during the life of a cell.

Eukaryotic organisms have sets of linear chromosomes located in the nucleus, and the number of chromosomes is constant in each species (except in the case of chromosomal abnormalities). At the stage they are normally viewed, chromosomes in eukaryotes consist of two strands called **chromatids** that are held together by a region called the **centromere** (Figure 6.1.2). Joined chromatids are known as **sister chromatids** and form the familiar 'X' shape of most chromosomes. Prokaryotes have a single circular chromosome. Prokaryotes may also contain smaller, circular DNA molecules called plasmids, which can move between cells.

A chromosome is a structure containing a single DNA molecule and associated proteins.

Chromosomes consist of two chromatids (DNA molecules) when a cell is undergoing mitosis or meiosis. At all other times, chromosomes consist of a single double-stranded DNA molecule.

Each chromosome carries a unique set of genes. In eukaryotes, chromosomes are passed on to daughter cells during mitosis and to gametes (sex cells) during **meiosis**. Mitosis and meiosis are both types of cell division—meiosis produces daughter cells that are genetically unique, whereas mitosis produces daughter cells that are genetically identical. You learnt about mitosis in Chapter 3 and you will learn more about meiosis in this chapter.

BIOFILE

Chromosome arms

Each chromosome has a constriction point called the centromere, which pinches the chromosomes into two sections. The regions on either side of the centromere are referred to as the chromosome arms. The shorter arm is called the p arm (from the French word 'petite'), and the longer arm is called the q arm simply because it is the next letter after p in the alphabet. Photographs or diagrams of chromosomes are always arranged so that the p arm is at the top.

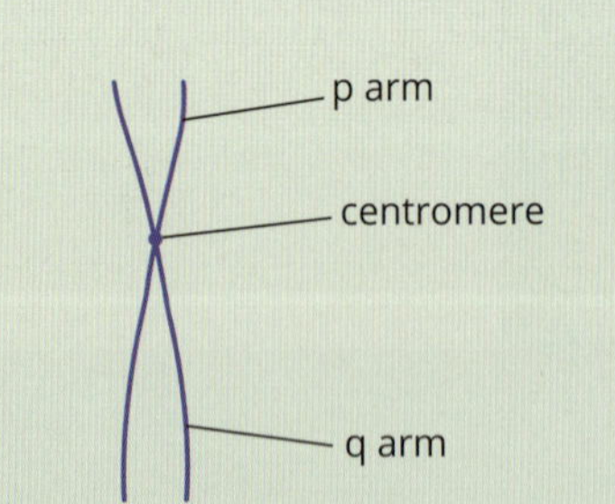

The regions on either side of a chromosome's centromere are called arms.

CHROMOSOME VARIATION BETWEEN ORGANISMS

The size and number of chromosomes can vary widely between organisms. This chromosomal variation is due to changes in genetic material over time, and species that are closely related will share more similarities in their chromosome size and number than species that are distantly related.

Chromosome size

Chromosome size varies between chromosomes and between different organisms. Chromosomes range in size from about 50 million to 300 million base pairs, with every chromosome carrying a different number of genes. There are 20 000 to 25 000 genes in the human genome. The longest human chromosome (chromosome 1) has about 2000 genes.

Each gene has a particular position, called a **locus** (plural loci), on a specific chromosome (Figure 6.1.2). The genes of each DNA molecule are separated by regions called spacer DNA (Figure 6.1.3). Spacer regions include DNA that does not encode a protein product. However, these regions may function in spacing genes far enough apart to enable enzymes or other molecules to interact easily with them. Chromosomes differ in size because of differences in the number of genes and the amount of spacer DNA between the genes.

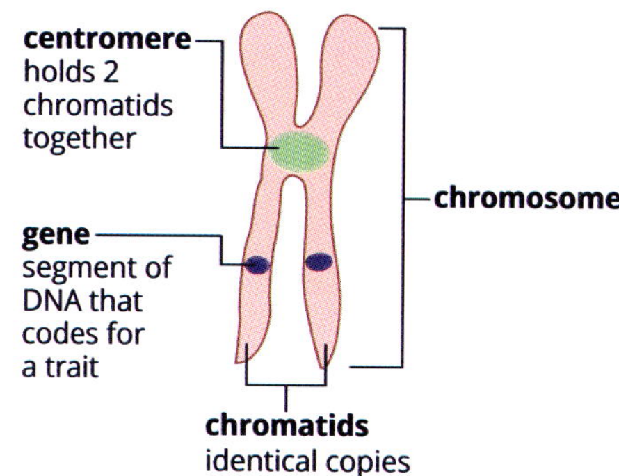

FIGURE 6.1.2 Each gene occupies a fixed locus, or position, on a chromosome. The gene indicated here is on the q arm at a precise distance from the centromere.

The position of a gene on a chromosome is called a locus.

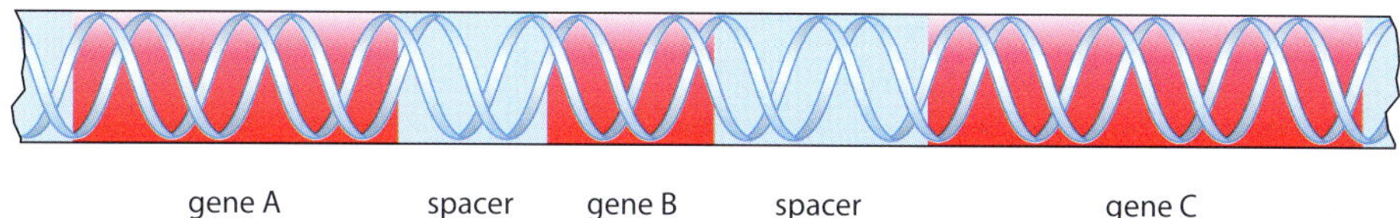

FIGURE 6.1.3 A short stretch of double-stranded DNA. Genes are highlighted in red and the spacer regions of DNA separating the genes are shown in blue.

CASE STUDY

Chromosome numbers of Australian plants

Chromosomes were first observed by Walther Fleming in 1882. While he was examining cells of a salamander larva under a light microscope, he saw minute threads in the nucleus. Since Fleming's time, the chromosome numbers of many plants and animals have been documented. This information is important for crossing species to breed new varieties of plants for horticulture and for understanding evolutionary patterns.

The family Proteaceae is a conspicuous and important part of the Australian bush. It includes banksias, grevilleas, waratahs and the macadamia nut tree. Waratahs (Figure 6.1.4) are characterised by a diploid number of 22. In contrast, all grevilleas (Figure 6.1.5) and their relatives such as hakeas have one pair of chromosomes less, with a diploid number of 20. Chromosome number provides important evidence about the evolutionary relationships of these plants.

FIGURE 6.1.4 Members of the family Proteaceae include waratahs (*Telopea* species), which have a diploid chromosome number of 22.

FIGURE 6.1.5 Grevilleas (*Grevillea* species) have a diploid chromosome number of 20.

Chromosome number

In eukaryotic organisms (plants, fungi, animals), cells that have a nucleus contain a fixed number of chromosomes. The number of chromosomes in somatic cells is characteristic of a species.

The number of sets of chromosomes in a cell is called the ploidy level. Haploid cells have one set, diploid cells have two sets, and polyploid cells have three or more sets.

The **ploidy** level of a cell is the number of chromosome sets that it carries. Gametes have nuclei that contain only one set of chromosomes and they are called **haploid** (designated as n). In most species, somatic cells are **diploid** ($2n$) because they contain two sets of chromosomes: one from each parent.

The diploid chromosome numbers of organisms vary widely, as shown in Tables 6.1.1 and 6.1.2. In humans the diploid number is 46. In some species of Australian ants (*Myrmecia* species) the diploid number is 2; each ant has only one pair of chromosomes ($n = 1$). Some ferns have more than a thousand chromosomes in each somatic cell.

TABLE 6.1.1 Variation between different organisms in genome size, the number of genes and the number of chromosomes

Organism	Genome size (base pairs)	Number of genes	Number of chromosomes
bacteria, *Escherichia coli*	4.6 Mbp	4300	1
corn plant, *Zea mays*	2.3 Gbp	33 000	20 ($2n$)
chimpanzee, *Pan troglodytes*	3.3 Gbp	19 000	48 ($2n$)
fruit fly, *Drosophila melanogaster*	1.65 Mbp	13 000	8 ($2n$)

TABLE 6.1.2 Diploid numbers of chromosomes in various species

Organism	Diploid number ($2n$)
Animals	
horse nematode worm, *Parascaris equorum*	2
koala, *Phascolarctos cinereus*	16
cat, *Felis catus*	38
human, *Homo sapiens*	46
platypus, *Ornithorhynchus anatinus*	52
dingo, *Canis lupus dingo*	78
Plants and algae	
garden pea, *Pisum sativum*	14
all eucalypts, *Eucalyptus* spp.	22
single-celled alga, *Euglena gracilis*	90
coconut palm, *Cocos nucifera*	596
fern, *Ophioglossum reticulatum*	1260
Fungi	
mould, *Penicillium* species	2
rust fungus, *Puccinia graminis*	6
oyster mushroom, *Pleurotus ostreatus*	22
brewer's yeast, *Saccharomyces cerevisiae*	32

HOMOLOGOUS CHROMOSOMES

Humans have 46 chromosomes, comprising 23 inherited from their mother and 23 inherited from their father. All the chromosomes together constitute the genome. Forty-four of these chromosomes form 22 matching pairs. The same genes are found at the same locations (loci) on the two chromosomes in a matching pair. They are referred to as **homologous chromosomes** or homologues. Each chromosome pair has different genes. The genes found on chromosome 1 are different to those found on chromosome 2 and so on.

Matching pairs of chromosomes are called homologous chromosomes or homologues.

Figure 6.1.6 shows metaphase chromosomes, which are classified according to the position of the centromere, in this case as either metacentric (the centromere is centrally positioned) or acrocentric (the centromere is close to one end). These are homologues because they contain the same gene sets. The sex chromosomes for all females are also homologues because they have a matching pair of X chromosomes. The sex chromosomes for males are not homologous because they have an X and a Y chromosome, which contain different gene sets.

The two chromatids of a chromosome contain matching DNA molecules.

Nevertheless, in most mammals the X and Y chromosomes behave as a homologous pair during meiosis because some small regions of these chromosomes are homologous.

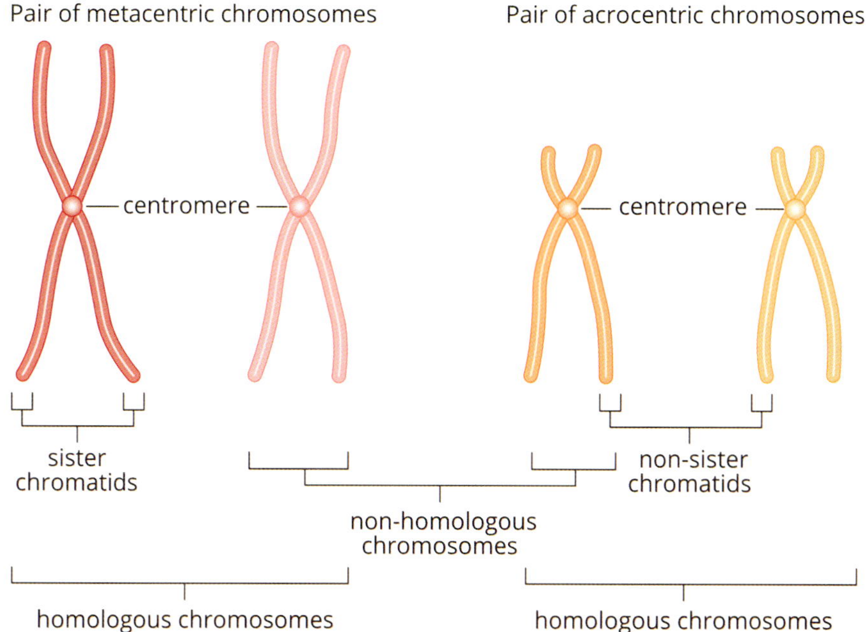

FIGURE 6.1.6 Metaphase chromosomes are classified according to the position of the centromere, in this case as either metacentric or acrocentric.

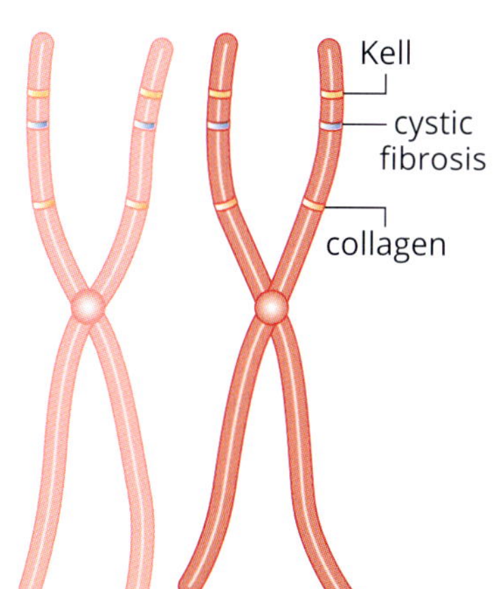

FIGURE 6.1.7 Two human chromosome homologues during metaphase. The loci for three genes are shown: a collagen gene, the cystic fibrosis gene and the Kell gene.

In Figure 6.1.7, two human homologues are shown during metaphase. The loci for three genes are shown: a collagen gene, the cystic fibrosis gene and the Kell gene, which produces a protein involved in determining the Kell blood groups. The three genes are in this same position in all cells in most individuals.

Non-dividing cells have two copies of each gene, so they have two copies of alleles (which may be identical or different) for each gene. However, in actively dividing cells there is a period during the cell cycle, after chromosome replication and before cytokinesis, when the diploid cell contains four copies of each gene (Figure 6.1.7), but there can be no more than two alleles in total (one from each parent).

Chromosome replication occurs during the S phase (DNA synthesis) in mitosis.

Sex chromosomes

> Sex chromosomes (also called allosomes) are chromosomes involved in sex determination.

> An autosome is any chromosome that is not a sex chromosome. Autosomes are represented by numbers (e.g. chromosome 12).

Sex chromosomes (also called allosomes) are chromosomes involved in sex determination. **Autosomes** are the numbered chromosomes. They are numbered in order of their size, with chromosome 1 being the largest chromosome. These chromosomes carry the genes for all of an organism's characteristics, except the sex-linked ones. Individuals with two sex chromosomes that are the same are the **homogametic** sex. Individuals with two different sex chromosomes are the **heterogametic** sex.

In humans and all other mammals, a pair of chromosomes known as the X and Y chromosomes determine the biological sex of an individual. Other types of organisms may have different types of sex-determining chromosomes. For example, in birds and strawberries the females are heterogametic (ZW) and males are homogametic (ZZ). The Z and W notation is used to distinguish this system from the XX/XY system.

In grasshoppers there is only one sex chromosome. Females are XX and males are X0, where the 0 refers to the absence of a matching sex chromosome. The diploid chromosome number in grasshoppers is therefore even in females and odd in males.

Some organisms (such as fungi and algae) do not have sex-determining chromosomes and therefore do not have sexes; instead they have 'mating types'. Table 6.1.3 shows some of the different sex chromosome combinations that are involved in sex determination.

TABLE 6.1.3 Different types and combinations of sex chromosome involved in determining the biological sex of an organism. '0' indicates the absence of a sex chromosome.

Examples of organisms	Female	Male
humans, other mammals, fruit flies	XX	XY
birds, butterflies, strawberries	ZW	ZZ
grasshoppers, moths	XX	X0
plants	XX	XY

In humans, females have two X chromosomes, i.e. XX. This is described as homogametic. Males have one X and one Y chromosome, i.e. XY. This is described as heterogametic. Sex is determined by the presence or absence of the Y chromosome. The gametes of females (eggs) carry an X chromosome. They also have one copy of each of the other 22 chromosomes, the autosomes, which are not involved in sex determination.

Figure 6.1.8 shows that human females are genotypically XX and therefore produce gametes carrying an X chromosome. Human males are genotypically XY, so 50% of the sperm produced carry an X chromosome and 50% carry a Y chromosome. Therefore you could expect the male to female ratio in human populations to be approximately 1:1.

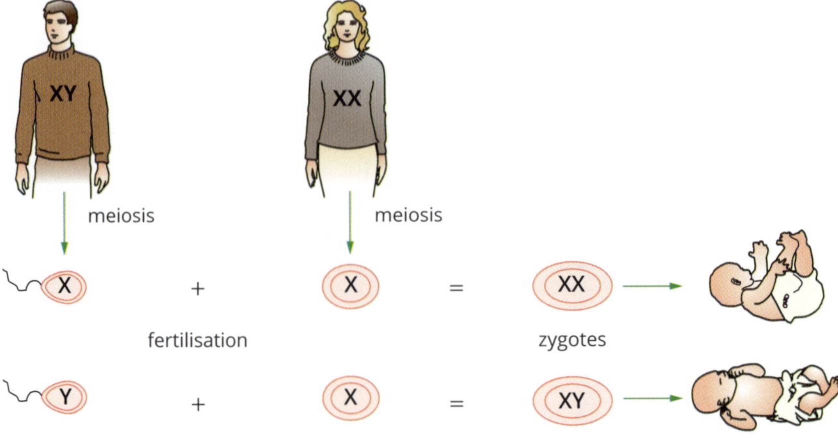

FIGURE 6.1.8 In humans, the inheritance of an X chromosome from both parents results in female offspring (XX). The inheritance of an X chromosome from the mother and a Y chromosome from the father results in male offspring (XY). The biological sex of a baby is determined by the sperm (X or Y) that fertilises the egg.

PRODUCING GENETIC DIVERSITY

Children are not identical to their mother or father, and they are not identical to their sisters or brothers (except in the case of identical twins). The variation seen between individuals is a result of the **genetic diversity** (also known as genetic variation) produced during meiosis and the combination of sex cells during fertilisation.

Gametes

Multicellular organisms are composed of two main types of cells—somatic cells and germ cells. **Somatic cells** are all the cells in the body of an organism apart from the sex cells (gametes). Examples of somatic cells include skin cells, muscle cells and nerve cells. **Germ cells** are the cells that give rise to **gametes**, which are the specialised sex cells that combine in sexual reproduction.

Male gametes (sperm) and female gametes (eggs) are often different in appearance. Gametes are formed by a type of cell division called meiosis, and this occurs in specialised reproductive organs, called gonads. The sperm or eggs formed as a result of this cell division are haploid, which means the number of chromosomes in the gametes is halved. Most normal eukaryotic organisms are composed of diploid cells (represented as $2n$), or one set of chromosomes (n) from each parent.

Female gametes (eggs or ova) are large, immobile cells. They contain the food stores needed for the development of the embryo. The male gametes (spermatozoa or sperm) contain limited food reserves and usually have a tail (or flagellum) for motility, which enables them to move towards an egg.

The process of **fertilisation** involves two haploid gametes fusing to form a diploid zygote (Figure 6.1.9). The zygote then divides by mitosis repeatedly to produce a large number of cells, which differentiate to form the tissues that make up the new organism. The organism continues to develop by mitotic divisions and becomes an adult. The reproductive cycle may then begin again.

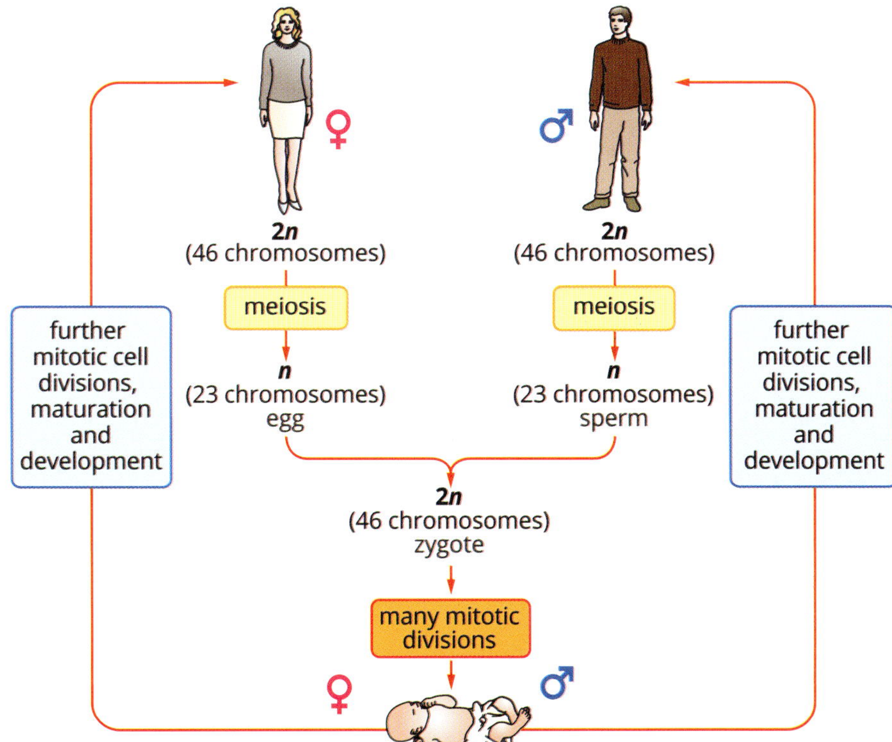

FIGURE 6.1.9 Meiotic cell divisions in females and males give rise to haploid (n) gametes (eggs and sperm). Two haploid gametes fuse to form a diploid ($2n$) zygote. The zygote develops into a new organism after many mitotic divisions and cellular differentiation.

In meiosis two successive cell divisions produce four daughter cells, each with half the number of chromosomes of the parent cell.

Identical (monozygotic) twins develop from one fertilised egg, so they have identical genes. But because of non-genetic factors that affect the way the embryos develop, identical twins are actually not exactly identical.

Meiosis

Gametes are produced by meiosis. Meiosis occurs only in eukaryotes and only in the gametes. The process of meiosis is essential to sexual reproduction and the creation of new genetic variation. Although similar to mitosis, the outcomes of meiosis are quite different. Table 6.1.4 and Figure 6.1.10 highlight the key differences between mitosis and meiosis.

TABLE 6.1.4 Key differences between mitosis and meiosis

	Mitosis	Meiosis
Genetic recombination	Mitosis does not involve recombination of alleles, and it creates daughter cells that are genetically identical.	Meiosis rearranges alleles between chromosome pairs, creating daughter cells that are genetically unique.
Number of cells	two daughter cells	four daughter cells
Number of chromosomes	The daughter cells produced from mitosis have the same number of chromosomes (diploid, 2*n*) as the parent.	The daughter cells produced from meiosis have half the number of chromosomes (haploid, *n*) of the parent.

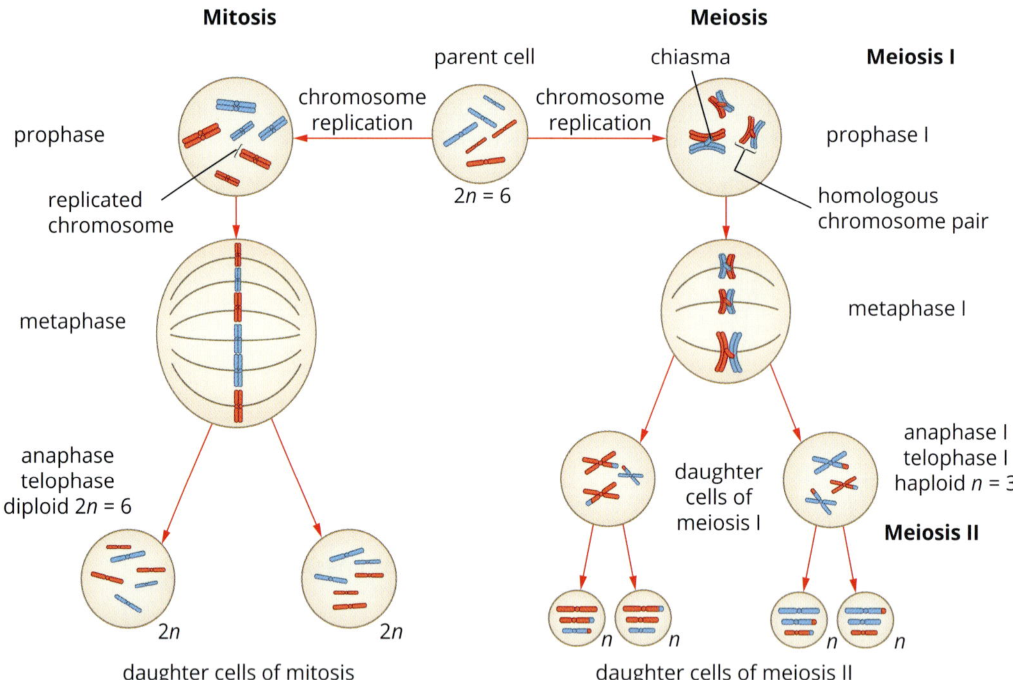

FIGURE 6.1.10 Mitosis and meiosis are both processes of cell division, but they are different in a number of important ways.

Haploid gametes

In order to maintain the chromosomal number of a species, the number of sets of chromosomes in somatic cells and gametes differs. In humans, for example, somatic cells have 46 chromosomes (23 pairs) whereas human gametes have 23 chromosomes. This means that when the haploid gametes combine during fertilisation, the resulting zygote will have a full set of chromosomes (23 chromosomes from the sperm + 23 chromosomes from the ovum = 46 chromosomes in human somatic cells).

Somatic cells in most animals are diploid ($2n$) because they contain two sets of homologous (matching) chromosomes, one from each parent. Meiosis is called a **reduction division** because it reduces the number of chromosomes in gametes (daughter cells) to half (n) of that in somatic cells. Cells with n chromosomes are called haploid cells. Gametes receive only one copy of each pair of homologous chromosomes (23 chromosomes in human gametes). Compare this to mitosis, where each daughter cell receives a copy of every chromosome (they are genetically identical).

Humans have 22 pairs of homologous chromosomes and two sex-determining chromosomes, called X and Y chromosomes. After meiosis in males, four haploid gametes (sperm) are formed from the original diploid parent cell. Each sperm cell contains 23 chromosomes (one of each homologous chromosome and either an X or Y chromosome). However, in females, only one haploid gamete results (the ovum) and the other three haploid cells degenerate (Figure 6.1.11). This occurs because of the uneven distribution of cytoplasm in cytokinesis, so that one daughter cell is very large and contains most of the cytoplasm. Each ovum also contains 23 chromosomes (one of each homologous chromosome and one X chromosome).

Chromosomes passed from a mother to her offspring are referred to as maternal chromosomes, whereas chromosomes passed from a father to his offspring are referred to as paternal chromosomes.

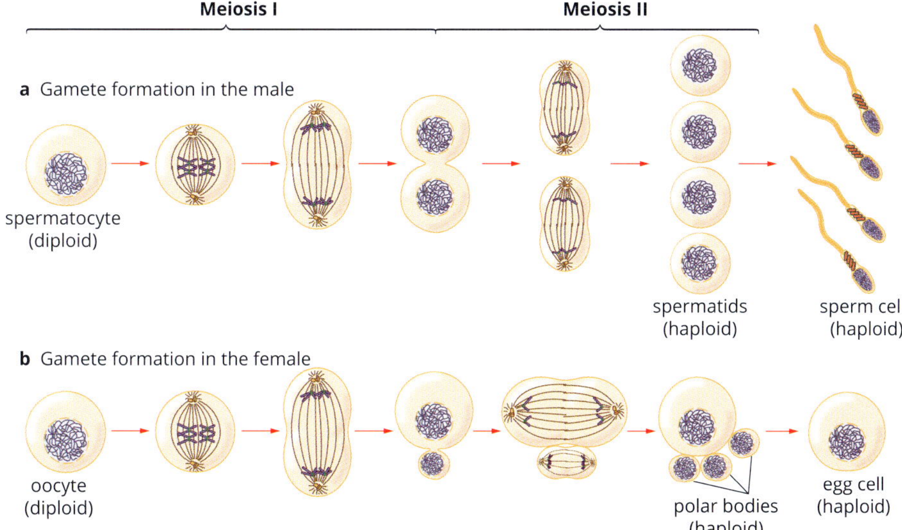

FIGURE 6.1.11 (a) Haploid male and (b) haploid female gametes are produced by meiosis.

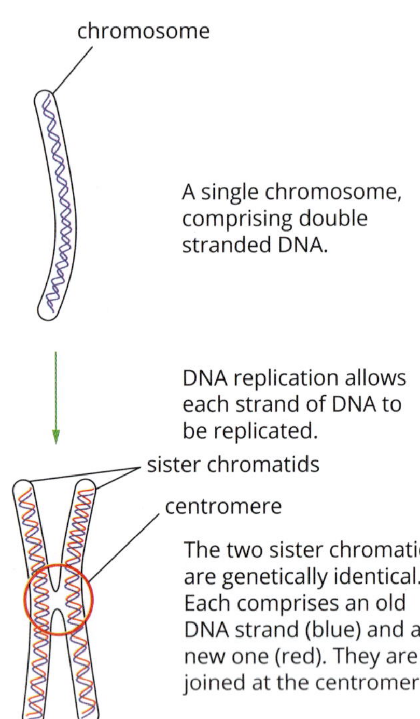

FIGURE 6.1.12 When a single chromosome replicates during the DNA synthesis of interphase, two sister chromatids are formed, joined at the centromere. This new structure is also known as a chromosome.

Meiotic cell division

Like mitosis, meiosis is a form of cell division that involves prophase, metaphase, anaphase, telophase and cytokinesis. However, in meiosis there are two sequential rounds of division, called meiosis I and meiosis II. During meiosis I, homologous chromosomes are separated, reducing the chromosome number by half (reduction division) and producing two haploid daughter cells. By the end of meiosis II, the sister chromatids are separated (**separation division**) and four genetically different haploid daughter cells are produced.

The stages prior to and during meiosis occur in the following order:

1 Interphase—Before meiosis, the DNA is replicated, meaning the chromosomes duplicate and there are now two identical chromatids held together at the centromere (Figure 6.1.12). A cell spends most of its time in interphase, carrying out cellular functions and preparing for cell division.

2 Prophase I—During prophase the chromosomes condense and become more visible. The nuclear envelope begins to disappear and spindle fibres begin to form, moving to opposite poles of the cell. Crossing over of homologous chromosomes occurs in late prophase. Sections of DNA are exchanged between homologous chromosomes that are nearby.

3 Metaphase I—Homologous chromosomes align along the equatorial plate. The arrangement of each pair occurs independently of one another in a process referred to as **independent assortment**. Spindle fibres attach to the centromeres of each chromosome.

4 Anaphase I—The spindles shorten and draw each chromosome within the pair to opposite poles of the cell, reducing the number of chromosomes (reduction division). Homologous chromosomes are now separated from one another.

5 Telophase I and cytokinesis—The nuclear envelopes form and the chromosomes decondense, forming thinner strands. A **cleavage furrow** is formed, pinching the plasma membrane inwards, and the cytoplasm divides. Cytokinesis is the division of the cytoplasm and occurs after meiosis I. Two haploid daughter nuclei are formed, but each chromosome is still in the replicated state.

6 Prophase II—The nuclear envelope breaks down again and the spindle fibres are recreated. Crossing over does not occur again.

7 Metaphase II—The duplicated chromosomes move and align along the equatorial plate, while the spindles on both sides of the cell attach to the chromosomes' centromeres.

8 Anaphase II—The spindles shorten, which cause the centromeres to split, separating the sister chromatids. Single-stranded chromosomes move to opposite poles of the cell (separation division).

9 Telophase II and cytokinesis—Four nuclear envelopes form and the chromosomes decondense. A cleavage furrow is formed, pinching the plasma membrane inwards, and the cytoplasm divides. Cytokinesis is the division of the cytoplasm and occurs after meiosis II. Four haploid (n) genetically unique daughter cells are produced.

The stages of meiosis are summarised in Figure 6.1.13.

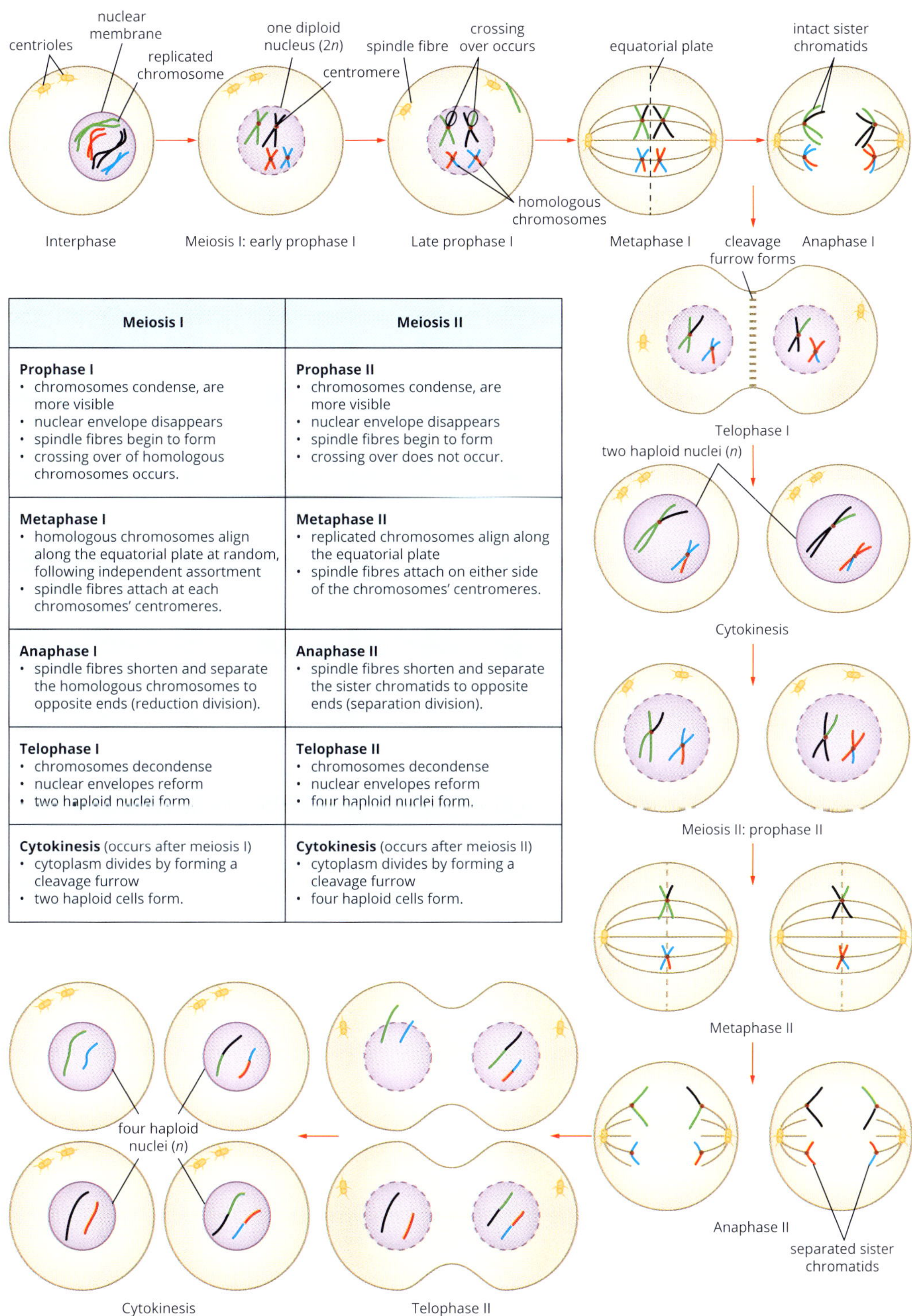

Meiosis I	Meiosis II
Prophase I • chromosomes condense, are more visible • nuclear envelope disappears • spindle fibres begin to form • crossing over of homologous chromosomes occurs.	**Prophase II** • chromosomes condense, are more visible • nuclear envelope disappears • spindle fibres begin to form • crossing over does not occur.
Metaphase I • homologous chromosomes align along the equatorial plate at random, following independent assortment • spindle fibres attach at each chromosomes' centromeres.	**Metaphase II** • replicated chromosomes align along the equatorial plate • spindle fibres attach on either side of the chromosomes' centromeres.
Anaphase I • spindle fibres shorten and separate the homologous chromosomes to opposite ends (reduction division).	**Anaphase II** • spindle fibres shorten and separate the sister chromatids to opposite ends (separation division).
Telophase I • chromosomes decondense • nuclear envelopes reform • two haploid nuclei form.	**Telophase II** • chromosomes decondense • nuclear envelopes reform • four haploid nuclei form.
Cytokinesis (occurs after meiosis I) • cytoplasm divides by forming a cleavage furrow • two haploid cells form.	**Cytokinesis** (occurs after meiosis II) • cytoplasm divides by forming a cleavage furrow • four haploid cells form.

FIGURE 6.1.13 The process of meiosis, including all the steps involved. The black and blue chromosomes are paternal whereas the red and green chromosomes are maternal.

Meiosis and genetic diversity

FIGURE 6.1.14 The physical differences between family members are due to the unique combination of genes that are packaged into gametes during meiosis, along with random selection of gametes during fertilisation.

Each gamete has its own unique combination of alleles. There will be similarities in genetic content between parents and offspring, but the offspring are always genetically different from the parents (Figure 6.1.14). This genetic diversity arises from two features of meiosis—crossing over and independent assortment.

Meiosis ensures that a wide range of genetic combinations occurs during the formation of gametes. Variability is further increased when different genetic combinations are brought together at fertilisation, with the random fusion of gametes. During fertilisation, only one sperm contributes its mixed set of chromosomes to form homologous pairs with the chromosomes in the egg. The randomness with which specific sperm fertilises the egg introduces yet more variation to the offspring from one parental pair and helps explain why siblings are different (Figure 6.1.14). Due to these processes, populations of organisms that reproduce sexually have a considerable genetic range, which enables species to survive and reproduce in varied and changing environments.

Crossing over and recombinant chromosomes

> The probability of two genes on the same chromosome exchanging alleles is related to the distance between these genes. The greater the distance, the greater the probability that crossing over between the genes will occur.

The significance of **crossing over** is that it produces chromosomes with new combinations of genetic information. During the first division of meiosis (prophase I), each chromosome pairs up precisely along its length with its matching (homologous) chromosome. The maternal chromosome 1 pairs up with the paternal chromsome 1; the maternal chromosome 2 pairs up with the paternal chromsome 2, and so on. This pairing is called **synapsis**.

When chromosomes pair up, chromatids of homologous chromosomes may exchange portions of their genetic information, including one or more genes, during the process of crossing over. DNA strands from the chromatids of two homologous chromosomes are cut at the equivalent point, a segment is exchanged, and the strands are reconnected. The point where crossing over occurs is called a **chiasma** (plural chiasmata). A long chromosome may have several chiasmata (Figure 6.1.15). The chromosomes produced from the crossing over of the maternal and paternal chromsomes have new combinations of genes and are referred to as **recombinant chromosomes** as shown in Figure 6.1.16. This ensures genetic variation in the daughter cells.

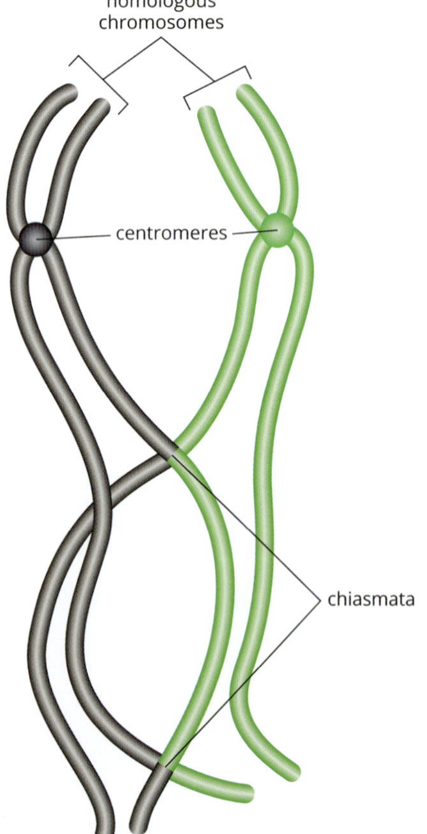

FIGURE 6.1.15 The points of crossing over, called chiasmata

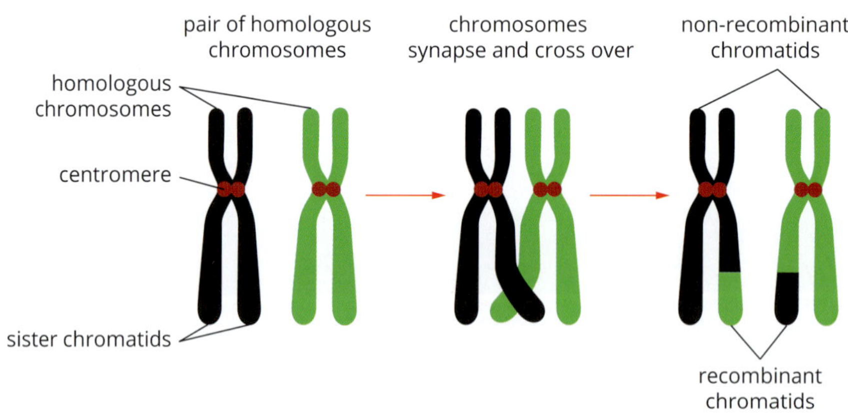

FIGURE 6.1.16 Crossing over occurs between the chromatids of homologous chromosomes. The chromosomes with exchanged DNA material are called recombinant chromosomes.

Independent assortment

When crossing over is finished, the homologous chromosome pairs align along the midline (equatorial plate) of the cell (metaphase I). They do this randomly, meaning the maternal and paternal chromosomes do not have to line up on the same side of the midline (Figure 6.1.17), a process called independent assortment. The homologues then separate and move to opposite poles. These two steps result in the random assortment of maternal and paternal chromosomes and their alleles in the gametes.

The formula 2^n can be used to calculate the number of possible chromosome combinations due to independent assortment, where n represents the organism's haploid number. In humans, the haploid number is 23 ($n = 23$), hence there are $2^{23} = 8\,388\,608$ possible combinations for sets of homologues in meiosis I. Variability is further increased when different genetic combinations are brought together at fertilisation. For humans, this is 2^{23} (egg) $\times$ 2^{23} (sperm) possible combinations, which equals over 70 trillion possible genetic variations (not including further variation introduced from crossing over).

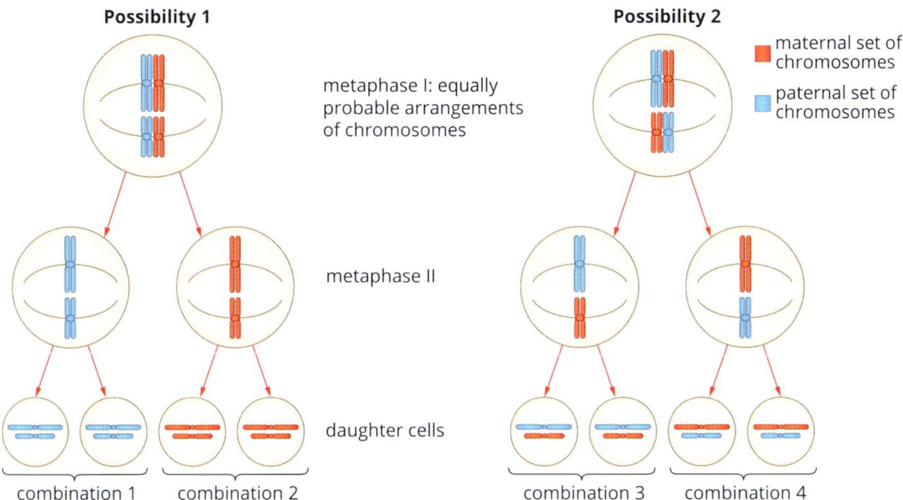

FIGURE 6.1.17 Independent assortment. Homologous chromosomes align randomly, regardless of the origin of the chromosome, along the cell's midline. Crossing over is not shown here.

KARYOTYPES

Scientists examine eukaryotic chromosomes when they are most visible in the cell—at the metaphase stage of the cell cycle. The chromosomes are stained so that characteristic patterns of light and dark bands (G bands) appear along the arms of the chromosomes. The bands reflect regional differences in the amounts of bases A and T versus G and C. Banding patterns are specific and consistent. They can be used to distinguish between chromosomes and to identify subtle changes in chromosome structure that may be associated with genetic conditions.

Figure 6.1.18 shows the striking similarity in banding patterns in a chromosome from human, chimpanzee, gorilla and orangutan.

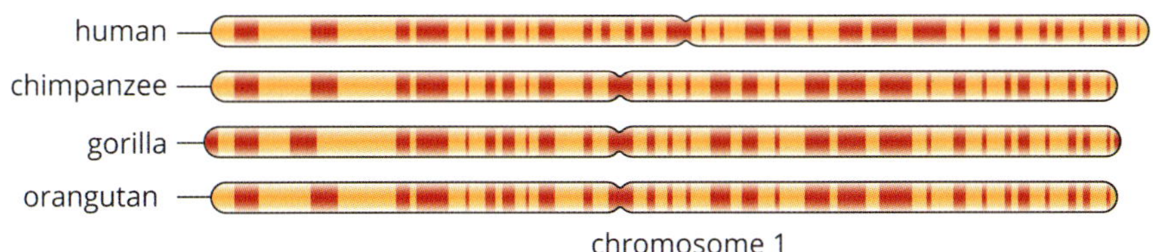

FIGURE 6.1.18 The banding patterns in chromosome 1 from human, chimpanzee, gorilla and orangutan show striking similarities.

A **karyotype** is the image or picture of the full set of chromosomes from an individual's cell. A karyotype is represented by photographs or diagrams of the chromosomes arranged in pairs according to their length and the position of the centromere. Karyotypes allow scientists to compare the chromosome sets of related species. Karyotypes also allow scientists to identify changes that may be associated with genetic conditions such as:

- changes in chromosome number (the loss or gain of whole chromosomes)
- changes in structure (such as the duplication, inversion or deletion of part of a chromosome).

The human karyotype

In humans (and some other organisms), sex chromosomes are distinguished from the remaining chromosomes (autosomes). Human somatic cells have a diploid chromosome number of $2n = 46$. A karyotype for a human male shows that there are 22 pairs of autosomes and two sex chromosomes, XY (Figure 6.1.19). The autosomal pairs are numbered 1–22 and are ordered from largest to smallest. The sex chromosomes are usually shown after the autosomes.

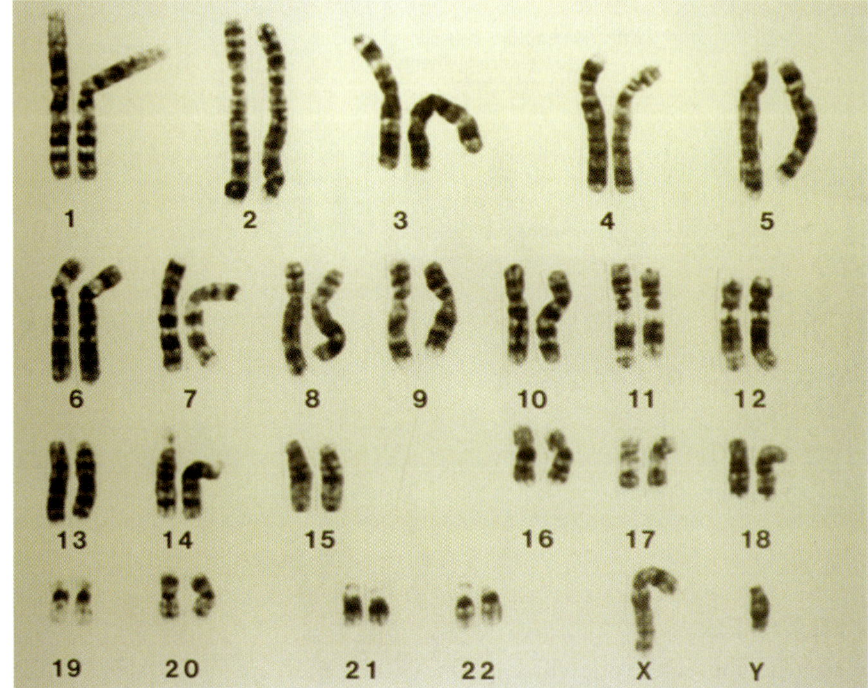

FIGURE 6.1.19 To construct a karyotype, metaphase chromosomes are stained and photographed. The chromosome images are arranged by size and centromere position, with the shorter arm (p arm) at the top.

Using karyotypes to identify chromosomal number abnormalities

A number of syndromes result from an increase or decrease in chromosome number. Down syndrome is a typical example of a syndrome that results from having one extra chromosome. It results from one extra copy of chromosome 21, which will be visible on a karyotype (Figure 6.1.20). This type of condition is called a **trisomy**, because there are three copies of the chromosome.

Aneuploidy refers to an abnormal number of chromosomes, such as 45 or 47 chromosomes in humans instead of 46. In monosomy one chromosome is missing. In trisomy there is an extra copy of a chromosome.

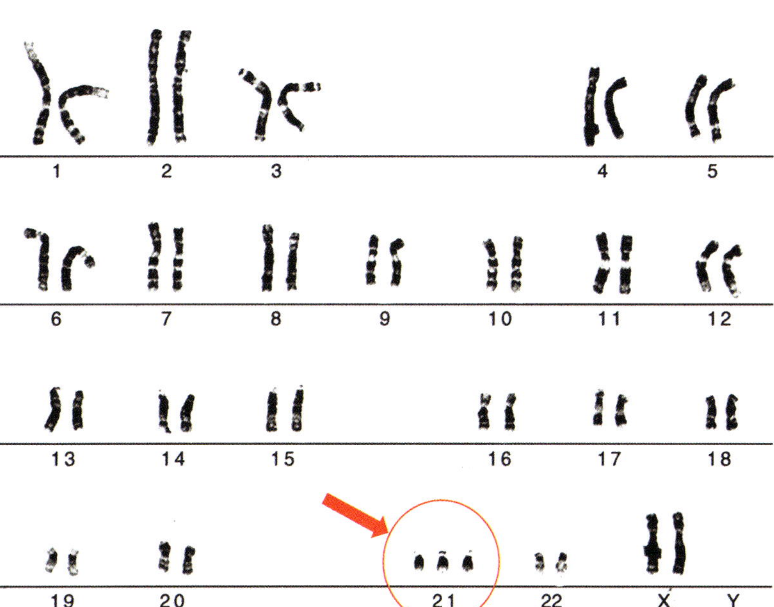

FIGURE 6.1.20 Down syndrome karyotype, showing an extra copy of chromosome 21

Two other syndromes that are the result of an abnormal chromosome number are Klinefelter syndrome and Turner syndrome. In Klinefelter syndrome (also called XXY syndrome), males have two X chromosomes and one Y chromosome instead of one X and one Y chromosome (Figure 6.1.21). As a result they have 47 chromosomes. People with Klinefelter syndrome are infertile and have other characteristics such as breast development and a tall stature.

In Turner syndrome (also called monosomy X), which occurs only in females, there is one X chromosome instead of two (Figure 6.1.22). People with Turner syndrome are infertile and have a short stature.

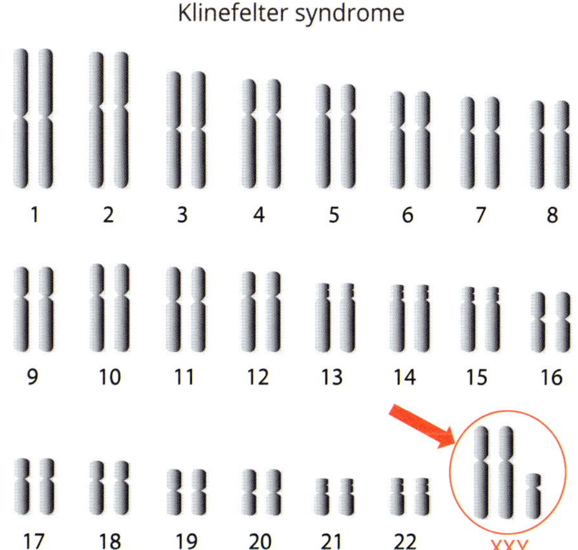

FIGURE 6.1.21 Klinefelter syndrome karyotype, showing an extra copy of the X chromosome

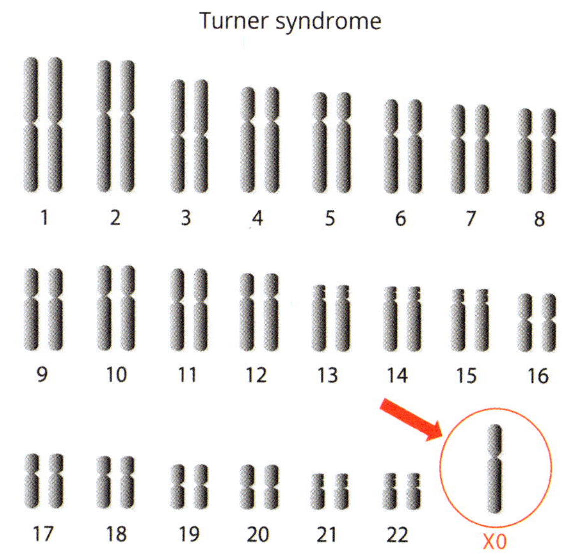

FIGURE 6.1.22 Turner syndrome karyotype, showing only one copy of the X chromosome. The other sex chromosome is absent, denoted by the '0' next to the 'X' in the karyotype.

Table 6.1.5 shows some of the consequences in humans of an extra or a missing member of a chromosome pair. This is the result of abnormal meiosis in one of the parents of the person with the condition.

TABLE 6.1.5 Conditions in humans that are a result of errors during meiosis and genetic recombination

Condition	Chromosome change	Traits of person with condition
Down syndrome	Three copies of chromosome 21 present (trisomy 21) (47 chromosomes)	Male or female, some intellectual disability, characteristic palm prints and facial features, may be infertile.
Klinefelter syndrome	Extra X (XXY) (47 chromosomes)	Male, sterile, often some intellectual disability, with female secondary sex traits (e.g. breast enlargement).
Patau syndrome	Three copies of chromosome 13 present (trisomy 13) (47 chromosomes)	Male or female, small skull, intellectual disability, cleft lip, cleft palate, usually has heart defects, seldom survives more than four months after birth.
Turner syndrome	All or part of one X chromosome is altered or missing (monosomy) (45 complete chromosomes)	Female, short stature, infertile, fluid retention and puffiness in hands and feet, kidney and heart problems, some learning difficulties but most people with Turner syndrome have normal intelligence.

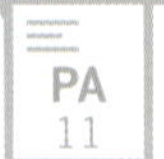

6.1 Review

SUMMARY

- Chromosome size is affected by the number and length of genes and the length of spacer DNA.
- Each chromosome carries different genes, so that together, all the genes and chromosomes make up the genome.
- Chromosome number varies between species but is consistent within species (except in the case of chromosomal abnormalities).
- Homologous chromosomes are matching pairs of chromosomes (one from each parent) that have the same genes.
- Genes on homologous chromosomes occur at the same loci but may have different alleles.
- Sex chromosomes (allosomes) are chromosomes that are involved in sex determination.
- In humans the sex chromosomes are X and Y.
- Other organisms may have other sex chromosomes, or no sex chromosomes at all.
- In humans, females have two X chromosomes and males have an X and Y.
- Autosomes are chromosomes that are not involved in sex determination.
- Gametes are haploid cells formed by meiosis and which combine in sexual reproduction.
- Meiosis is a division of the nucleus that halves the normal diploid number of chromosomes to produce four genetically unique haploid daughter cells.
- Variation in gametes arises from the exchange of alleles through crossing over (recombination) and independent assortment of chromosomes during meiosis.
- The great advantage of meiosis and sexual reproduction is the production of genetic diversity in a population.
- A karyotype is an image showing the number and appearance of a cell's chromosomes.
- A karyotype can identify:
  - the number of chromosomes in a cell
  - the biological sex of the individual
  - whether an individual has an extra chromosome, such as in Down syndrome or Klinefelter syndrome
  - where an individual is missing a chromosome, such as in Turner syndrome
  - the position of the centromere
  - the size of the chromosome.

KEY QUESTIONS

Knowledge and understanding

1 Distinguish between the following terms:
  - a autosomes and sex chromosomes (allosomes)
  - b sex chromosomes and homologues.

2 Identify the number of homologous pairs of chromosomes you would expect to find in the cells of most:
  - a human females
  - b human males.

3 Select the phase of meiosis in which crossing over occurs.
  - **A** prophase I
  - **B** anaphase I
  - **C** metaphase II
  - **D** prophase II

4 Select the process that does not occur during meiosis.
  - **A** formation of two haploid nuclei
  - **B** crossing over
  - **C** pairing of homologous chromosomes
  - **D** formation of two diploid daughter cells

5 For each of the conditions listed below, state which chromosome is affected and whether the chromosome is in excess or missing.
  - a Down syndrome
  - b Turner syndrome
  - c Klinefelter syndrome

Analysis

6 Complete the table with the number of chromosomes in the haploid or diploid cells of these organisms.

Organism	Haploid number (n)	Diploid number ($2n$)
koala (*Phascolarctos cinereus*)	8	
spinach (*Spinacia oleracea*)	6	
common potato (*Solanum tuberosum*)		48
goat (*Capra aegagrus hircus*)		60

7 a When a cell with chromosome number $n = 24$ undergoes mitosis, how many daughter cells are produced, and what is their chromosome number?
  - b When a cell with chromosome number $2n = 24$ undergoes meiosis, how many daughter cells are produced, and what is their chromosome number?

8 Refer to Figure 6.1.13 on page 255 to answer the following questions about meiosis.
  - a State the colours of each of the homologous pairs of chromosomes.
  - b Describe the events that occurred in prophase I and metaphase I, and their significance for genetic diversity.
  - c Using a cell with a diploid number of 6, draw two separate cells to show any of the two possible alignments of chromosomes during metaphase I. You do not need to display any crossing over that may have occurred.
  - d Outline how:
    - i prophase I differs from prophase II
    - ii metaphase I differs from metaphase II
    - iii anaphase I differs from anaphase II
    - iv telophase I differs from telophase II.

9 The figure below shows a human karyotype.
  - a State whether this individual is a male or female.
  - b Is there any evidence of trisomy in this person? Explain your answer.

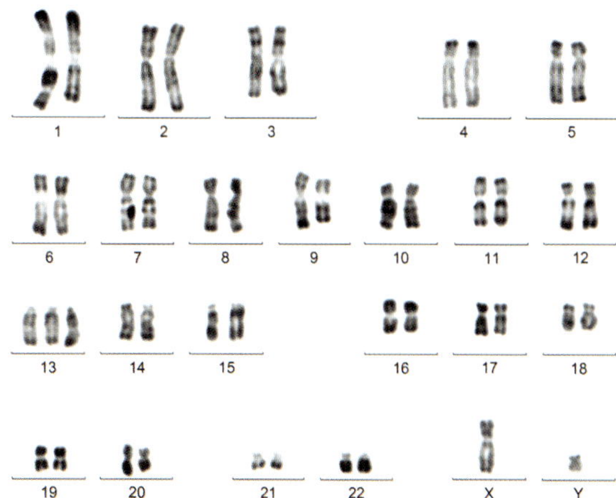

6.2 DNA and genes

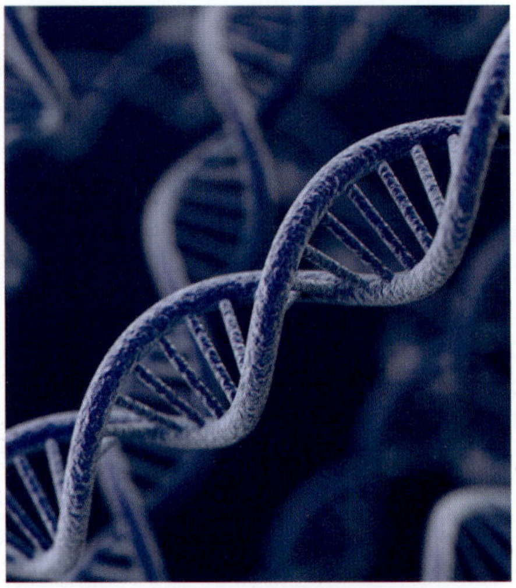

In all organisms, inherited characteristics are determined by genes located on DNA (Figure 6.2.1).

In this section you will learn about the difference between a gene, genome and allele. You will also learn about the genome as the complete set of genetic material (DNA) present in an organism.

FIGURE 6.2.1 Representation of part of the molecule of DNA (deoxyribonucleic acid), showing its double helix structure

THE STRUCTURE OF DNA

All cells contain genetic material in the form of **deoxyribonucleic acid (DNA)** packaged in the chromosomes (Figure 6.2.2). DNA carries **hereditary** information, directs the cell's activities, and is passed on from generation to generation. The DNA molecule has certain regions known as **genes**, which contain the genetic information that determines what you look like and the function of each cell in your body. To understand how genes work, it is important to understand the basic structure of DNA.

DNA (deoxyribonucleic acid) is a complex molecule that contains all the genetic information necessary to build and maintain an organism.

Hereditary information refers to genetic material that is passed on from parent to offspring (or from one generation to the next).

A gene is a unit of heredity that determines the characteristics of an organism. At the molecular level, a gene is a section of DNA with a unique sequence.

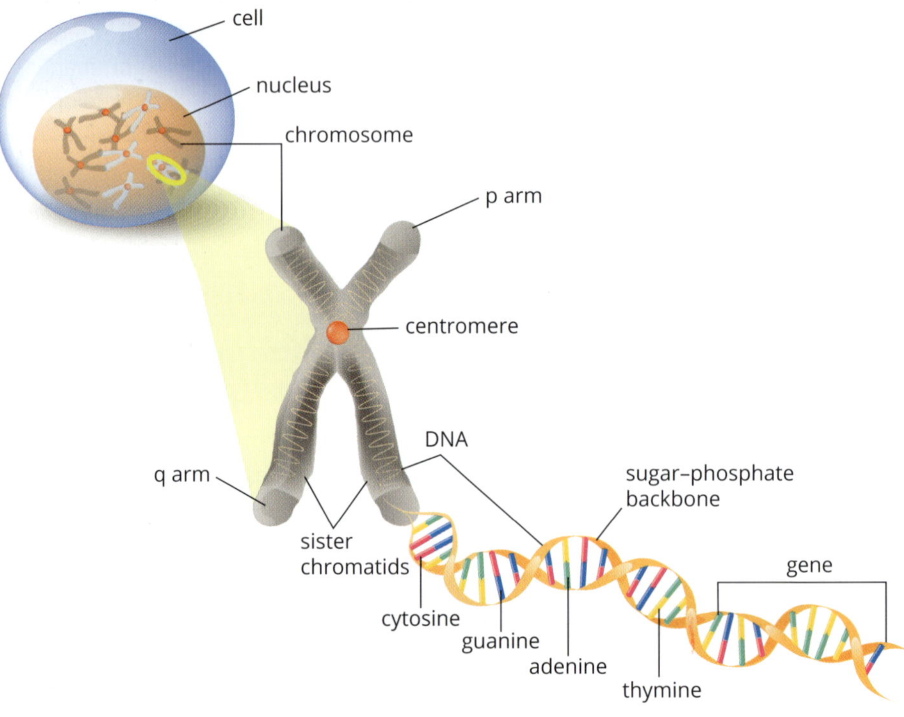

FIGURE 6.2.2 The relationship between a cell, chromosomes and the DNA molecule. The double helix structure and chemical sub-units of DNA are also depicted.

Nucleotides—building blocks of DNA

DNA is a large (macro) molecule, which is made up of a series of chemical building blocks called nucleotides. As shown in Figure 6.2.3, a **nucleotide** is composed of a phosphate molecule, the sugar deoxyribose, and one of four nitrogen-containing (nitrogenous) **bases** (adenine, cytosine, guanine or thymine).

There are two types of nucleotide: **purines**, with a double ring structure, and **pyrimidines**, with a single ring. The purine bases are **adenine (A)** and **guanine (G)**, and the pyrimidine bases are **thymine (T)** and **cytosine (C)**. In ribonucleic acid (RNA) thymine (T) is replaced by uracil (U), which also pairs with adenine (A).

In DNA, the four nitrogenous bases are A, T, G and C. In RNA, the four nitrogenous bases are A, U, G and C. A pairs with T (U in RNA) and G pairs with C.

Nucleotides are distinguished from one another by the nitrogen-containing base (Figure 6.2.3). One nucleotide is joined to the next nucleotide by a covalent phosphodiester bond between the phosphate group on the 5' carbon of one nucleotide and the 3' carbon of the other nucleotide. When many nucleotides are joined together, a polynucleotide chain which runs from 5' to 3' is formed (Figure 6.2.3). Different nucleotides can occur in any order within a strand: if a particular base is A, the next base in the sequence could be A, G, T or C.

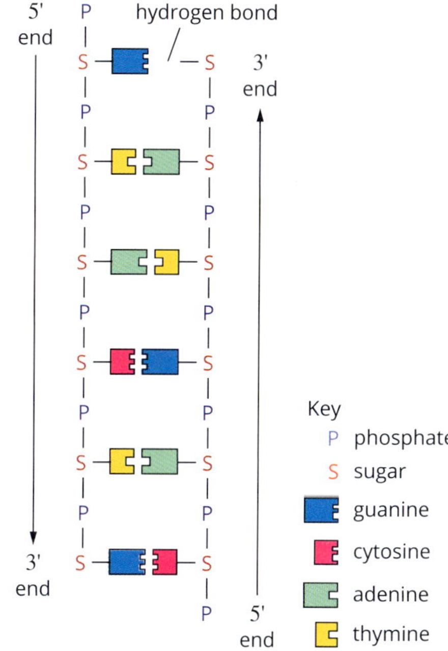

FIGURE 6.2.3 Two-dimensional representation of the DNA molecule. Individual nucleotides within the single polynucleotide chains are joined together by phosphodiester bonds. Hydrogen bonds between the complementary base pairs hold the two antiparallel polynucleotide strands together.

DNA—a double-stranded helix

A DNA molecule is made up of two polynucleotide chains. The two polynucleotide chains of DNA are held together by hydrogen bonds between **complementary base pairs** (Figure 6.2.4). There is a direct pairing between A and T, and between G and C in the DNA molecule. This complementary base pairing results in the two polynucleotide strands joining together to form the double-stranded DNA molecule. Given the base sequence of one strand you can determine the sequence of the other by this complementary base-pairing rule.

The two polynucleotide strands will then spiral around an imaginary axis, forming a **double helix** (Figure 6.2.4).

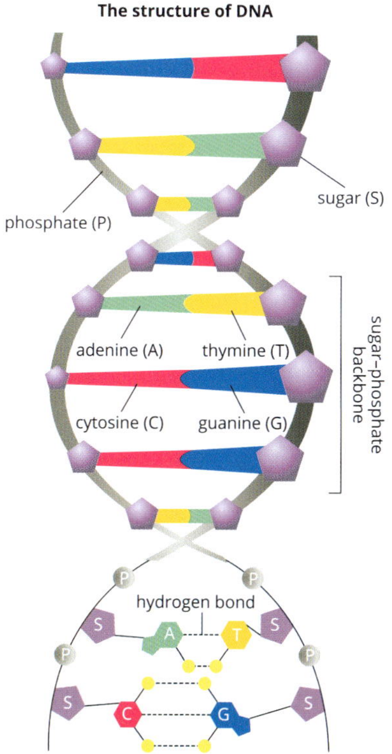

FIGURE 6.2.4 The double helix structure of DNA is formed by two strands of complementary nitrogenous bases that are joined by hydrogen bonds. Each side of the helix is comprised of deoxyribose sugar and phosphate molecules, known as the sugar–phosphate backbone.

A genome is an organism's complete set of DNA.

GENOMES

The somatic cells (all cells in a body except gametes) of most organisms are diploid ($2n$). This is because they contain two sets of chromosomes (one set from each parent). Because the sets of chromosomes are almost identical, the **genome** of an organism is the total of an organism's DNA measured in the number of base pairs contained in a haploid (n) set of chromosomes. A genome includes the DNA found in cell organelles, the mitochondria and chloroplasts. A typical human cell has two similar sets of chromosomes and each set has DNA totalling 3234 million base pairs. In other words, the human genome has approximately 3234 million DNA base pairs.

FIGURE 6.2.5 Genes carry instructions for a cell to make proteins. A spider's web is made of the structural protein silk (a fibroin protein).

GENES

DNA is the molecule of life that encodes the information from which organisms are built. The DNA molecule consists of many genes, and these genes determine the characteristics of an organism. At a molecular level, a gene is a unique sequence of DNA. Each gene carries a particular instruction for a cell; for example, how to make silk for a spider web (Figure 6.2.5). The web is made of silk fibroin, which is a structural protein. The genetic information on how to make this protein is in one particular gene, called the silk fibroin gene. The process by which the information in the gene is decoded to assemble silk fibroin is called **gene expression**.

A gene is a unit of heredity, made up of a unique sequence of DNA that determines a characteristic of an organism. Nearly all genes specify the production of polypeptide chains or **proteins**, which perform essential functions in our body's cells. For example, enzymes are proteins that catalyse chemical reactions within the body, such as the reaction that produces sugars from starch. Carrier proteins found in the plasma membrane control the movement of essential elements and molecules into and out of the cell. The protein haemoglobin in red blood cells carries oxygen.

Proteins control cellular functions and genes control the production of proteins; therefore inherited genes ultimately govern the functions of organisms. Genes vary in size from about 100 to 2.5 million base pairs. The length of the sequence of DNA and the precise order of the base pairs in a gene are the critical factors that determine what the gene product will be like and what it will do in a cell.

Diploid organisims ($2n$) have two copies of each gene, one on each chromosome. As each parent contributes one chromosome in each chromosome pair, so too does each parent contribute one copy of all genes on autosomes.

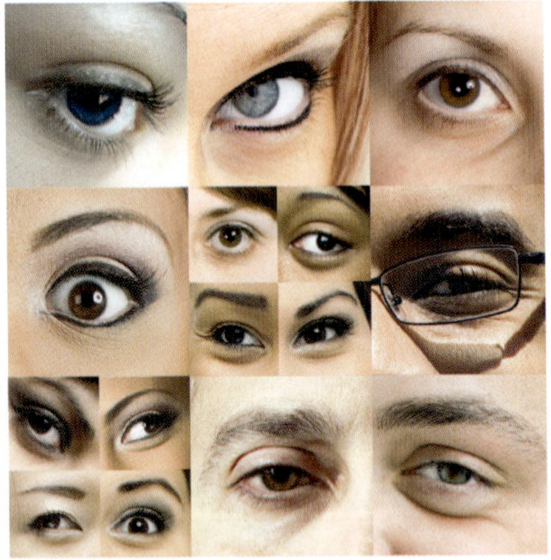

FIGURE 6.2.6 Eye colour is a complex trait: many genes are involved in producing the variety of eye colours in humans. Each of these genes has different alleles.

ALLELES

Physical characteristics or **traits** such as skin colour, eye colour and hair colour all vary within populations and even within families. Although an individual gene may be responsible for a specific trait, that gene can exist in different forms known as **alleles**. For example, there is a gene that codes for the amount of pigment in the eye, and therefore eye colour. Alternative forms (alleles) of the gene exist, including one for blue eye colour and another for brown eye colour (Figure 6.2.6). This means that the DNA sequence of bases for these two alleles is slightly different. Both alleles still code for the eye colour gene and are found in the same place (locus) on the chromosome (DNA strand). Somatic cells of a diploid organism contain two alleles for every gene, with one allele for every gene in an organism inherited from each parent.

A trait is a particular characteristic or feature of an organism.
An allele is one form of a gene.

CASE STUDY **DATA ANALYSIS**

DNA sequencing

DNA sequencing is a procedure that determines the order of nucleotides in a gene. It is an automated process involving the polymerase chain reaction (PCR) using nucleotides that are tagged with coloured fluorescent markers. Each type of nucleotide (A, T, G or C) is tagged with a different coloured marker. In the final stage of the sequencing process, a laser beam causes the final nucleotide of each fragment to fluoresce. The colour of each nucleotide is recorded and compiled into a data file.

The data file is then analysed by software that arranges the detected bases in the correct sequence. The output from the software is the sequence of bases and a matching chromatogram (graph) showing the strength of the fluorescence and the base that produced it (Figure 6.2.7).

Sequences are always obtained in both directions (5' to 3', and 3' to 5'), and the two sequences are matched up by the software. The order of bases in the original sample of DNA can be determined from the order of the peaks. Figure 6.2.7 shows how the terminating nucleotides on DNA fragments are read and converted into a chromatogram.

DNA sequences can be analysed to screen for various genetic conditions in humans. Variation in the DNA sequences of genes can have different outcomes for the phenotype—it may have no effect, be of uncertain significance, be likely to cause disease, or be disease-causing. For example, most people have an allele that is involved in the production of a protein called melanin. If a lot of melanin is produced, the person will have brown eyes, brown hair and/or brown skin. If a person has the allele that does not allow a functional melanin protein to be produced, then their hair, skin and eyes will appear white. This is called oculocutaneous albinism (Figure 6.2.8). The eyes of a person with oculocutaneous albinism may look red because light reflects off blood vessels in the back of the eye and the red blood vessels can be seen through the pale iris.

One of the genes where mutations are known to cause oculocutaneous albinism is the *OCA2* gene on chromosome 15. A section of this gene is shown in Figure 6.2.7.

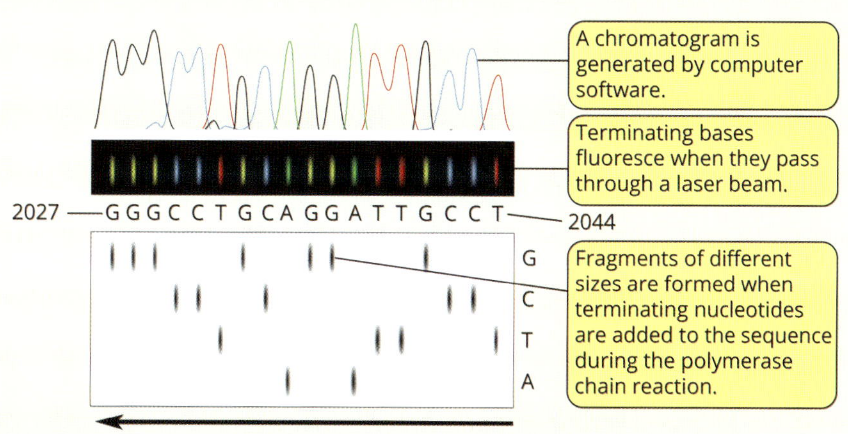

FIGURE 6.2.7 Capillary electrophoresis separates the DNA fragments according to size. The fragments pass through a laser beam, causing the nucleotides to fluoresce. Computer software then translates the data into a sequence of bases and a matching chromatogram. The fragment shown is from a section of the *OCA2* gene.

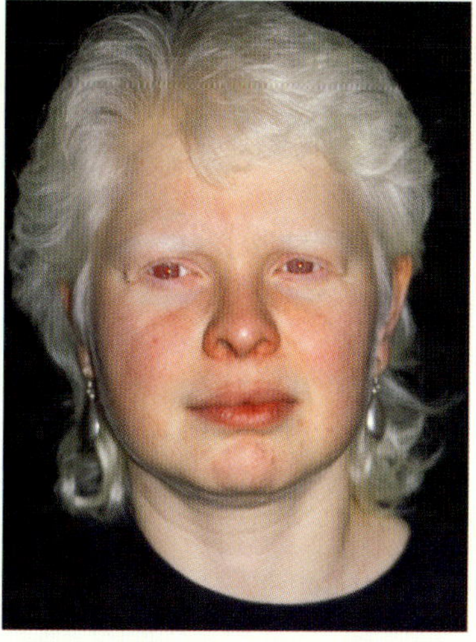

FIGURE 6.2.8 Albinism is a condition caused by an allele that does not produce a functional melanin protein. This results in white hair and pale skin, and eyes that look red.

Analysis

1 Analyse the sequence of the *OCA2* gene fragment in Figure 6.2.7 and write down the sequence of the complementary DNA strand.

2 Study the sequence in Figure 6.2.7 and compare it to the sequence shown below:

GGGCCTGCAGCATTGCCT

Use the position numbers to help you describe the difference between the sequences.

3 What conclusion can you draw about this difference when it is identified in an individual that has oculocutaneous albinism?

6.2 Review

OA ✓✓

SUMMARY

- Deoxyribonucleic acid (DNA) is the genetic material that contains the instructions for cells to make proteins.
- DNA is a double-stranded molecule made up of two polynucleotide strands, held together by hydrogen bonds.
- The two polynucleotide strands of DNA are made up of complementary nucleotides: adenine (A) and thymine (T), and guanine (G) and cytosine (C).
- The genome of an organism is the total of an organism's DNA, measured in the number of base pairs contained in a haploid (n) set of chromosomes.
- A gene is a unit of heredity made up of a unique sequence of DNA that determines characteristics of an organism.
- Genes can have many alternate forms, known as alleles.
- There are two alleles for a particular trait: one allele is inherited from each parent.

KEY QUESTIONS

Knowledge and understanding

1 Outline the structure of the DNA molecule. In your answer, include:
- the name and components of the unit (building block) of DNA
- the names of the four bases found in DNA
- how nucleotides are joined to build a single-stranded DNA molecule
- how a double-stranded DNA molecule is formed.

2 Distinguish between DNA, genome, gene and allele.

3 How many alleles are involved in the determination of a physical trait controlled by a single gene? Explain your answer.

4 Distinguish between the number of gene copies in diploid and haploid sets of chromosomes.

Analysis

5 A strand of DNA has the sequence ATTCCGTA. Write this out, and under it write the sequence of the complementary strand.

6 The following table shows the composition of bases in the cells of a variety of species.

Composition of bases in cells of selected species.

Species	Thymine (%)	Adenine (%)	Guanine (%)	Cytosine (%)
humans	30.1	30.4	19.6	19.9
cattle	28.7	29.0	21.2	21.2
salmon	29.1	29.7	20.8	20.4
wheatgerm	27.4	28.1	21.8	22.7
E. coli	23.6	24.7	26.0	25.7
sea urchin	32.1	32.8	17.7	17.3

Discuss whether or not the evidence supports the theory that there are complementary base pairs.

7 Complete the table by assigning 'purine' or 'pyrimidine' to each nitrogenous base and filling in their complementary bases.

Base	Purine or pyrimidine	Complementary base	Purine or pyrimidine
adenine (A)			
guanine (G)			
cytosine (C)			
thymine (T)			

6.3 Genotypes and phenotypes

You inherited many of your physical features or traits from your parents. You share certain traits with your mother and others with your father. You might even appear to have a totally different version of a trait.

In this section you will look at what determines the traits and characteristics that humans have and how they are passed to children from parents. You will also learn the difference between genotype and phenotype, the use of symbols for writing genotypes by stating the alleles present at a particular gene locus, and the distinction between dominant and recessive phenotypes. You will also learn about the influences on phenotypes of genetic material, environmental factors (Figure 6.3.1) and interactions of DNA with other molecules.

FIGURE 6.3.1 Hydrangeas are a well-known example of the effects the environment can have on phenotype. Differences in soil pH result in differences in flower colour.

GENOTYPE

A **genotype** is the set of alleles present in the DNA of an individual organism. It is the result of inheritance. In Section 6.2 you learned that an allele is an alternative form of a gene. Each individual usually only has two alleles for each trait: one inherited from their mother and one inherited from their father. But one gene may have many alleles, and this is what leads to variation in a population.

> The genotype is the set of alleles present in the DNA of an individual organism. The genotype is the result of inheritance.

You will recall from Section 6.1 that a diploid cell has two sets of chromosomes (one set from each parent). You have two sets of autosomes, which are denoted by the numbers 1 to 22. Each pair of these chromosomes is homologous because they are identical and carry the same genes.

Consider a gene, which might be called gene *A*, that has two alleles. Different alleles are represented by different symbols. One allele can be represented by an upper-case *A*, and the other by a lower-case *a*. The names of genes and alleles are always italicised.

If gene *A* is in a homologous chromosome, there will be two copies of the gene. If you inherited the allele *A* from both parents, your genotype for gene *A* will be *AA*. On the other hand, if you inherited the *A* allele from one parent and the *a* allele from the other parent, you will have the genotype *Aa*. If you inherited the allele *a* from both parents, you will have the genotype *aa*.

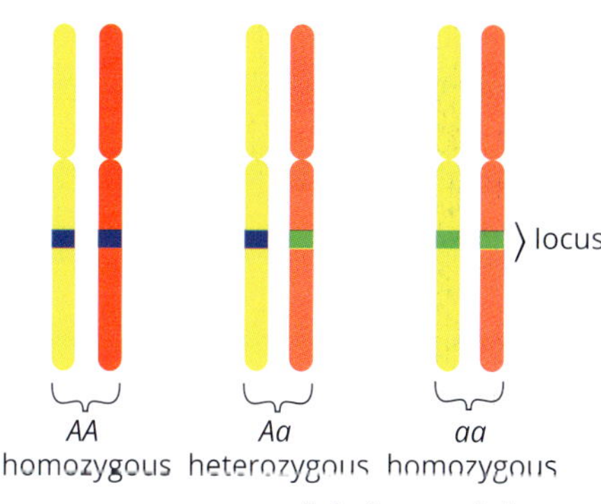

FIGURE 6.3.2 On homologous chromosomes, alleles of a gene occur at the same locus. If a gene has two alleles, there can be three different combinations or genotypes. Two of these combinations are homozygous, and one is heterozygous.

Therefore there are three different combinations or genotypes for gene *A*: *AA*, *Aa* or *aa* (Figure 6.3.2). Genotypes *AA* and *aa* contain only one type of allele, so the individual is said to be **homozygous** for that gene and is called a **homozygote**. Genotype *Aa* contains two different alleles, so the individual is said to be **heterozygous** for that gene and is called a **heterozygote**.

Now consider a gene, which might be called gene *B*, on the X chromosome. This gene also has two alleles, *B* and *b*. Figure 6.3.3 shows the symbols used to record genotypes of autosomal and X-linked genes. To show that this gene is on the X chromosome, the symbol X (an X-linked gene) is used, with a superscript to represent the gene. So the two alleles are named X^B and X^b.

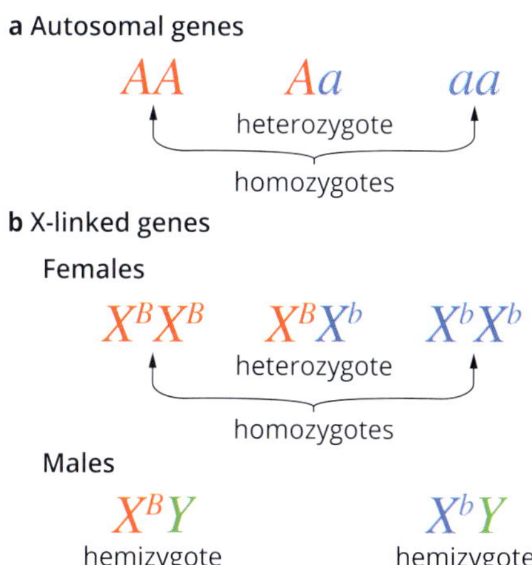

FIGURE 6.3.3 These are the symbols used to record genotypes of (a) autosomal and (b) X-linked genes.

An organism that has two copies of the same allele of a gene is said to be homozygous for that gene; 'homo' means 'the same'.

An organism that carries two different alleles of a gene is said to be heterozygous for that gene; 'hetero' means 'different'.

Because females have two copies of the X chromosome they can be homozygous (X^BX^B or X^bX^b) or heterozygous (X^BX^b) for gene *B*. Males only have one copy of the X chromosome, so they are referred to as being **hemizygous** ('hemi' = 'half'). This term is used to indicate that the human male has only half the number of copies of those genes that occur on the X chromosome, compared with a female. The Y chromosome is used in describing the genotype to emphasise that the individual is a male.

The two possible genotypes of gene *B* for a male are X^BY and X^bY.

Multiple alleles at a single gene locus

A single gene locus may have more than two alleles. The human ABO blood group system is based on such alleles (Figure 6.3.4). In this case there are three alleles, represented as I^A, I^B and *i*. Allele I^A produces the A antigen, I^B produces the B antigen, and *i* produces no antigen. Each person carries two copies of these three possible alleles. There are therefore six possible genotypes and four phenotypes, as shown in Figure 6.3.4.

	ABO blood groups			
Blood type	**Type A**	**Type B**	**Type AB**	**Type O**
Possible allele combinations	$I^A I^A$ $I^A i$	$I^B I^B$ $I^B i$	$I^A I^B$	$i\ i$
Antigen (on RBC)	A antigen	B antigen	AB antigens	no antigens

FIGURE 6.3.4 The ABO blood group system is based on three alleles. The production of antigens A and B depends on the combination of alleles present.

BIOFILE

Naming genes

There are internationally accepted names for genes and their abbreviated forms. For example, the gene that codes for phenylalanine hydroxylase, an enzyme involved in the inherited disorder phenylketonuria, is abbreviated to *PAH*. The gene name is always italicised, to distinguish them for the proteins they encode. For example, BRCA1 is an enzyme expressed in the cells of breast and other tissue, where it helps repair damaged DNA or destroy cells if DNA cannot be repaired. The gene that codes for this enzyme is known as *BRCA1*.

PHENOTYPE

An organism's **phenotype** is all of its observable characteristics. It is the result of inheritance and the effects of the organism's environment. An example of a phenotype is skin colour. Your skin colour depends on how much skin pigment (melanin) is produced. But skin colour also depends on environmental factors such as exposure to sunlight, especially in pale-skinned people. The greater the exposure, the more melanin is produced (Figure 6.3.5).

An organism's phenotype is all of its observable characteristics. It is the result of inheritance and the effects of the organism's environment.

FIGURE 6.3.5 The effect of sun exposure on skin colour. The darker part of the skin has been exposed and has produced more melanin, causing it to darken. Unexposed skin (the ankle) does not produce extra melanin.

When studying inheritance it is important to know the genotypes of parents and offspring. However, it is equally important to know the specific observable characteristics that can result from a given genotype; that is, the phenotype. The phenotype includes any distinct property of an organism: physical, chemical, physiological or behavioural. In experimental crosses (matings), phenotypes are observed to determine the underlying genotypes.

Importantly, environmental conditions and epigenetic factors can affect the phenotype of certain genotypes. In these cases, the genotype determines the possible range of phenotypes for a particular characteristic or trait, and the environment and epigenetics influence where in that range the actual phenotype will be.

For example, in Arctic foxes (*Vulpes lagopus*), two fur colour genotypes occur, called 'white morph' and 'blue morph'. The fur of the blue morph remains dark blue-grey throughout the year, but the fur of the white morph varies from dark brown or grey to pure white. In summer the fur is dark, but as winter approaches the fur gradually changes to white in response to the increasing cold and shorter day length (Figure 6.3.6). At the end of winter the fur gradually returns to its summer colour.

Although the phenotype is relevant to the functioning of an individual organism, the genotype is what is passed on to the next generation.

FIGURE 6.3.6 The fur of the white morph genotype of Arctic fox (*Vulpes lagopus*) changes from dark brown or grey (a) to pure white (b) as winter approaches, in response to the increasing cold and shorter day length.

CASE STUDY

Gregor Mendel, the founder of genetics

Modern genetics began in an abbey garden, where a monk named Gregor Mendel proposed a mechanism for inheritance (Figure 6.3.7a). Mendel developed his theory of inheritance before chromosomes were observed under the microscope or the significance of their behaviour was understood.

Mendel's work began with breeding garden peas to study inheritance. His choice of breeding peas was fortunate. There were a number of easily observable phenotypes, and the plants were self-fertile but could also be outcrossed (that is, crosses could be carried out within and between pure-breeding lines). Furthermore, large numbers of offspring could be counted for each cross, the generation time was short, and the peas were easy to maintain.

Figure 6.3.7b shows some of the discrete traits that Mendel observed in his breeding experiments with peas: flower colour, pod colour, pod shape, seed colour and seed shape. Each of these traits was important in the choice of peas used for crossings. It was likely that Mendel devised his model of inheritance theoretically and used the data from peas to confirm the model.

Mendel presented his results to the Brunn Natural Science Society in 1865 and they were published by the society in 1866. But the results were ignored until three other scientists independently produced similar data in 1900. Only then was Mendel's scientific contribution recognised—16 years after his death.

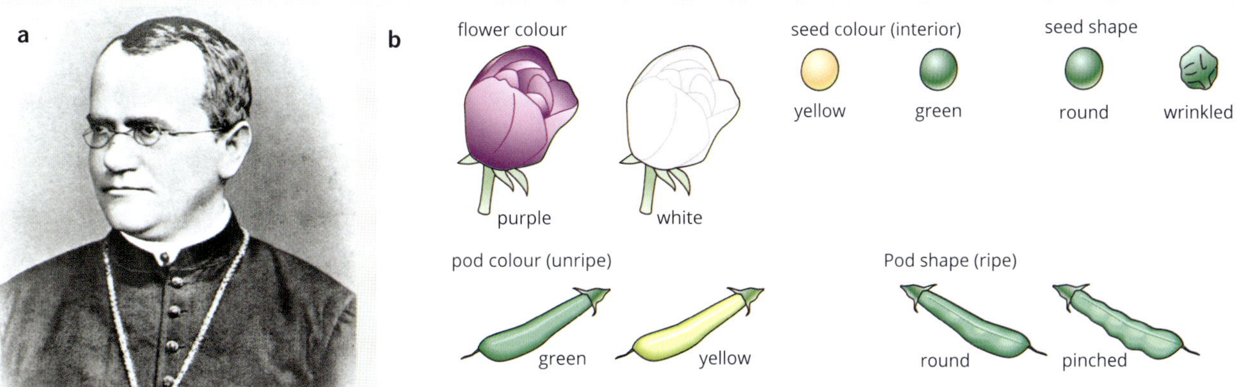

FIGURE 6.3.7 (a) Gregor Mendel, the founder of genetics. (b) Some of the discrete traits that Mendel observed in his breeding experiments with peas: flower colour, pod colour, pod shape, seed colour and seed shape

GENOTYPIC INFLUENCES ON PHENOTYPE

Dominance and recessiveness are properties of alleles, not genes. They are expressed as either dominant or recessive phenotypes.

The relationship between genotype and phenotype gives an insight into an important property of phenotypes known as **dominance**. For a given gene, the phenotype of the heterozygote compared with the appearances of each homozygote allows us to determine whether a phenotype is completely dominant, incompletely dominant, codominant or **recessive**. It is important to understand that dominance (and recessiveness) are properties of alleles, not genes. Genes are expressed as dominant or recessive phenotypes but the genes are neither dominant nor recessive.

Complete dominance

To understand **complete dominance** it is useful to consider the white eye gene in blowflies. There are two alleles for the white eye gene, *W* and *w*. Individuals with genotype *WW* have red eyes, while individuals with genotype *ww* have white eyes (Figure 6.3.8). Individuals with genotype *Ww* do not show an in-between trait such as pink eyes but instead have red eyes, making them indistinguishable from those of the *WW* genotype.

Blowflies with genotype *Ww* display red eyes because the *W* allele makes enough membrane transporter protein to give the eye normal red pigment levels. The red eye colour phenotype is referred to as the **dominant phenotype**, because it only needs one *W* allele for that phenotype to be displayed. The white phenotype is referred to as the **recessive phenotype** because it is not observed in the heterozygote. It needs two copies of the *w* allele for it to be observed in the phenotype. This example shows that scientists can determine which phenotype, if any, is dominant only by examining the heterozygote.

By convention, the allele associated with the dominant phenotype is represented by an upper-case symbol (e.g. *W*). The allele associated with the recessive phenotype is represented by a lower-case symbol (e.g. *w*). The blowfly's white eyes is an example of complete dominance. *Ww* individuals have the same eye colour as *WW* flies.

FIGURE 6.3.8 Blowflies with the *WW* and *Ww* genotypes have the red-eye phenotype (left). Blowflies with the *ww* genotype have the white-eye phenotype (right).

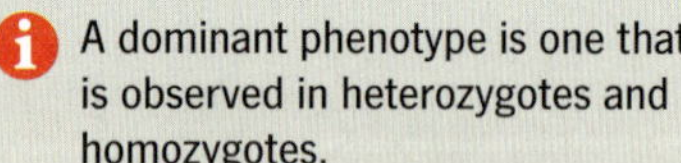

A dominant phenotype is one that is observed in heterozygotes and homozygotes.

A recessive phenotype is one that is observed only in homozygotes.

Incomplete dominance

Not all phenotypes are completely dominant or recessive. When neither phenotype is completely dominant and intermediate phenotypes occur, it is known as **incomplete dominance**. Flower colour in snapdragons is a trait that shows incomplete dominance. In heterozygous individuals, neither allele is completely expressed and the result is a blending effect of the two phenotypes. In snapdragon flowers, R_1 represents the red colour allele and R_2 represents the white colour allele. In this case, because neither is completely dominant, upper-case letters and subscripts are used to distinguish the alleles.

Crossing red-flowering snapdragons (R_1R_1 genotype) with white-flowering snapdragons (R_2R_2 genotype) will yield an F1 generation in which all individuals have the genotype R_1R_2 and have pink flowers.

Codominance

When both alleles are equally expressed, it is known as **codominance**. Human blood type is an example of autosomal codominant inheritance. In this case, depending on the allele inherited, the expression of the genotype differs. There are three alleles for blood type at the same locus, and individuals can have A, B, AB or O phenotypes. Those with the less common AB blood type are heterozygotes carrying one allele that produces an A antigen and one allele that produces a B antigen. Because both the A and B antigens are present on the surface of red blood cells, which can be detected using antibodies, neither phenotype is fully dominant. So A and B phenotypes are codominant, whereas the O phenotype is recessive.

ENVIRONMENTAL INFLUENCES ON PHENOTYPE

The examples discussed so far show that the phenotype is determined by the genotype but may also be affected by the environment. So if an individual with a given genotype develops in one environment, its phenotype may be different from that it would have developed in another environment. For example, silver banksia plants growing in inland areas have a greater average height than silver banksias growing in coastal areas due to environmental differences, such as soil type.

Phenylketonuria

The inherited disorder phenylketonuria (PKU) is a consequence of the build up of an amino acid called phenylalanine in the blood. This is toxic to developing neurons, leading to abnormal development of the nervous system and intellectual disability. PKU is caused by a mutation in the *PAH* gene, which codes for an enzyme that converts phenylalanine into another amino acid, tyrosine. If an individual inherits two copies of the mutant allele (that is, they are homozygous for the gene), they will develop PKU. Fortunately, development of the symptoms can be prevented by modifying the diet (environment) of babies that test positive for PKU shortly after birth. If homozygous individuals reduce their intake of dietary phenylalanine, particularly during childhood, they show normal brain development. Newborns in Victoria are routinely tested for genetic disorders such as PKU.

Fur colour in Himalayan rabbits

Coat colour of Himalayan rabbits provides an example of the effect of environmental temperature on phenotype. The Himalayan rabbit is homozygous for a mutant allele, c^h, that encodes a heat-sensitive enzyme called tyrosinase. Tyrosinase is produced but it is inactivated at normal body temperature, resulting in no melanin being produced and hence a white coat. At low temperatures, the tyrosinase enzyme is activated and results in the formation of melanin, causing black fur to form. When a small section of fur is shaved from a white region on the back, the fur grows back black if the animal is kept at low temperatures, but white if the animal is kept at high temperatures (Figure 6.3.9). Both rabbits bred at high temperatures and those bred at low temperatures have the same genotype, c^h c^h.

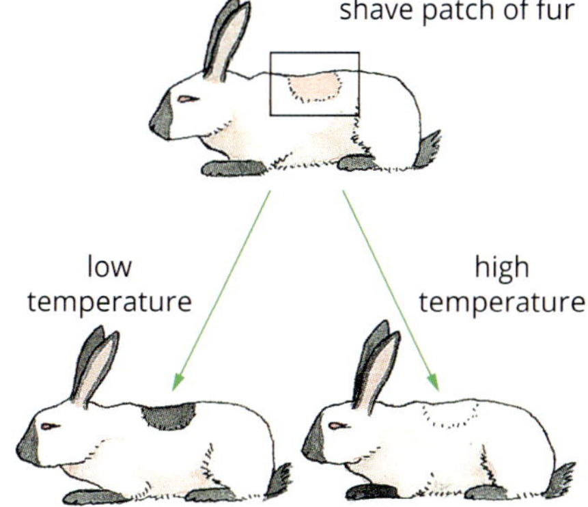

FIGURE 6.3.9 The relationship between temperature and fur colour in Himalayan rabbits. The rabbits with patches of different fur colour have the same genotype—c^h c^h.

Flower colour in hydrangeas

FIGURE 6.3.10 Flower colours of cuttings of the same hydrangea plant grown in an acidic soil (top) and an alkaline soil (bottom)

Hydrangeas are a commonly seen example of environmental effects on phenotype. If cuttings of a single hydrangea plant are grown in very acidic soil (pH 5.5 or less), the flowers produced are blue; if the cuttings are grown in weakly acidic or alkaline soil (pH 6.5 or more), the flowers are pink (Figure 6.3.10). The cuttings are of identical genotype, so it must be the environment (the pH of the soil) that affects the phenotype of the hydrangea.

This effect is caused by the relationship between soil pH, a pigment called anthocyanin, and the availability of aluminium in the soil for uptake by the plant. At a soil pH of 5.5 or less, aluminium is free to be taken into the plant. Anthocyanin is normally red, but at low pH it binds to aluminium in the plant to form a blue pigment called metalloanthocyanin, resulting in blue flowers. At a soil pH of 6 or more the aluminium binds to soil particles and is less available to the plants. This leaves most of the anthocyanin in the plant in its red form, resulting in pink flowers.

EPIGENETIC INFLUENCES ON PHENOTYPE

Phenotypes can sometimes be affected by the interaction of DNA with other molecules. For example, the queen and worker honeybees (*Apis mellifera*) are genetically identical, but their behaviour, physiology and appearance are different (Figure 6.3.11). The phenotype differences are due to the differences in the diet of the bees. Queen bees are fed royal jelly while worker bees are fed nectar.

Royal jelly contains ingredients that inhibit an enzyme which adds a methyl group ($-CH_3$) to cytosine bases in honeybee DNA, allowing for certain genes to be expressed (Figure 6.3.12). When scientists mimicked the effects of royal jelly on worker bees, worker bees exhibited characteristics of queen bees.

FIGURE 6.3.11 All the honeybees in a colony are genetically identical to each other. The queen bee (marked with blue paint on her head) looks different to the worker bees because of epigenetics.

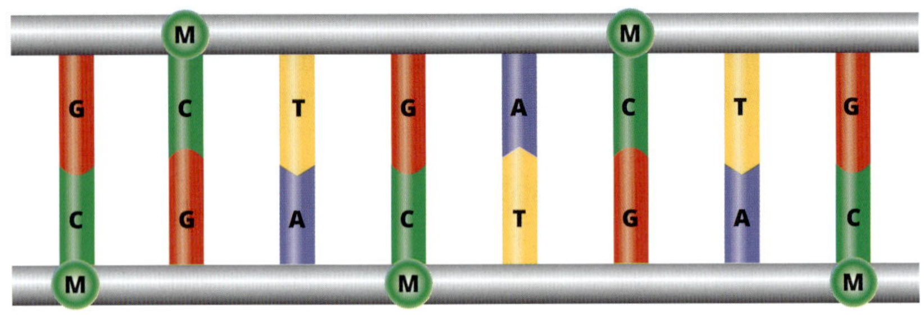

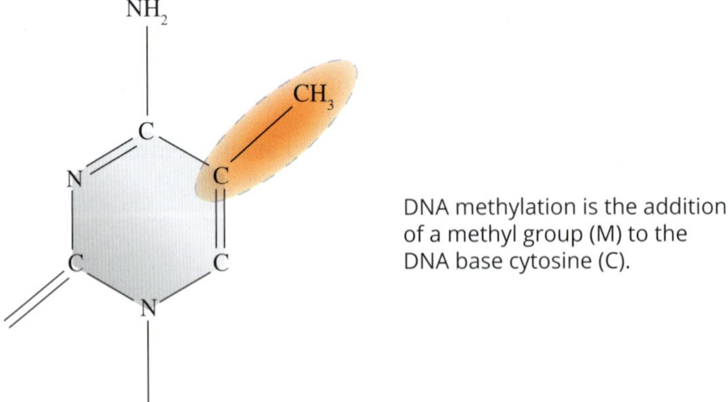

FIGURE 6.3.12 When a methyl group ($-CH_3$) is attached to cytosine bases, it prevents the expression of genes.

This effect of royal jelly in the honeybee is an example of epigenetics. **Epigenetics** is a term used to describe molecular events, such as adding methyl groups, that influence the expression of DNA without altering the DNA sequence. Such modifications are called 'epigenetic marks' and result in changes in gene expression and variations in phenotype. Other forms of epigenetic modification include addition of methyl or phosphate groups to histones, which affect how DNA is coiled and whether particular genes are expressed. Another example of epigenetics is X-inactivation.

X-inactivation occurs in females, who all have two copies of the X chromosome. It is a process in which one of the two copies of the X chromosome is inactivated. One of the X chromosomes must be inactivated to ensure that females do not end up with twice as many X chromosome gene products compared to males. The inactive X chromosome, Xi, is silenced by the addition of methyl groups to DNA and histones, resulting in Xi being coiled in such a way that it has an inactive structure called heterochromatin.

During the formation of gametes, epigenetic tags are usually erased during meiosis to ensure the growth of a healthy embryo. This process is known as **reprogramming** (Figure 6.3.13). Certain epigenetic changes can, however, be inherited. For example, feeding pregnant laboratory rats vinclozolin (a fungicide used on grape plants) results in lifelong epigenetic changes to the offspring. The male offspring have low sperm count and the sperm have abnormally high levels of methyl tags. The great-grandsons of the exposed male offspring also have low sperm count with abnormally high levels of methyl tags.

Researchers are amassing more and more evidence for the importance of epigenetic information in the regulation of gene expression. Epigenetic variations might help explain why one identical twin may acquire a genetically based disease, while the other twin does not.

A methyl group is a carbon with three hydrogen atoms bound to it. The attachment of a methyl group to DNA, referred to as DNA methylation, can affect gene expression; the gene might not be 'turned on' to produce the protein, or it might be silenced.

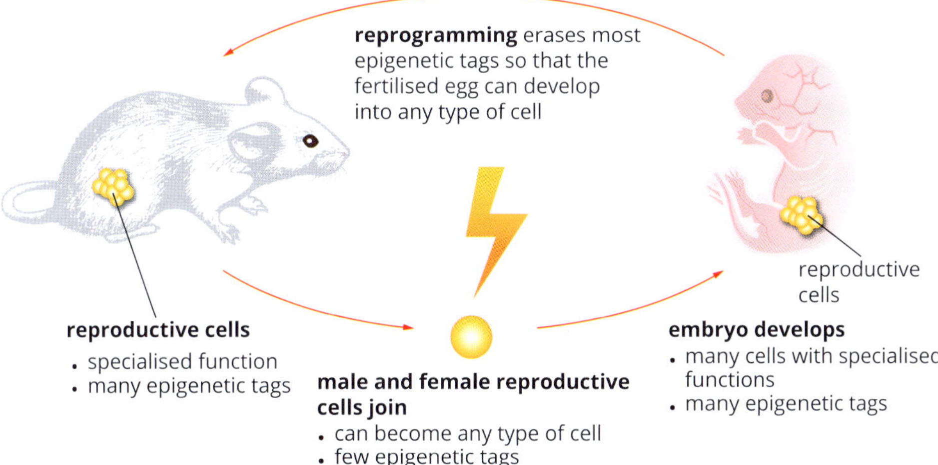

FIGURE 6.3.13 Reprogramming removes the epigenetic tags of the early embryo so that it can form every type of cell in the body.

6.3 Review

OA ✓✓

SUMMARY

- The genotype is the combination of alleles at a particular locus or loci.
- An organism that has two copies of the same allele is homozygous for that allele.
- An organism that carries two different alleles is heterozygous.
- Phenotype is an observable characteristic or trait that results from the genotype under the influence of the environment.
- Dominance and recessiveness are properties of alleles and are expressed as dominant or recessive phenotypes.
- A phenotype can be dominant, incompletely dominant, codominant or recessive depending on its appearance in the heterozygote.
  - A dominant phenotype is one that is visible in the heterozygote and homozygote.
  - When neither phenotype is completely dominant and intermediate phenotypes occur, it is known as incomplete dominance.
  - When both alleles are equally expressed, it is known as codominance.
  - A recessive phenotype is only observed in the homozygote.
- An italic upper-case letter is used to signify the allele for a dominant phenotype.
- An italic lower-case letter is used to signify the allele for a recessive phenotype.
- Phenotype is influenced by:
  - genotype
  - interaction between genotype and the environment
  - interaction between DNA and other molecules (epigenetic factors).
- Epigenetics refers to molecular events that affect the expression of genes without altering the DNA sequence. These events usually involve switching genes on or off.

KEY QUESTIONS

Knowledge and understanding

1 Explain the difference between the genotype and phenotype of an individual.

2 What factors contribute to an individual's phenotype? Give an example.

3 Use an example to explain how two organisms can have the same phenotype but different genotypes.

4 Use an example to distinguish between dominant and recessive phenotypes.

5 Explain the role of codominance in human blood types.

6 Describe an example where an organism's phenotype can be affected by the environment in which it is raised.

7 Mutations are changes in a DNA sequence, and can result in new alleles for a gene. Outline the difference between epigenetic events affecting phenotypes and mutations affecting phenotypes.

Analysis

8 Cystic fibrosis (CF) is a genetic disorder that affects the cells that produce mucus, sweat and digestive juices. The gene responsible for causing CF is located on chromosome 7. More than 1500 CF mutations have been identified. For an individual to express CF, they must inherit two mutated CF alleles. The table below shows the frequency of the four most common CF mutations in people living with CF in Australia in 2014.

Frequency of *CFTR* mutations in people living with CF in Australia, 2014

CFTR mutation	Number of alleles
F508del	4321
G551D	238
R117H	122
G542X	91
other	1298

a Construct a graph from the data in the table.

b CF is a recessive genetic condition. Do affected individuals have a homozygous or a heterozygous genotype?

Chapter review

KEY TERMS

adenine (A)
allele
autosome
base
centromere
chiasma (pl. chiasmata)
chromatid
chromosome
cleavage furrow
codominance
complementary base pair
complete dominance
crossing over
cytosine (C)
deoxyribonucleic acid (DNA)
diploid
dominance
dominant phenotype
double helix
epigenetics
fertilisation
gamete
gene
gene expression
genetic diversity
genome
genotype
germ cell
guanine (G)
haploid
hemizygote (adj. hemizygous)
hereditary
heterogametic
heterozygote (adj. heterozygous)
homogametic
homologous chromosomes
homozygote (adj. homozygous)
incomplete dominance
independent assortment
karyotype
locus (pl. loci)
meiosis
nucleotide
phenotype
ploidy
protein
purine
pyrimidine
recessive
recessive phenotype
recombinant chromosome
reduction division
reprogramming
separation division
sex chromosome
sister chromatids
somatic cell
synapsis
thymine (T)
trait
trisomy

REVIEW QUESTIONS

Knowledge and understanding

1 State the number of autosomes and homologous chromosomes in the following types of human cells.

a female somatic cell

b male somatic cell

2 Select the correct statement about chromosomes in mammalian gametes.

A They are all identical to those in the parent cell.

B They are different to those in the parent cell but only because of mutation.

C They are all identical to those in the parent cell because crossing over and recombination between homologues does not create new combinations of alleles.

D They are different to those in the parent cell partly because of the effects of independent assortment.

3 Why is meiosis a necessary process in living organisms?

A It happens in the reproductive organs.

B It replicates the parents' DNA to produce genetically identical offspring.

C It produces new cells to replace dead or dying cells.

D It enables each parent to contribute genetic information to the offspring.

4 Consider a cell with a diploid number ($2n$) of 8.

a Draw an illustration of a chromosome during the stage of prophase I. Label the sister chromatids and centromere.

b State how many chromosomes and chromatids are present in a cell during the following stages.

i prophase I **ii** prophase II

c Calculate how many possible chromosome combinations may result due to independent assortment.

5 Arrange the following stages of meiosis in the correct order, from first to last.

metaphase II

telophase II

prophase I

anaphase I

metaphase I

anaphase II

6 The diploid number of chromosomes in a horse is 64. State the number of chromosomes in a horse's:

a fertilised ovum

b sperm cell

c somatic cell

d cell during telophase I

e cell during prophase I

f cell during telophase II.

7 Describe karyotyping and one application of its use.

8 The three parts of a nucleotide are:

A sugar, phosphate, base

B phosphate, base, protein

C thymine, sugar, protein

D protein, sugar, phosphate

9 Identify the correct statement regarding the structure of DNA.

A Deoxyribose is a six-carbon sugar.

B The base C always pairs with the base G.

C The building blocks are called nucleosomes.

D The DNA molecule is a single-stranded helix.

10 Sketch a labelled diagram to symbolise a:

a nucleotide with adenine as its base

b strand of DNA, showing the base pairing.

11 Each DNA molecule contains genes and regions called spacer DNA. Distinguish between genes and spacer DNA.

12 What is the difference between the alleles of a gene?

A their locus on the chromosome

B their amino acid sequence

C the type of sugar on the nucleotides

D the sequence of bases

13 Recall the name given to an individual that has two different alleles for the one gene.

14 *P* and *p* are alleles of a particular autosomal gene. Identify the possible combinations of alleles and name them appropriately. List the possible genotypes of the dominant and recessive phenotypes.

Application and analysis

15 In mice, coat colour is controlled by a single gene. Black coat colour is dominant to white coat colour.

a Assign allele symbols for the gene responsible.

b How many genotypes are possible with respect to these alleles? State the genotypes and phenotypes.

16 The diagram below shows a pair of chromosomes during meiosis to form a human sperm. The position of the alleles of some of the genes is shown.

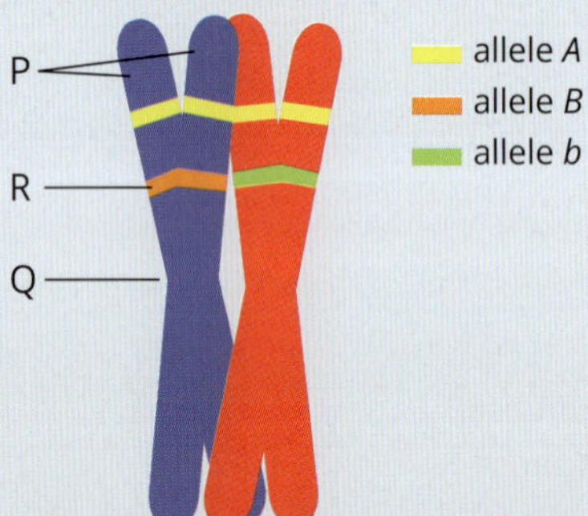

a Identify the chromosome structures labelled P, Q and R.

b Suggest, with reasons, whether the chromosomes are:

i sex chromosomes or autosomes

ii homologous or non-homologous

iii homozygous, hemizygous or heterozygous with respect to the *B* gene locus.

17 The figure below shows six chromosomes belonging to three homologous pairs. Conclude, with justification, which chromosomes are homologous pairs.

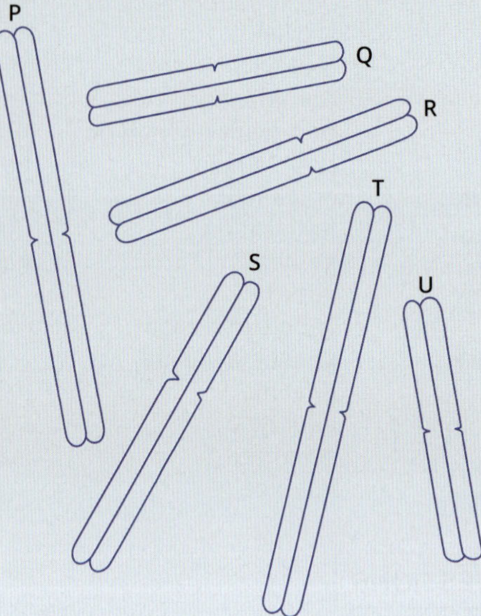

18 a Does the cell division diagram below represent mitosis or meiosis? Justify your answer.

b Write a short description for each of the stages 1, 2 and 3 shown on the diagram.

c Based on your answer to part **a**, which important process is not highlighted on this diagram?

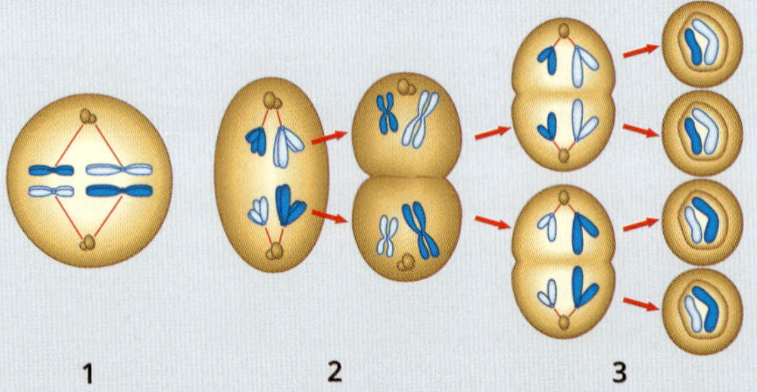

19 A student argued that all chromosomes consist of two chromatids. Assess the student's understanding.

20 A student drew the following diagram to show her understanding of meiosis. Critique the diagram.

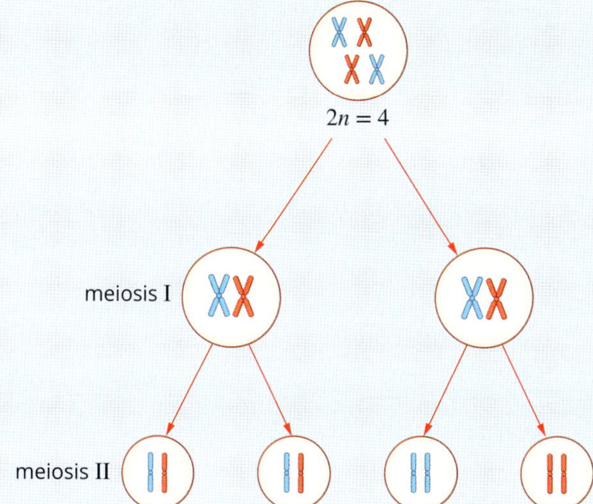

21 The figure below shows a human karyotype.

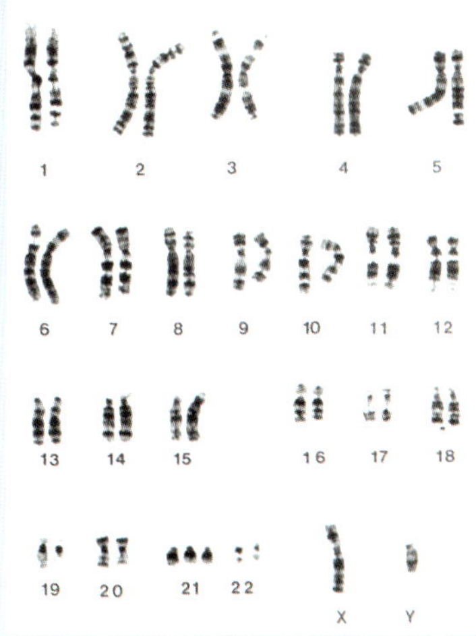

What conclusions can be reached from the karyotype?

22 Consider the following karyotypes. In each case identify the sex of the individuals and the conditions that they have.

a

1 2 3 4 5
6 7 8 9 10 11 12
13 14 15 16 17 18
19 20 21 22 X Y

b

1 2 3 4 5
6 7 8 9 10 11 12
13 14 15 16 17 18
19 20 21 22 X Y

c

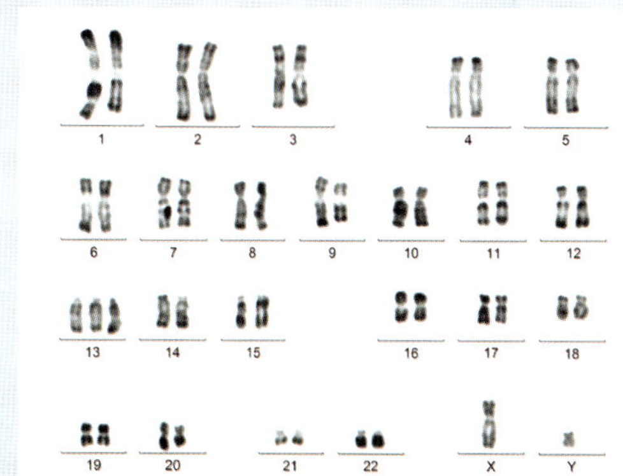

23 The diagram illustrates a proposed model to help students understand the concept of genes and alleles as part of chromosome structure.

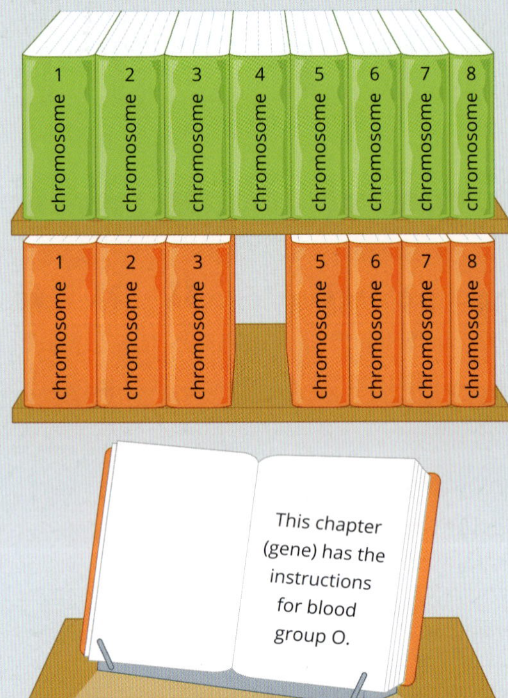

a Interpret how this model depicts the concept.

b Evaluate if such a model helps your own understanding of this area of biology.

24 The table below shows the DNA base composition of thymine (T) and cytosine (C) for a number of species.

Species	Thymine (%)	Cytosine
human	30	20
wheatgerm	27	23
E. coli	24	26
sea urchin	32	17

a Draw a graph plotting the percentage of cytosine against the percentage of thymine using the data from the table above.

b Identify the relationship that exists between the percentage of thymine and the percentage of cytosine.

c Explain the relationship that you have observed from the data.

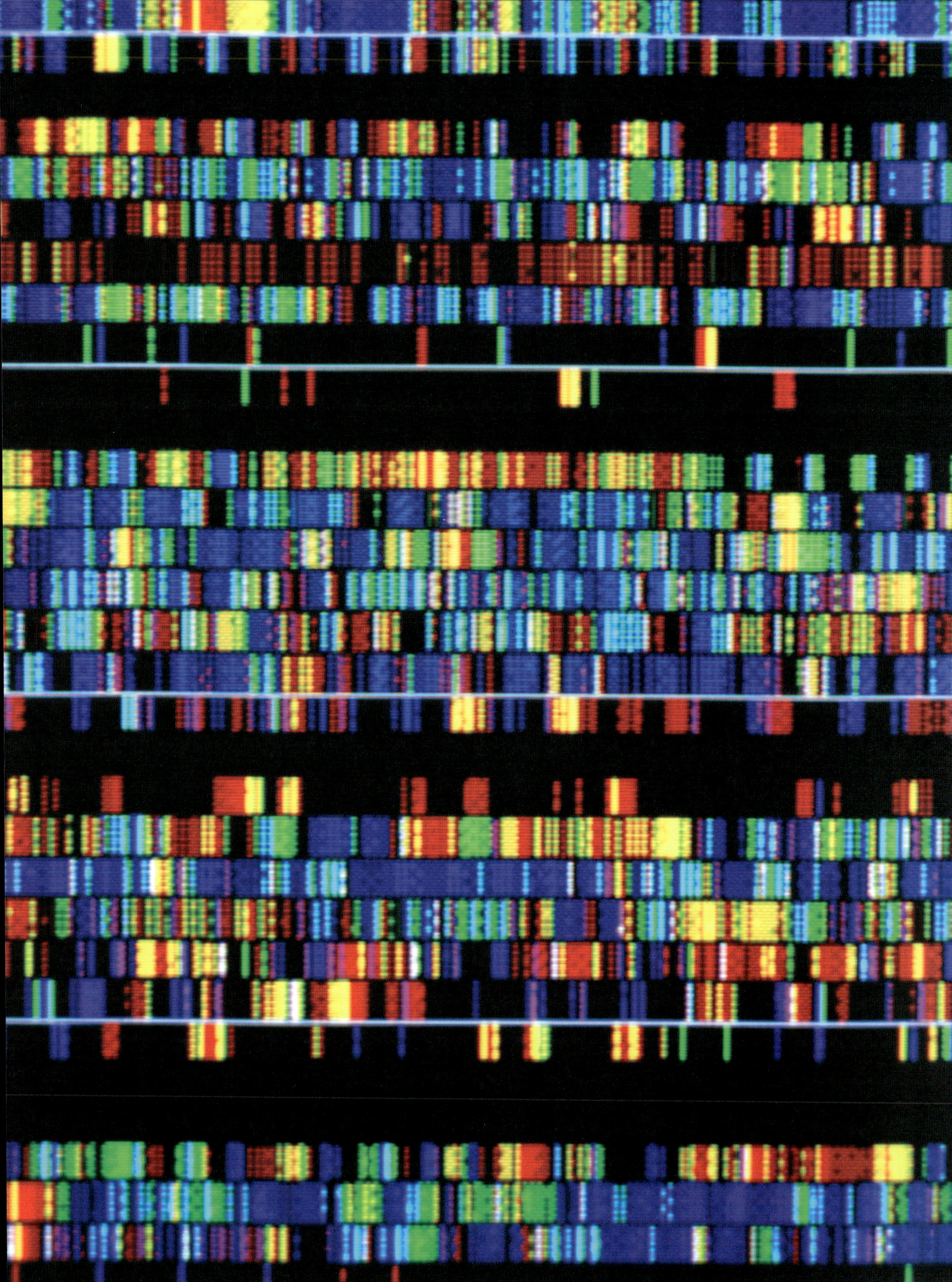

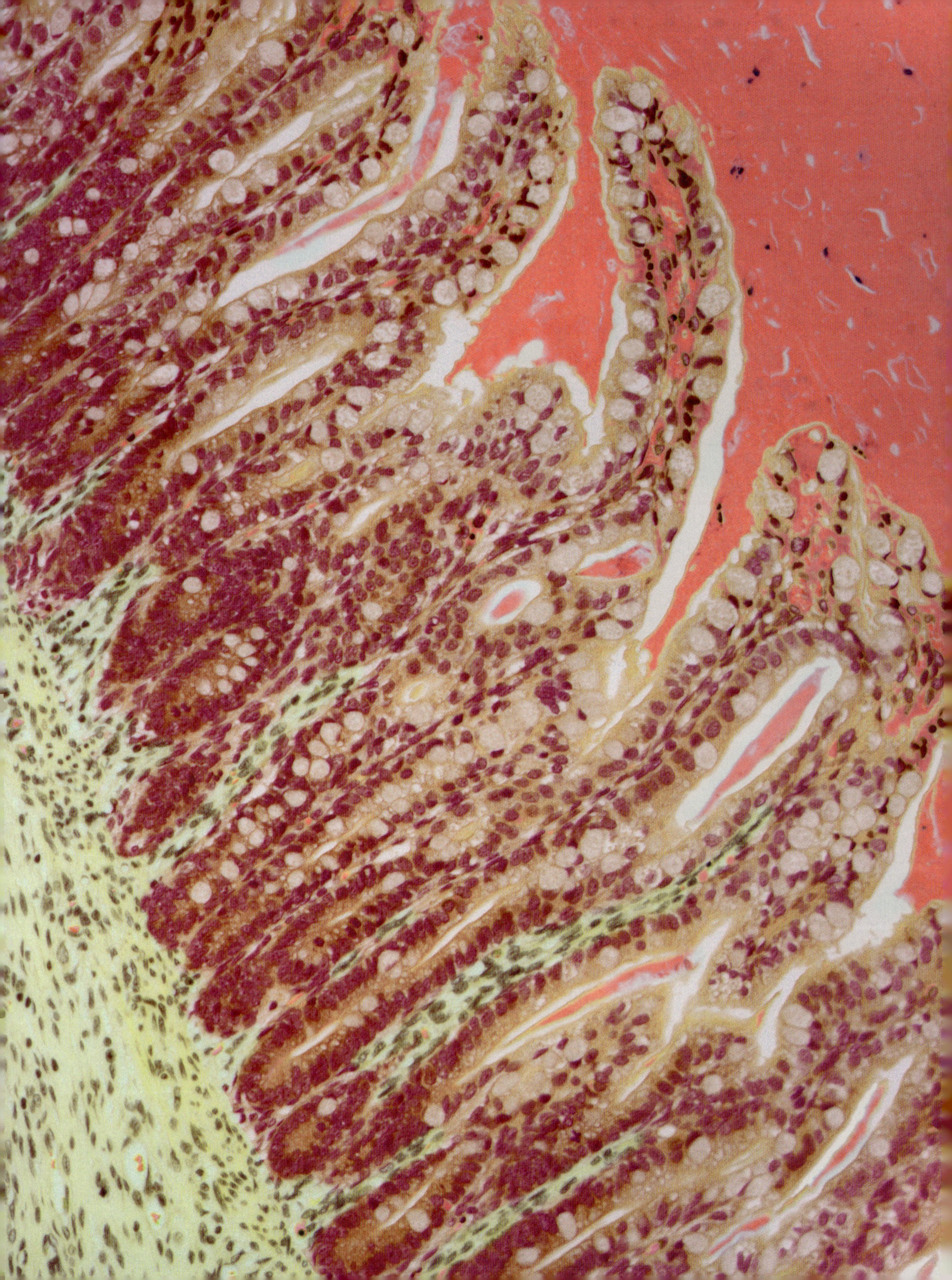

CHAPTER

07 Patterns of inheritance

In this chapter you will learn about the transmission of biological information from generation to generation. You will use chromosome theory and terminology from genetics to explain the inheritance of characteristics, analyse patterns of inheritance, learn about the differences between linked and non-linked dihybrid crosses, interpret pedigree charts and predict the outcomes of genetic crosses.

Key knowledge

- pedigree charts and patterns of inheritance, including autosomal and sex-linked inheritance **7.2**
- predicted genetic outcomes for a monohybrid cross and a monohybrid test cross **7.1**
- predicted genetic outcomes for two genes that are either linked or assort independently. **7.3**

7.1 Monohybrid crosses

Much of what is now understood about natural variation and patterns of inheritance in sexually reproducing organisms was originally gained through the work of Gregor Mendel in the 1860s. Mendel accurately deduced the basic principles of inheritance by studying several heritable traits in pea plants (*Pisum sativum*; Figure 7.1.1), using precise experimentation and careful observations over many years.

Mendel used monohybrid crosses to investigate the inheritance of traits in pea plants. In this section, you will learn about how the outcomes of monohybrid crosses can be used to determine different types of inheritance, including autosomal and sex-linked inheritance.

FIGURE 7.1.1 After carefully studying the results of crossing different pea plants (*Pisum sativum*) in his garden over many years, Gregor Mendel deduced the basic principles of inheritance.

> A genetic cross is the intentional breeding of two genetically different organisms to determine the inheritance pattern of particular traits.

> Phenotype refers to the observable characteristics of an organism. The phenotypic features of an organism are an expression of their genes (genotype) and their interaction with the environment.

MENDEL'S STUDY OF PATTERNS OF INHERITANCE

Mendel crossed different varieties of pea plants and cultivated their seeds to determine their traits and understand patterns of inheritance. He performed these crosses over many generations, recording data for over 20 000 pea plants. Mendel made several important observations from his pea plant experiments:

- Traits are passed from parents to offspring and these traits form specific patterns over generations of crossbreeding.
- Organisms have heredity units—Mendel called these units 'factors', and they are now known as **genes**.
- The heredity units (genes) occur in pairs, now known as **alleles**.
- The alleles separate during the formation of the gametes (sperm and eggs) and each parent passes one allele for each gene to their offspring.
- The offspring of two **true-breeding** (homozygous) parents are not a mix of their parents' traits but express only one trait. The expressed trait is known as the dominant trait or phenotype (e.g. purple flowers are dominant over white flowers).
- **Dominant** phenotypes are expressed if the individual carries at least one allele for the dominant phenotype. **Recessive** phenotypes are expressed only if the individual carries two alleles for the recessive phenotype.
- The first generation of offspring from a true-breeding cross produce offspring with two different traits in the ratio of 3 : 1 (e.g. three plants with purple flowers : one plant with white flowers).

From these observations, Mendel's laws of inheritance were developed:

- **law of segregation** (the first law)—alleles are separated (i.e. segregated) when gametes form so each gamete carries one allele for each gene.
- **law of independent assortment** (the second law)—the segregation of alleles for one gene is independent of the segregation of alleles for any other gene.
- **law of dominance** (the third law)—alleles for recessive traits are masked by alleles for dominant traits.

GENETIC CROSSES

A **genetic cross** is the deliberate mating (or crossing) of two organisms to determine the inheritance pattern of particular traits. There are different types of genetic crosses, including monohybrid crosses, dihybrid crosses, test crosses and backcrosses. You will learn about dihybrid crosses in Section 7.3.

A **monohybrid cross** is a cross between two individuals with different alleles at a single **locus** (position on a chromosome). In the standard monohybrid cross, **homozygous** parents (P) with different phenotypes of the same trait (e.g. purple flowers and white flowers) are crossed to produce heterozygous offspring (the first filial generation, or F1). These **heterozygous** offspring are then crossed with each other to produce the second filial generation (F2). The phenotypic ratios in the offspring of F1 and F2 generations indicate which phenotypes are dominant or recessive.

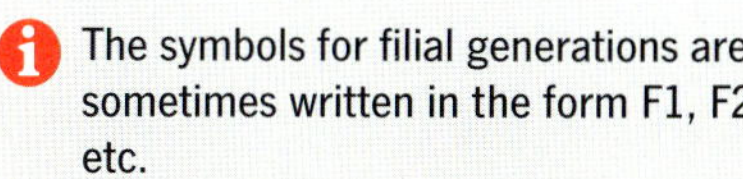

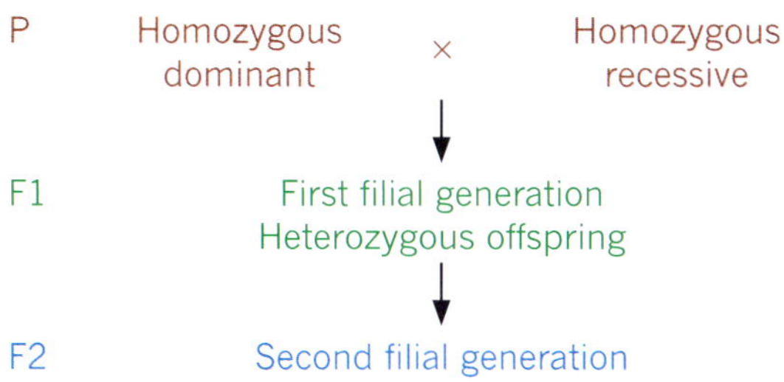

PUNNETT SQUARES

In 1905, geneticist Reginald Punnett devised a simple method for showing the random combination of gametes and the genotypes of the resulting offspring. In a **Punnett square** the alleles of each parent are first written in the top and side cells. Then by going down each column and across each row, the alleles are combined and written into the remaining cells (Figure 7.1.2).

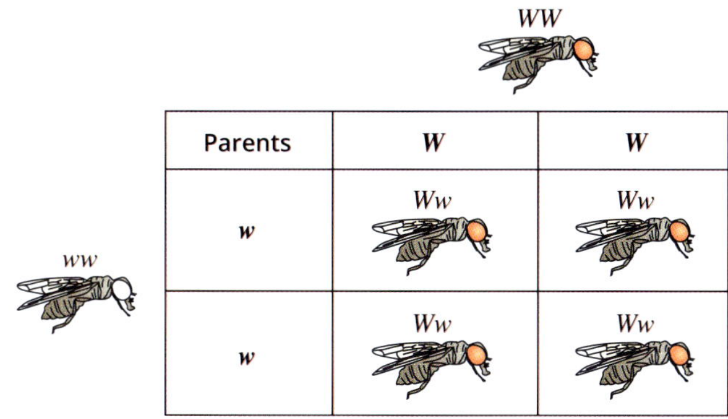

FIGURE 7.1.2 A Punnett square for a cross between two homozygous parents to produce an F1 generation. All F1 individuals are heterozygous.

Punnett squares make it easy to establish all of the possible combinations of alleles carried by the parents' gametes and are used to determine the expected outcomes of a genetic cross (i.e. the possible genotypes and phenotypes of the offspring). This is useful in fields such as animal husbandry and horticulture because it allows breeders to select individuals to cross according to the desired traits of the offspring.

Genotypic and phenotypic ratios

Genotypic and phenotypic ratios are used to express the expected frequency of genotypes and phenotypes in the offspring from a genetic cross.

The **genotypic ratio** in the offspring is written in the following order: homozygous dominant : heterozygous : homozygous recessive

The **phenotypic ratio** observed in the offspring is written as: dominant phenotype : recessive phenotype.

BIOFILE

Punnett squares vs experimental data

Punnett squares provide only the theoretical results of a cross; the actual results from an experiment may be different. Fertilisation can be compared to tossing a coin—for most genes there are two possible outcomes. If a coin is tossed, there is a 50% chance of getting heads and a 50% chance of getting tails. If the coin is tossed 10 times, you might not get 5 heads and 5 tails, but if it is tossed 1000 times, a heads : tails ratio very close to 1 : 1 would be observed.

Similarly, the more fertilisation events (data) there are in a breeding experiment, the closer the results will be to the theoretical ratio.

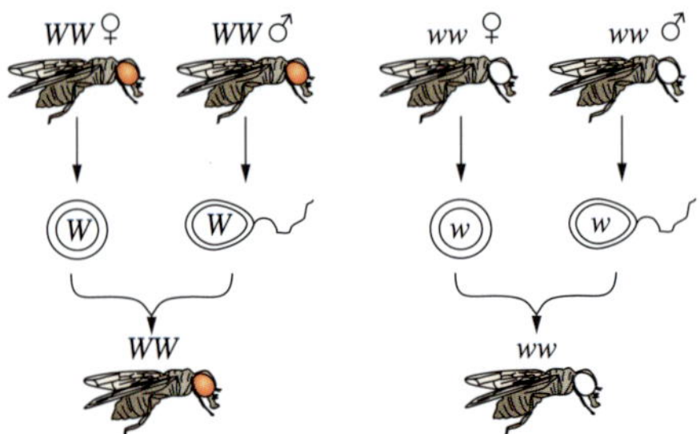

FIGURE 7.1.3 Homozygous genotypes produce only one type of gamete. By crossing homozygotes of the same genotype together, a true-breeding strain can be established.

AUTOSOMAL INHERITANCE

Autosomal inheritance refers to the inheritance of alleles carried on autosomes. **Autosomes** are chromosomes that are not involved in sex determination. In humans, all chromosomes are autosomes except the X and Y chromosomes.

Autosomal dominant inheritance

Autosomal dominant inheritance (complete dominance) refers to a dominant trait that is passed on to offspring via an autosomal gene. Only one copy of the allele from one parent is needed to express a dominant phenotype.

Parent (P) generation

The inheritance of eye colour in the Australian sheep blowfly is an example of a single gene with two alleles (found on an autosomal chromosome) coding for the trait. In the blowfly, red eye colour is dominant over white eye colour. The homozygous genotypes are *WW* and *ww*. Homozygous genotypes produce only one type of gamete. *WW* individuals produce only *W* gametes and *ww* individuals produce only *w* gametes.

In a cross between two red-eye homozygous (*WW*) individuals, all the offspring would be homozygous *WW* (red eye). As long as *WW* individuals were crossed together, it would be a true-breeding strain. Similarly, as you can see on the right side of Figure 7.1.3, crosses between two homozygous white-eye (*ww*) individuals would yield a true-breeding white eye (*ww*) strain.

The F1 generation

To test the principle of dominance, two true-breeding parents with two different traits can be crossed. This type of cross is known as hybridisation, and the offspring are known as **hybrids**.

In the blowfly example, two true-breeding strains (one with red eyes, *WW*, and one with white eyes, *ww*) can be crossed to produce an **F1 generation**. The results of the cross can be shown in a Punnett square, as shown in Figure 7.1.2 on page 283.

Each of the offspring in the F1 generation has the heterozygous genotype *Ww*. The phenotype resulting from this genotype is red eyed. From this, it can be deduced that the red-eye phenotype is dominant over the white-eye phenotype.

The term **wild type** is used to describe the typical or standard form of an organism, gene or characteristic that is the most common in natural or normal populations. It is distinguished from forms that may result from selective breeding. The wild type is not necessarily dominant to other forms, but it is the most prevalent within the population.

The F2 generation

The **F2 generation** is the result of crossing the individuals from the F1 generation. In this example, half of the gametes produced by an F1 individual (*Ww*) will be *W* and half will be *w*. So three different combinations of alleles are possible in the F2 generation: *WW*, *Ww* and *ww*, as shown by the Punnett square in Figure 7.1.4.

In the F2 generation the dominant phenotype is likely to occur in three out of four crossings, and the recessive phenotype only once.

Ww

Ww

Parents	W	w
W	WW	Ww
w	Ww	ww

FIGURE 7.1.4 Punnett square of a cross between two F1 individuals to produce the F2 generation

> **BIOFILE**
>
> **Choosing symbols for alleles**
>
> When choosing symbols for alleles, it is common practice to select one that relates to the dominant phenotype. For example, if the dominant phenotype is grey fur, the dominant allele would be given the symbol G and the recessive phenotype would be g.
>
> However, the symbols *W* and *w* are traditionally used for eye colour alleles in flies, even though red eye colour is dominant. This is because other genes are involved in eye colour in flies, and the discoverers of this gene named it 'white eye gene'.

The Punnett square shows that the F2 genotypic pattern is:

$WW : Ww : Ww : ww$ or $1WW : 2Ww : 1ww$

Because red eye colour is dominant over white eye colour, the F2 phenotypic pattern is:

3 red-eyed flies ($WW : Wr : Wr$) : 1 white-eyed fly (ww)

From this information it can be determined that the wild type is red-eyed.

The genotypic ratio 1 : 2 : 1 of the F2 generation resulting from a monohybrid cross occurs because of the following reasons.

In meiosis, heterozygous (Ww) individuals (both male and female) produce two gametes (a W gamete and a w gamete). This is because of the separation of pairs of alleles during the formation of reproductive cells.

Fertilisation occurs at random. For example, a W sperm has equal chance of fertilising a W egg or a w egg, because these eggs are produced in equal frequency. A w sperm also has an equal chance of fertilising a W egg or a w egg. So the four equally possible genotypic outcomes are WW, Ww, wW, ww.

The Punnett square accounts for both of these factors in demonstrating the possible outcomes of the cross.

CASE STUDY

The law of segregation

In the 1860s Gregor Mendel conducted breeding experiments on 34 different varieties of pea plants. During this time he carefully collected data and made many observations that would later lay the foundations for modern genetics and the study of inheritance.

One of his most significant observations was that the offspring of the pea plants did not always have the same phenotype as the parents, and that offspring from the same parents were often different from one another. Mendel hypothesised that hereditary units or 'factors' (now called genes) must have different forms (now called alleles) that separate randomly during the production of gametes. These forms would then unite after fertilisation, with each parent contributing one allele to the offspring. Mendel's hypothesis became known as the law of segregation or Mendel's first law.

With the advancement of cell biology, we now have a better understanding of the process of the law of segregation. During meiosis, each daughter cell (or gamete) receives one chromosome from each homologous pair. The alleles for each trait are separated into different gametes. Because gametes are haploid, they carry only one of the two alleles of a genotype. Offspring then receive one allele from each parent at fertilisation (Figure 7.1.5).

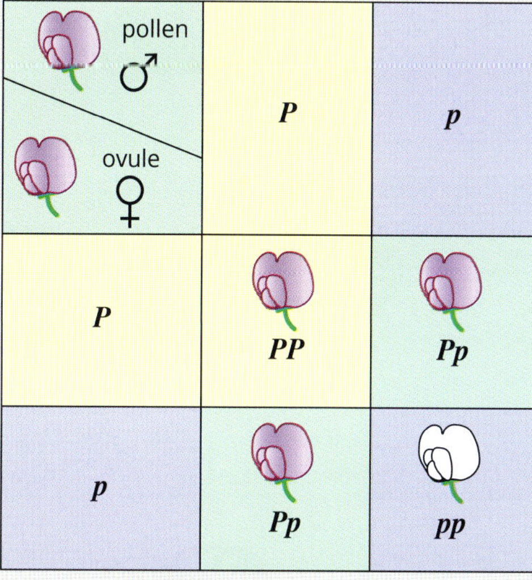

FIGURE 7.1.5 Punnett square for flower colour resulting from a cross between two heterozygous pea plants (P = purple flowers, p = white flowers). The resulting offspring have a 3 : 1 phenotypic ratio.

Test crosses

It is not immediately obvious whether an individual with a dominant phenotype is homozygous; it might be either *AA* or *Aa*. Apart from sequencing the gene involved (which is expensive and time-consuming), the only way to determine this is to do a **test cross**. This involves crossing the individual with another that has the recessive trait and is therefore homozygous. Homozygous individuals produce gametes with one type of allele, whereas heterozygous individuals can produce gametes with two types of alleles.

If the offspring from the test cross all have the dominant phenotype, then both the parents are likely to be homozygous. (It is not possible to be certain because of the random nature of fertilisation.) If the offspring have both dominant and recessive phenotypes, then the parent with the dominant phenotype must also carry a recessive allele and is therefore heterozygous.

Example: coat colour in guinea pigs

The coat colour of guinea pigs is determined by the alleles of one gene, and black fur is the dominant phenotype. If a true-breeding white guinea pig (*bb*) is crossed with a true-breeding black guinea pig (*BB*), the resulting F1 has black fur (*Bb*). But if the genetic history of a black guinea pig is unknown, its genotype can be determined by crossing the F1 black guinea pig with a white-coated guinea pig, which must be *bb* (homozygous recessive).

Figure 7.1.6 illustrates the test cross that would be carried out. Of the resulting offspring in this example, half are white and half are black. This ratio of about 1 : 1 is consistent with the results of a heterozygote crossed with a homozygote if the trait is determined by the alleles of one gene and one trait is dominant. So the black guinea pig is likely to be heterozygous. If all the offspring of the test cross were black-coated (all *Bb*), the F1 guinea pig in question would have been shown to be homozygous.

The predicted outcome for a cross between heterozygote black guinea pigs is 1 black (*Bb*) : 1 white (*bb*). However, as Figure 7.1.6 shows, the resulting ratio of the test cross was 27 black (*Bb*) : 23 white (*bb*) rather than, for example, 23 : 23 (which is equal to a 1 : 1 ratio). The difference between predicted and observed ratios is due to chance.

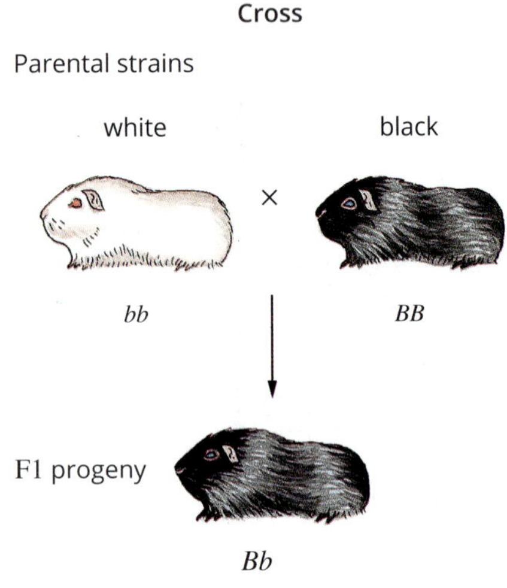

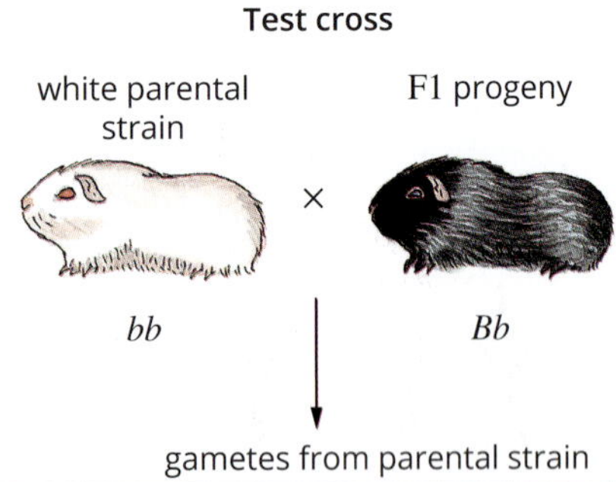

		b	*b*	
gametes from F1	*B*	*Bb*	*Bb*	27 black
	b	*bb*	*bb*	23 white

FIGURE 7.1.6 A cross and test cross between true-breeding strains of guinea pigs and their progeny (F1)

Autosomal incomplete dominance and codominance

Some traits do not show simple dominance or recessiveness. These are instances in which both alleles are expressed to varying degrees in the phenotype of heterozygous individuals. When neither phenotype is completely dominant and intermediate phenotypes occur, it is known as **incomplete dominance**. When both alleles are equally expressed in an individual, it is known as **codominance**. You learnt about these traits in Chapter 6. In this section, you will learn about how these traits are inherited and how this inheritance is represented in Punnett squares.

Autosomal incomplete dominance

Flower colour in snapdragons is an example of incomplete dominance. If red-flowering snap dragons (R_1R_1) are crossed with white-flowering snapdragons (R_2R_2), the F1 generation will all have the intermediate phenotype of pink flowers (R_1R_2).

Parents		Red flowers	
		R_1	R_1
White flowers	R_2	R_1R_2	R_1R_2
	R_2	R_1R_2	R_1R_2

If the F1 plants ($R_1R_2 \times R_1R_2$) are crossed, an F2 generation with a 1 : 2 : 1 genotypic ratio (1 R_1R_1 : 2 R_1R_2 : 1 R_2R_2) would be expected:

Parents		Pink flowers	
		R_1	R_2
Pink flowers	R_1	$R_1 R_1$	$R_1 R_2$
	R_2	$R_1 R_2$	$R_2 R_2$

The heterozygote pink-flowering snapdragon (R_1R_2) can be distinguished from the two homozygotes, red R_1R_1 and white R_2R_2, because of the incomplete dominance of both the red and white alleles resulting in a pink-flowering phenotype.

Genotypic ratio: 1 R_1R_1 : 2 R_1R_2 : 1 R_2R_2

Phenotypic ratio: 1 white : 2 pink : 1 red

This phenotypic ratio of 1 : 2 : 1 (Figure 7.1.7) is different to the 3 : 1 ratio of two phenotypes observed in complete dominance (Figure 7.1.5 on page 285).

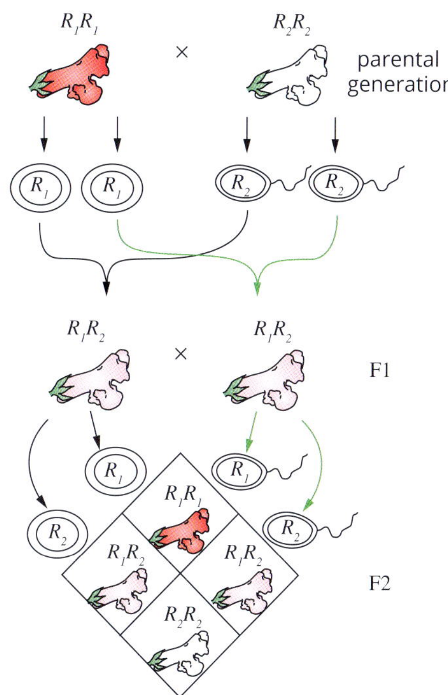

FIGURE 7.1.7 A cross between homozygous red and white snapdragons produces pink-flowering progeny in the F1 generation. If plants from the F1 are then crossed, a phenotypic ratio of 1 red : 2 pink : 1 white would be expected in the F2 generation.

Autosomal codominance

Human blood type is an example of codominant inheritance of multiple alleles at the same locus. In this case, the three alleles are represented as I^A, I^B and i. I^A codes for the A antigen, I^B codes for the B antigen while i does not produce either antigen. Therefore the effects of I^A and I^B dominate over i. Each person carries copies of one or two of these three possible alleles. Table 7.1.1 shows the possible genotypes and phenotypes for the ABO blood group system.

> Blood group antigens are molecules on the surface of red blood cells. There are two antigens in the ABO blood group system—A antigen and B antigen.

TABLE 7.1.1 Possible genotypes and phenotypes in the ABO blood group system

Genotype	Phenotype (blood group)	Diagram of red blood cell
I^AI^A I^Ai	A	
I^BI^B I^Bi	B	
I^AI^B	AB	
ii	O	

From this table it can be seen that there are six possible genotypes and four phenotypes, with the A and B blood groups each having two possible genotypes.

The possible genotypes and phenotypes of the offspring of a parent with blood type O and a parent with blood type AB can be determined using a Punnett square, as shown below. The F1 generation in this example would be either blood type A or B, but all would be heterozygous.

Parents		Blood type AB	
		I^A	I^B
Blood type O	i	I^Ai	I^Bi
	i	I^Ai	I^Bi

If a heterozygous individual for blood type A and a heterozygous individual for blood type B were to have children, four possible combinations of blood type are possible, as shown in the following Punnett square.

Parents		Blood type B	
		I^B	i
Blood type A	I^A	I^AI^B	I^Ai
	i	I^Bi	ii

The F1 generation would show all of the phenotypes possible: AB, A, B and O. The important principle illustrated by this example is that phenotypes are not always dominant or recessive. The dominance of a phenotype is always in relation to another phenotype. Thus, phenotype A is codominant with B, but dominant to O.

SEX-LINKED INHERITANCE

So far you have examined the inheritance of genes located on autosomes. However, the patterns of inheritance are not the same for genes located on either of the two sex chromosomes. Phenotypes inherited through genes on sex chromosomes are said to be 'sex-linked' and they show **sex-linked inheritance**. It is important to remember that sex chromosomes also carry other genes which are not related to sex determination.

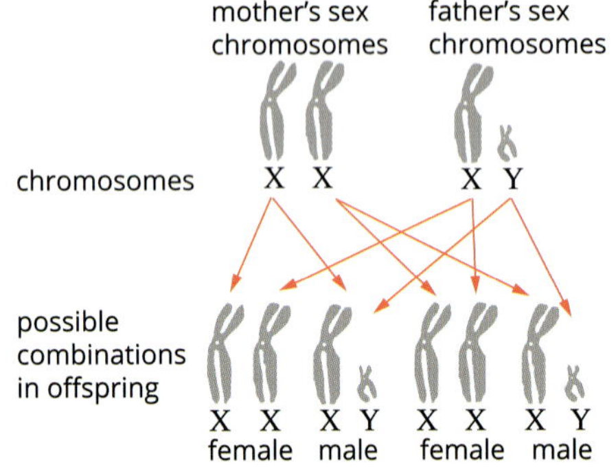

FIGURE 7.1.8 Inheritance of the sex chromosomes in the XY sex-determination system. The outcome of this inheritance is two possible arrangements—XX or XY, with half the offspring being female and half being male.

Figure 7.1.8 shows how sex chromosomes are transferred to the offspring, with an equal probability of the offspring being female or male. The XY system determines sex in humans, most other mammals, some insects and some plants. In this system, females are **homogametic** (XX) and males are **heterogametic** (XY). The female passes one X chromosome on to her offspring, while the male can pass on either an X or Y chromosome; an X chromosome produces female offspring and a Y chromosome produces male offspring. It is therefore the father's genetic contribution that determines the sex of the offspring.

Males are **hemizygous** for all sex-linked genes—that is, they only carry one copy of each sex-linked gene because they only have one X chromosome and one Y chromosome.

X-linked recessive inheritance

In humans, **X-linked** recessive traits are predominantly expressed in males, because males carry only one X chromosome. Females carrying an X-linked recessive allele might not express the trait, or show only mild expression. This is because the second X chromosome that females carry could mask the recessive trait. The probability in humans of a female carrying two X-linked recessive alleles is very low.

> Males only carry one X chromosome and therefore always express the phenotype of the alleles on the X chromosome. The second X chromosome in females can mask the recessive phenotype.

The pattern of sex-linked inheritance is evident when a **reciprocal cross** is performed. This is an experiment to investigate the role of parental sex on the inheritance of genotypes. Two crosses are performed: one crossing a male with the trait of interest with a female not expressing the trait (usually homozygous wild type), and another crossing a female with the trait of interest (homozygous) with a male that does not express the trait (usually wild type). If the trait is sex-linked (carried on the X chromosome), the phenotypic ratios of the male and female offspring will be different.

Paralysis in Drosophila

The temperature-sensitive paralytic gene, named after the **mutant** phenotype, is on the X chromosome of the fruit fly (*Drosophila melanogaster*). A mutant phenotype arises from a genetic mutation that causes phenotypic change from the normal wild type phenotype. Individuals with the mutant allele are paralysed when incubated to a temperature of 29°C, whereas wild type flies show normal behaviour at this temperature. The paralytic phenotype is recessive to wild type. For this trait, wild type flies move around normally, are not paralysed, when the temperature is 29°C.

The alleles are defined differently for sex-linked traits. An X is used to indicate that the trait is carried on the X chromosome, and the allele is written in superscript next to the X. In the example of the fruit flies, the alleles can be written as:

X^P = wild type

X^p = paralysis

As the paralytic phenotype is recessive, females that are homozygous for the mutant allele (X^pX^p) express the mutant paralysis phenotype; females that are homozygous dominant (X^PX^P) and females that are heterozygous (X^PX^p) are both wild type phenotype.

As males have only one X chromosome there are only two male genotypes: X^pY males are paralytic and X^PY males are wild type.

If paralytic females (X^pX^p) are crossed with wild type males (X^PY), all of the F1 male offspring will be paralytic (X^pY) and all of the F1 female offspring will be wild type phenotype (X^PX^p) (Figure 7.1.9a). This pattern of transmission of the mutant phenotype from the female parent to male offspring is characteristic of X-linked recessive inheritance.

The reciprocal (reverse) cross, shown in Figure 7.1.9b, produces a different outcome. If a wild type homozygous female (X^PX^P) is crossed with a paralytic male (X^pY) all of thc offspring (malc and fcmalc) arc wild typc (X^PX^p and X^PY).

These different outcomes of the reciprocal crosses are characteristic of X-linked recessive inheritance. In contrast, reciprocal crosses give the same outcome for autosomally inherited genes.

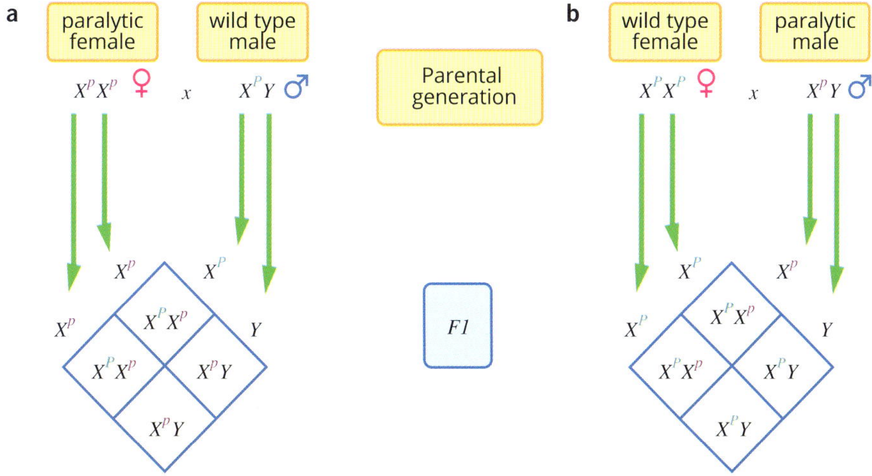

FIGURE 7.1.9 The characteristics of X-linked inheritance are evident in a reciprocal cross. (a) A male receives an X chromosome from the female parent, so males are paralytic (X^pY) and females are wild type (X^PX^P or X^PX^p). (b) In the reciprocal cross, both the male and female offspring are wild type.

Haemophilia in the British royal family

Figure 7.1.10 shows part of the family tree of the British royal family, including Queen Victoria, whose eighth child Leopold was born with haemophilia. Haemophilia is a blood disorder in which blood clotting is slow, resulting in excessive bleeding. It results from a mutation in a gene on the X chromosome that is involved in the production of a blood clotting protein that controls bleeding.

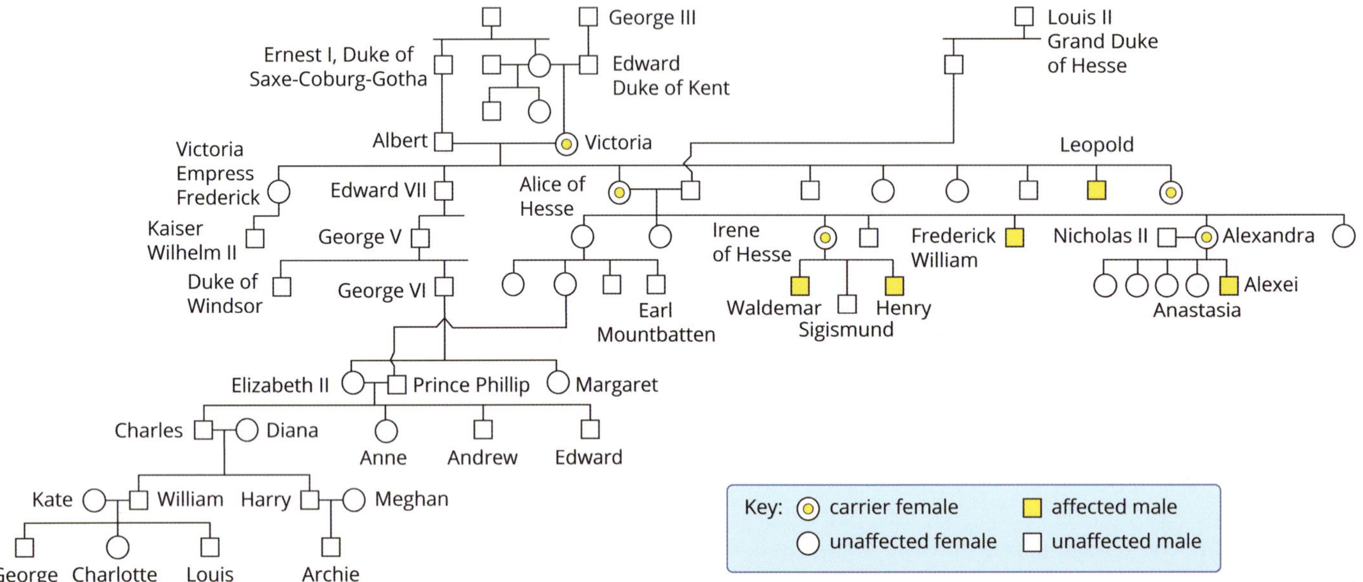

FIGURE 7.1.10 Queen Victoria was a carrier of a mutation that causes haemophilia. Some of Queen Victoria's female descendants have been carriers, and some male descendants have had the disease.

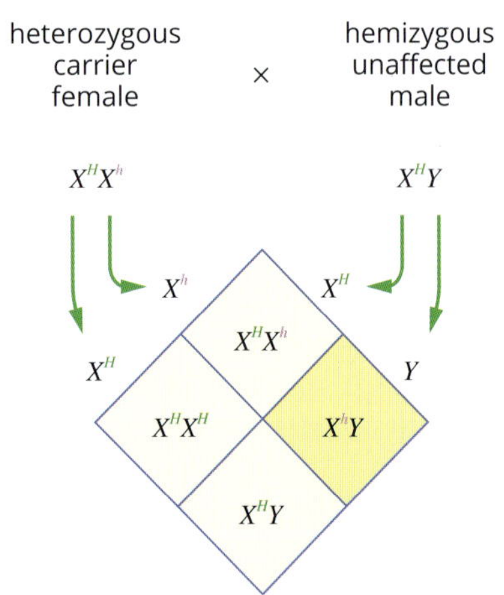

FIGURE 7.1.11 Inheritance pattern of an X-linked recessive condition. Males only need one copy of the X-linked allele to be affected. Females with one copy of the X-linked allele are unaffected carriers.

The incidence of haemophilia in the descendants of Queen Victoria shows the hallmarks of X-linked recessive inheritance. All of the individuals with haemophilia shown in the tree are male. The female **carriers** of the disease are heterozygous, carrying one haemophilia allele and one normal allele. Given that the haemophilia phenotype is recessive, carrier females are generally phenotypically normal. However, because females produce eggs carrying the normal and haemophiliac alleles with equal frequency, and males receive their single X chromosome from the egg, there is a 50% chance that the son of a carrier will have haemophilia.

Through marriage, some of Victoria's phenotypically unaffected daughters who carried this mutation spread haemophilia to other royal families in Europe. For example, Irene of Hesse transmitted the haemophilia allele to her eldest and youngest sons Waldemar and Henry, and the normal allele to her other son, Sigismund. This form of haemophilia occurs at a frequency of 1 in 10 000 males and 1 in 100 million females.

Pedigree charts are discussed further in Section 7.2.

In general, X-linked recessive disorders occur at much higher frequencies in males than females because to be affected females need to inherit a copy of the allele from both parents (that is, the mother must be a carrier and the father must be affected by the disorder). Males, however, need only inherit one copy of the X-linked allele from their carrier mother (Figure 7.1.11).

X-linked dominant inheritance

X-linked disorders may also display a dominant phenotype. Consider the inheritance of vitamin D-resistant rickets disorder, which causes bone deformities. This is shown in the pedigree chart in Figure 7.1.12. The mother of the first generation is heterozygous and affected by the condition. Her children had a 50% chance of having vitamin D-resistant rickets, regardless of whether they were male or female.

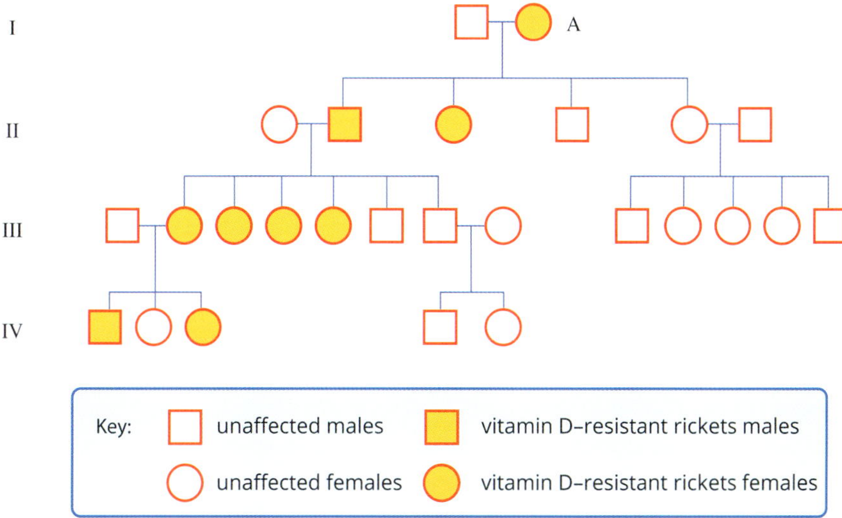

FIGURE 7.1.12 Pedigree chart showing the inheritance of the X-linked dominant condition vitamin D-resistant rickets. The mother (A) of the first generation is heterozygous for the gene that controls the condition.

When a father is affected and a mother is unaffected (as in the second generation), all female offspring will show the condition, and all male offspring will be unaffected. This is because female children all receive an X chromosome, which carries the allele for the disease, from their father (Figure 7.1.8 on page 288).

The alleles for this trait could be shown as:

X^D = vitamin D-resistant (rickets) allele

X^d = normal allele

The following Punnett square shows the pattern of inheritance of offspring of a heterozygous female affected by the vitamin D-resistant rickets allele and a male with the normal allele.

Parents		Mother	
		X^D	X^d
Father	X^d	X^DX^d	X^dX^d
	Y	X^DY	X^dY

The possible genotypes of the offspring are:

$X^DX^d : X^dX^d : X^DY : X^dY$

1 : 1 : 1 : 1

The possible phenotypes are therefore:

female with condition : unaffected female : male with condition : unaffected male

1 : 1 : 1 : 1

BIOFILE

Spontaneous mutations

Spontaneous mutations are mutations in DNA that have not been inherited. In the case of Queen Victoria there is no history of haemophilia in her ancestry, so it seems likely that she (or possibly her mother, Victoria, Duchess of Kent) was the source of the mutation. Spontaneous mutations are believed to cause about one third of all haemophilia occurrences.

The following Punnett square shows the pattern of inheritance of offspring between a homozygous unaffected female and a male with vitamin D-resistant rickets.

Parents		Mother	
		X^d	X^d
Father	X^D	X^DX^d	X^DX^d
	Y	X^dY	X^dY

The possible genotypes of the offspring are:

$X^DX^d : X^dY$

1 : 1

The possible phenotypes are therefore:

all females have vitamin D-resistant rickets : all males are unaffected

Y-linked inheritance

Compared to the X chromosome, the Y chromosome has few genes: it has only about 72 protein-coding genes, compared to 800–900 on the X chromosome. Most of these genes are involved in male sex determination and fertility. Therefore there are far fewer **Y-linked** traits than X-linked traits.

If a trait is passed from father to son and never observed in females, it is likely to be Y-linked, meaning the gene for that trait is on the Y chromosome. Until recently, hairy ears were thought to be controlled by a Y-linked gene, but recent studies suggest there are also autosomal genes involved in the trait.

Sex-limited inheritance

The Y-linked pattern of inheritance is sometimes confused with **sex-limited inheritance**. Sex-limited traits can only occur in one sex because the phenotype affected is unique to that sex, even though the gene involved is present in males and females. Therefore males and females have different phenotypes. For example, complete androgen insensitivity syndrome, in which the fetus is unresponsive to male hormones, can only occur in males, because only males carry the Y chromosome. This means that even if females have the genotype for this syndrome, they cannot express the condition.

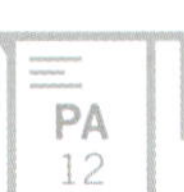

CASE STUDY

Male pattern baldness—sex-limited inheritance

It is estimated that 80% of hair loss is genetic, and though the causes are not yet well understood, it is known that several genes are involved (that is, it is a polygenic trait).

Male pattern baldness is the most common type of baldness. It affects around 40% of men by the age of 40 and around 60% by the age of 60.

Affected males gradually start losing their hair, until eventually they have hair only on the sides and back of the head (Figure 7.1.13).

You may have heard that baldness is inherited from your mother's father. This is because one of the key genes associated with balding is on the X chromosome.

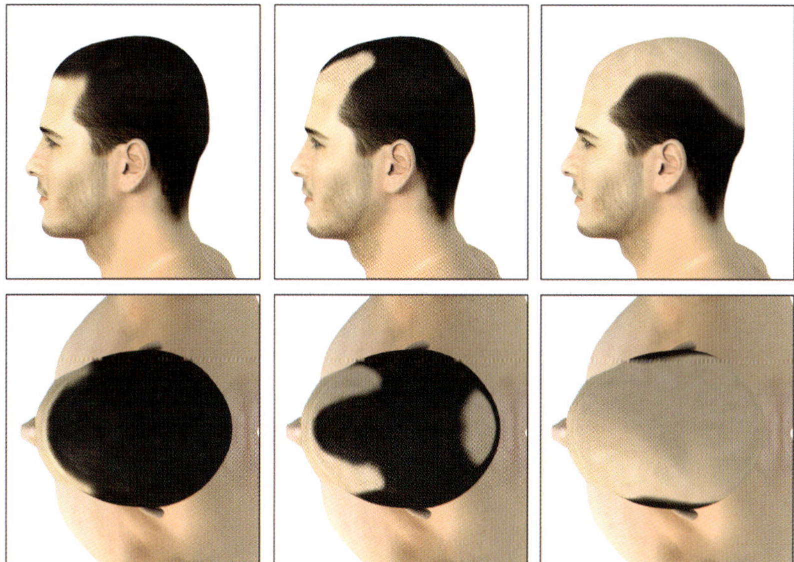

FIGURE 7.1.13 The change over time of the hairline of a man with male pattern baldness. This phenotype can be caused by several genes located on autosomes and the X chromosome.

If your mother's father has male pattern baldness, then your mother will carry the allele for this characteristic on the X chromosome that she inherited from her father. Because you inherited one of your X chromosomes from your mother, there is a 50% chance that you will inherit the affected X chromosome. However, males are much more likely to express the balding phenotype because females have a second X chromosome to mask the expression of the gene.

Balding can also be passed from fathers to offspring, indicating that autosomal genes must be involved. Two genes on chromosome 20 have been found to contribute to balding. The effects of these genes are neither dominant nor recessive, but have an additive effect the more copies of the alleles you have, the more likely you are to go bald. Even though these genes are found on autosomes, males are affected more than females. This is because some of the genes are associated with male hormone receptors. This is an example of sex-limited inheritance—males and females may have the same genotype but express different phenotypes.

7.1 Review

OA ✓✓

SUMMARY

- Much of what is now understood about inheritance has been gained through Mendel's experiments with pea plants in the 1800s. Mendel's observations led him to develop of the laws of inheritance.
- True-breeding strains are homozygous at the locus of interest and produce genetically identical progeny when crossed with each other.
- A phenotypic ratio approaching 3 : 1 will be observed in the F1 generation of a monohybrid cross between two heterozygous individuals for any trait controlled by a single autosomal gene, with two different alleles, controlling a dominant trait.
- A test cross involves crossing an individual displaying the dominant phenotype but unknown genotype with an individual displaying the recessive phenotype(s).
- Test crosses are used to determine whether an individual of dominant phenotype is homozygous or heterozygous.
- A phenotypic ratio approaching 1 : 2 : 1 will be observed in the F2 generation of a monohybrid cross for any trait controlled by a single autosomal gene, with two different alleles, displaying codominance.
- Incomplete dominance results when neither phenotype is completely dominant and intermediate phenotypes occur (e.g. pink snapdragon flowers).
- Codominance results when both alleles are equally expressed in an individual (e.g. ABO blood groups).
- Phenotypes inherited through the action of genes located on either the X or Y chromosomes show sex-linked inheritance.
- X-linked recessive inheritance shows a pattern of transmission of the mutant phenotype from the female parent to male offspring.
- X-linked dominant inheritance shows a pattern of transmission of the dominant trait from an affected male parent to all female offspring and from an affected heterozygous female parent to 50% of all offspring.
- Y-linked inheritance shows a pattern of transmission of the trait from father to son, and it is never observed in females.

KEY QUESTIONS

Knowledge and understanding

1 What is meant by the term 'monohybrid cross'?

2 Explain what the following ratios of phenotypes in the F2 generation suggest about inheritance patterns for a particular trait.

a 3 : 1 **b** 1 : 2 : 1

3 Freckles are an inherited trait that results in the formation of spots of pigment on fair skin. The gene for this trait is found on chromosome 4 and an individual only needs one allele to show the trait.

a State whether the type of inheritance is autosomal or sex-linked, and dominant or recessive.

b Using suitable symbols, draw a full genetic diagram to show how a mother and father who have freckles can have a child that does not have freckles. State the probability that the child does not have freckles.

4 Define sex-linked inheritance.

Analysis

5 A gardener is growing snapdragons, which exhibit incomplete dominance for flower colour. They want to have a hedge of white, red and pink flowers. They start with two plants, one with red flowers and the other with pink flowers. They begin to get frustrated when no white flowers are produced.

a Explain why no white flowers are being produced by the offspring of these plants. Use a Punnet square to support your answer.

b If only two plants could be selected to start the hedge, which flower colours would be needed to ensure all three colour flowers were produced? Use a Punnet square to support your answer.

6 The figure below, which is incomplete, shows the inheritance of haemophilia in a family. Haemophilia is a recessive X-linked disorder.

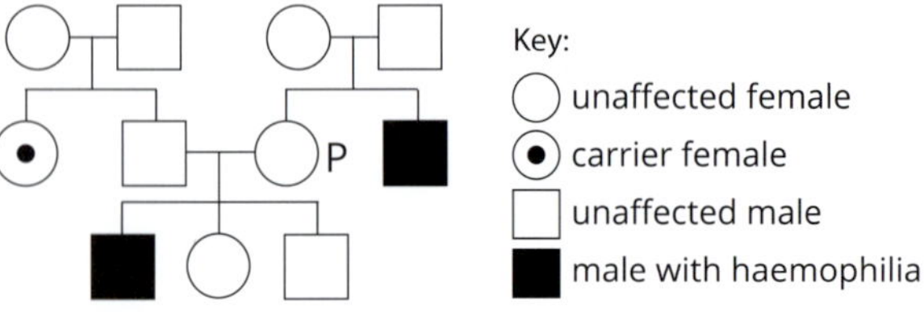

What is the genotype of P? Show your workings using a Punnett square and appropriate symbols.

7.2 Pedigree charts and inheritance patterns

Studying the patterns of inheritance in humans has its challenges. In this section you will learn how patterns of inheritance of alleles across generations of families (Figure 7.2.1) can be analysed using pedigree charts.

FIGURE 7.2.1 Pedigree charts help to determine patterns of inheritance for different traits in families.

PEDIGREE ANALYSIS

Pedigree analysis is a technique involving studying a family tree for the occurrence of a particular character or trait in a family over a number of generations. **Pedigree charts** can be used to determine how traits are inherited, as well as the presence of particular alleles within a family and the chances of the allele occurring in offspring. Given sufficient data, the likely mode of inheritance can be determined; for example, dominance patterns and whether inheritance is autosomal or sex-linked. In practice it may be necessary to combine the pedigree data of several families to determine the most likely mode of inheritance of a particular character.

When it comes to studying the genetics of humans, pedigree analysis is a useful method for the following reasons:

- Genetic crosses cannot be set up as required.
- The environment in which humans live cannot be controlled experimentally.
- There are strict legal and ethical laws concerning human experimentation.
- Humans tend not to have large families, so there are rarely large numbers of offspring to score.
- Each generation of humans takes many years to reach sexual maturity and produce offspring.

Knowledge of the mechanisms of inheritance in humans and other organisms continues to advance through research using model organisms, and DNA- and genome-sequencing technology.

A pedigree chart is a record of the ancestry (also called the lineage) of an individual or a group of related individuals.

Pedigree charts

Pedigree analysis makes use of pedigree charts to track and organise data. When analysing a pedigree chart, key features in the pattern of inheritance can be used to distinguish between one type of inheritance and another.

Symbols and conventions used for pedigree charts

Pedigree charts use a number of standard symbols and conventions (Figure 7.2.2). The main ones you need to know are as follows:

- Circles represent females and squares represent males.
- Shapes are shaded or unshaded to represent the presence of a phenotype for a particular trait.
- A horizontal line represents a cross between the individuals.
- A vertical line represents a link from parents to offspring.
- Individuals are numbered from left to right (if required).
- Generations are represented with roman numerals (if required), with the first generation in the pedigree chart being generation I.
- A carrier of an X-linked trait is shown with a dot in the centre of the symbol.

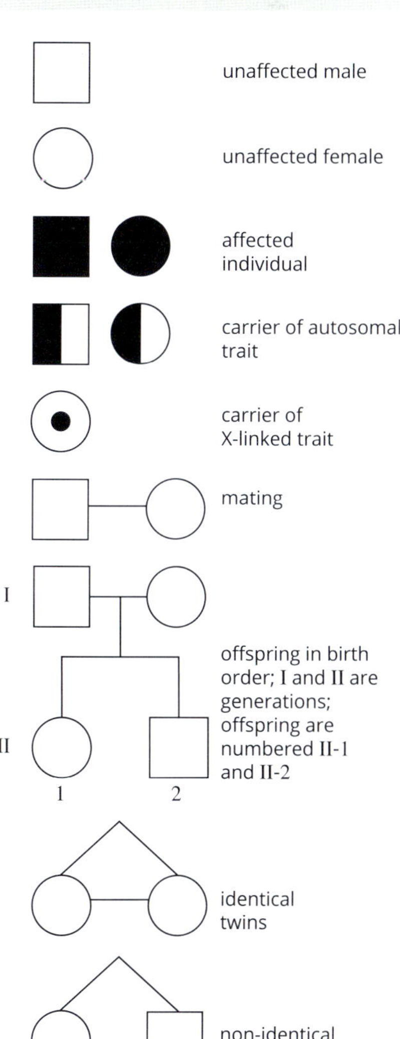

FIGURE 7.2.2 Conventions for pedigree charts

RECOGNISING INHERITANCE PATTERNS IN PEDIGREE CHARTS

Pedigree analysis can be used to understand how particular genetic traits are inherited, but as you will have learnt by now, there are many possible modes of inheritance. Fortunately, these can be recognised if you are aware of a few simple methods. Once you know these methods, you should be able to examine a comprehensive pedigree chart and then work out the inheritance pattern for a trait.

Autosomal inheritance

When the inheritance of a trait is just as likely in males as in females, it is an autosomal trait—that is, the gene(s) associated with the trait are not located on the sex chromosomes, but on the autosomes. This is not always immediately obvious, particularly if a mating produces only male or only female offspring.

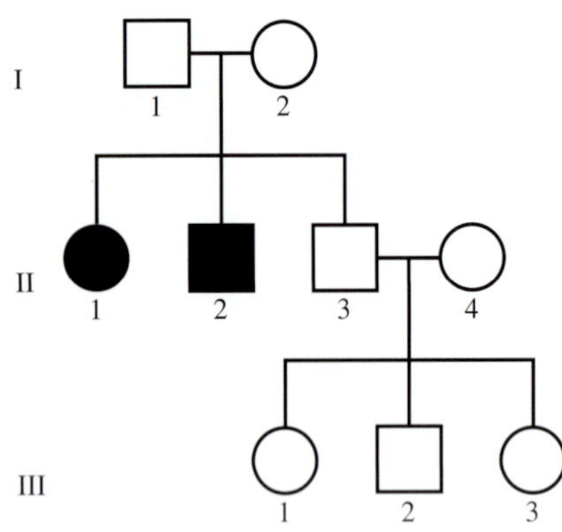

FIGURE 7.2.3 Example of autosomal recessive inheritance

Autosomal recessive inheritance

Autosomal recessive inheritance of a trait is likely if two parents do not have a particular phenotype but one or more of their offspring does. All the affected individuals in a family are usually in a single sibling group—multiple generations are not typically affected. The autosomal recessive condition represented in the pedigree chart in Figure 7.2.3 is haemochromatosis, a disorder in which too much iron accumulates in the body, leading to tissue damage. Shaded individuals are affected (that is, they express the trait), and unshaded individuals are not affected.

For this exercise it may be assumed that the inheritance of the trait is not sex-linked. Figure 7.2.3 shows that:

- individuals I-1 and I-2 are both unaffected but a cross between them produced two affected offspring
- males and females are equally affected—individual II-1 is female and affected, individual II-2 is male and affected and individual II-3 is male and unaffected
- none of the individuals in the third generation are affected.

The parents in generation I (I-1 and I-2) both contribute one allele each for the trait to II-1 and II-2, so both parents must carry the allele for the trait. Since the parents are both unaffected, both must be heterozygous. The trait is therefore autosomal recessive.

Another indicator of autosomal recessive inheritance is that the trait skips generations (that is, it does not appear in every generation). However, not skipping a generation does not rule out autosomal recessive inhritance.

Once the form of inheritance is determined, it is possible to work out the genotypes of some of the individuals in the pedigree chart. First, a symbol should be allocated for the two alleles. In this particular example:

- *H* can represent the allele for unaffected (dominant trait)
- *h* can represent the allele for haemochromatosis (recessive trait).

In generation I, individuals I-1 and I-2 are unaffected, but are heterozygous (*Hh*) as they have two affected offspring, II-1 and II-2. Individual II-3 could either be either homozygous (*HH*) or heterozygous (*Hh*). There is not enough information in this pedigree chart to be able to determine his genotype, or that of the remaining individuals.

Autosomal dominant inheritance

Autosomal dominant inheritance of a trait is likely if the trait is seen in both sexes, occurs in all (or most) generations and affected individuals have at least one affected parent. The pedigree chart in Figure 7.2.4 shows the inheritance pattern of Huntington's disease, a condition that causes the progressive breakdown of brain nerve cells. This condition generally presents later in life, with affected individuals experiencing signs and symptoms in their 30s and 40s.

It may be assumed for this example that the trait is not sex-linked. Figure 7.2.4 shows that:

- II-3 and II-4 produce three offspring
- II-4 is affected so must have at least one allele for the Huntington's disease phenotype
- offspring III-1 is female and unaffected, and III-2 and III-3 are male and affected.

Because both II-3 and II-4 contribute one allele each to III-1, each must carry the allele for the unaffected phenotype. This means that II-4 must be heterozygous and their phenotype must be dominant.

Other examples of autosomal dominant inheritance in humans include Marfan syndrome and neurofibromatosis Type 1.

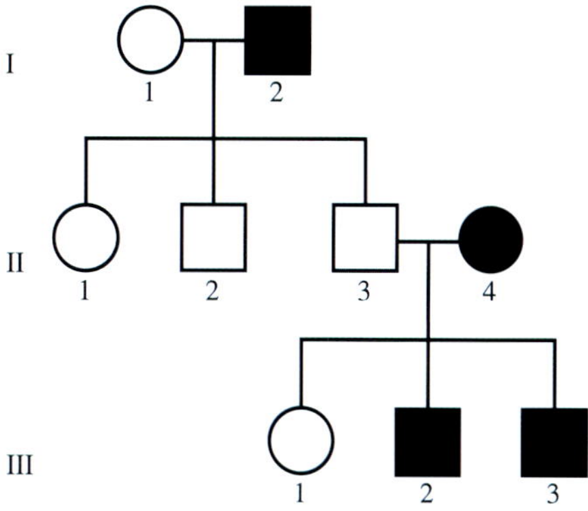

FIGURE 7.2.4 Example of autosomal dominant inheritance

CASE STUDY

The principle of penetrance

Researchers in France in the 1990s conducted one of the largest pedigree analyses to date when they traced the inheritance of a particular form of blindness (juvenile glaucoma) in over 30000 living descendants of a 15th century couple. The analysis accounted for almost half of the known cases of this disease in France. The massive pedigree chart clearly showed that juvenile glaucoma resulted from autosomal dominant inheritance. As a result, it might be expected that every individual who had at least one parent with the disease would also have the disease. However, the researchers noted that this was not always true; sometimes offspring of a parent who developed juvenile glaucoma did not develop the disease (Figure 7.2.5). This illustrates the principle of penetrance.

Complete penetrance of a phenotype means that all individuals with an affected genotype will have the affected phenotype. Incomplete penetrance describes the situation that occurred in the juvenile glaucoma pedigree chart where a proportion of a population with an affected genotype does not show the expected phenotype. In Figure 7.2.5, individual V-2, whose father had the disorder, did not show the glaucoma phenotype yet passed it on to his daughter (VI-2).

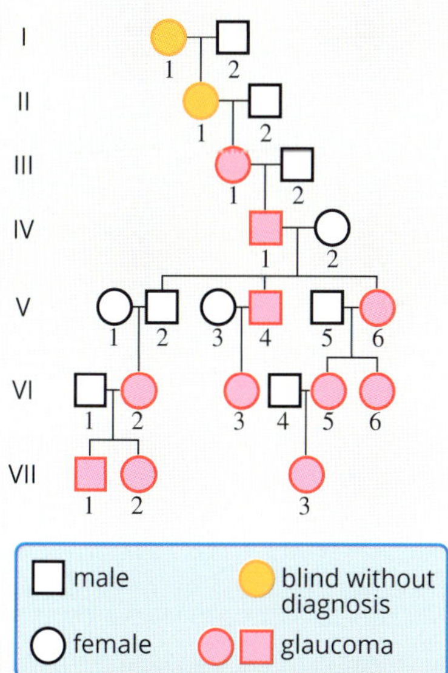

FIGURE 7.2.5 Part of a pedigree chart showing the pattern of transmission of juvenile glaucoma over several generations. Incomplete penetrance is demonstrated by the lack of juvenile glaucoma in V-2.

Sex-linked inheritance

When an inheritance pattern is 'sex-linked', that means the trait in question affects males and females differently. We therefore conclude that one or more of the gene(s) responsible for the trait is located on a sex chromosome. If enough male and female offspring are produced in each generation, sex-linked inheritance should be obvious when looking at the pedigree chart.

Distinguishing between autosomal and sex-linked inheritance

In the previous examples of autosomal inheritance it was assumed that inheritance was autosomal. However, when investigating inheritance it is important to check whether the inheritance might be sex-linked, because similar patterns can occur in both types.

The inheritance pattern of colour vision deficiency (colour blindness) in humans can be examined by studying the pedigree chart of a family in which some individuals have a colour vision deficiency. By convention, if a mating partner is not shown in the pedigree chart, he or she is not affected by the phenotype.

Differences in the incidence of the trait between males and females, and the frequency of the trait across generations, are two inheritance patterns that can help you determine whether a trait is sex-linked or auosomal.

In the pedigree chart shown in Figure 7.2.6 only the males have a colour vision deficiency. This demonstrates that the colour vision deficiency trait is not dominant and is potentially sex-linked (X-linked). Y-linked alleles affect only males and are very rare, and the phenotype will be present in every male offspring. Because of these factors, colour vision deficiency cannot be a Y-linked trait.

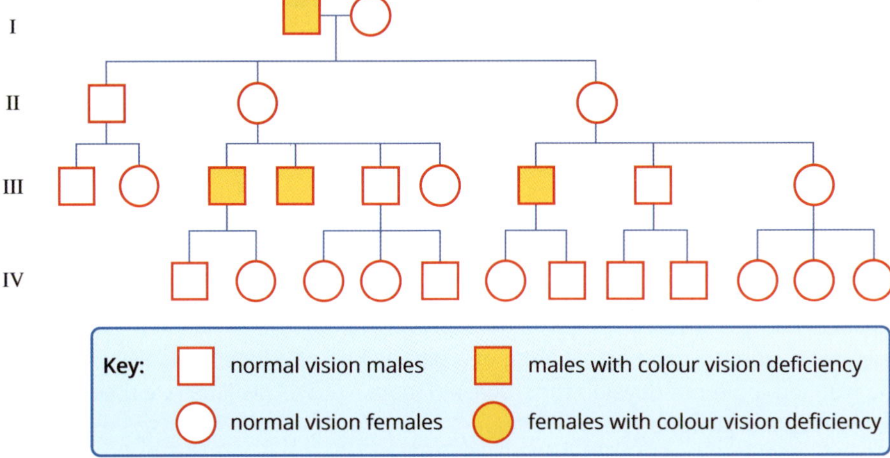

FIGURE 7.2.6 Pedigree chart of a family in which colour vision deficiency is present. The inheritance of colour vision deficiency is X-linked recessive and so males are affected more frequently than females. No females with colour vision deficiency were observed in this particular pedigree chart.

X-linked recessive inheritance

The pedigree chart in Figure 7.2.6 also indicates that the inheritance of colour vision deficiency is more likely to be X-linked recessive than autosomal recessive, because X-linked recessive traits affect many more males than females. This is because males only inherit one X chromosome, while females inherit two. This second X chromosome has a masking effect on recessive alleles, resulting in females carrying the affected allele but not expressing the phenotype. In order for females to be affected by X-linked recessive traits, they must carry two copies of the allele, one on each X chromosome.

Given the low number of offspring, it is possible that all the individuals with colour vision deficiency are males purely by chance. However, in reality, many such pedigree charts have been studied and show similar patterns, confirming that the inheritance of colour vision deficiency is X-linked recessive.

X-linked dominant inheritance

Traits that are X-linked dominant are rare and affect more females than males. This is because females inherit two X chromosomes and so have twice the chance of inheriting an affected X chromosome compared to males, who inherit only one.

Evidence of X-linked dominant inheritance is seen in a pedigree chart in which affected males have daughters who are all affected and sons who are not affected. This is because daughters inherit their father's only X chromosome, while sons inherit their father's Y chromosome. If the X chromosome carries an allele for a dominant trait, then the daughter will express its phenotype. X-linked dominant inheritance can be seen in Figure 7.2.7.

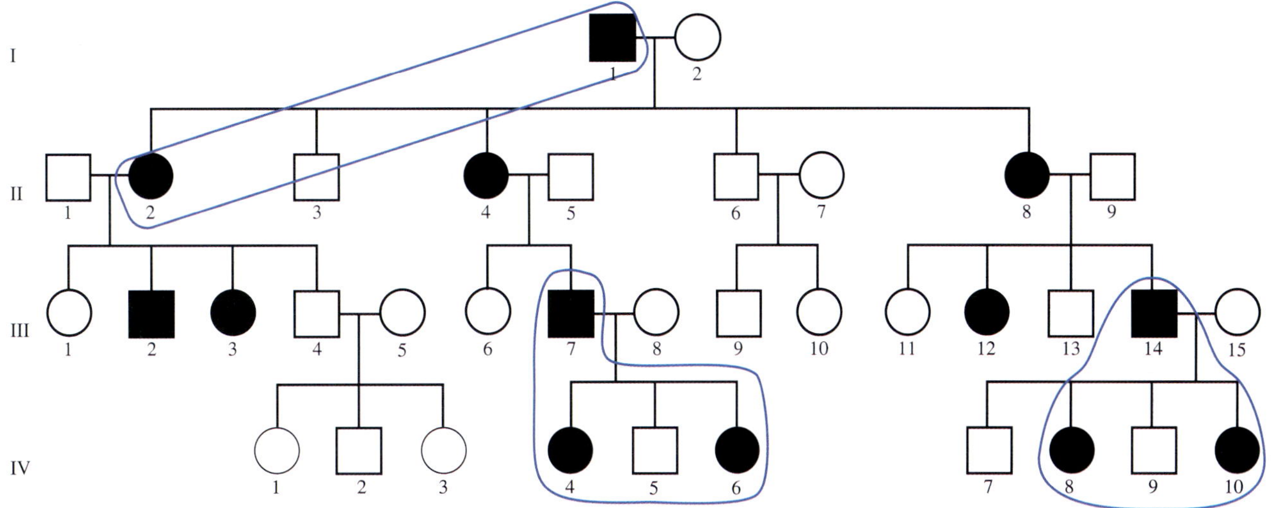

FIGURE 7.2.7 An example of X-linked dominant inheritance

Males I-1, III-7 and III-14 have daughters who are all affected and sons who are unaffected (see circled parts of Figure 7.2.7).

Because females pass on one of their two X chromosomes to both their daughters and sons, there is a 50% chance that they will pass on an X-linked dominant trait to their offspring. This pattern of inheritance is seen in female individuals II-2, II-4 and II-8, who have both affected and unaffected daughters and sons. An example of X-linked dominant inheritance in humans is Fragile X syndrome, caused by an abnormal gene located on the X chromosome, which results in developmental issues including learning disabilities and impaired cognitive function. Another example is vitamin D-resistant rickets (Figure 7.1.12, page 291, and Figure 7.2.8).

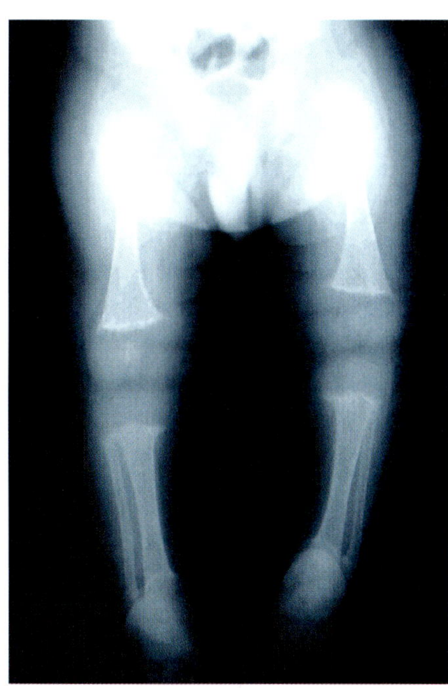

FIGURE 7.2.8 X-ray of the legs of a young child with vitamin D-resistant rickets

Y-linked inheritance

Male offspring inherit their father's Y chromosome, so any alleles carried on this chromosome will be passed on from father to son. The phenotype of Y-linked disorders is, therefore, seen in fathers and all their sons. Females are never affected by Y-linked traits because they do not inherit a Y chromosome. As there is only one Y chromosome (hemizygous), and thus only one allele present, the general principles of dominant and recessive inheritance do not apply.

A Y-linked trait is likely if:

- only males are affected
- all male offspring are affected
- the trait is observed in every generation in which males are born.

The Y-linked pattern of inheritance can be seen in Figure 7.2.9. Because the Y chromosome has far fewer genes that code for proteins compared to the X chromosome, Y-linked inheritance of traits is relatively rare.

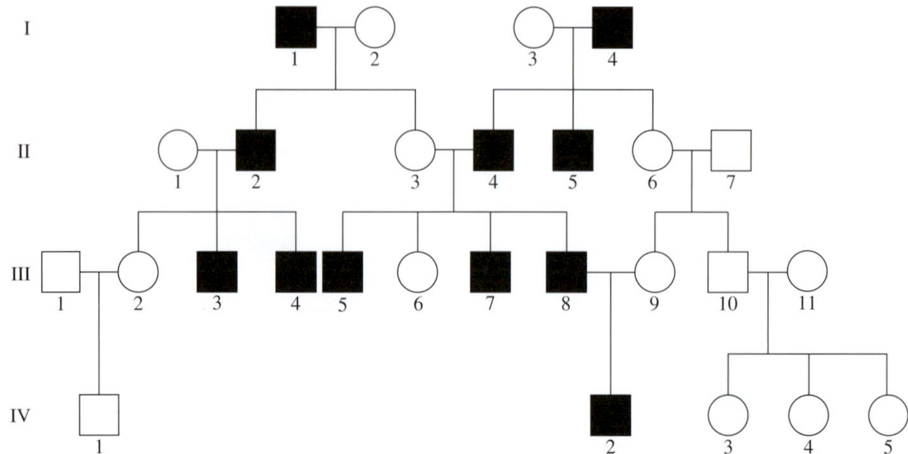

FIGURE 7.2.9 Pedigree chart for a trait with Y-linked inheritance. This pattern of inheritance is evident because only males are affected and the trait is present in males in every generation.

Ruling out sex-linked inheritance

Consider the following pedigree chart for a trait in humans. It is possible to rule out sex-linked inheritance in this pedigree chart (Figure 7.2.10) by looking for a pattern that will rule out each type of sex-linked inheritance in turn. (If no such pattern can be found, then that type of inheritance cannot be ruled out.)

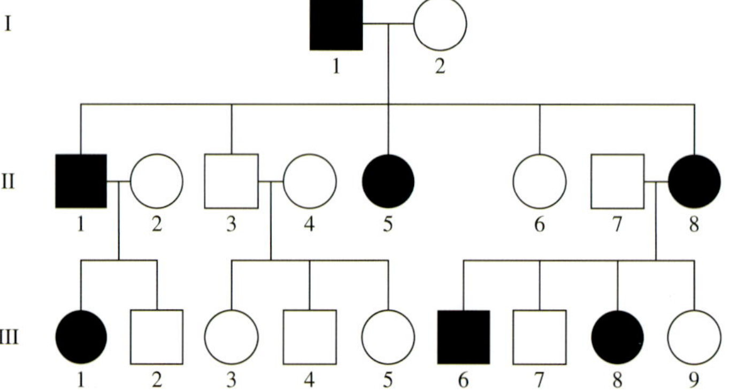

FIGURE 7.2.10 A pedigree chart for an inherited trait in humans. At this stage it is not known whether the inheritance is sex-linked or autosomal.

The following annotated versions of the pedigree chart in Figure 7.2.10 illustrate how each type of sex-linked inheritance can be ruled out.

Ruling out X-linked recessive inheritance: In order for a trait to be X-linked recessive, affected mothers must have affected sons. In Figure 7.2.11, we can see that individual II-8 has two sons, one affected by the trait and the other unaffected. For the trait to be X-linked recessive, the mother would have to carry the allele on both her X chromosomes (X^hX^h) to be affected. The mother contributes one of these chromosomes to her offspring, so both sons would have to receive a chromosome with the affected allele. But one son does not show the trait (and therefore did not receive the affected allele), so X-linked recessive inheritance cannot be involved.

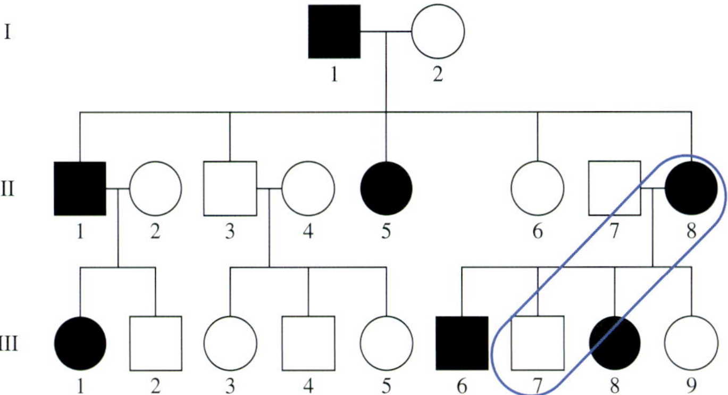

FIGURE 7.2.11 Ruling out X-linked recessive inheritance

Ruling out X-linked dominant inheritance: For X-linked dominant inheritance to be involved, every daughter of an affected male must be affected. This is because the daughters must receive an X chromosome from their father, and the father has only one X chromosome. In the pedigree chart shown in Figure 7.2.12, male I-1 has the trait but one of his daughters (II-6) does not. X-linked dominant inheritance therefore cannot be involved in this pedigree chart.

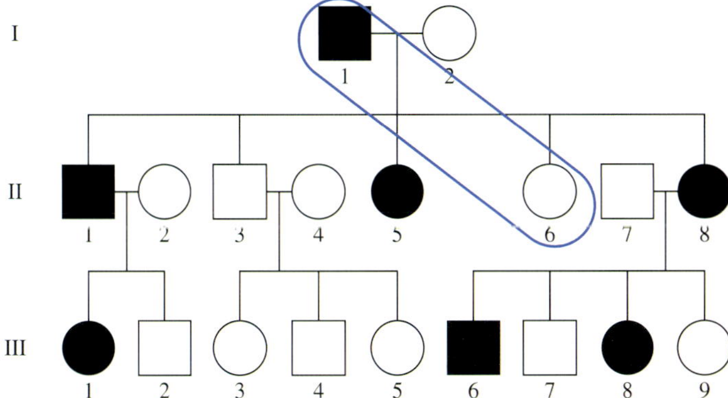

FIGURE 7.2.12 Ruling out X-linked dominant inheritance

Ruling out Y-linked inheritance: If a trait is Y-linked, only males are affected (because females lack a Y chromosome) and affected fathers pass the trait on to all their sons. In the pedigree chart shown in Figure 7.2.13, some females have the trait, and not all fathers with the trait passed it on to their sons (e.g. male I-1 and son II-3; also II-1 and III-2). Y-linked inheritance is therefore not involved in this pedigree chart.

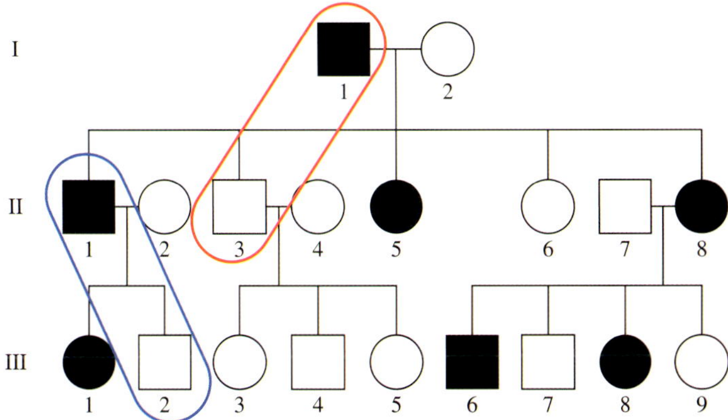

FIGURE 7.2.13 Ruling out Y-linked inheritance

After ruling out X-linked recessive, X-linked dominant and Y-linked inheritance, it can be concluded that the trait shown in the previous figures must be the result of autosomal inheritance. Although the trait is observed in every generation, which often indicates dominant inheritance, in this case the trait could be autosomal dominant or autosomal recessive inheritance. Further information would be needed to determine which is involved in this pedigree chart.

STEPS IN PEDIGREE ANALYSIS

A pedigree analysis should be carried out in a methodical series of steps to determine the pattern of inheritance. These steps are outlined below and illustrated in Figure 7.2.14.

Step 1 Determine whether the condition is sex-linked.

a Are males and females affected in equal proportions? If yes, then it is likely an autosomal condition. Go to step 2.

b Are mostly or only males affected? If no, and more females are affected, then it is likely an X-linked dominant condition. Confirm this by checking that transmission occurs from fathers to daughters but not to sons.

If yes to mostly or only males affected, then either the condition is X-linked recessive or Y-linked. Go to part c.

c Is there male-to-male transmission? If yes, then it is likely to be Y-linked inheritance. If no, then it is likely to be X-linked recessive.

Step 2 Determine the type of autosomal inheritance.

WS 28 PA 13

Are there multiple generations of affected individuals? If yes, then it is likely to be an autosomal dominant condition. If no, then it is likely to be an autosomal recessive condition.

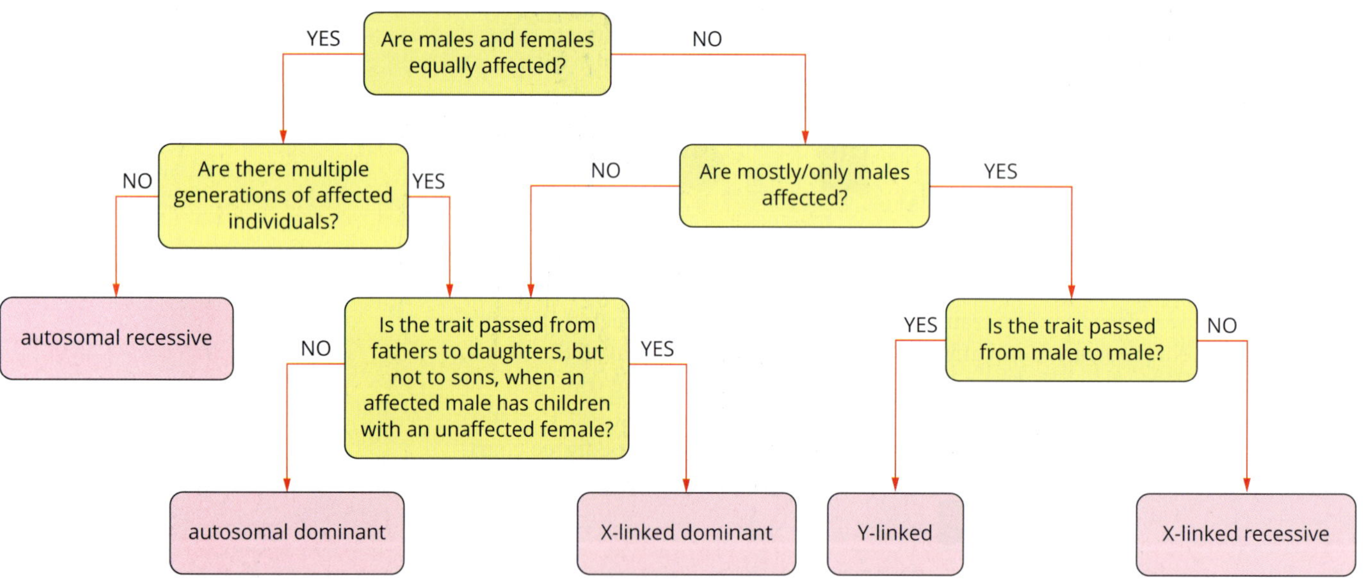

FIGURE 7.2.14 Flow diagram for pedigree analysis of simple modes of inheritance

7.2 Review

SUMMARY

- Pedigree analysis is a technique involving studying a family tree for the occurrence of a particular character or trait in a family over a number of generations.
- Pedigree charts can be used to determine the likely mode of inheritance, such as dominance patterns and whether inheritance is autosomal or sex-linked.
- When a trait is autosomal recessive:
  - Both sexes display the trait in equal numbers in a pedigree chart.
  - Offspring of unaffected parents have a 25% chance of being affected.
  - Affected individuals are homozygous.
- When a trait is autosomal dominant:
  - Both sexes usually display the trait in equal numbers in a pedigree chart.
  - One parent must be affected to have an affected offspring.
  - The trait is observed in each generation.
- When a trait is X-linked recessive:
  - The trait is rare within the pedigree chart, but males are more affected than females.
  - Affected fathers do not pass the trait on to their sons, so the condition can skip generations.
  - Females can be carriers and not show the condition; female carriers can pass the trait on to their sons.
- When a trait is X-linked dominant:
  - Males and females are affected (often more females than males).
  - All affected sons have an affected mother.
  - The trait is observed in each generation.
- When a trait is Y-linked:
  - Only males are affected, not females.
  - Fathers pass the trait on to their sons.
  - The trait is observed in each generation.

KEY QUESTIONS

Knowledge and understanding

1 Why is pedigree analysis often the easiest way to investigate inheritance patterns in humans? Give three reasons.

2 The figure at right shows part of a family pedigree chart. If individual III-3 was shaded, which of the following best describes the trait?

A dominant
B sex-linked
C recessive
D codominant

3 If individual III-5 in the pedigree chart at right was shaded, what could the genotypes of the parents be? Choose from options A to D and draw a Punnett square to show your reasoning.

A *BB*, *BB*
B *Bb*, *Bb*
C X^BX^B, X^BY
D X^BX^B, X^bY

4 Explain why Y-linked disorders are rare.

Analysis

5 Blue eyes in humans is a homozygous recessive trait. All other eye colours are dominant to blue. Determine the probability of two non–blue eyed individuals, who each have one blue-eyed parent, having a blue-eyed child.

6 What type of inheritance is shown in the pedigree chart below? Give three reasons for your choice of inheritance pattern.

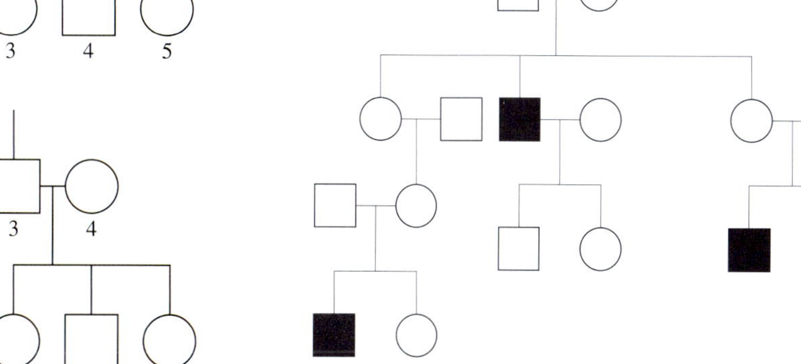

7.3 Dihybrid crosses

The crosses discussed so far relate to one trait; for example, the colour of a flower. In this section, you will learn how to predict the outcome of dihybrid crosses for both linked and non-linked traits, and the biological consequence of crossing over for linked genes.

FIGURE 7.3.1 This fruit fly (*Drosophila melanogaster*) has red eyes and a brown body.

NON-LINKED DIHYBRID CROSSES

Mendel's second law of inheritance, the law of independent assortment, states that the alleles of a gene controlling one trait assort independently of alleles of another gene controlling a different trait. This can be illustrated by considering crosses involving two genes that affect two distinct traits.

You can cross true-breeding strains of flies (*Drosophila*) that differ for two traits: eye colour and body colour (Figure 7.3.1) and then conduct a **dihybrid cross** ('di' meaning two). The two traits in this example are eye colour and body colour.

The eye-colour gene in this example is the yellow eye gene. (This is a different gene from the white eye-colour gene discussed in Section 7.1, and is located on a different chromosome). The 'yellow eye' gene is autosomal; the alleles are *Y* and *y*. The wild type 'red eye' phenotype (genotypes *YY* and *Yy*) is dominant and the 'yellow eye' phenotype (genotype *yy*) is recessive.

The second trait is body colour and the gene in this example is called 'black body'. It is an autosomal gene with two alleles *B* and *b*. The wild type brown body phenotype (genotypes *BB* and *Bb*) is dominant and black body phenotype (genotype *bb*) is recessive.

The eye-colour and body-colour genes are on different chromosomes. These characteristics are an example of non-linked traits as they are inherited independently of one another—they are said to assort independently.

F1 generation

In a dihybrid cross, a cross is first set up between, for example, a true-breeding red eye, brown body (*YY*, *BB*) strain and a yellow eye, black body (*yy*, *bb*) strain. The homozygous *YY*, *BB* strain will produce gametes carrying only the *Y* and *B* alleles, and the homozygous *yy*, *bb* strain will produce only *y* and *b* alleles. The gametes fuse to produce F1 progeny, which are heterozygous for both genes (*Yy*, *Bb*). Following meiosis, a gamete will end up with any of four possible combinations of alleles: *YB*, *yb*, *yB* or *Yb* (Figure 7.3.2).

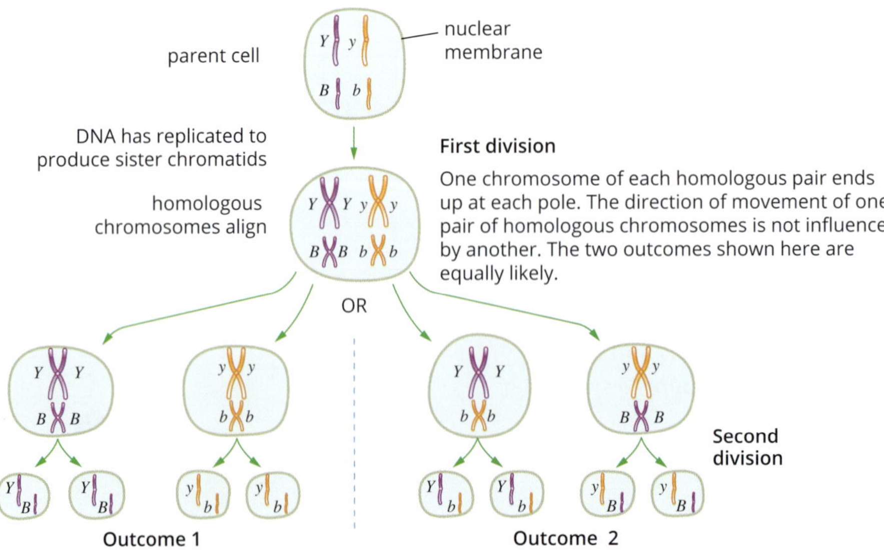

FIGURE 7.3.2 New combinations of alleles result from cells dividing by meiosis. Two genetic loci are shown, each on separate chromosomes within the cell nucleus. Daughter cells produced by meiosis may have any of four possible combinations of alleles.

During gamete formation in F1, the chance of a sperm or egg cell containing a Y allele is 0.5 (because half of the gametes contain a Y allele). The chance that a gamete will contain a B allele is also 0.5. Therefore, the chance of the gamete being YB is $0.5 \times 0.5 = 0.25$. The probability of each of the other three possible gametic combinations is also 0.25.

This is because the segregation of alleles of a gene on one chromosome in meiosis is independent of the segregation of alleles of a gene on another chromosome. The **homologous chromosome** (matching chromosome) carrying the Y allele can move to either pole, as can the homologous chromosome carrying the y allele. These homologous chromosomes move independently of the chromosomes that carry the B and b alleles (Figure 7.3.2).

The law of independent assortment states that the alleles of genes that code for different traits are inherited independently from each other.

F2 generation

The heterozygotes generated in the F1 can be crossed together (a dihybrid cross) to produce an F2 generation. Figure 7.3.3 shows that the expected ratio of phenotypes in the F2 generation is:

- 9 red eye, brown body
- 3 red eye, black body
- 3 yellow eye, brown body
- 1 yellow eye, black body.

If these crosses were actually performed, the phenotypic ratio in the F2 generation should be close to the 9 : 3 : 3 : 1 ratio. There would be some difference between the expected and observed phenotypic ratios due to sampling error. The larger the number of F2 progeny scored, the closer the result will be to the 9 : 3 : 3 : 1 ratio.

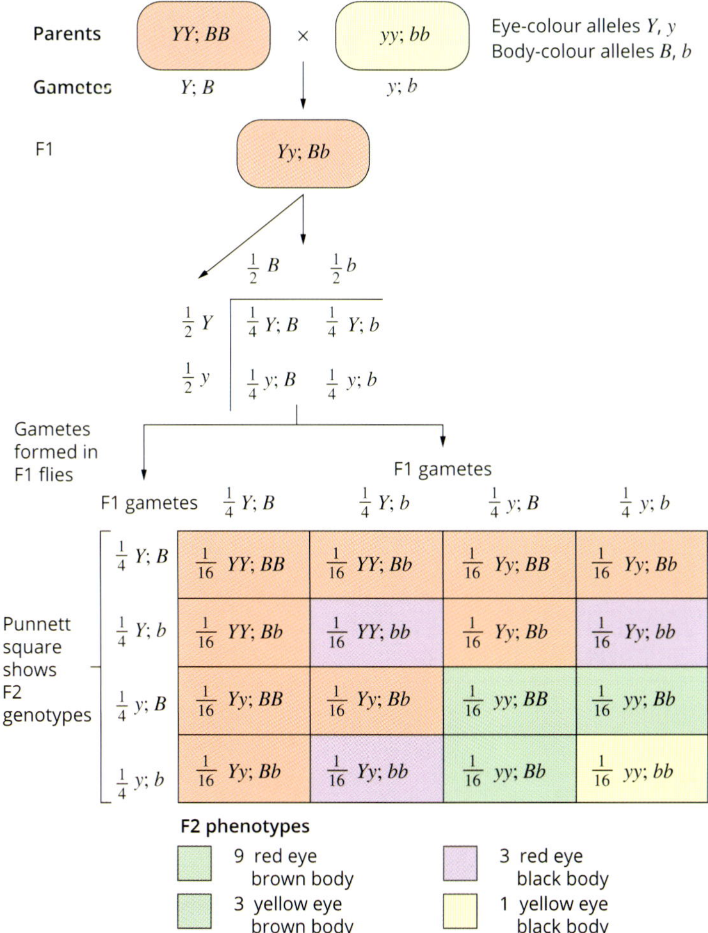

FIGURE 7.3.3 A Punnett square showing the genotypes and phenotypes of the F2 progeny. The F1 generation produces *YB*, *yB*, *yB* and *yb* gametes in equal frequency. When F1 individuals are crossed, the resulting F2 shows a 9 red eye, brown body : 3 red eye, black body : 3 yellow eye, brown body : 1 yellow eye, black body phenotypic ratio.

Non-linked dihybrid cross summary

In summary, a phenotypic ratio approximating 9 : 3 : 3 : 1 will be observed in the F2 generation of a dihybrid heterozygous cross if the following five conditions apply:

- the two genes control two distinct traits
- there are two alleles for each of the genes
- one phenotype is dominant for each trait
- both genes are on autosomes
- the two genes assort independently.

In this example, independent assortment occurs because the two genes are on different chromosomes. However, you will learn later in this section that independent assortment can occur via another mechanism.

GENE LINKAGE

Gregor Mendel was a truly outstanding scientist. His law of segregation and law of independent assortment are the cornerstones upon which our current understanding of inheritance is built. Since the rediscovery of Mendel's work 16 years after his death, scientists have continued to refine and extend these laws or principles to explain new and unexpected aspects of heredity, and more complex patterns of inheritance.

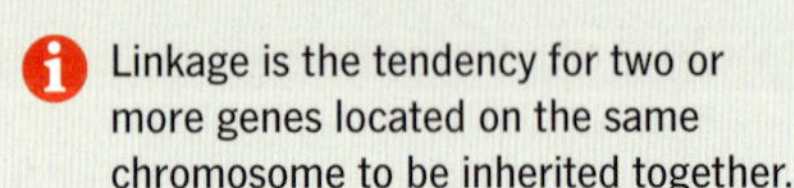

Linkage is the tendency for two or more genes located on the same chromosome to be inherited together.

Although many traits are inherited in accordance with Mendel's laws, this is not always the case. The exceptions occur when two or more genes are located on a single chromosome and are inherited together. This is known as **linkage**, and is another key principle of inheritance. The closer the genes are, the more likely they are to be inherited together, or 'linked'. A group of genes located on the same chromosome that is inherited as one unit is known as a **linkage group**. The alleles in a linkage group do not assort independently. However, linkage is never complete because of **crossing over**, which occurs during meiosis.

The consequences of gene linkage for phenotypes in offspring can be seen in the following example of maize seed shape and colour. These phenotypes are encoded by autosomal genes, each with two alleles:

- seed colour—orange (genotypes *OO* and *Oo*) is dominant to white (genotype *oo*).
- seed shape—round (genotypes *RR* and *Rr*) is dominant to flat (genotype *rr*).

A standard dihybrid cross was conducted, starting with two true-breeding parents, one with orange round seeds (genotype *OORR*) and the other with white flat seeds (genotype *oorr*). The F1 offspring all showed the dominant phenotypes (phenotype: orange, round seeds; genotype: *OoRr*)

If the two genes assort independently, the expected genotypes and phenotypes in the F2 generation can be predicted with a Punnett square for a cross between two F1 plants with orange, round seeds (*OoRr* × *OoRr*). The predicted phenotypic ratio is:

9 orange, round seed : 3 orange, flat seed : 3 white, round seed : 1 white, flat seed

The expected genotypes in the F2 generation are shown in the Punnett square below.

Parents	***OR***	***Or***	***oR***	***or***
OR	*OO RR*	*OO Rr*	*Oo RR*	*Oo Rr*
Or	*OO Rr*	*OO rr*	*Oo Rr*	*Oo rr*
oR	*Oo RR*	*Oo Rr*	*oo RR*	*oo Rr*
or	*Oo Rr*	*Oo rr*	*oo Rr*	*oo rr*

However, when this cross was actually carried out, only orange, round seeds and white, flat seeds were obtained, with a phenotypic ratio of 3 : 1. This phenotypic ratio is not in accordance with Mendel's law of independent assortment. It occurs because both genes are located on the same chromosome and are inherited together. They are linked genes.

Because the alleles are inherited together, there is no crossing over and so only two types of gametes are produced (*OR* and *or*). In cases where genes are linked, the correct Punnett square to use is shown below.

Parents	***OR***	***or***
OR	*OO RR*	*Oo Rr*
or	*Oo Rr*	*oo rr*

Linkage and recombination

Crossing over is a normal event that results in genetic exchange between non-sister **chromatids**. It will occur in most germ line cells going through meiosis. The probability of at least one cross-over event occurring somewhere on the chromosome is high because there is usually at least one **chiasma** (point of crossing over between chromosomes) for each homologous pair in meiosis (Figure 7.3.4).

Crossing over is the exchange of chromosomal material between members of a homologous pair of chromosomes during meiosis.

Non-sister chromatids are chromatids of paired homologous chromosomes, one from each parent. Paired non-sister chromatids form chiasmata (crossing points) during prophase I of meiosis.

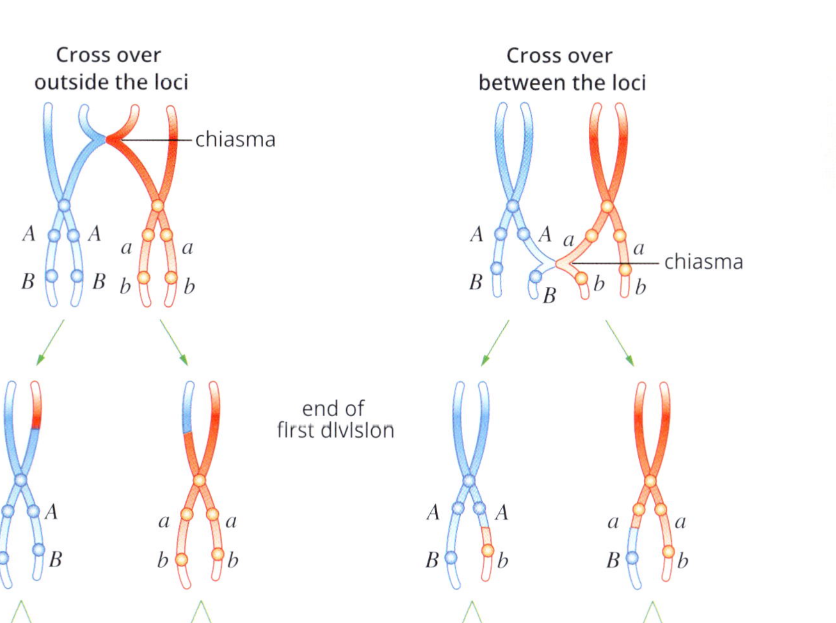

FIGURE 7.3.4 Possible outcomes of crossing over between chromosomes. When two genes, A and B, are located on the same chromosome, crossing over may occur outside the loci (outcome 1) or between the two loci (outcome 2). The gametes produced in these two situations (outcome 1 and outcome 2) are very different.

Figure 7.3.4 features two hypothetical genes, *A* and *B*, with their pairs of alleles *A*, *a* and *B*, *b*. The A and B loci are on the same chromosome.

If a cross is initiated between *AA*, *BB* and *aa*, *bb* parents, an F1 of *Aa*, *Bb* (all heterozygote) individuals will be produced. The diagram shows two possible outcomes of meiosis in *Aa*, *Bb* heterozygotes.

Genes are said to be linked when the percentage of recombinant gametes falls below 50%.

The percentage of recombination between two linked genes is correlated with their physical distance apart along the length of the chromosome.

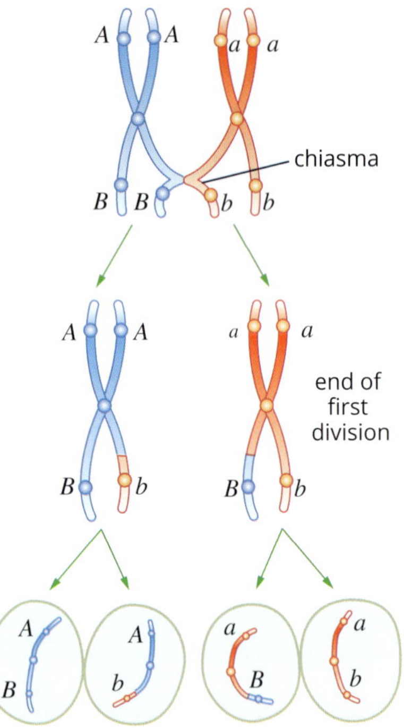

FIGURE 7.3.5 If the genes *A* and *B* are far enough apart on the same chromosome, there will be an average of one cross-over event between the genes in every cell, and 50% of the gametes will be recombinant. Therefore, the genes and their alleles assort independently.

WS 27

- Outcome 1 is that crossing over does not occur between the A and B loci, but occurs elsewhere on the chromosomes. In this case:
  - only gametes of allelic combinations *AB* and *ab* are formed, and they are formed in equal frequencies
  - these gametes are referred to as being of the **parental type** because they are the gametes that the *AA*, *BB* and *aa*, *bb* parents of the F1 would have produced.
- Outcome 2 is when crossing over occurs between loci A and B. Gametes containing allelic combinations *AB*, *Ab*, *aB* and *ab* are observed in equal frequency.
  - Some parental type gametes (*AB* and *ab*) are formed.
  - **Recombinant gametes**, *Ab* and *aB* are also formed.
  - Recombinant gametes carry a combination of alleles not observed in the *AA*, *BB* and *aa*, *bb* parents.

If the A and B loci are very close together, the probability of a random cross-over event occurring between them (outcome 2) is very low. If genes are close together, there will be fewer recombinant gametes and more parental gametes produced (outcome 1). The closer the two genes are together, the more rare the recombinant gametes will be.

If genes are so far apart (on the same chromosome) that close to 50% of the gametes are recombinants, then independent assortment is observed. If the percentage of recombinant gametes is less than 50%, the two genes are considered to be linked (Figure 7.3.5).

Recombination and distance between linked genes

By measuring the percentage of recombinant gametes produced by an F1 heterozygote when genes are linked, it is possible to estimate the distance between the two genes. The farther apart two genes are on the chromosome, the more frequently crossing over will occur and the higher the observed percentage of recombination. By repeating such measurements for different pairs of genes, the position of any identifiable gene on a particular chromosome can be found. This process is called **gene mapping**.

CASE STUDY **ANALYSIS**

Cystic fibrosis and linkage mapping

Cystic fibrosis (CF) is an inherited disorder that affects the respiratory and digestive systems. It can significantly shorten the lifespan of people with the condition. In a person with cystic fibrosis, the mucus glands secrete thick, sticky mucus, which clogs the airways, leading to breathing difficulties, respiratory infections and lung damage (Figure 7.3.6). The mucus also affects the pancreas, inhibiting the release of important digestive enzymes, which causes a range of nutritional problems. There is currently no cure for the disorder, but modern treatments are continuing to improve life expectancy for those with cystic fibrosis.

The symptoms of cystic fibrosis were first identified in 1938. Finding the gene responsible was a difficult task because its protein product was not known at the time and the gene could have existed on any of the 23 human chromosomes.

In the 1980s, researchers conducting linkage analysis tracked and mapped five genes that were linked to the gene that causes CF (the *CFTR* gene). The data showed that the *CFTR* gene was located on the long arm of chromosome 7. The gene was subsequently cloned in 1989 and its gene product (protein) was identified as a membrane chloride channel protein in 1992. This protein regulates the movement of salt in and out of cells. Because the gene is faulty, the regulation of salt movement is inefficient and leads to a build up of salt in the cells, which causes the production of thick mucus.

It is now known that cystic fibrosis is an autosomal recessive disorder (Figure 7.3.7). It is the most common genetic life-threatening disorder in Australia; more than one million Australians carry a copy of the faulty *CFTR* gene.

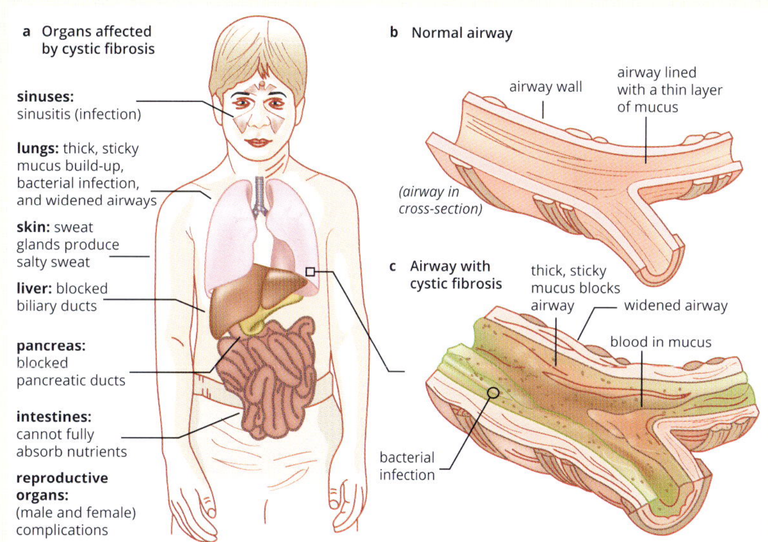

FIGURE 7.3.6 (a) The effects of cystic fibrosis on the systems of the human body and (b) a normal airway compared to (c) the airway of someone with cystic fibrosis

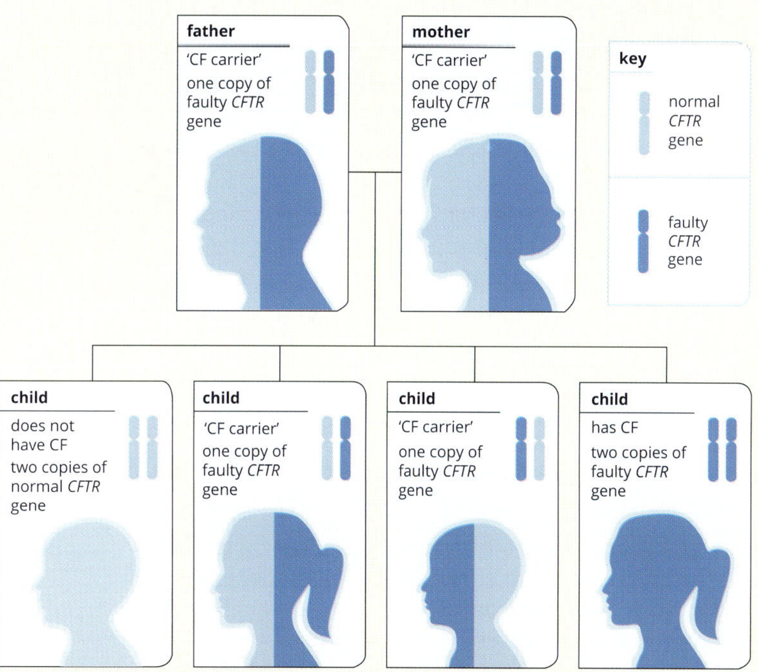

FIGURE 7.3.7 Autosomal recessive inheritance of cystic fibrosis. Regardless of their biological sex, an individual has a 25% chance of inheriting cystic fibrosis if both their parents are carriers of the *CFTR* gene.

Analysis

1 A couple decide that they would like start a family. They have both undergone genetic testing and are both carriers of the faulty *CFTR* gene. What is the chance that their child will be born with CF? Draw a Punnett square to support your answer.

2 The couple's first child is born with CF. After adjusting to life with a child with CF, they decide to have another baby.
 a What is the probability that their second child will have CF?
 b What is the probability that both children will have CF?

7.3 Review

SUMMARY

- A dihybrid cross is a cross between two individuals that carry alleles for different traits at two genetic loci.
- A phenotypic ratio approximating 9 : 3 : 3 : 1 will be observed in the F2 generation of a dihybrid cross if the two traits are each separately controlled by a single autosomal gene with two alleles, where for each trait one phenotype is dominant and the genes assort independently.
- Independent assortment occurs because the segregation of one pair of homologous chromosomes (and the alleles they carry) in meiosis does not influence the segregation of other homologous pairs of chromosomes.
- Linkage is the tendency for two or more genes located on the same chromosome to be inherited together.
- Genes are 'linked' when the percentage of recombinant gametes falls below 50%.
- Recombinant gametes carry a combination of alleles not observed in the parents.
- The percentage of recombinant progeny can be used to estimate the distance between two genes on chromosomes.

KEY QUESTIONS

Knowledge and understanding

1 What is the name of Mendel's second law, and what does it state?

2 Determine the possible gamete genotypes produced by meiosis in a heterozygote with the genotype *AaBb*, without linkage.

3 Complete the sentences below about the principle of linkage.
The tendency for two or more ________ located close together on the same _______ is that they are inherited _______. The _______ the _______ are to each other, the more likely they are to be _______ together—this is known as _______.

4 State whether each of the following statements is true or false.
   a Genes are linked when the percentage of recombinant gametes is above 50%.
   b The effects of crossing over can be seen in offspring.
   c If A and B loci are very close together on the same chromosome, the probability of a random cross-over event is very low.

5 Sheep blowfly chromosome 5 carries genes for resistance to the insecticide dieldrin (gene *D*). The same chromosome carries a gene called furrowed eyes (*F*).

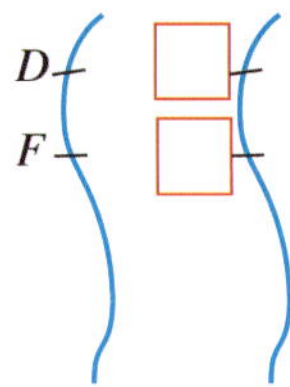

   a Add allele symbols to the unlabelled chromosome to show a fly that is heterozygous at both loci.
   b What combinations of alleles will be present in gametes if crossing over does occur?
   c Construct a Punnett square for a cross between a fly that is heterozygous at both loci and a fly that is homozygous recessive at both loci.
   d If dieldrin resistance is a dominant trait and furrowed eye is recessive, what proportion of the offspring from part **c** with normal (wild type) eyes are resistant to the chemical dieldrin?

Analysis

6 A domestic fruit plant has phenotypes of tall or dwarf, and variation in fruit texture of smooth or furry. The tall phenotype is dominant to the dwarf, and the smooth fruit is dominant to the furry. A fruit grower is trying to establish a pure-breeding stock of tall plants with smooth fruit. To identify the homozygous specimens in her crop, she crosses dwarf, furry-fruited plants with tall, smooth-fruited ones. The results of one cross are set out in the table.

Fruit plant cross results

Phenotype	Frequency
tall plant, smooth fruit	88
tall plant, furry fruit	10
dwarf plant, smooth fruit	8
dwarf plant, furry fruit	94

The fruit grower was surprised at the proportions of the different offspring, as she was expecting a 1 : 1 : 1 : 1 ratio.

a Account for the results of the cross.

b Explain in what circumstances the expected 1 : 1 : 1 : 1 ratio might have been achieved.

c In terms of the fruit grower's aim in conducting this cross, what valuable information has been gained from these results?

Chapter review

07

OA ✓✓

KEY TERMS

allele
autosomal dominant
autosomal recessive
autosome
carrier
chiasma (pl. chiasmata)
chromatid
codominance
crossing over
dihybrid cross
dominance (adj. dominant)
F1 generation
F2 generation
gene
gene mapping
genetic cross
genotypic ratio
hemizygote
(adj. hemizygous)
heterogametic
heterozygote
(adj. heterozygous)
homogametic
homologous chromosome
homozygote
(adj. homozygous)
hybrid
incomplete dominance
law of dominance
law of independent assortment
law of segregation
linkage
linkage group
locus (pl. loci)
monohybrid cross
mutant
parental type
pedigree analysis
pedigree chart
phenotypic ratio
Punnett square
recessive
reciprocal cross
recombinant gamete
sex-limited inheritance
sex-linked inheritance
test cross
true-breeding
wild type
X-linked
Y-linked

REVIEW QUESTIONS

Knowledge and understanding

1 Explain the difference between the genotype and phenotype of an individual.

2 Use an example to explain how two organisms can have the same phenotype but different genotypes.

3 Explain the difference between a monohybrid and a dihybrid cross.

4 Explain what a test cross is, outlining its purpose.

5 Recall the term given to the first generation of offspring from a test cross.

6 What do 'carried on the X chromosome' and 'occurs more in males than females' suggest?

A a monohybrid cross
B a dihybrid cross
C Mendel's experiments
D sex-linked inheritance

7 Explain why males are more likely to exhibit the phenotype of an X-linked trait than females.

8 Explain why Y-linked traits are not considered to be recessive.

9 In mice, black coat colour is dominant to white coat colour. Calculate the expected genotypic and phenotypic ratio for a cross between two heterozygotes. Use appropriate notation.

10 Draw the symbol that is commonly used to represent the following individuals within a pedigree chart.

a affected female
b male carrier of an autosomal trait
c female carrier of an autosomal trait
d female carrier of an X-linked trait
e male identical twins

11 Recall the name given to the point at which two chromosomes cross during meiosis.

A histone
B chiasma
C homologous chromosome
D centromere

Application and analysis

12 The diagram below represents a linkage group on a chromosome from a common crop plant. R, S and T represent different loci on the chromosome. MU is map units (the distance between the loci).

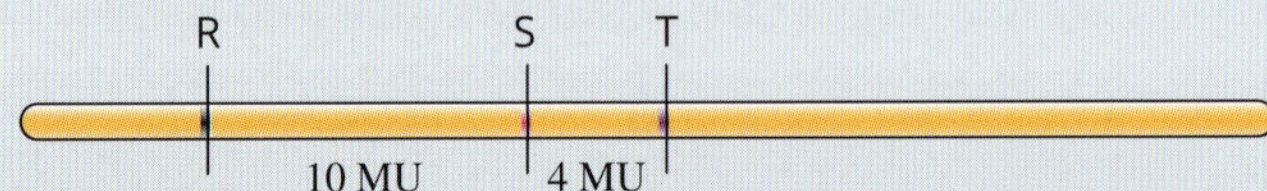

Would you expect the greatest percentage of recombination to occur between R and S, S and T, or R and T? Explain your reasoning.

13 Use the key terms that relate to different sorts of crosses (test cross, monohybrid cross and other crosses) to make a poster that distinguishes between them.

14 The shape of a human earlobe is determined by a single autosomal gene. Free lobe is dominant to attached lobe.

a Write appropriate allele symbols for this gene.
b How many genotypes are possible with respect to these alleles? How many phenotypes are possible?
c A homozygous man with free lobes and a heterozygous woman have children together. Show the genotypes and phenotypes possible in their children.
d Can two people with free lobes have a child with attached lobes? If so, what is the probability of their child having attached lobes? Explain your answer using a Punnett square.

15 Determine the probability of Robert, who has blood type A, and Lee, who has blood type B, having a baby of blood type O. Assume Robert and Lee are heterozygous. Include a Punnett square in your answer.

16 In cats, the allele for orange coat and the allele for black coat show a codominant pattern of inheritance. The phenotype of heterozygous individuals is called tortoiseshell. The gene for coat colour is found on the X chromosome. Sex determination in cats is the same as in humans. Draw a Punnett square to determine the genotypes and phenotypes of the offspring of a tortoiseshell female and a black male.

17 The following pedigree chart illustrates the pattern of inheritance for a particular characteristic.

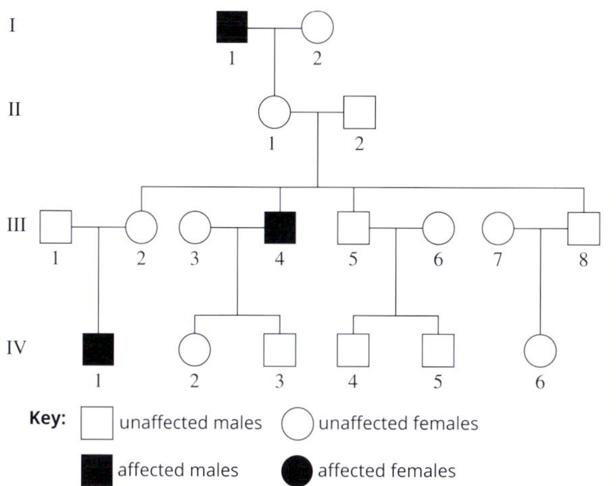

a i State the mode of inheritance for this characteristic.
 ii Describe the evidence that suggests this mode of inheritance.

b Use appropriate notation to assign genotypes to individuals I-1, II-1, III-2, III-4, III-5 and IV-2.

c State the possible genotypes and phenotypes in the children fathered by individual IV-1 if his partner has no family history of the trait.

18 The following pedigree chart shows the inheritance of albinism.

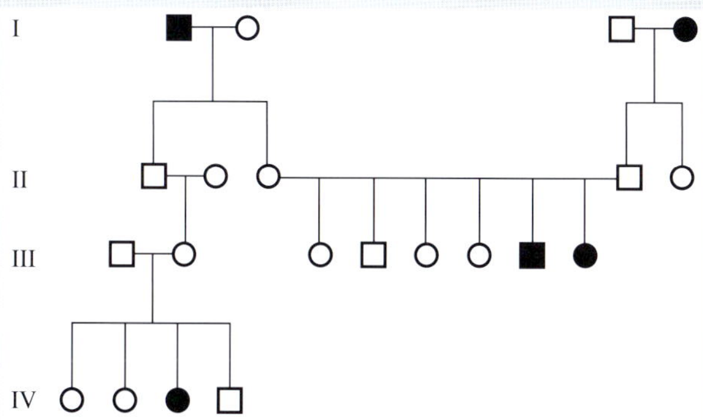

What is the most likely mode of inheritance of the condition? Explain your choice.

19 A genetics student undertakes a study of inheritance patterns for feather colour in domestic chickens. The student observes the following:

- Matings between black-feathered adults always results in black-feathered offspring.
- Matings between white-feathered adults always results in white-feathered offspring.
- Matings between black-feathered adults and white-feathered adults produces only blue-grey–feathered offspring.
- Matings between blue-grey–feathered adults results in black, blue-grey and white offspring in a ratio of 1 : 2 : 1.

a Describe the mode of inheritance of this trait. Outline the evidence that leads you to this conclusion.

b How many genes and alleles control this trait? Outline the evidence that leads you to this conclusion.

c Use appropriate notation to set up a model that explains this student's observation.

20 The graph below represents data from an F1 generation. Use the graph to determine the genotypes of the parental population. Dashes in the genotypes represent more than one allele (for example, *B* or *b*).

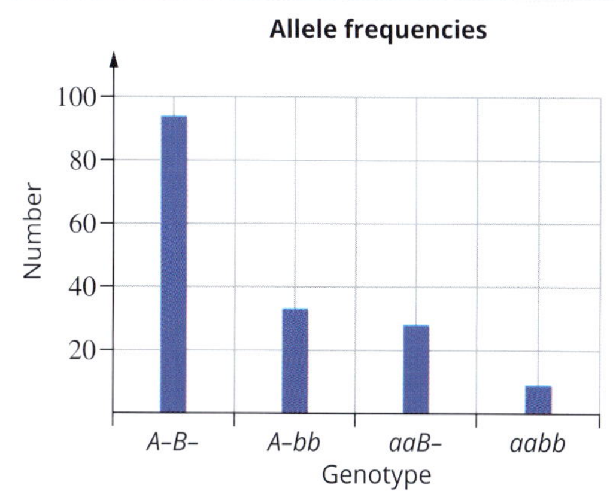

UNIT 2 • Area of Study 1

REVIEW QUESTIONS

How is inheritance explained?

Multiple-choice questions

1 How many autosomes are there in a human sperm?
- A 1
- B 22
- C 23
- D 44

2 What is the difference between the X chromosome and Y chromosome in humans?
- A The X chromosome is much shorter.
- B Many genes found on the X chromosome are absent from the Y chromosome.
- C Both chromosomes carry the same genes but the loci of the genes are different.
- D Only the X chromosome determines biological sex.

3 Identify which one of the following statements about genes or alleles is correct.
- A Alleles randomly segregate during meiosis.
- B Genes randomly segregate during meiosis.
- C Alleles are alternative codes for genetic information at defined loci on homologous chromosomes.
- D Gene and allele mean the same thing, with genes being an older term and alleles the more modern term.

4 Which of these correctly lists a sequence of events in meiosis?
- A crossing over occurs → chromosomes replicate → cytokinesis occurs
- B chromosomes replicate → haploid daughter cells form → chromatids separate
- C chromatids separate → chromosomes line up at the equator → cytokinesis occurs
- D pairing of homologous chromosomes → chromosomes replicate → crossing over occurs

5 Select the genotype that shows alleles for a heterozygous trait.
- A *Bb*
- B *AA*
- C *CD*
- D *Cd*

6 To make a karyotype, which phase of cell division is photographed?
- A metaphase of mitosis
- B metaphase I of meiosis
- C anaphase of mitosis
- D anaphase I of meiosis

7 What information is evident in a karyotype?
- A only the size of the chromosomes
- B only the gene mutations of the chromosomes
- C the size of the chromosomes and the gene mutations of the chromosomes
- D the size of the chromosomes, the gene mutations of the chromosomes and the age of the individual

Use the information below and in the pedigree chart to answer questions 8, 9 and 10.

The ability to taste a particular chemical, PTC, is controlled by one gene. The pedigree chart shows the transmission of this gene in a family.

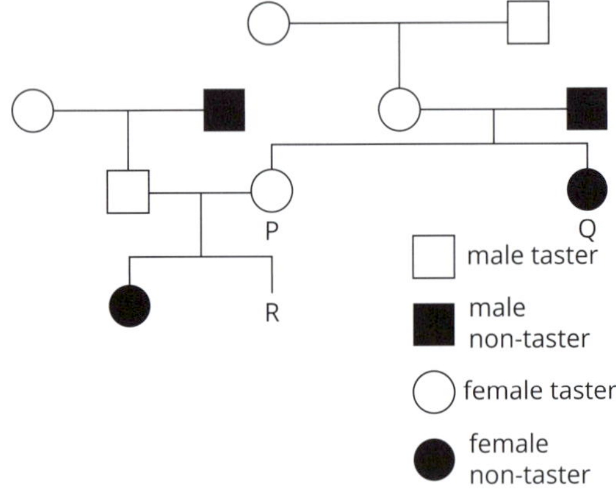

8 Determine the mode of inheritance of the tasting allele.
- A Y-linked
- B X-linked recessive
- C autosomal dominant
- D autosomal recessive

9 Determine the probability that child R is a taster.
- A 0.13
- B 0.25
- C 0.38
- D 0.75

10 Infer the genotypes of females P and Q.

	P	Q
A	*tt*	*tt*
B	*tt*	*Tt*
C	*Tt*	*tt*
D	*Tt*	*Tt*

11 The cross-over percentage between linked genes *P* and *Q* is 40%, between *Q* and *R* is 20%, between *R* and *S* is 10%, between *P* and *R* is 20%, and between *Q* and *S* is 10%. Identify the sequence of genes on the chromosome.

A *P, Q, R, S*

B *P, S, R, Q*

C *P, Q, S, R*

D *P, R, S, Q*

Short-answer questions

12 Distinguish between the terms 'genome', 'gene' and 'allele'.

13 The figure below shows the karyotype of a young adult.

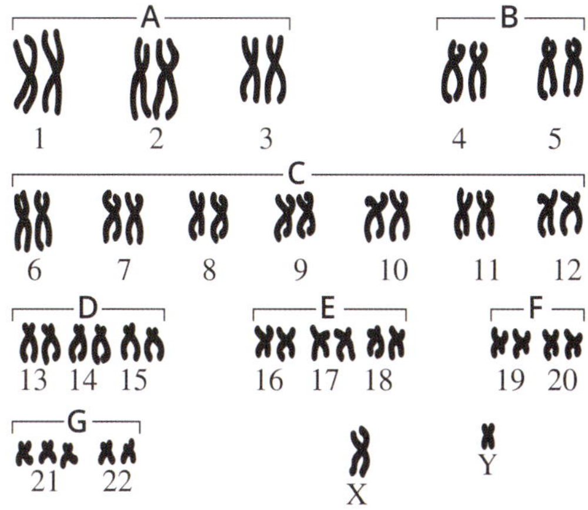

a **i** Identify the abnormality present in this karyotype.

ii Name the term or terms used to describe all similar conditions.

b Explain how such a condition could arise.

14 The figure below shows a karyotype of a human baby.

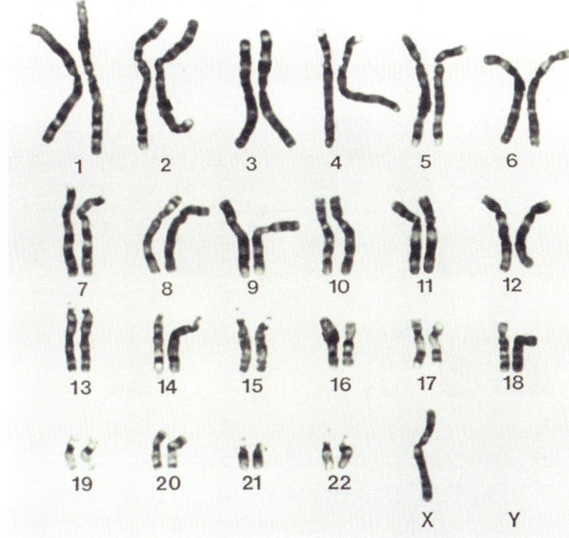

a The chromosomes are arranged in pairs. State the name given to chromosomes that contain the same gene loci.

b Explain the term 'diploid' and why all human somatic cells are diploid.

c Is there any chromosome abnormality with this baby? Explain your answer.

15 Describe the types of variation seen in chromosome structure and number between different species.

16 Meiosis is a type of cell division that results in the production of gametes. Meiosis is divided into phases, but before meiosis can start, a special phase called interphase is required. The following diagram is a summary of interphase and meiosis.

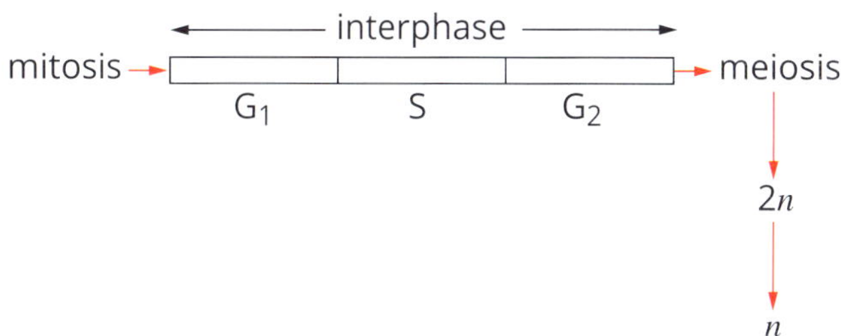

a The first cell division of meiosis is called the reduction division. Explain why.

b The following graph represents the changes in the amount of DNA in a cell as it goes through meiosis.

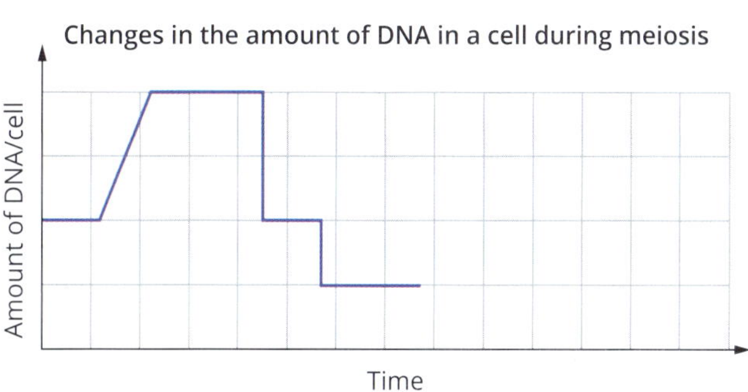

i On the graph, write the letter D where DNA replication is occurring, and the letter C where cytokinesis is occurring.

ii On the graph, extend the line to show the amount of DNA if fertilisation occurred and then the cell underwent one mitotic cell division.

17 The following diagram shows a stage during meiosis. The circles represent the cell and the structures within represent a homologous pair of chromosomes.

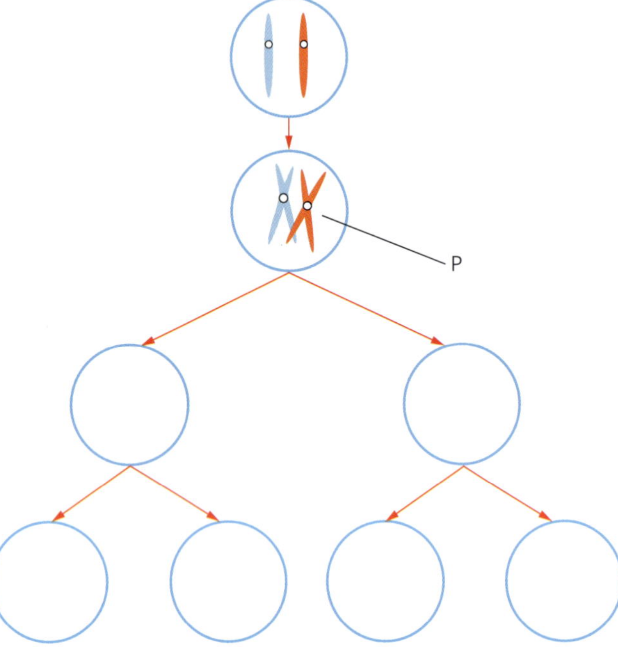

a Copy the diagram and then complete it by drawing the chromosomes in all of the cells.
b Explain what is happening at P.
c Distinguish between mitosis and meiosis.
d Explain how meiosis promotes genetic diversity in a species.

18 In bees and wasps, males of the species are haploid. The males develop from unfertilised eggs, yet the male offspring of the same female are not all identical.
a Explain, with reference to the processes occurring during egg formation, how the male offspring can be genetically different.
b Draw diagrams to show the different genotypic outcomes for two genes next to each other on the same chromosome during the process of egg formation, where crossing over occurs and when it does not occur.
c Explain whether the sperm produced by a male bee has genetic variation.

19 The snowshoe hare's fur is brown in summer and white in winter, as shown below. The change in fur colour is caused by epigenetic factors.

a Outline what is meant by epigenetic factors, using the snowshoe hare as an example.
b i List three factors that can cause epigenetic effects in humans.
ii Alzheimer's disease results from damage to brain tissue caused by the formation of abnormal proteins. It has been shown through familial studies to have a genetic component. However, studies of identical twins have shown that one twin may develop Alzheimer's while the other twin does not.
Explain how epigenetics can account for this.
c Besides epigenetic factors, what other factors can influence an organism's phenotype?

20 The pedigree chart below shows the hypothesised inheritance of oligospermia. Oligospermia results in low sperm count and sometimes, morphological changes to sperm, which reduces fertility. Shaded individuals have the oligospermia condition.

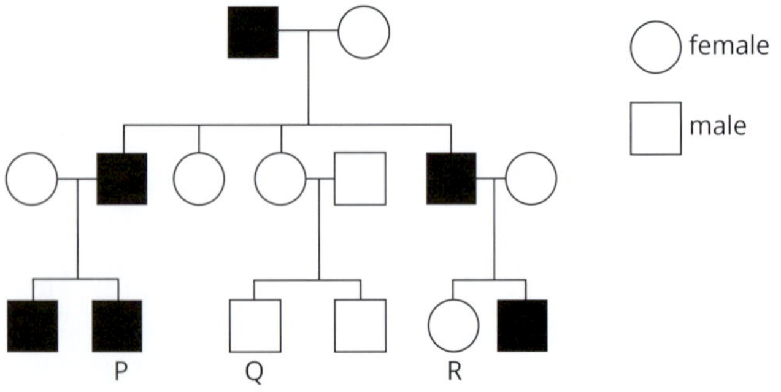

a State the type of inheritance shown.
b Assign an appropriate symbol and state the genotype for individuals P and R.
c Calculate the probability that any of individual Q's future children would have oligospermia. Use a Punnett square to show your working.

21 In the garden pea, *Pisum sativum*, the phenotype for tall plants (allele represented as *T*) is dominant over the phenotype for short plants (*t*). The phenotype for round seeds (*R*) is dominant over the phenotype for wrinkled seeds (*r*). The alleles are unlinked. Pure-breeding tall plants with round seeds were crossed with pure-breeding short plants with wrinkled seeds.

a State the dihybrid genotypes and the phenotypes of the F1 individuals.

The F1 plants were then crossed with plants that had the genotype *ttrr*. The table below shows the results obtained in the F2 generation.

Phenotype	Frequency
tall plants with round seeds	22%
short plants with round seeds	26%
tall plants with wrinkled seeds	25%
short plants with wrinkled seeds	27%

b Draw up a Punnett square to show the expected ratio of dihybrid phenotypes in the F2 generation.

c Are the results listed in the table exactly as you expected? If not, suggest an explanation for the differences.

d Outline an experiment to investigate the genotype of a tall *P. sativum* whose genetic history is unknown.

22 Pedigree analysis is a useful tool for determining patterns of genetic inheritance. In pedigree charts, a standard set of symbols and conventions is applied.

a Name what each symbol below represents in a pedigree chart.

i 

ii 

iii 

iv 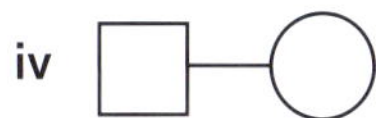

v 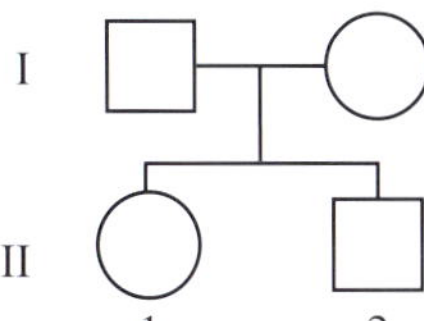

vi 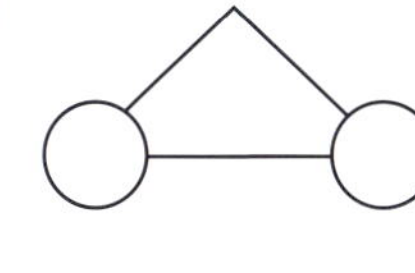

vii 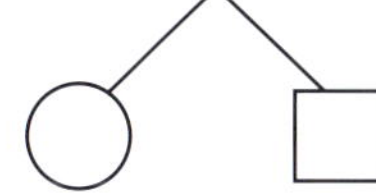

The pedigree chart below is for a family in which colour vision deficiency is present.

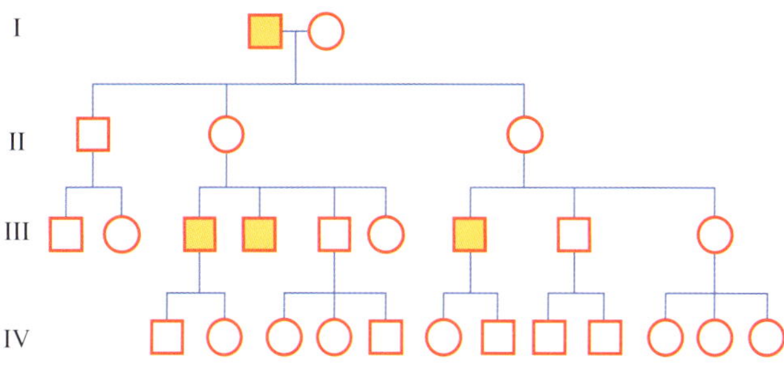

b Draw a key for the chart so that each individual can be identified by their sex and if they are affected by colour vision deficiency.

c What type of inheritance pattern can be inferred from the pedigree chart? Support your answer by giving at least two reasons for your choice.

23 Boys can inherit a sex-linked allele carried on the X chromosome (X^c) that causes the recessive phenotype red–green colour vision deficiency.

a Explain the following terms:

i X-linked allele

ii recessive phenotype.

b Write down the possible genotypes for red–green colour vision deficiency in:

i men

ii women.

c A boy inherited red–green colour vision deficiency from one of his grandfathers. Which of the boy's grandfathers (maternal or paternal) also had a colour vision deficiency? Explain your reasoning.

d A woman with red–green colour vision deficiency and an unaffected man had five children: three boys and two girls. The three sons and the elder daughter did not have children. The younger daughter married a man with normal colour vision, and they had four children: two boys and two girls. Draw a pedigree chart to illustrate the inheritance of the X-linked condition in this family. Use conventional symbols.

24 In humans, the adenomatous polyposis coli (*APC*) gene is located on chromosome 5. *APC* controls cell division and is also known as a tumour suppressor gene. Mutations of *APC* will cause a genetic disease called familial adenomatous polyposis (FAP).

a Clarify whether FAP is a sex-linked genetic disease.

b Half of the gametes produced by a person with FAP have an *APC* gene with the mutation. State whether FAP is a dominant or recessive phenotype. Justify your choice.

c A male who is heterozygous for FAP and an unaffected female are planning to have children together. Predict the possible phenotypes and genotypes of their children, showing your working.

25 The fruit fly (*Drosophila melanogaster*) is commonly used for genetic studies, one reason being that sex determination in *Drosophila* is the same as humans. Females are XX and males are XY. One common mutation in these fruit flies results in offspring that have an ebony (black) body and another mutation results in small (vestigial) wings, as shown below.

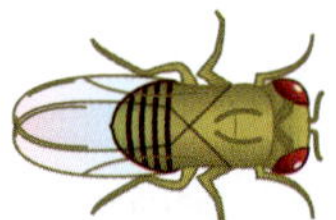
wild type

ebony body

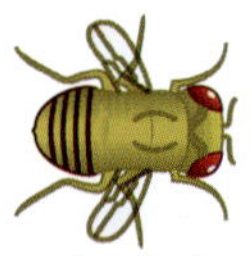
vestigial wings

Grey body phenotype (allele *G*) is dominant to ebony body (*g*), and normal wing shape (*N*) is dominant to vestigial wings (*n*). The genes for these traits are linked. The normal version of the fruit fly, with grey body colour and normal wing shape, is called the wild type.

Male flies, heterozygous for both grey body and normal wings, were mated with ebony-bodied and vestigial winged females. Five thousand offspring were examined for body colour and wing type. The following table shows the results obtained.

Offspring	Frequency
grey body, normal wings	36%
ebony body, vestigial wings	38%
grey body, vestigial wings	12%
ebony body, normal wings	14%

a Contrast the two terms: linked genes and unlinked genes.

b The genotype for the female parental flies can be represented as *ggnn*. Suggest a possible genotype designation for the male parentals.

c i Describe the process of recombination.

ii Does the experiment provide evidence for recombination of linked genes? Explain your answer.

iii Identify which offspring are the result of recombination in this cross.

A geneticist took a fly homozygous for wild type at both the body colour and wing shape loci, and crossed it with a fly that was homozygous for ebony body and vestigial wings. All of the resulting F1 offspring were wild type.

d Explain how this establishes that ebony body and vestigial wings are dominant or recessive.

e i Design a test cross involving one of the F1. Clearly identify the genotypes and phenotypes of the parents and offspring.

ii Identify the proportion of the offspring which will possess only one recessive phenotype.

26 In the 1860s, Gregor Mendel had no knowledge of DNA and chromosomes, nor did he use the word 'gene' in reporting his genetic experiments. Yet Mendel accurately deduced the basic principles of inheritance.

a i Name the experimental species used by Mendel for his breeding trials.

ii Outline how Mendel was able to determine an inheritance pattern from observing phenotypes without detailed knowledge of their genotypes.

b Sixteen years after his death, Mendel's work was rediscovered and its significance realised. Now he is known as the 'father of modern genetics'. Discuss two experimental principles used by Mendel that led to his success in understanding inheritance.

c In the early 1900s, T. H. Morgan followed Mendelian principles in test crosses with fruit flies and discovered a different form of inheritance when he crossed mutated white-eyed males with the wild type of homozygous red-eyed females. Eventually this was determined to involve genes not located on the autosomes.

i Identify this type of inheritance.

ii The XY system that determines sex in humans also operates in fruit flies. Explain why gene linkage on these allosomes does not produce Mendelian ratios in a test cross.

iii The diagram below was developed much later to depict the inheritance pattern in fruit flies discovered by T. H. Morgan. Interpret how the diagram explains this type of inheritance.

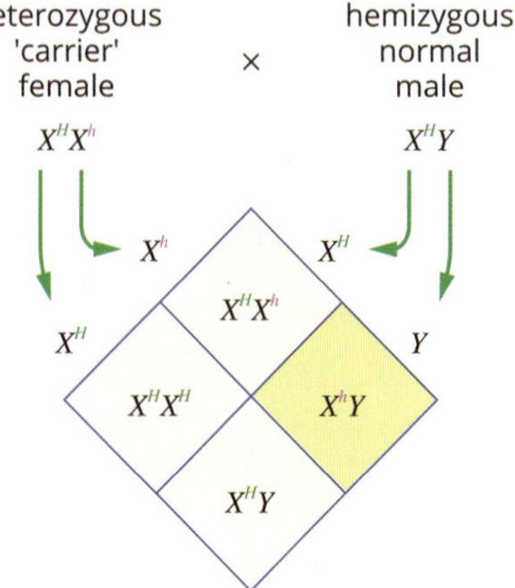

27 Genotypes and phenotypes do not always display a simple dominant and recessive pattern of inheritance.

  a An inheritance pattern called incomplete dominance is displayed in the flower colour of snapdragon plants. The flower phenotype can be red, white or pink in colour as shown below. Draw a Punnett square diagram using the symbols R_1 and R_2 to support your explanation of incomplete dominance for flower colour in snapdragons.

  b The human blood groups are an example of autosomal codominant inheritance.
    i Clarify the meaning of this statement in relation to human blood types. Use specific genetic terms in your answer.
    ii Use a table to identify the six possible genotypes and four phenotypes of the ABO blood group system.

  c Distinguish between codominance and incomplete dominance by referring to the examples of human blood types and snapdragon flower colour.

28 The language of genetics is very specialised. Check your understanding by matching each term in the table below with its correct definition.

1	complementary base pairs		A	adenine with thymine and cytosine with guanine in a DNA molecule
2	chromosome		B	sex cell that carries a different combination of alleles from the parent
3	chromatid		C	chromosome that determines biological sex
4	centromere		D	chromosome that is not a sex chromosome
5	chiasma		E	diploid individual with different alleles for a gene
6	phenotype		F	diploid individual with only one copy of some alleles
7	genotype		G	diploid individual with the same alleles for a gene
8	autosome		H	exchange of sections between chromatids of homologous chromosomes
9	allosome		I	genes close together on a chromosome that are usually inherited together
10	haploid		J	matching chromosomes with the same genes in the same loci
11	diploid		K	observable traits of an individual organism
12	homozygous		L	offspring of a cross between members of F1 generation
13	heterozygous		M	offspring of a cross between members of the parent generation
14	hemizygous		N	one of two copies of a chromosome duplicated during cell division
15	homologous		O	having one set of chromosomes
16	F1 generation		P	point where crossing over occurs between chromatids at prophase I
17	F2 generation		Q	point where sister chromatids link together and attach to the spindle
18	linkage group		R	sequence of genes on a DNA strand coiled around histone proteins
19	crossing over		S	combination of alleles at a locus
20	reciprocal cross		T	used to determine if inheritance is sex-linked
21	recombinant gamete		U	having two sets of chromosomes

WS 29 EQ U201

Reproductive strategies

Organisms need to reproduce in order for their species to survive. In this chapter you will learn how organisms can reproduce asexually and sexually, as well as the advantages and disadvantages of each method of reproduction.

You will learn about the methods used to clone plants and animals and some of the applications of these methods.

Issues associated with cloning will also be investigated.

Key knowledge

- biological advantages and disadvantages of asexual reproduction **8.1**
- biological advantages of sexual reproduction in terms of genetic diversity of offspring **8.1**
- the process and application of reproductive cloning technologies. **8.2**

WS 30

OA ✓✓

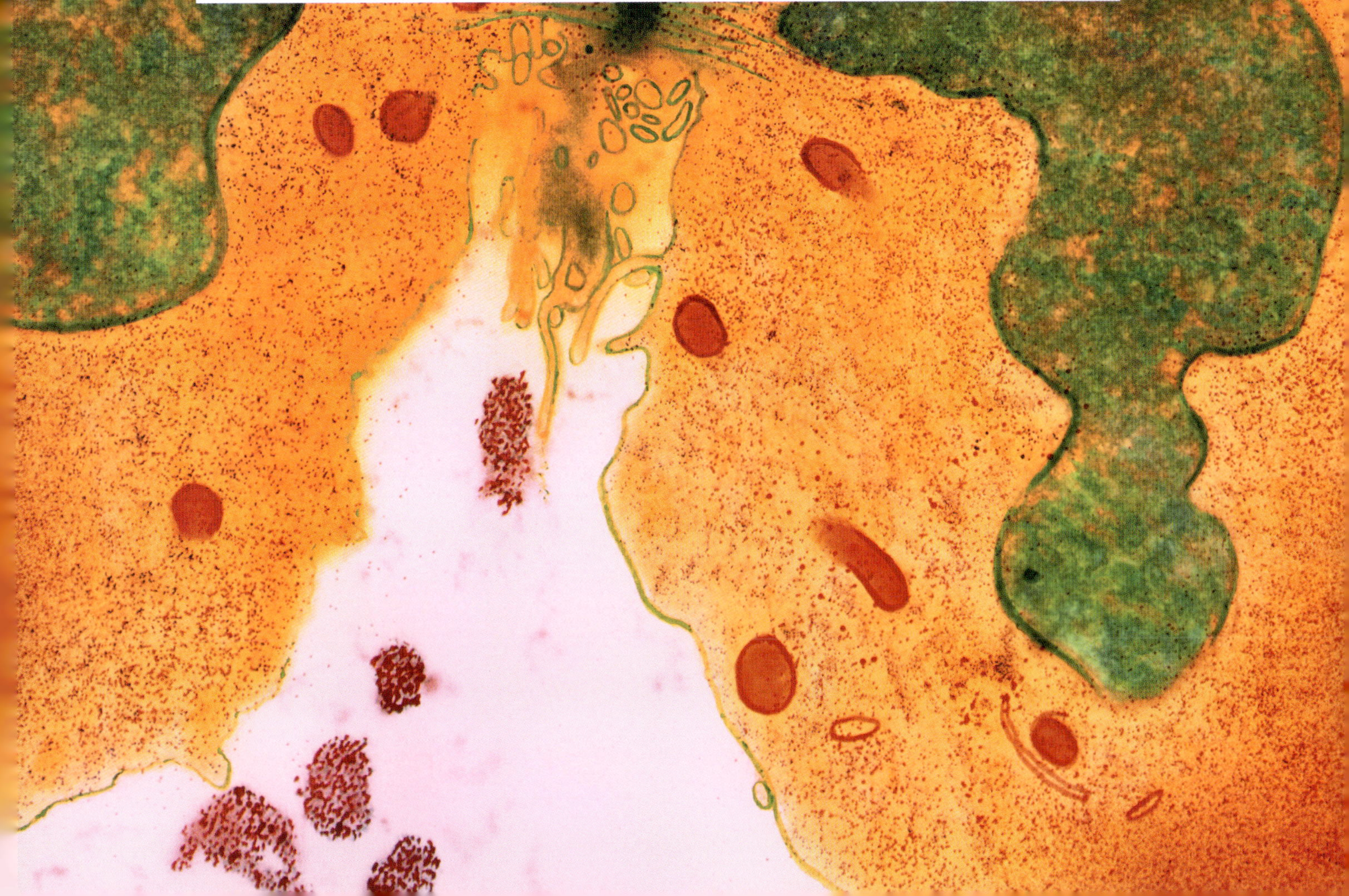

8.1 Sexual and asexual reproduction

Individual organisms do not live forever. The continuity of life from generation to generation is the result of reproduction. Reproduction is one of the distinctive characteristics of living organisms. Reproduction may be sexual, asexual or a combination of both.

Sexual reproduction involves the union of male and female sex cells to form a unique individual. Most animals, including humans, reproduce sexually, although some animals can reproduce asexually as well. Some unicellular organisms are also capable of sexual reproduction. Many plants and most fungi have a complicated alternation of generations, which sees them reproduce using both sexual and asexual reproduction.

Asexual reproduction is the production of offspring from just one parent. Asexual reproduction occurs in bacteria and fungi as well as in many plants and some animals.

The different types of reproduction and the conditions under which each method is most likely to assist a species to survive are explored throughout this chapter.

SEXUAL REPRODUCTION

In this section you will learn about how an offspring of two parents (Figure 8.1.1) has a unique genetic identity, and about the biological advantages of sexual reproduction.

FIGURE 8.1.1 Blue ringtail damselflies (*Austrolestes annulosus*) form a mating 'wheel' when mating. The male (top) is holding the female's neck, while the female has moved her abdomen towards the male's genitalia to receive his sperm.

> Haploid cells contain one set of chromosomes.
>
> Diploid cells contain two sets of chromosomes.

Features of sexual reproduction

All sexually reproducing organisms go through a cycle involving both haploid (n) and diploid cells ($2n$). Diploid cells divide to produce haploid cells and then haploid cells fuse, resulting in a new diploid cell. In many species, including humans, this process takes place within the one individual.

In some plants, such as ferns, the haploid and diploid stages occur in different individuals. There is a **gametophyte**, which is haploid and produces gametes by mitosis, and a **sporophyte**, which is diploid and produces spores by meiosis (Figure 8.1.2). This alternation of haploid and diploid individuals in the life cycle is known as **alternation of generations**.

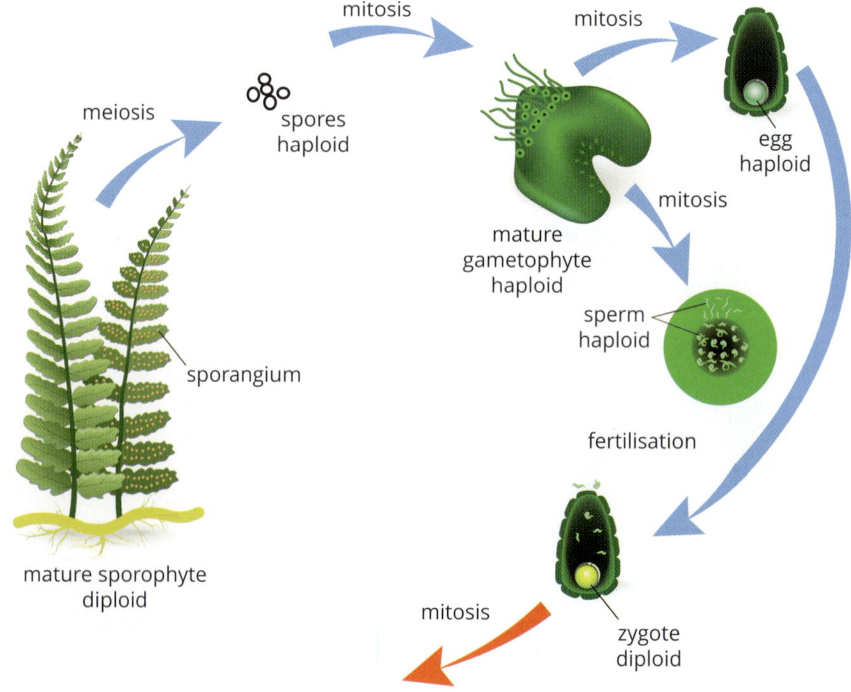

FIGURE 8.1.2 Ferns undergo true alternation of generations, with separate spore-producing and gamete-producing individuals.

Note that although the spores produced by ferns are haploid, they are not identical because crossing over and independent assortment during meiosis makes all of the spores different. As a result, each of the gametophytes is genetically different but all the ova (eggs) and sperm produced by a single gametophyte are genetically identical, as they are produced by mitosis. For species using this method, such as ferns and many mosses and liverworts, the major challenge to reproducing sexually is that the gametes may struggle to meet each other as they require a moist environment for the sperm to swim to the ova and for the gametes not to dry out before fertilisation occurs. For this reason, these types of plants can only reproduce sexually in moist environments, such as rainforests, so this limits their spread.

Introducing variation—sexual reproduction

Sexual reproduction involves the fusion of two cells, called gametes, to produce a new cell called a zygote. Meiosis is critical to sexual reproduction as it forms genetically unique gametes. When the gametes fuse to form the zygote, the individual that results will have similarities in genetic content to the parents, but the offspring are always genetically different from the parents and from each other (except for identical twins). Here lies much of the advantage and disadvantage for any species using this form of reproduction. However, the widespread occurrence of sexual reproduction in almost all eukaryotic organisms shows that the long-term benefits to the species far outweigh any costs to the individuals.

Advantages and disadvantages of sexual reproduction

The considerable benefit of sexual reproduction is evident from its widespread occurrence in almost all eukaryotic organisms. The most beneficial aspect of sexual reproduction is the genetic diversity that is introduced through gamete production and genetic recombination.

Crossing over and recombination during gamete production results in the formation of new combinations of alleles. When these new combinations of alleles are beneficial, the individuals possessing them will have a higher chance of surviving and reproducing, and the frequency of these alleles will increase in the population. Equally, if combinations of alleles are harmful, the individuals possessing them are less likely to survive and reproduce, and those alleles will decrease in frequency or be removed from the population. You will learn more about the role of genetic diversity in populations in Chapter 9.

Ultimately, genetic diversity within a population enables a species to survive and reproduce in varied and changing environments. If a population has high levels of diversity it is more likely that if a new selection pressure, such as a new disease or a new predator, is encountered there will be some individuals that have traits that enable them to cope with this new pressure. We can see this when a new disease affects human populations: some individuals have few or no symptoms, others have mild symptoms and for others the disease is serious, possibly resulting in death. This occurs because of the variation in human populations due to sexual reproduction.

In the long term, increased genetic diversity provides greater adaptability and evolutionary potential in changing conditions. The pool of genetic diversity in a population also facilitates the selection of beneficial traits and elimination of unfavourable traits, according to the survival and reproductive success of individuals. This process ultimately benefits the population, as those individuals that are most successful will reproduce, increasing beneficial genetic variants in the population.

However, sexual reproduction usually involves changes in an organism's way of life, and these changes are not without cost. Finding and competing for a mate can be energetically costly and risky. Some reproductive behaviours, such as the calling used by frogs to attract the attention of potential mates, might also attract the attention of predators. In some animals, mating leads to considerable, and potentially harmful, competition between males. For example, bighorn sheep not only risk injury but also expend considerable energy in the mating battles which they must undertake in order to reproduce (Figure 8.1.3).

BIOFILE

Hermaphrodites

Sexual reproduction usually requires two parents for the production of offspring. However, this is not the case for all species. Many plants and some animals, such as tapeworms, snails, earthworms and some fish, have both male and female reproductive organs in the same individual—they are hermaphrodites. Some species are able to self-fertilise, whereas others require a partner. Even when hermaphrodites self-fertilise, the recombination of genes and random assortment of chromosomes that occurs during meiosis means that the offspring are always genetically unique.

Earthworms are hermaphrodites. They have both male and female reproductive systems in their bodies.

Sexual reproduction produces genetic diversity in populations, which allows them to adapt to changing environments.

FIGURE 8.1.3 Big horn sheep fight each other for the right to mate. These fights can lead to injury or death of the competitors and involve considerable expenditure of energy.

For reproduction, some of the food resources of each parent must be used to produce gametes and to ensure that mature gametes are brought together at the right time of year. In other words, not all the food eaten by the parents is used to maintain their own body systems.

ASEXUAL REPRODUCTION

In this section you will learn about the different forms of asexual reproduction (Figure 8.1.4), and the biological advantages and disadvantages of this type of reproduction.

FIGURE 8.1.4 Pin moulds (*Mucor* species), like all fungi, reproduce asexually. This pin mould is growing on a tomato.

Genetically identical offspring

Clones are offspring that are genetically identical. Clones reproduced by asexual reproduction are also genetically identical to the parent.

An organism's phenotype is its overall appearance. Phenotype is influenced by an organism's genotype and its environment.

Asexual reproduction when used on its own produces new individuals in which each daughter cell receives a copy of every chromosome of the parent cell. The offspring are therefore **clones**—individuals that are genetically identical. Although offspring from asexual reproduction are genetically identical, they are not necessarily identical in appearance; environmental conditions also affect their growth and development. Remember that phenotype is influenced by genotype and environment.

The potato is an example of a plant that can reproduce asexually. A shoot grows from the buds or 'eyes' on the tuber (the starch-rich part of the potato that you eat). When planted, a shoot grows into stems above and below the soil. Roots grow from the underground stem, which eventually produces new tubers. Each tuber can then grow into a new potato (Figure 8.1.5).

FIGURE 8.1.5 Potato plants reproduce asexually through tubers.

Methods of asexual reproduction

Asexual reproduction is the most common method of reproduction for unicellular organisms. This is because there is no cell specialisation or differentiation in unicellular organisms, so there are no sex cells or reproductive organs. Many multicellular organisms also have the capacity to reproduce asexually. In multicellular organisms, the new individual arises from ordinary body cells, known as **somatic cells**. Asexual reproduction can be by:

- fission
- budding
- fragmentation
- spore formation
- vegetative propagation
- parthenogenesis.

Fission

Fission is the most common form of asexual reproduction among unicellular organisms, such as bacteria and protozoans. **Fission** occurs when a single parent cell divides into two or more approximately equal parts, each of which develops into a new organism. Fission can be binary fission (division of a single organism into two parts) or multiple fission (division of a single organism into more than two parts). Fission in unicelluar organisms (e.g. prokaryotes) mostly occurs by binary fission (Figure 8.1.6a) but some species also undergo multiple fission. You learnt about binary fission in prokaryotes in Chapter 3. Asexual reproduction by fission is also seen in some multicellular organisms, such as sea anemones (Figure 8.1.6b).

a

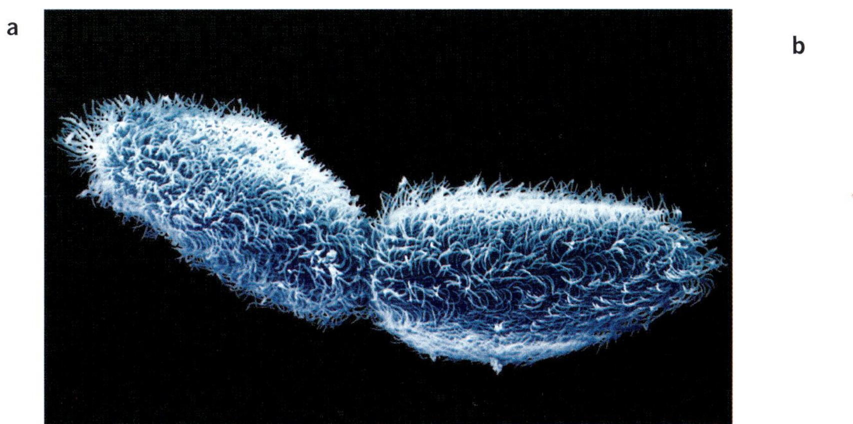

b

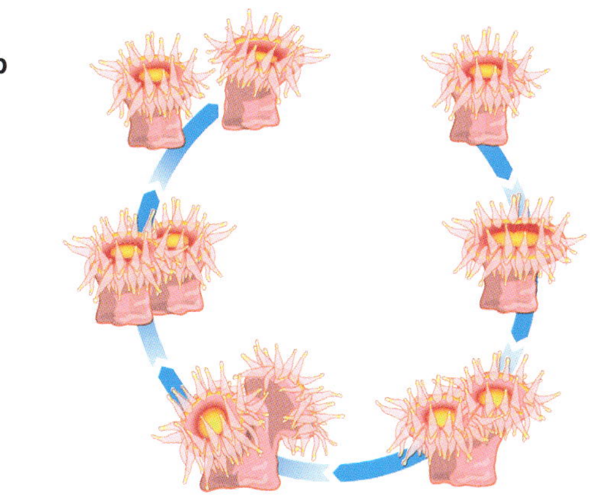

FIGURE 8.1.6 Asexual reproduction by fission is observed in (a) prokaryotes (binary fission) and (b) eukaryotes such as sea anemones

Budding

Budding of unicellular organisms such as yeasts is similar to fission, except that the division of the cytoplasm is unequal. The new individual arises as an outgrowth, or bud, from the parent. Budding also occurs in small multicellular animals such as hydra (Figure 8.1.7).

FIGURE 8.1.7 A hydra undergoing budding. The new individual is the bud on the left. When it detaches, it will be considered an individual organism.

Fragmentation

Fragmentation is similar to fission, but it happens in multicellular organisms. The body of the organism breaks into two or more parts, each of which regenerates the missing pieces to form a new, complete individual (Figure 8.1.8). Fragmentation is common in some flatworms, marine worms and echinoderms.

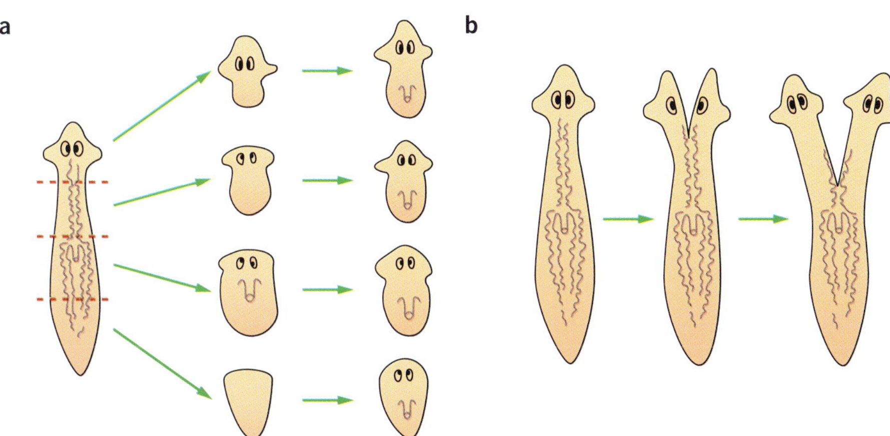

FIGURE 8.1.8 Fragmentation occurs when a part of the parent organism breaks off and grows into a new individual. Planarians (free-living flatworms) have a remarkable ability to regenerate in this way.

Spore formation

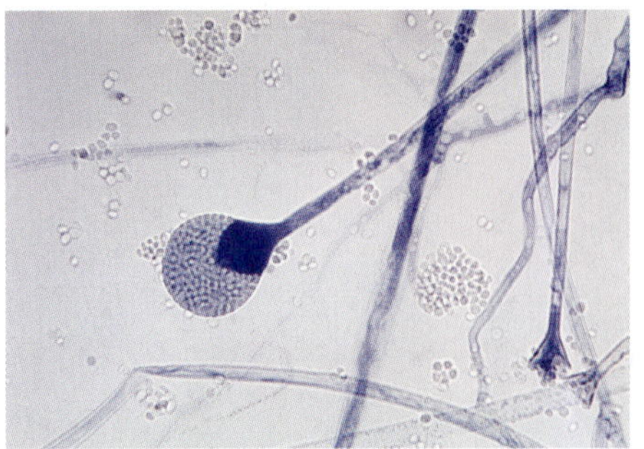

FIGURE 8.1.9 Light micrograph (LM) of sporangium of a mature *Mucor* fungus containing spores. When the wall of the sporangium disintegrates, the spores will be released. Broken sporangia can be seen in the lower right corner of the image, along with released spores in the background.

Spores are reproductive cells that are produced by bacteria, fungi, algae and plants. **Spore formation** can be a form of asexual or sexual reproduction. Spores that are produced asexually are often called **mitospores** and are formed by mitosis. The spores are cells that are encased in a protective coating that enables them to survive in unfavourable environments.

Some fungi produce a cluster of spores inside a structure called a **sporangium** (plural sporangia) (Figure 8.1.9). The spores are released when the sporangium wall disintegrates and are dispersed by wind or water. When a spore lands in a suitable environment it germinates, forming a new fungus. Spore formation and dispersal can rapidly increase the population of a species.

The clouds of green powder that come off the surface of a mouldy lemon are a particular type of asexually reproduced spore of blue mould (*Penicillium expansum*). These spores, produced by budding, are called **conidia** (Figure 8.1.10). In *Penicillium expansum* these spores are formed in long chains.

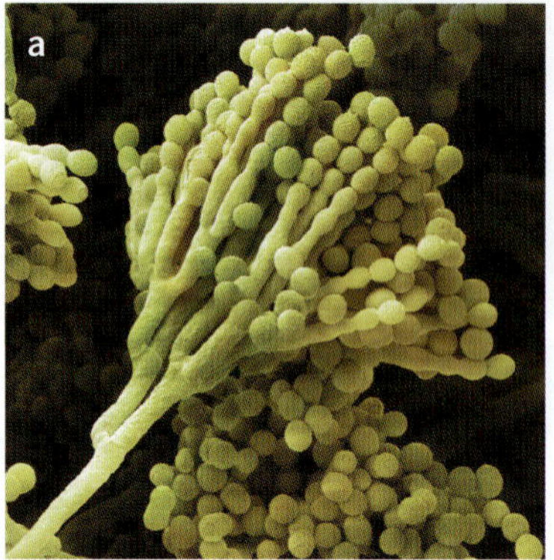

FIGURE 8.1.10 (a) Coloured scanning electron micrograph (SEM) of conidiophores of blue mould (*Penicillium expansum*), which produce spores called conidia by budding. (b) Blue mould on a rotting lemon. The spores are visible on the surface of the fruit; they are white at first but later turn blue.

Vegetative propagation

Many plants, including flowering plants, can reproduce asexually by **vegetative propagation**—the growth of specialised plant tissues that form new plants when separated from the parent plant. Naturally occurring vegetative propagation may arise from many parts of the plant, such as the leaves and underground stems. Some types of vegetative propagation in plants are described below.

Rhizomes are underground stems that branch and give rise to new shoots and roots. Well-known examples of plants with rhizomes are couch grass, irises and ginger (Figure 8.1.11a).

Stolons are like rhizomes, but they grow above ground. Examples include spider plants and strawberry plants (Figure 8.1.11b).

Tubers are swollen underground stems with buds ('eyes') that easily grow into new plants, such as sweet potatoes and cassava (Figure 8.1.11c).

FIGURE 8.1.11 Some types of vegetative propagation. (a) Rhizome formation in ginger (*Zingiber officinale*). (b) Stolon (runner) formation in the garden strawberry (*Fragaria* species). (c) Tuber formation in cassava (*Manihot esculenta*), an important vegetable in Africa and Asia

Bulbs and **corms** produce lateral buds that also develop into new plants. Examples include daffodil and hyacinth bulbs (Figure 8.1.12a), taro and gladioli corms (Figure 8.1.12b).

Plants are able to reproduce sexually by flowering and producing seeds, or forming spores, as well as asexually by vegetative propagation. An advantage of vegetative propagation is it enables a rapid increase in the number of plants growing in a favourable area so that they outcompete, or displace, neighbouring species. In contrast, seeds produced by sexual propagation may land in unfavourable conditions and fail to germinate. Potential disadvantages of vegetative propagation are competition from sister and parent plants for resources, and lack of genetic diversity to protect the population against disease or changing environmental conditions.

a

FIGURE 8.1.12 (a) A hyacinth bulb (*Hyacinthus orientalis*) undergoing vegetative propagation. If the growing bulb on the right is removed from the parent bulb, it can grow into a new individual. (b) This sword lily (*Gladiolus* species) has produced multiple cormlets, which can be separated from the corm to form many new individual plants.

Parthenogenesis

The development of an egg in the absence of fertilisation is an unusual form of asexual reproduction known as **parthenogenesis**, a Greek term meaning 'virgin birth'. Because it involves the development of an egg, parthenogenesis can occur only in females. Parthenogenesis is a normal part of the life cycle of some lizards, birds, insects (bees, wasps and ants), rotifers and nematodes.

In spiny leaf insects (Figure 8.1.13), eggs develop by parthenogenesis into females; no males are present in these populations. In these species there is an extra doubling of the chromosomes during egg development, resulting in a full clone of the mother. However, genetic recombination (see Section 6.1) can occur, increasing genetic diversity.

FIGURE 8.1.13 Female spiny leaf insects (*Extatosoma tiaratum*) can reproduce by parthenogenesis.

Advantages and disadvantages of asexual reproduction

Given the abundance of organisms that reproduce by asexual means, there must be significant advantages to asexual reproduction. Asexual reproduction is an efficient way to reproduce when environmental conditions are ideal. Asexually reproducing organisms are therefore commonly found in relatively stable and uniform environments to which they are well suited.

Asexual reproduction allows populations to rapidly expand and is beneficial in stable environments.

However, when environmental conditions are variable, asexually reproducing populations are at a disadvantage. Because these organisms are genetically the same, they will all respond to change in the same way. Lack of genetic diversity in a population means that there will be no unusual individuals that may be able to tolerate changed environmental conditions. As a group, they will either survive or die. Clones are offspring that are genetically identical. Clones produced by asexual reproduction are also genetically identical to their parent. Asexual reproduction has short-term benefits, enabling rapid population expansion, but the lack of genetic diversity in asexual populations limits their adaptability and evolutionary potential in the long term.

CASE STUDY **ANALYSIS**

Reproductive methods of the New Zealand mudsnail

The New Zealand mudsnail (*Potamopyrgus antipodarum*; Figure 8.1.14) is an invasive gastropod that has spread from New Zealand to Australia, Europe and North America by attaching itself to cargo ships. Once in a country, mudsnails are easily spread to new waterways by humans—transported on items such as recreational watercraft and camping equipment. In New Zealand, mature mudsnails can reach up to 12 mm in length, whereas mudsnails found in North America are usually 4–6 mm in length when mature (Figure 8.1.15). The mudsnails' ability to adapt to a wide range of environmental conditions make it a highly invasive species—they can tolerate temperatures of 0–34°C, depths of 4–45 m and a wide range in salinity. Outside its natural range, the mudsnail has no natural predators. As a result, numbers have increased dramatically and the mudsnail has disrupted river ecosystems by competing with native aquatic snails and insects, causing declines in the populations of native species.

Female mudsnails can reproduce sexually and asexually, using parthenogenesis. Offspring produced by parthenogenesis have been found to reach reproductive maturity faster than offspring produced by sexual reproduction. The ability of mudsnails to reproduce asexually gives them an advantage when colonising a new habitat—a single snail can start an infestation, and in some places their density may be over 500 000 individuals per square metre. Each female mud snail produces between 20 and 120 offspring in each brood. Development of offspring to maturity is rapid and up to six generations can be produced in a year. Females are born with embryos already developing inside them. This rapid reproduction has resulted in the mudsnail becoming a significant pest in the many countries to which it has accidently been transported.

Experiments have shown that when a bacterial or viral disease is introduced into an asexually reproducing population of mudsnails, the entire population may be killed. But when the same disease is introduced into a sexually reproducing population of the snails, the number of snails initially drops but the population soon recovers.

Analysis

1 Assuming a female produces an average of 60 offspring when she reproduces, how large could a population of mudsnails begun by a single female be in one year?

2 Explain why the New Zealand mudsnail has been labelled an invasive species in many countries.

3 Suggest two ways of preventing the New Zealand mudsnail from spreading to other countries or waterways.

4 The female population of *P. antipodarum* mudsnails is much larger than the male population. Give two reasons why.

FIGURE 8.1.14 The New Zealand mudsnail (*Potamopyrgus antipodarum*)

FIGURE 8.1.15 Although the New Zealand mudsnail is only a few millimetres long, it can cause major environmental problems in river systems outside its natural range.

ADVANTAGES AND DISADVANTAGES OF SEXUAL AND ASEXUAL REPRODUCTION SUMMARY

The advantages and disadvantages of sexual and asexual reproduction are summarised in Table 8.1.1.

TABLE 8.1.1 Summary of the advantages and disadvantages of sexual and asexual reproduction

	Advantages	Disadvantages
Sexual reproduction	• It gives greater long-term evolutionary potential. • Unfavourable (deleterious) alleles are removed from the population more efficiently. • It generates genetic diversity and allows selection for beneficial phenotypes more efficiently. • Populations are better able to adapt to changing environmental conditions.	• The slower reproductive rate means fewer offspring are produced in a specific time frame. • Recombination can break apart beneficial genomic combinations and introduce deleterious alleles to populations • There is potential for the spread of sexually transmitted diseases throughout a population. • It is energetically costly; it requires a lot of ongoing energy input from the parent.
Asexual reproduction	• It is an efficient form of reproduction. • The amount of time and energy to produce offspring is minimal. • Population sizes can increase rapidly in optimal environments. • There is no need to find a sexual partner, thus energy is conserved. • Offspring are genetically identical to the parent, so they are well suited to a stable environment.	• Rapid population growth can lead to overcrowding and increased competition for resources. • There is a lack of genetic diversity. • The lack of genetic diversity in a population means the entire population is vulnerable if conditions change (e.g. the climate changes or a new pathogen is introduced).

8.1 Review

SUMMARY

Sexual reproduction

- in multicellular organisms, involves the fusion of gametes from two different individuals to form a zygote that is genetically different to either parent
- involves equal genetic contributions from male and female parents
- is most advantageous in changing environments
- produces variation between individuals and increases genetic diversity within a population
- increases the chances that some individuals of a population will have traits that allow them to cope in the face of a new selection pressure, e.g. a new disease
- can require significant effort for finding a mate, mating and producing offspring, and may even expose the prospective parents to physical danger.

Asexual reproduction

- involves a single parent producing a new individual from part of itself
- usually involves mitosis and produces offspring that are genetically identical to their parent
- is suited to organisms living in relatively stable and uniform environments
- is a disadvantage in changing environmental conditions
- provides many benefits, including the ability to reproduce rapidly
- is more energy efficient as there is no need to find a mate or undergo mating rituals
- produces offspring that lack genetic variability
- includes fission, budding, fragmentation, spore formation, and vegetative propagation and parthenogenesis
  - parthenogenesis is an unusual form of cloning in which an egg develops without fertilisation to form a new individual.

KEY QUESTIONS

Knowledge and understanding

1 What is meant by the term 'alternation of generations'?

2 List three advantages and three disadvantages of sexual reproduction.

3 Explain why there is more variability in the offspring of sexually reproducing organisms than asexually reproducing organisms.

4 Why is a sexually reproducing species better able to survive in a changing environment than an asexually reproducing species?

5 In the table below, match each type of asexual reproduction to its correct description.

6 Define 'asexual reproduction' and state what type of nuclear division is usually involved.

7 What are the ideal environmental conditions for asexual reproduction?

8 List three advantages and three disadvantages of asexual reproduction.

Analysis

9 In ideal conditions, some bacteria species reproduce asexually via fission every 20 minutes. Calculate how many bacteria there will be after 2 hours if there is only one bacterium to begin with.

budding		separation of structures from a parent plant to form a new, independent plant, without the formation of seeds or spores
fission		form of asexual reproduction in which the new organism arises as an outgrowth or bud from the parent
fragmentation		development of an egg in the absence of fertilisation by sperm; a normal part of the life cycle of some insects and crustaceans
spore formation		form of asexual reproduction of unicellular organisms where the parent cell divides into two approximately equal parts
parthenogenesis		formation of structures that are resistant to adverse environmental conditions and can give rise to complete organisms when conditions become favourable
vegetative propagation		form of asexual reproduction of multicellular organisms in which an organism breaks into two or more parts, each of which regenerates the missing pieces to form a complete new organism

8.2 Reproductive cloning technologies

Cloning is the production of new individuals that contain the same genetic information as the parent organism. A clone is produced by asexual reproduction when a single parent cell divides to produce two new daughter cells. The daughter cells carry identical genetic information to the parent cell—they are clones.

A clone is an individual that is genetically identical to its parent.

Artificial cloning is used to increase the number of individuals in a population having a particular trait or traits. Many organisms have been selectively bred or genetically engineered to express traits desired by humans, but allowing the normal processes of meiosis, random assortment and crossing over can result in offspring that fail to express the desired trait. Using cloning overcomes this problem.

In this section you will learn about the processes used to artificially clone organisms. You will also learn about the applications of artificial cloning and will explore some of the issues associated with its use.

THE PROCESSES OF REPRODUCTIVE CLONING

Although cloning occurs naturally in some cells and organisms, the term cloning is also used to refer to the artificial processes which produce genetically identical organisms. Artificial processes of cloning currently used in horticulture and agriculture include:

- cuttings and grafts
- tissue culture
- embryo splitting
- nuclear transfer.

Cuttings and grafts

Cuttings and **grafts** are both forms of asexual propagation in plants. Growing plants from cuttings is one of the oldest and simplest forms of artificial cloning. To clone a plant in this way, a section of the plant (stem, root or leaf) is removed from the parent and planted in soil or water (Figure 8.2.1). Under suitable conditions the cutting will develop new roots, stems and leaves, because cells in the cutting re-enter the cell cycle, replicate, and differentiate into new structures. As in all forms of asexual reproduction, new plants grown from cuttings are clones of the parent plant.

FIGURE 8.2.1 Plant cuttings are the simplest method of cloning in horticulture and agriculture.

Grafting is a more complex form of artificial cloning. It involves transferring part of the stem of the desired plant variety (plant A) on to the cut stem of another plant (plant B). Plant B has well-developed roots and is referred to as the **rootstock**. Plant A is referred to as the **cultivar** and is usually selected for fruiting, flowering or aesthetic qualities.

Figure 8.2.2 shows how a cutting from the desired plant cultivar, Bing cherry, is spliced into the rootstock, *Prunus avium*. If the grafting successfully connects the vascular tissues of the plants, the cutting will begin to grow. The plant that develops is a clone of the plant from which the graft was taken.

FIGURE 8.2.2 Grafting a Bing cherry (cultivar) on the stem of the wild cherry, *Prunus avium* (rootstock)

There are many advantages to grafting, including an increase in yield, an increased tolerance to cold, disease resistance, early fruiting and production of new varieties of plants. Grafting enables much more efficient and rapid growth of desired plant varieties because they do not need to be grown from seeds. The qualities of the cultivar have been carefully selected over many generations and grafting ensures these qualities are preserved for future propagation. Grafting is commonly used in horticulture to grow fruiting and ornamental trees.

Many varieties of important food crops are propagated by cloning through cuttings or grafts. These include sugarcane, pineapple and onion. The advantage of this is that as long the conditions under which they are growing do not change, the plants produced will be exactly as expected. The food crops will have the right size, shape and taste, and will be suitable for sale.

Tissue culture

Tissue culture is a cloning technique used to grow large numbers of plants rapidly. It is used commercially to develop large crops with ideal characteristics, such as wheat with large grains. The technique is also being used in research and recovery programs for endangered plant species.

To clone plants using plant tissue culture, fragments or single cells from a parent plant are selected and grown in a culture medium. Plant fragments are treated with a sterilising solution to kill any fungi and bacteria that could contaminate the culture. The culture medium contains nutrients and plant hormones (auxin and cytokinin) to encourage plant growth and differentiation, as shown in Figure 8.2.3. During the initial stage of growth, the tissue samples form a mass of undifferentiated cells called a **callus**. The callus cells are then separated and each cell grows into a clone of the parent plant.

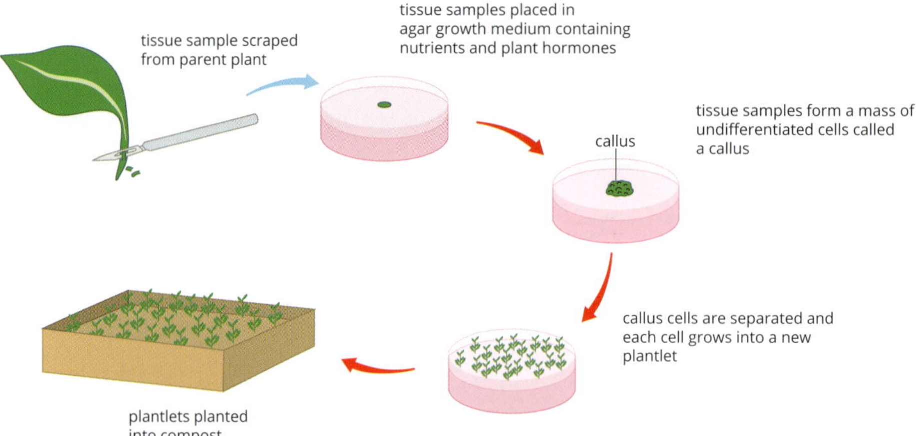

FIGURE 8.2.3 The process of plant cloning by tissue culture involves collecting cells from a plant that has the desired characteristics and then growing the cells in nutrient media with hormones. The hormones cause the cells to develop into new individual plants that are genetically identical to the parent plant.

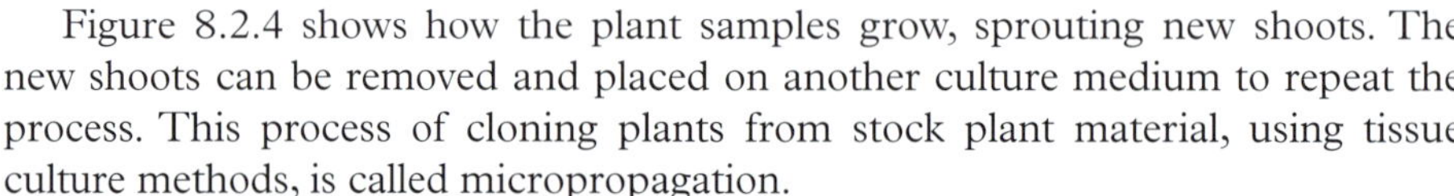

FIGURE 8.2.4 Round-leaved sundews (*Drosera rotundifolia*) growing from tissue cultures in a Petri dish

Figure 8.2.4 shows how the plant samples grow, sprouting new shoots. The new shoots can be removed and placed on another culture medium to repeat the process. This process of cloning plants from stock plant material, using tissue culture methods, is called micropropagation.

Plant tissue culturing can produce thousands of genetically identical plants very quickly, but it is very labour intensive and expensive. Table 8.2.1 outlines the main advantages and disadvantages.

TABLE 8.2.1 Advantages and disadvantages of plant tissue cultures

Advantages	Disadvantages
• A large number of plants can be produced in a short time. • The technique provides the opportunity to control growth conditions and to synchronise the processes of growth and development, making it possible to obtain plants with preferred characteristics. • New genes can be introduced into the plants, which boost crop yield or confer resistance to pests and infections.	• All the plants that are produced have the same genetic material, so they are equally vulnerable to environmental factors, infections and pests. • A lack of new combinations of traits, which only result from meiotic division and genetic recombination (see Section 6.1). • The genetic diversity of the plants is reduced and some gene variants (alleles) can be irreversibly eliminated from the gene pool.

Embryo splitting

During the early stages of embryonic development, each cell is capable of developing into a complete organism. This is possible up until the 16-cell stage of development, because all the cells of the embryo are undifferentiated. At the 32-cell stage, cells begin to undergo cell specialisation.

If an embryo splits during the early stages of embryonic development, identical twins, triplets or even quadruplets can result. When this happens, all of the offspring are genetically identical, but they are not phenotypically identical because of the influence of environment. Figure 8.2.5 shows identical twins that, while very similar in appearance, do have differences; the most obvious is that one twin is taller than the other.

Embryo splitting can be used in livestock breeding programs to increase the number of offspring born in each breeding season. Rather than allowing livestock to breed naturally, farmers use **in vitro fertilisation (IVF)** techniques. By fertilising eggs in a Petri dish, scientists are able to split embryos in the early stages of embryonic development (up to the 16-cell stage) to create multiple genetically identical embryos, which are then implanted into surrogate mothers (Figure 8.2.6). This method can be used to increase the genetic contribution from a single prize cow into the next generation. She can effectively have many more genetic offspring than could be possible if she was allowed to calve naturally.

FIGURE 8.2.5 Identical twins are formed as a result of natural embryo splitting. They are genetically identical but are not always identical in appearance.

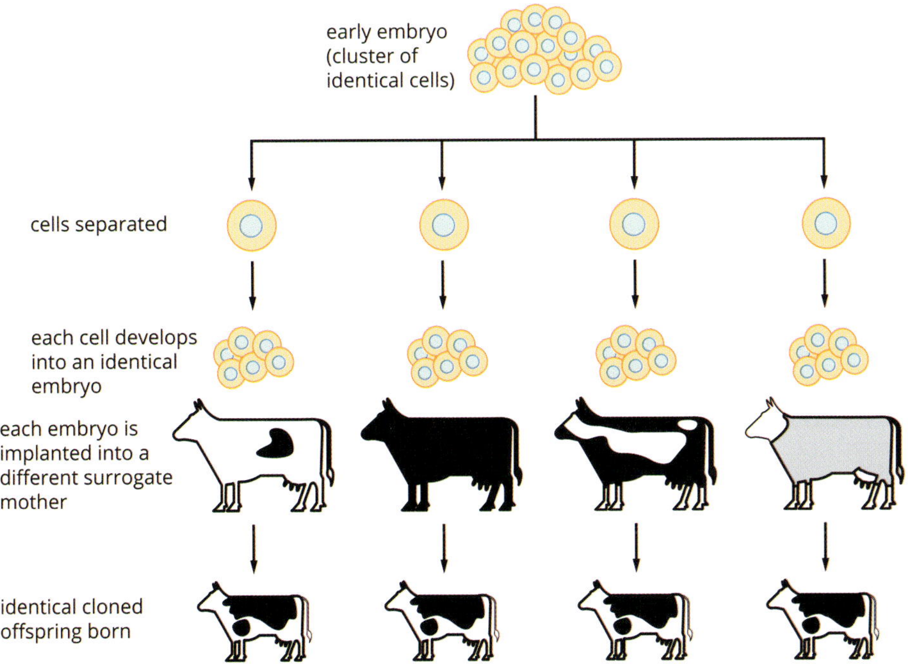

FIGURE 8.2.6 Cloning by embryo splitting is used in livestock breeding to increase the number of desirable offspring produced each season.

BIOFILE

Wollemi pine—a living fossil

The Wollemi pine (*Wollemia nobilis*) was thought to be extinct until 1994, when it was rediscovered in a remote region of the Blue Mountains in New South Wales. In nature, this species reproduces sexually in a typical conifer reproductive mode of separate female and male cones with transfer of pollen through the air. Many people were interested in having one in their garden. Because it is so rare and its one known location needed protection, it was cloned for commercial sale using tissue culture and micropropagation. The asexually cultivated pines have proved to be remarkably easy and fast to grow. Funds raised from the sale of the propagated specimens are used for protection of the rare pine in its natural habitat.

The Australian Wollemi pine (*Wollemia nobilis*) was thought to be extinct. After it was rediscovered, tissue culture was used to mass-produce pines for commercial sale. This enabled protection of the original location.

Genetic engineering has been used to create a variety of transgenic animals. These are animals that have genes from another species inserted into their genome. Goats were found to be suitable mammals to use for genetic engineering. Once a suitable and viable embryo has been produced by genetic engineering, cloning techniques such as embryo splitting enable the maximum number of animals to be produced. Goats have been engineered so that their milk produces at least 10 different proteins that are used in medical applications. Among these proteins are human growth hormone, human alpha-fetoprotein (which is used to treat autoimmune diseases) and malaria antigens suitable for vaccine production.

Nuclear transfer

The most advanced cloning technique is used in agriculture and is known as nuclear transfer. This technique has been used in many organisms, including sheep, dogs, cats and horses. It involves removing the nucleus from an unfertilised egg and replacing it with a nucleus from an adult somatic cell using a syringe (Figure 8.2.7).

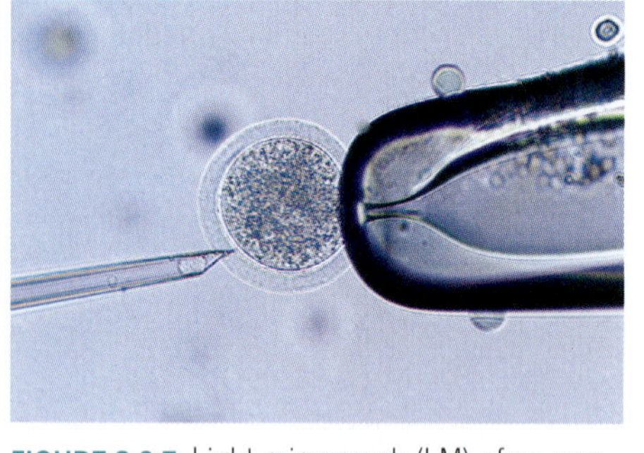

FIGURE 8.2.7 Light micrograph (LM) of an egg from a cow (centre) held in place by a pipette (right) during nuclear transfer. The egg's genetic material has been removed and now an adult cell nucleus is being injected in its place with a syringe (left).

The egg is then transplanted back into a host mother, or surrogate, where it will develop into a new individual. The new individual is genetically identical to the donor of the somatic cell. This technique is also known as **somatic cell nuclear transfer (SCNT)** (Figure 8.2.8).

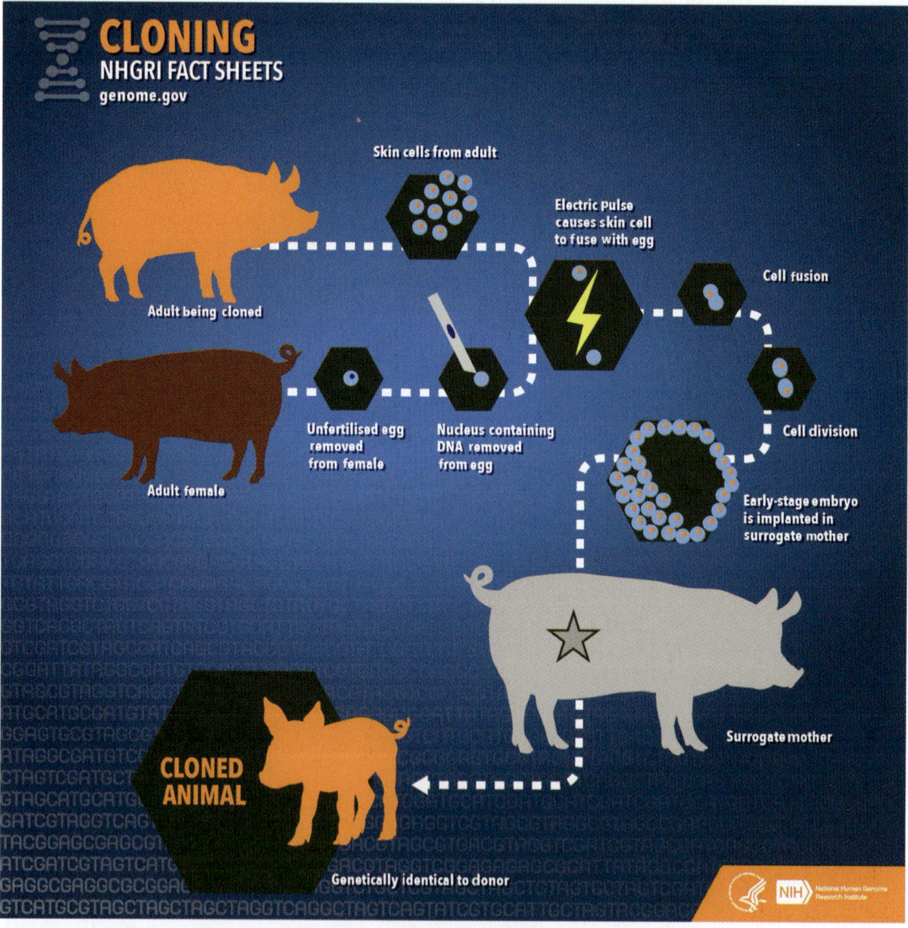

FIGURE 8.2.8 The process of somatic cell nuclear transfer. Dolly the sheep is perhaps the most famous example of this cloning technique.

THE APPLICATION OF REPRODUCTIVE CLONING

Scientists are also experimenting with cloning in order to increase the population sizes of endangered species and to even bring back extinct species. In 2001 scientist cloned a gaur. Gaurs are a type of wild ox, native to south east Asia. They are highly endangered. The clone was born healthy but died after two weeks due to an infection which is common among young calves. Later, a banteng (another type of ox) and three African wild cats were cloned successfully.

In 2013, scientists involved in the Lazarus Project, a research team based at the University of New South Wales, attempted to bring an extinct species back to life. The gastric brooding frog, *Rheobatrachus silus* (Figure 8.2.9), was declared extinct in 1983, but tissue samples had been stored frozen and these were used in an attempt at resurrection of the species. Eggs of a related species, the great barred frog, *Mixophyes fasciolatus*, were enucleated (had the nucleus removed) and nuclei from the frozen brooding frog cells were inserted. A number of embryos were produced, but none survived to maturity.

FIGURE 8.2.9 The gastric brooding frog (*Rheobatrachus silus*) is the subject of cloning attempts in order to resurrect this extinct species. The mother frog incubated her eggs in its stomach.

CASE STUDY

Dolly the sheep

In 1996 the 'most famous sheep in the world', Dolly, was born (Figure 8.2.10). Dolly was the first mammal to be cloned from the cells of an adult individual using somatic cell nuclear transfer, or SCNT. In SCNT the nucleus of an egg cell from one animal is removed, and the nucleus of a somatic cell from another animal is extracted. The somatic nucleus is inserted into the ovum, which then reprograms the nucleus. An electric shock applied to the ovum then causes the ovum to enter the G_0 phase of meiosis and begin dividing. When the ovum reaches the embryonic stage, it is transplanted into a host organism to continue development.

In Dolly's case a number of embryos were produced from a Finn Dorset ewe and a Scottish blackface ewe. The embryos were transplanted into 13 surrogate Scottish Blackface ewes. Eventually one ewe gave birth to a healthy Finn Dorset lamb—Dolly. The cloning of Dolly showed that an adult somatic cell (in Dolly's case, a mammary cell) could be used to produce a whole organism.

Although the normal life expectancy for sheep is approximately 12 years, Dolly was euthanised at the age of six and a half because she was suffering from an incurable lung disease and arthritis. It has been speculated that Dolly may have been born with a genetic age of six, the same age as the donor sheep, after research found that the lengths of her telomeres were decreased. Telomeres are the ends of chromosomes, and they are linked to the ageing process.

During her lifetime Dolly gave birth to six healthy lambs. The knowledge gained from breeding Dolly has been applied to the cloning of other large mammals such as pigs and horses.

FIGURE 8.2.10 A team led by Professor Ian Wilmut cloned Dolly the sheep, the first mammal successfully cloned using somatic cell nuclear transfer (SCNT).

Therapeutic cloning

Another application of SCNT is therapeutic cloning. This involves using SCNT to create an embryo, not for reproductive purposes, but in order to harvest embryonic stem cells which have a genetic complement identical to the donor of the nucleus. This type of cloning could potentially be used to create human embryos. Early stage embryonic stem cells have the potential to become any kind of cell in the body. These embryonic stem cells could then be used to form tissues or organs genetically compatible with the donor of the nucleus so they would not stimulate an immune response. In the process of harvesting the cells the embryo is destroyed.

BIOLOGICAL ISSUES ASSOCIATED WITH CLONING

While cloning may provide benefits for farmers and industries in terms of increased production and thus increased profits, there are biological costs to its use both for the individual organisms and for the environment.

Susceptibility to disease

FIGURE 8.2.11 The Cavendish banana is under threat from the fungus *Fusarium oxysporum*.

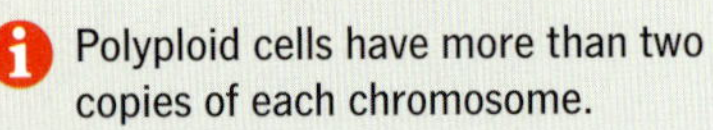

Polyploid cells have more than two copies of each chromosome.

The main concern about widespread cloning, especially for crops and animals raised for meat, is the risk that populations with less genetic diversity are more susceptible to disease and changes in environmental conditions. This in turn could leave human populations vulnerable to a wide-scale loss of food resources.

In Ireland between 1845 and 1851, disease caused potato crops to fail, resulting in a famine during which over a million people died of starvation or diseases that thrived in the malnourished population. Another one million people emigrated from Ireland to escape the famine, and many of these came to Australia. The potato crop failure was the result of a disease caused by the pathogen *Phytophthora infestans*. The entire crop failed because the potato plants were clones and were all susceptible to the same diseases.

Disease remains an issue with crops produced by cloning even today. Cavendish bananas (Figure 8.2.11) are the most popular variety in the world. This variety of plant is a polyploid. Instead of the normal two copies of each chromosome it has three, and is therefore triploid. Homologous pairing and meiosis are impossible for these plants, so they are all clones. They are currently under threat from a fungal disease, *Fusarium oxysporum,* which attacks their roots. Once infected the plants inevitably die.

In sexually reproducing wild populations of crop plants, the natural variation means that new diseases are unlikely to result in extinction of the population because there is a greater chance that some individuals will be resistant.

High failure rate

The current cloning technology associated with SCNT is highly inefficient; the rate of production of offspring is about 0.1 to 3%. The scientists who produced Dolly had 277 attempts in order to attain one successful outcome. Failure may occur at the nuclear transfer stage, the embryonic cell division stage, the implantation stage, or at many stages throughout fetal development. Studies in cattle have shown that all reproductive technologies have increased failure rates over natural methods, but that cloning has by far the highest rate of failure (Table 8.2.2).

TABLE 8.2.2 Rates of pregnancy failure in cattle for reproductive technologies and natural pregnancies

Type of reproduction	Percentage of pregnancies lost during gestation
natural	2–4
IVF	11
cloning	50+

Adverse health effects

A growing area of concern in industrial farm production is the welfare of animals. Many cloned animals have experienced adverse health effects, such as impaired immune systems and organ malformations. If the parent has any genetic defects these will be passed to all the offspring. Significant defects associated with cloning can lead to a wide range of medical problems throughout the animal's life. There are even associated health issues for the surrogate mothers of clones. In a study looking at the effects of cloning on the birth rate in cattle it was observed that the surrogate mothers of cloned calves were much more likely to develop a serious pregnancy condition called hydrops. Hydrops is a build up of fluid in the uterus. The fluid eventually fills the abdominal cavity and leads to the death of the mother. While this is normally a rare condition, it has a much higher incidence when reproductive technology—especially cloning—is used, as can be seen in Table 8.2.3.

TABLE 8.2.3 Incidence of hydrops in association with reproductive technologies

Type of technology used	Percentage of pregnancies with hydrops
none (natural reproduction)	0.0125
artificial insemination	0.7
IVF	3.3
cloning	17

Premature ageing

Associated with the welfare of cloned animals is the premature ageing that has been observed in some cloned animals. As a cell ages, the telomeres on the chromosomes shorten. **Telomeres** are repeated sequences at the ends of chromosomes. Using the genetic material from an adult organism for cloning means that the newborn cloned organism is already genetically old at birth. Dolly the sheep had a significantly shorter lifespan and suffered age-related illnesses such as arthritis. Her shorter lifespan is thought to be associated with the decreased telomere length of the donor sheep's chromosomes. Since the production of Dolly many species have been cloned. While some have shown shortened lifespans, others have had lifespans the same as animals that have reproduced naturally (Table 8.2.4).

TABLE 8.2.4 Comparison of normal life expectancy against lifespan of cloned species

Species	Normal life expectancy of species (years)	Reported maximum lifespan of cloned animals (years)
goat	15	12–18
cattle (Jersey)	15	11.8
cattle (Simmental Fleckvieh)	15	14.4
dog (Afghan hound)	10–12	10
sheep (Finn Dorset)	<10	10–12
mouse	2–3	3
cat	15	10
pig	15–17	6

Loss of genetic diversity

As more crops are produced through cloning, the loss of genetic diversity is a considerable concern. Reduced genetic diversity increases the likelihood of extinction for any species. For human populations, the loss of genetic diversity in crop species poses a risk for food security. Food crop varieties with low genetic diversity are at greater risk of crop failure or extinction, which has the potential to cause food shortages or widespread famine, as occurred in 1840s Ireland.

In the case of the cloning of endangered or extinct species, many scientists argue that the use of cloning to increase the numbers of individuals is futile as the populations will be so lacking in diversity that even if numbers increase, the species will still likely become extinct as they can all be wiped out by a single environmental change.

OTHER ISSUES ASSOCIATED WITH CLONING

Cloning can have social, ethical and legal implications. Although humans have been cloning plants for hundreds of years, modern cloning technology is much more complex and extends to the animal kingdom, including humans, and therefore has wider implications.

Ethical concerns

FIGURE 8.2.12 Animal health and welfare is a concern in cloning programs.

The formation of human embryos for reproductive or therapeutic cloning is controversial. The idea of selecting traits in humans or the creation of embryos to destroy them for their cells is seen by some people to be morally wrong. Cloning is also expensive: it costs over $10 000 to create a single clone. Some people question the morality of such spending when it is used to clone animals such as pet cats and dogs. There are also concerns by animal-rights activists over the health problems that appear more prevalent in cloned animals. Any cloning techniques should ensure the health and welfare of the animals are not negatively affected (Figure 8.2.12).

Cross-contamination

Some people have concerns about cross-contamination between clones that have been genetically modified and natural crops if they are planted close to each other. Farmers who have organic certification could potentially lose their organic status because of this, which would lead to a loss of income.

Cloning of food products

Another concern about cloning is the use of these products as food and the regulations around labelling to indicate the food has been produced via cloning. In 2006 the US Food and Drug Administration (FDA) approved the use of meat products from cloned animals for human consumption. Although the FDA states that cloned animal products are identical to those from animals bred conventionally and are safe for human consumption, there is some consumer concern about cloned products entering the food supply without sufficient tracking and labelling. As of 2020, cloned animal products have not yet entered the food supply in Australia.

Legal issues

In many countries cloning happens within a regulatory framework which encompasses laws and regulations about who owns the rights to the methods used and in the case of some plants, who owns the rights to the actual clones. In most countries, including Australia, cloning of humans for reproductive purposes is illegal; however, in Australia and several other countries human clones can be made for research and the production of stem cells. In Australia, such clones must not be permitted to develop beyond the blastocyst stage (5 to 9 days after fertilisation).

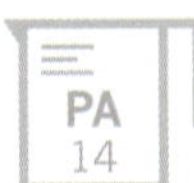

8.2 Review

OA ✓✓

SUMMARY

- Humans have been cloning organisms for hundreds of years, for example to achieve better crop yields.
- Clones are genetically identical to each other but are not phenotypically identical due to environmental influences during development.
- Cloning plants involves the following steps:
  - Cuttings are made by removing a part of a plant and encouraging it to grow roots.
  - Grafting is a process where the shoots of one plant are attached to the roots of a different plant.
  - Tissue culture involves taking cells from a plant, treating them with plant hormones and growing them in a nutrient solution on agar plates.
- Identical twins are a natural form of embryo splitting.
- Artificial embryo splitting involves taking an early stage embryo and separating it into more than one bundle of cells, which are then implanted into surrogate mothers.
- Somatic cell nuclear transfer (SCNT) is a modern form of cloning where the nucleus from one animal is implanted into an enucleated egg of another animal of the same or a related species. The embryo develops and is implanted in a surrogate mother.
- Cloning has issues such as a high rate of failure during embryonic and fetal development in animals, and adverse health effects, including increased susceptibility to disease in plants and animals, and premature ageing in animals.
- Widespread use of cloning could lead to the loss of species due to reductions in genetic diversity.
- There are social, ethical and legal issues that must be considered in relation to cloning.

KEY QUESTIONS

Knowledge and understanding

1 What is tissue culture used for, and what are the advantages of using this technique?

2 **a** What is grafting?

b How does grafting increase the economic benefits to the farmer?

3 Describe embryo splitting and SCNT.

4 Explain why clones will not necessarily have the same phenotype.

5 List at least three problems associated with cloning.

6 Why does the use of cloning lead to concerns about loss of genetic diversity?

Analysis

7 Cloning of mammals has been undertaken for some time. A literature review provided the data shown in the two graphs below.

What conclusions could be drawn about success rates in achieving live birth in each species?

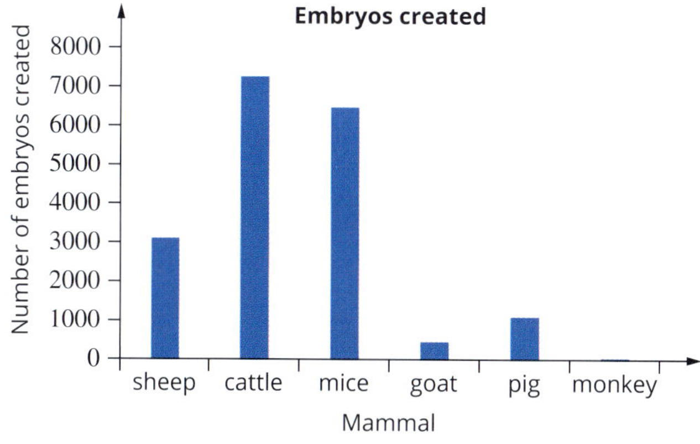

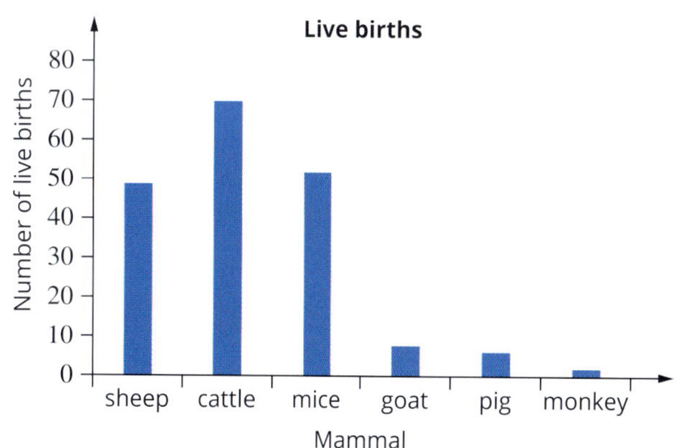

Chapter review

KEY TERMS

alternation of generations
asexual reproduction
budding
bulb
callus
clone
cloning
conidia
corm
cultivar
cutting
fission
fragmentation
gametophyte
graft
in vitro fertilisation (IVF)
mitospore
parthenogenesis
rhizome
rootstock
sexual reproduction
somatic cell
somatic cell nuclear transfer (SCNT)
sporangium
spore
spore formation
sporophyte
stolon
telomere
tissue culture
tuber
vegetative propagation

REVIEW QUESTIONS

Knowledge and understanding

1 Which of the following methods of asexual reproduction does not necessarily produce all genetically identical offspring?
  A vegetative propagation
  B parthenogenesis
  C fission
  D budding

2 Sexual reproduction has costs to a species, such as the expenditure of energy to find a mate, and yet sexual reproduction has persisted. Why?

3 Asexual reproduction is highly successful in a stable environment but less so when the environment is unstable. Explain why.

4 Asexual reproduction is common amongst invasive species. Discuss why this reproductive strategy helps make organisms successful invaders of a new habitat and what impact this has on the native species in that environment.

5 What type of reproduction and reproductive structures are responsible for mouldy bread?

6 Arrange the following stages of plant tissue culture in the correct order, from first to last.
  - New shoots are removed and placed on another culture medium.
  - Plant hormones promote the rapid growth of shoots and roots.
  - A sample of the ideal stock plant is removed.
  - The sample is sterilised and placed on a culture medium.

7 **a** What are four methods of cloning currently used in horticulture and agriculture?
  b Discuss one of these methods, using an example of its application in horticulture or agriculture.

8 Identify four ethical issues associated with cloning.

9 Do hermaphrodites reproduce via asexual or sexual reproduction? Explain your answer.

Application and analysis

10 A farmer grows a range of plants. The table below outlines the reproductive strategies of the plants.

Plant reproductive strategies

Plant	Type of reproduction
tulip	asexual
poppy	sexual
lily	asexual
strawberry	asexual and sexual

If a virus infects all of the plants, which plants are most likely to survive?

A poppy and strawberry
B tulip and lily
C tulip, lily and strawberry
D poppy only

11 Consider the steps in the process of somatic cell nuclear transfer (SCNT). Suggest two possible reasons why cloning animals using SCNT has such a high failure rate.

12 Is the embryo used in SCNT a perfect clone of the donor of the nucleus? Explain why or why not.

13 Sexual reproduction increases variation, which is beneficial to the population or species but is often not beneficial to the individual. Discuss how this might be the case.

14 The sheep that donated the egg used to clone Dolly the sheep had a light brown fleece and yet Dolly was born with a white fleece. Consider the process used to create Dolly and explain why this is the case.

15 Contrast the processes of asexual and sexual reproduction. Mention at least two points of difference.

16 Experiments have shown that when a disease is introduced into an asexually reproducing population of New Zealand mudsnails, the entire population may be killed. When the same disease is introduced into a sexually reproducing population of mudsnails, the number of snails initially drops but the population soon recovers.

Explain why an asexual population of mudsnails can be wiped out by a disease but a sexually reproducing population can recover after exposure to the same disease.

17 The pineapple mealybug, *Dysmicoccus brevipes*, is an insect related to aphids. It lives on pineapples and has been transported around the world as pineapple agriculture has spread. In the 1930s this bug was transported to Okinawa, an island in Japan. In Okinawa there are two lineages of mealybug, one which reproduces by parthenogenesis and the other which reproduces sexually. Observation of these lineages indicates that although they are both the same species, they only ever use the same method of reproduction as their mother. Surveys of the Okinawa population were undertaken on three separate occasions over the period of a year, the methods of reproduction being used was determined, and the following results were obtained.

Method of reproduction in *Dysmicoccus brevipes*, Okinawa

Survey	1	2	3
Percentage of population reproducing sexually	60	63	85
Percentage of population reproducing asexually			

a Complete the table by calculating the percentage of population reproducing asexually for each survey.

Researchers also found the mealybugs on Ishigaki Island. This island is further south than Okinawa and has a much gentler climate. On Ishigaki the mealybugs are all female.

b How could the population on Ishigaki be sustained if all individuals are female?

c Discuss whether the data from Okinawa and Ishigaki supports the idea that sexual reproduction provides greater survival benefits than asexual reproduction.

18 It has been observed that in some species the offspring born as a result of SCNT have a tendency to develop obesity and its associated health problems. Researchers wished to determine whether this problem continues into future generations. The researchers took groups of rats: group A consisted of rats born normally, group B were rats cloned using SCNT, and group C were an F2 generation formed from the mating and reproduction of previously cloned rats. Once born, the rats were weighed every 4 weeks until they were 24 weeks old. Their weights in grams are shown below.

Weights of rats (in grams) from birth to 24 weeks of age

Age (weeks) / **Group**	birth	4	8	12	16	20	24
A (born normally)	5	38.5	52.5	62.5	67.5	72.5	79
B (cloned using SCNT)	5	41.5	67.5	31.5	79	83.5	95
C (F2 generation from clones)	5	41	60	67.5	72.5	76	79

a What was the research question being investigated by the researchers?

b Plot the data on a correctly constructed graph.

c Describe the trends shown by your graph.

d What conclusions can be drawn from the data?

19 It has been theorised that reduction in telomere length is responsible for some of the poor health consequences experienced by cloned animals. An experiment was undertaken in cattle where four daughters of a cloned bull and five daughters of a bull produced naturally were observed. The relative telomere length of the calves was observed at 2, 6 and 12 months. In both the daughters of cloned and naturally produced bulls, telomeres had a range of lengths, as shown in the data. The range was wider in the daughters of the naturally produced bull. Natural variation introduces random errors into the results. Results are shown in the table below.

Telomere length in daughters of cloned and naturally reproduced bulls

Age (months) / **Relative telomere length**	2	6	12
Daughters of cloned bull	21.36 ± 2.05%	20.56 ± 2.72%	20.56 ± 1.49%
Daughters of naturally produced bull	23.78 ± 4.37%	23.53 ± 3.98%	22.43 ± 4.65%

a What is the control in this experiment?

b Do the results of the experiment indicate that there is a difference in telomere length in the second generation after cloning?

c What conclusions can be drawn for future generations, in regard to telomere length, if cloning becomes widespread in animal farming?

20 Navel oranges first appeared as a result of a mutation in a single plant which occurred during the early 1800s. Today's navel oranges are all cloned descendants of this original plant. How may this affect the long-term viability of this variety?

21 Alternation of generations is particularly beneficial for organisms such as ferns, which are unable to move around to find a mate. Discuss.

OA ✓✓

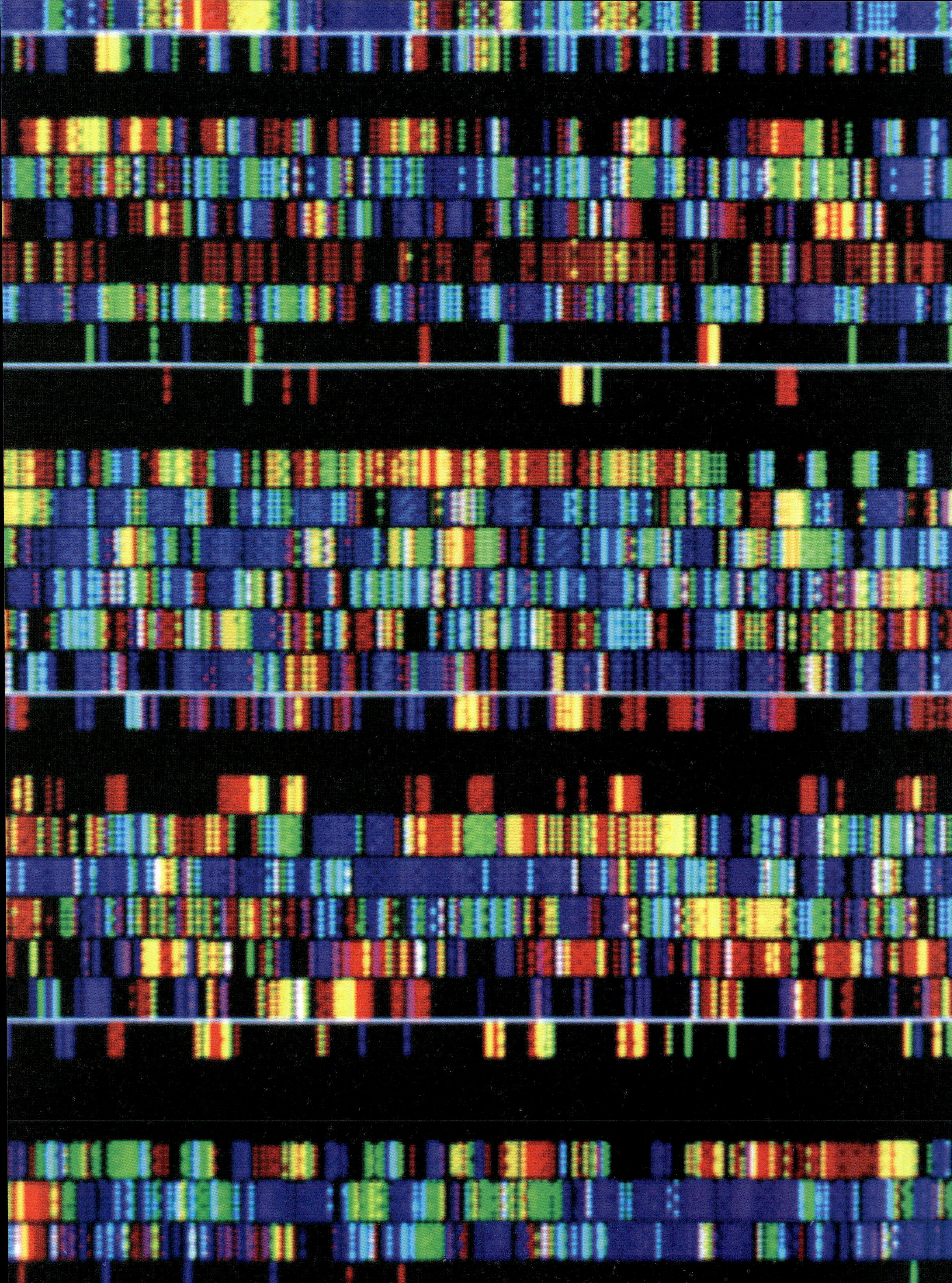

CHAPTER 09 Adaptations and diversity

This chapter examines the extraordinary variety of adaptations that species have evolved to survive in a wide range of environments. You will learn about the important role genetic diversity plays in enabling populations and species to adapt to changing environments and some of the different types of adaptations that plants and animals have evolved in response to their environment.

You will also learn about interdependencies between species in ecosystems and the contribution of Aboriginal and Torres Strait Islander peoples' knowledge and perspectives to our understanding of adaptations and interdependencies between species in Australian ecosystems.

Key knowledge

- the biological importance of genetic diversity within a species or population **9.1**
- structural, physiological and behavioural adaptations that enhance an organism's survival and enable life to exist in a wide range of environments **9.2**
- survival through interdependencies between species, including impact of changes to keystone species and predators and their ecological roles in structuring and maintaining the distribution, density and size of a population in an ecosystem **9.3**
- the contribution of Aboriginal and Torres Strait Islander peoples' knowledge and perspectives in understanding adaptations of, and interdependencies between, species in Australian ecosystems. **9.3**

OA ✓✓

9.1 Genetic diversity

FIGURE 9.1.1 Harlequin ladybirds (*Harmonia axyridis*) vary in colour and number of spots. This variation is due to the genetic diversity in the species.

Evolution is change in the genetic composition of populations over time. This can be observed as changes in alleles (gene variants) and phenotypes (physical traits) in a population. New species can evolve in response to changes in environmental conditions or after populations become isolated and accumulate genetic differences. Biodiversity increases as genetic changes result in new genetic diversity. In this sense, evolution promotes biodiversity. However, evolution can also lead to the loss of biodiversity, through the extinction of alleles, populations and species.

In Chapter 8 you learnt about sexual and asexual reproduction and the importance of sexual reproduction for the production of genetic diversity in offspring (Figure 9.1.1). In this section you will learn about the biological importance of genetic diversity within populations and species.

SOURCES OF GENETIC DIVERSITY

You inherited many of your physical features or traits from your parents. You share certain traits with your mother and others with your father. You might even appear to have a totally different version of a trait. If you look at the people in the population you live in there may be different versions of the trait again, and then even more variation across the human species. These differences between individuals are due to the inheritance of different combinations of **alleles** (gene variants). The variety of genes or alleles in a population or species is known as **genetic diversity** (also known as genetic variation). The collection of alleles in a population is known as a **gene pool**.

There is always variation between individuals within a sexually reproducing population. This is due to there being different combinations of alleles that have arisen from crossing over of chromatids and independent assortment during gamete formation, and the random fusion of gametes at fertilisation. Genetic mutations that occur during gamete formation can also introduce new genetic variation into a population. These factors, along with differences in gene expression and the influence of environmental factors, result in a wide variety of phenotypes in a population.

The key sources of genetic diversity in a population or species are:

- crossing over of chromatids during meiosis
- independent assortment of chromosomes during meiosis
- mutation
- random fusion of gametes at fertilisation
- **gene flow** (the exchange of alleles between populations).

Mutation is the only source of new alleles in a species, whereas crossing over, independent assortment, fertilisation and gene flow reshuffle alleles that already exist in a species. The reshuffling of alleles gives offspring combinations of alleles that are different to their parents and siblings, producing the genetic diversity that can be seen within families, populations and species (Figure 9.1.2).

> Meiosis and sexual reproduction produce new combinations of alleles, but mutations are the only source of new alleles in a species.

> Gene flow occurs when individuals from different populations interbreed and when individuals enter or leave a population.

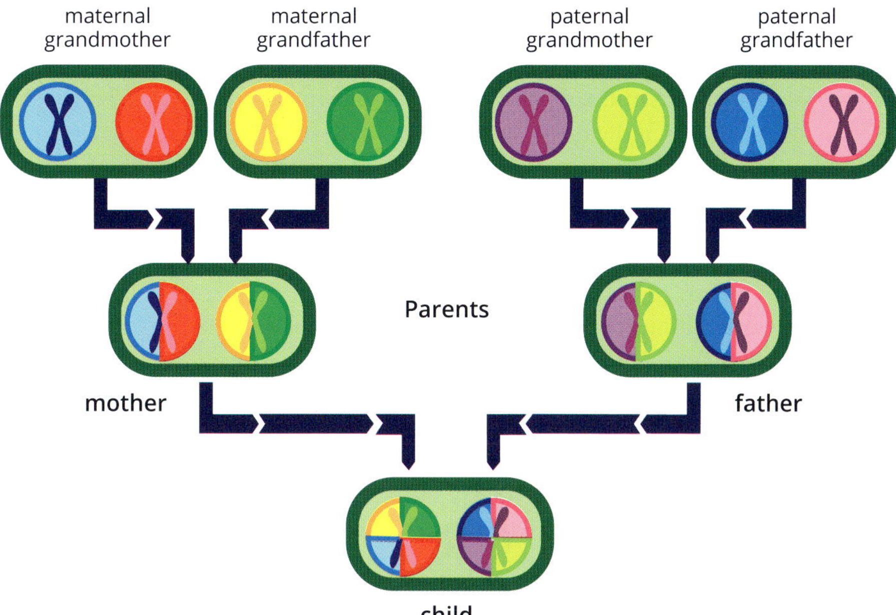

FIGURE 9.1.2 The child in the pedigree chart has inherited a unique mix of genetic information from grandparents and parents.

IMPORTANCE OF GENETIC DIVERSITY

For populations and species to remain viable, they must have genetic diversity. Genetic diversity allows populations to adapt to changing environments, provides resistance to disease and reduces the likelihood of inbreeding and the inheritance of disadvantageous recessive alleles. Low genetic diversity increases the risk of extinction, particularly in species with small, fragmented populations. For these reasons, genetic diversity is an important aspect of conservation programs for endangered species.

Genetic diversity in populations and species is important for the following reasons.

- It is the basis for adaptation—genetic diversity results in a variety of phenotypes, providing populations with the flexibility to adapt to changing conditions.
- It maintains the health and stability of a population—a population with high genetic diversity is more likely to have alleles that provide resistance to stressors such as disease. In a population with low genetic diversity, it is likely that individuals will all be vulnerable to the same stressors, putting the whole population at risk of collapse.
- It improves **biological fitness** (the ability to survive and produce viable offspring)—small populations with low genetic diversity are more likely to have individuals with reduced fitness due to inbreeding and the inheritance of disadvantageous recessive alleles.
- It improves the long-term **evolutionary potential** of populations—a population or species will experience many environmental changes and their ability to adapt to these changes will determine their evolutionary potential (how long they can adapt and survive). Species or populations with high genetic diversity are more likely to adapt to changing environments and persist for longer periods of time than species or populations with low genetic diversity.

BIOFILE

The Irish potato famine

In the 1800s the Irish imported a variety of potato from South America to feed their growing population. The potatoes were grown asexually through vegetative propagation, so all the potato plants were genetically identical to one another. When the potato crops were infected by the fungal pathogen *Phytophthora infestans* in the 1840s, the entire crop was destroyed because the lack of genetic diversity meant every potato was susceptible to the disease. The crop failure occurred across Europe but Ireland was the country worst affected. At that time, an Irish working-class family grew their own crop, and an adult labourer would eat up to 5 kg of potatoes a day as their main food.

The Irish potato famine is a reminder of the importance of maintaining genetic diversity in crops. The famine was responsible for the death of over one million Irish people between 1845 and 1851. Another one million emigrated to other countries.

Potatoes infected by the fungus *Phytophthora infestans*.

A population's evolutionary potential is determined by its ability to adapt and survive in the long term. Populations with high genetic diversity are more resilient to change and have greater evolutionary potential.

Natural selection

The genetic diversity between individuals in a population results in different phenotypes (traits) that have varying advantages for survival and reproduction. Individuals with phenotypes that are well-suited to their environment are more likely to survive and reproduce—this process is known as **natural selection**.

> Allele frequency refers to the proportion of the gene pool that carries a particular allele, relative to other alleles for that gene.

The alleles of the individuals that survive and reproduce will persist in the population and increase in frequency. This is the mechanism by which natural selection changes the allele frequencies and phenotypes in populations. The conditions or factors that influence which phenotypes are most successful in a population—and therefore, influence allele frequencies in that population—are known as **selection pressures**. Examples of selection pressures include climatic conditions, competition for resources (e.g. food and shelter), mate availability and predator abundance.

Over many generations, a population will undergo genetic and phenotypic changes in response to selection pressures in the environment. If populations of the same species experience different selection pressures, natural selection can result in different phenotypes becoming dominant in the populations. If there is no interbreeding between the populations, they may become different species over time.

Finches from the Galápagos Islands show how genetic diversity provides the basis for adaptation and evolution of new phenotypes and species. There are more than 14 species of Galápagos finches recognised today, each with different beak shapes, body size and feeding behaviours that are advantageous for the conditions on the island on which they live (Figure 9.1.3). All of these species evolved from a common ancestor with genetic diversity that allowed the populations colonising new islands to survive and adapt to the different environmental conditions.

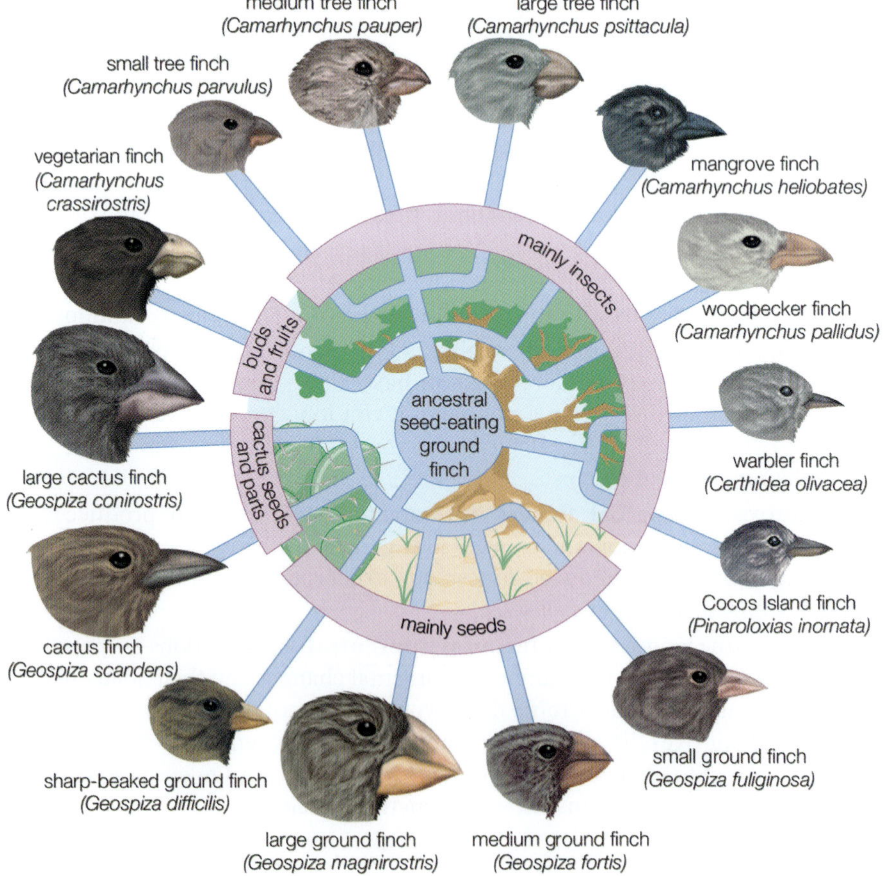

FIGURE 9.1.3 Galápagos finches. All of these species have evolved from a common ancestor which had genetic diversity that allowed populations to adapt to new environments.

CAUSES OF LOW GENETIC DIVERSITY

Genetic diversity can naturally change in a population over time, but rapid declines in diversity can be difficult for a population or species to recover from. Once alleles are lost from a population or species, it is unlikely that the same alleles will recur by chance. Loss of genetic diversity can be permanent.

Low genetic diversity can be caused by a variety of factors but is usually associated with population decline, fragmentation or isolation.

> Genetic diversity allows populations to adapt to changing environments.

Population reduction

The number of individuals in a population can be rapidly reduced because of a random event, such as a bushfire, or can decline over time due to ongoing selection pressures. Population reduction can lead to a loss of alleles from a population and can significantly reduce genetic diversity. Ongoing population decline is a concern for many species, with human activities such as hunting and land clearance greatly reducing genetic, species and ecosystem diversity.

Genetic drift

In the case of rapid population reduction, the phenotypes within a population are unlikely to significantly increase its chance of survival because the population does not have time to adapt to the change. The individuals that survive will do so by chance. The allele frequencies of the remaining population are unlikely to reflect those of the original population as alleles are randomly lost from the population. Random changes in allele frequencies due to chance events (e.g. births and deaths) is called **genetic drift**. Genetic drift has a greater impact on small populations as there are fewer alleles, and the death of one individual can significantly alter the allele frequencies of the population.

> Genetic drift is random changes in alleles in a population. Small populations are at greater risk of losing alleles via genetic drift.

Genetic drift can occur when populations decrease for a period of time (a bottleneck effect) or in small founding populations (the founder effect).

Bottleneck effect

A 'population bottleneck' is a sudden reduction in the size of a population due to an environmental event, such as a natural disaster (e.g. bushfire, earthquake) or human activity (e.g. habitat clearance). The **bottleneck effect** describes the impact of the substantial population reduction on the remaining population. The smaller the population, the greater the bottleneck effect and impact of genetic drift. Alleles may be lost from the gene pool immediately after the bottleneck or lost by chance in only a few generations. The small population size and low genetic diversity makes bottlenecked populations more vulnerable to environmental change, disease and inbreeding.

Cheetah (*Acinonyx jubatus*) populations underwent a bottleneck around the end of the last glacial period (approximately 12 000 years ago) and consequently have unusually low genetic diversity (Figure 9.1.4). Such low genetic diversity has caused sperm cell defects, low fertility and increased susceptibility to disease, putting cheetah populations at increased risk of extinction.

FIGURE 9.1.4 Cheetahs (*Acinonyx jubatus*) have experienced a population bottleneck in the past, resulting in low genetic diversity in their populations today.

Founder effect

The **founder effect** occurs when a small group of individuals is genetically isolated from a larger population, either by migration, new geographic barriers or habitat fragmentation. The smaller population only has a small portion of the alleles of the original population and therefore has lower genetic diversity.

In the new environment, the selection pressures on the founder population are likely to be different from those experienced by the original population. These differences in selection pressures drive further changes in allele frequencies and genetic diversity.

> Low genetic diversity increases the risk of extinction, as seen when species or populations go through a genetic bottleneck.

FIGURE 9.1.5 Habitat fragmentation due to deforestation. The fragmentation of habitat reduces connectivity and gene flow between populations, increasing the likelihood of inbreeding and loss of genetic diversity.

Isolated populations

When populations are isolated, there is little or no gene flow to introduce new alleles into the population. The isolation of small populations often results in breeding between genetically similar individuals (inbreeding), causing the genetic characteristics of the population to become concentrated over multiple generations. The overall genetic diversity in isolated populations is reduced, and disadvantageous traits that are rare in the wider population are more likely to occur.

Small, fragmented populations are commonly seen in endangered species. Human-induced changes such as habitat clearance and urbanisation have caused habitat fragmentation and increased the chance of isolation (Figure 9.1.5). Habitat fragmentation and isolation reduces gene flow between populations, potentially leading to inbreeding and further loss of genetic diversity.

CASE STUDY

Low genetic diversity in Tasmanian carnivores

Fossils and ancient Indigenous art in northern Australia indicate that the thylacine (*Thylacinus cynocephalus*), also known as the Tasmanian tiger, once inhabited most of Australia (Figure 9.1.6). During the course of human expansion across Australia, its range decreased until it was confined to Tasmania. It faced competition from the introduced dingo and was close to being extinct in the wild, when people hunted it for a bounty. The last individual died at Hobart Zoo in 1936. It is inferred from the complete genome sequence of a preserved thylacine fetus that population numbers and genetic diversity started decreasing 70000 years ago or earlier, before human habitation of Australia 50000–60000 years ago. The cause may have been a cooling climate and the consequent habitat change.

Low genetic diversity may have meant that populations were less able to adapt to changing environments, including the effects of human habitation such as hunting and habitat loss, putting the species at greater risk of extinction.

The genome of another Tasmanian marsupial, the Tasmanian devil (*Sarcophilus harrisii*), shows that it also has very low genetic diversity. Such low genetic diversity increases its susceptibility to the facial tumour disease that is currently threatening its survival as a species and its susceptibility to other extinction risk factors, such as habitat fragmentation.

FIGURE 9.1.6 Thylacine at London Zoo around 1910.

CONSEQUENCES OF LOW GENETIC DIVERSITY

Low genetic diversity has a number of negative consequences for populations and species. The most significant consequence is loss of evolutionary potential and increased risk of extinction. While some species are able to recover from population decline and loss of genetic diversity, many others will decline further when faced with environmental change, disease and reduced biological fitness from inbreeding. Each of these factors puts populations at risk of collapse and increases the risk of extinction for species.

Inability to adapt

Natural populations are regularly faced with environmental change and stressors in the form of disease, competition for resources, introduced species and climate change. In a changing environment, having low genetic diversity can lead to population decline or collapse if the traits within the population are not well suited to the new environmental conditions. Because genetic diversity is required for populations to adapt to environmental change, populations with low genetic diversity are at greater risk of population decline and extinction than populations with high genetic diversity.

Disease

Genetic diversity in genes associated with immune response is important for both individuals and populations to resist or recover from disease. Individuals with low genetic diversity in genes associated with immune response have been found to have higher parasite loads and greater susceptibility to infectious disease than individuals with high genetic diversity. At the population level, genetic diversity creates a greater variety of individuals that will respond to disease differently. For this reason, populations with high genetic diversity are more likely to be resistant to disease than populations with low genetic diversity. In populations with low genetic diversity, alleles that provide disease resistance are less likely to occur, leaving the entire population vulnerable in the event of a disease outbreak.

Inbreeding depression

Inbreeding is the production of offspring from parents that are closely related or genetically similar. While many species have evolved mechanisms to avoid inbreeding, it may become unavoidable in small populations of individuals that are genetically similar. Inbreeding results in an increase in the number of homozygous individuals (individuals with two copies of the same allele) in a population and further loss of genetic diversity.

The reduction in biological fitness and survival due to inbreeding is known as **inbreeding depression**. Over time, inbreeding allows unhealthy mutations to accumulate in a population. The more closely related individuals are, the more likely it is that they share the same recessive alleles, which may be disadvantageous or even lethal if both alleles are inherited. This is not a concern if closely related individuals do not breed, but in small populations inbreeding is likely to concentrate recessive alleles, increasing the chance that offspring will express disadvantageous traits.

BIOFILE

Maintaining genetic diversity in captivity

Zoos that participate in global captive breeding programs must keep very careful records, known as studbooks, to prevent inbreeding and maintain genetic diversity. Studbooks are records of parent and offspring phenotypes kept for many years for animals for which ancestry is important. Many zoos participate in captive breeding programs for endangered animals such as the Sumatran tiger (*Panthera tigris sumatrae*). The estimated population of Sumatran tigers in the world is less than 700, a dangerously low number resulting in a very limited gene pool. When genetic diversity is this low, there is a high risk of inbreeding, further loss of genetic diversity and extinction.

Genetic analysis gives conservationists insight into the alleles in populations. Zoos can use this information to select mating partners that have different alleles, improving the genetic diversity of the offspring and in the captive population. Zoos may introduce genetic diversity to a captive population by selecting mating partners from other zoos' captive breeding programs or by collecting sperm from wild individuals for artificial insemination. These practices introduce new genetic diversity to the captive population and allow gene flow to occur in an otherwise isolated population.

Sumatran tiger cub Achilles and mother Melati at London Zoo in 2017

CASE STUDY

Genetic diversity and conservation

Genetic diversity is recognised by the International Union for Conservation of Nature (IUCN) as one of the three forms of biodiversity worthy of conservation (species diversity and ecosystem diversity are the other forms). Genetic diversity is an important factor to consider when developing conservation programs for endangered species.

Siberian tigers

The largest of all the tigers, the Siberian tiger (*Panthera tigris altaica*) (Figure 9.1.7), is estimated to have only around 500 individuals remaining in the wild. This is a very low population in terms of genetic diversity, but unfortunately the story is more alarming than that—when the Siberian tiger was first protected in 1947, there were only an estimated 50 individuals in the wild. The remaining 500 individuals are descended from those 50 individuals. Although the population size has increased, the genetic diversity within the population has not. This is referred to as a 'population bottleneck' and is often a cause of low of genetic diversity in endangered species. This lack of genetic diversity means that the Siberian tiger population is extremely vulnerable to inbreeding, disease and environmental change.

FIGURE 9.1.7 The Siberian tiger (*Panthera tigris altaica*) has low genetic diversity due to a population bottleneck in the past.

FIGURE 9.1.8 Despite being endangered, the giant panda (*Ailuropoda melanoleuca*) has high genetic diversity in the genes associated with immunity.

Giant pandas

Not all endangered species have such low genetic diversity. There are only an estimated 1500 giant pandas (*Ailuropoda melanoleuca*) (Figure 9.1.8) still breeding in the wild, confined to mountain ranges in northern China. Although pandas have been difficult to breed in captivity, scientists have discovered that they have surprisingly high variation in genes associated with immune response. This means that, while the giant pandas are still considered endangered, they are less susceptible to disease than many other endangered species.

Koalas

The koala (*Phascolarctos cinereus*) (Figure 9.1.9) is under threat from habitat loss, urbanisation and disease, with many populations declining and concerns of decreasing genetic diversity. The 2019–20 Australian bushfires decimated important koala habitat and resulted in a high loss of life, adding further pressure to their populations.

Researchers from the Australian Museum have sequenced the koala's genome, providing insight into their historical and present-day genetic diversity.

The genome sequencing has revealed that koala populations have undergone substantial and rapid declines in the past, most likely due to climate changes and hunting. These population declines significantly reduced the genetic diversity in populations, leading to high levels of inbreeding. Prior to the 2019–20 bushfires, researchers found that populations in New South Wales and Queensland had retained higher levels of genetic diversity and connectivity between populations. Continuing to assess and monitor genetic diversity will be an important part of the recovery and conservation of koala populations in the future.

FIGURE 9.1.9 Koalas (*Phascolarctos cinereus*) are facing threats from habitat loss, urbanisation, disease and low genetic diversity.

9.1 Review

SUMMARY

- Genetic diversity is the variety of genes or alleles in a population or species.
- The key sources of genetic diversity in a population or species are:
  - crossing over of chromatids during meiosis
  - independent assortment of chromosomes during meiosis
  - mutation
  - random fusion of gametes at fertilisation
  - gene flow.
- Genetic diversity in populations and species is important because it:
  - is the basis for adaptation
  - maintains the health and stability of a population
  - improves biological fitness
  - improves the long-term evolutionary potential of populations.
- The genetic diversity between individuals in a population results in different phenotypes (traits) that have varying advantages for survival and reproduction. Individuals that are well-suited to their environment are more likely to survive and reproduce—this process is known as natural selection.
- Low genetic diversity is caused by population decline, fragmentation or isolation.
- Genetic drift is random changes in alleles in a population. Small populations are at greater risk of losing alleles through genetic drift.
- Isolated populations have little or no gene flow to introduce new alleles into the population. Small, isolated populations often have low genetic diversity and are at greater risk of extinction than large, connected populations.
- Loss of genetic diversity can be permanent—once alleles are lost from a population or species, it is unlikely that the same alleles will recur by chance.
- Small populations with low genetic diversity are more likely to lose alleles by chance (genetic drift), experience inbreeding and inherit disadvantageous recessive alleles, increasing the risk of extinction.
- Genetic diversity is required for populations to adapt to environmental change. Populations with low genetic diversity are at greater risk of population decline and extinction than populations with high genetic diversity.
- Populations with low genetic diversity are less likely to have alleles that provide resistance to disease and are more vulnerable in the event of a disease outbreak.
- Reduced biological fitness due to inbreeding is known as inbreeding depression. Populations with low genetic diversity are at increased risk of inbreeding depression.
- Genetic diversity is recognised by the World Conservation Union (IUCN) as one of the three forms of biodiversity worthy of conservation and is an important factor to consider when developing conservation programs for endangered species.

KEY QUESTIONS

Knowledge and understanding

1. List the main sources of genetic diversity for a population.
2. Explain why inbreeding is problematic for populations.
3. The thylacine, also known as the Tasmanian tiger, is an extinct marsupial that once roamed the Australian continent. During the course of human expansion across Australia, its range decreased until it was confined to Tasmania. Suggest why the thylacine population was reduced to Tasmania only.

Analysis

4. Propose two ways in which genetic diversity can be increased in an isolated population.
5. Both the Siberian tiger and the giant panda are endangered species, but they have different levels of genetic diversity.
   a. Explain the difference between the genetic diversity in their populations.
   b. Based on our understanding of their populations and genetic diversity, which species potentially faces greater risk of extinction? Justify your response.

9.2 Adaptations

FIGURE 9.2.1 Panther chameleons (*Furcifer pardalis*) of Madagascar have evolved a spectacular coloured skin that can change to signal aggression, and territorial and mating behaviour, as well as mood.

Organisms have different features that enable them to survive and reproduce in different environments. These features have evolved in response to various environmental factors and are known as **adaptations**. Adaptations can enable animals and plants to live in extreme environments, access resources and mates, defend themselves and their territory, and communicate and interact with their own and other species (Figure 9.2.1).

Adaptations can be classified into three broad categories:

- structural (morphological or anatomical)
- physiological (functional)
- behavioural.

Adaptations are characteristics that increase the likelihood of survival and reproduction of an organism in a particular environment. They have a genetic basis and are passed on from generation to generation. Adaptations are the result of the evolutionary process of natural selection, in which those organisms that are best suited to their environment survive and reproduce, passing on their advantageous adaptations to their offspring. While adaptations help individual organisms survive and reproduce, they are also critical for the long-term evolutionary potential of populations and species.

In this section you will learn about the different types of adaptations in plants and animals, and how theses adaptations enable organisms to exist in a wide range of environments.

ENVIRONMENTAL FACTORS

To live in a particular habitat, an organism must have access to the basic requirements necessary for growth. These requirements are usually met by the organism's environment, which consists of:

- **abiotic factors**—non-living components of an environment such as water, temperature, pH, and salinity
- **biotic factors**—living components of an environment, such as bacteria, fungi, plants and animals.

TOLERANCE RANGE

Consider what it feels like when you are ill and have a fever, usually indicated by a body temperature of above 38°C. You start feeling unwell, lightheaded and uncomfortable as your body is not functioning within its optimum range. The **optimum range** is the range of an abiotic factor, such as temperature, within which an organism can thrive and function at its best. For a human, optimum body temperature is in the range 36–37°C. **Tolerance range** includes the optimum range and values on either side at which the organism is still active but does not thrive. At 38°C, a human's body temperature is within their tolerance range as they can still be active, but they will not thrive at this temperature. Environmental changes outside the optimum range of an organism can lead to **physiological stress** and can reduce an organism's functional capacity.

> The range of environmental conditions in which an organism can survive is called its tolerance range.

Typically, a species' tolerance range forms a bell-shaped curve on a graph (Figure 9.2.2)—that is, the closer the environmental conditions are to the optimum range of a species, the more likely it is that individuals will survive and reproduce, leading to population growth. The optimum range is where abiotic conditions are ideal and where organisms survive and reproduce most effectively. Outside the optimum range, organisms may be physiologically stressed and be less likely to survive and reproduce. Species have different optimum and tolerance ranges depending on their adaptations and the environment they are in.

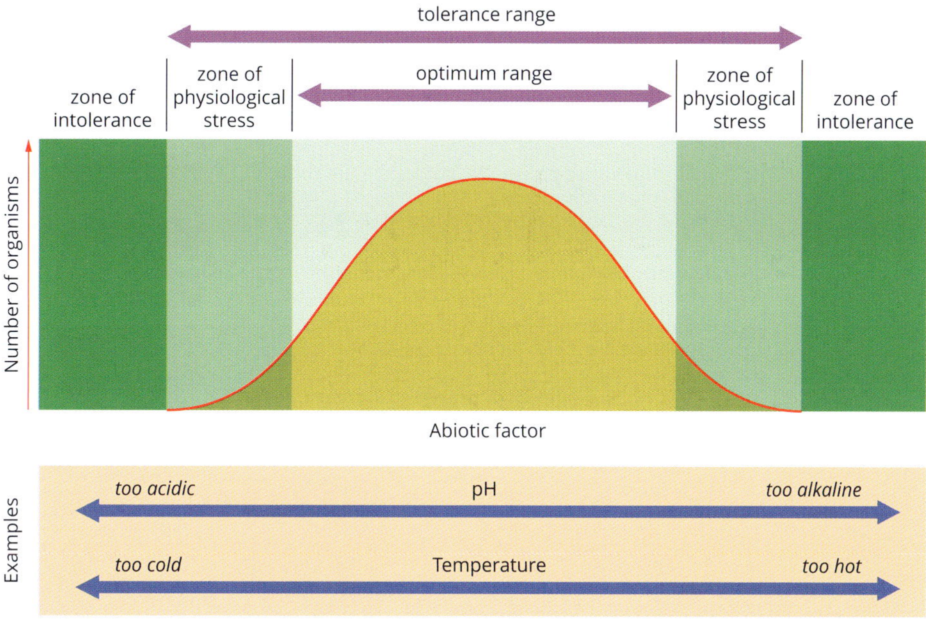

FIGURE 9.2.2 A bell curve representation of the pH and temperature tolerance range and optimum range of an organism. Organisms function best when abiotic factors are within their optimum range—outside this range they are in physiological stress and are less likely to survive and reproduce. Organisms cannot survive outside their tolerance range.

STRUCTURAL ADAPTATIONS

Structural adaptations are anatomical or morphological features that help organisms survive in a specific environment. They are the physical characteristics relating to body size and features that increase the likelihood of the organism's survival in a particular environment.

Morphology relates to the structure and form of an organism.

Structural adaptations of plants

Plants can be classified based on the type of environments they live in and the structural adaptations they have to survive in these environments. **Hydrophytes** (from the Greek *hydro-*, meaning water, and *phyton*, plant) are plants which grow in water or on the surface of water, such as water lilies. **Mesophytes** form the largest group and include the terrestrial plants that survive in moderate climates—not too dry and not too wet. The plants you see around you in public spaces and household gardens are mesophytes. Plants that grow in dry, hot environments are known as **xerophytes**, from the Greek *xeros* (dry). Cacti are well-known examples of xerophytes.

Water is essential for photosynthesis. Therefore, a large number of structural adaptations that we observe in plants reduce water loss caused by salinity, heat and wind in the environment. Some of these adaptations include:

- reduced leaf surface area
- fewer stomata
- stomatal hairs
- sunken or protected stomata
- thick, waxy cuticle
- extensive root systems
- rolled leaves
- leaves orientated away from sunlight.

Cacti—adaptations to dry environments

Xerophytes, such as cacti, have adaptations that conserve moisture and prevent the leaf temperature from rising too much. They also have an increased tolerance to desiccation (drying). Some of the adaptations of xerophytes are shown and described further in Figure 9.2.3.

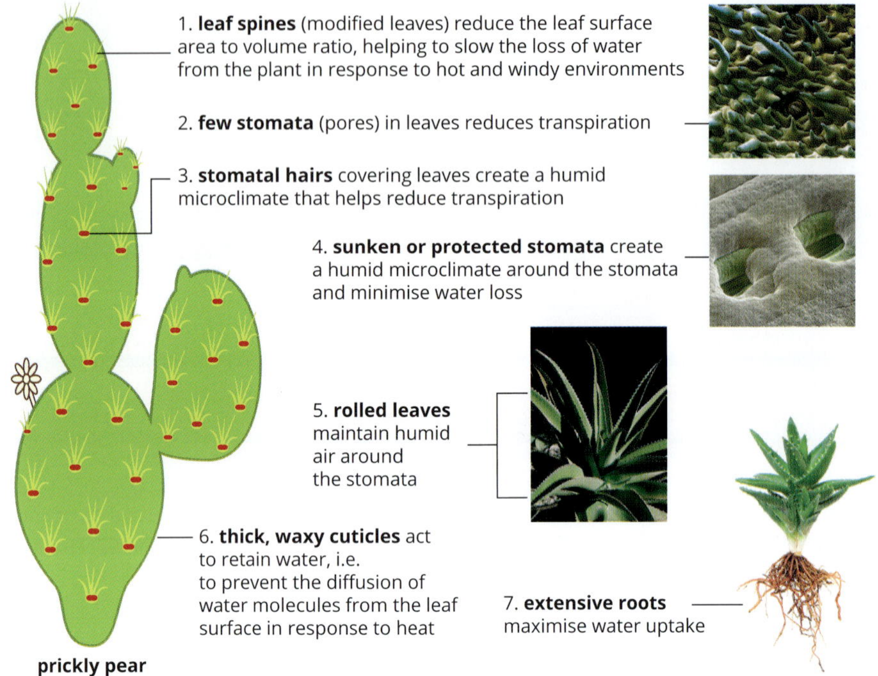

FIGURE 9.2.3 Plants that live in harsh, dry environments such as deserts have evolved adaptations that enable them to conserve water.

Marram grasses—rolled leaves

Marram grasses (*Ammophila* species) are xerophytes that grow well in the salty, sandy soils of coastlines (Figure 9.2.4a). When conditions are hot and dry, thin-walled bulliform (bubble-shaped) cells partially collapse, causing the leaves to roll inwards, reducing water loss. Hairs on the inside of the rolled-up leaf trap moisture, creating a humid microclimate (Figure 9.2.4b). This humidity reduces the concentration gradient between the outside and inside of the leaf, which reduces transpiration. Because of this and other adaptations, these grasses have been used to stabilise sand dunes that are prone to erosion.

FIGURE 9.2.4 (a) Marram grasses (*Ammophila* species) grow well in the salty, sandy soils of coastlines. They have adaptations that allow them to survive in this dry, salty environment. (b) One of these adaptations is leaf rolling. This enables the plant to trap moisture and reduce water loss.

Eucalypts—leaf orientation

Eucalyptus trees also have structural features that enable them to survive in hot, dry environments. Many *Eucalyptus* species have leaves that hang vertically to reduce the amount of direct sunlight they receive (Figure 9.2.5). They also have hard leaves with waxy cuticles on both sides. Both of these features of *Eucalyptus* leaves are structural adaptations that reduce the amount of water that the plant loses via transpiration.

FIGURE 9.2.5 The leaves of this adult *Eucalyptus* tree are hanging downwards, reducing their exposure to sunlight. This adaptation reduces transpiration in hot, dry climates.

BIOFILE

Lithops

Lithops is a genus of succulent plants that live in dry, rocky environments in southern Africa. They are called stone plants or pebble plants because of their stone-like appearance. This appearance helps them to blend in with their surroundings, so they are less likely to be eaten by herbivores. Another unusual feature of *Lithops* is that most of the leaves grow underground. While this adaptation helps to reduce water loss, it also makes it difficult for the leaves to access vital sunlight for photosynthesis. To overcome this problem, *Lithops* have evolved translucent tissue on their leaf tips that allows sunlight to be magnified to the chloroplasts that are deep within their underground leaves. *Lithops* can also tolerate a diverse temperature range from a temperature of −16.4°C to a maximum of 68.7°C. The enzymes in this plant have a broad functional temperature range to enable it to survive in these extremes.

Lithops or stone plants have a unique structure to cope with their dry environment.

Structural adaptations of animals

All animals have evolved structures that enable them to survive in their environment. Adaptations to abiotic factors such as temperature and water availability, as well as biotic factors such as predators, prey and competitors, are critical for survival in any environment.

One of the main abiotic factors to which animals need to be able to adapt is temperature. Animals can be classified into two groups based on how they regulate their body temperature. **Ectotherms** are animals that depend on external heat sources to regulate their body temperature. An ectotherm's body temperature changes with the temperature of the external environment. Lizards are one example of an ectotherm (Figure 9.2.6). **Endotherms** are animals that internally generate heat to maintain a constant body temperature. Examples include birds and mammals. As the environmental temperature changes throughout the day, the internal body temperature of an endotherm remains steady (e.g. approximately 37°C for humans). Both ectotherms and endotherms have various structural adaptations that allow them to regulate their body temperature.

Some examples of structural adaptations in animals include:

- thick fur and blubber (fat) to insulate against cold
- bright feathers to help attract mates
- large ears to increase heat loss
- small ears to reduce heat loss
- webbed feet and flippers for swimming
- spines for protection against predators
- body coverings for camouflage.

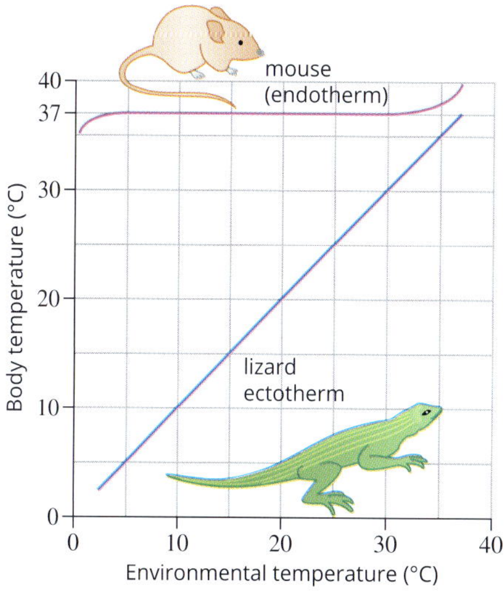

FIGURE 9.2.6 An endotherm is able to maintain a steady body temperature as the environmental temperature changes. The body temperature of an ectotherm changes as the environmental temperature changes.

Ecto- is a prefix that indicates that something is external. *Endo-* is a prefix that indicates that something is internal.

Size and proportion

An organism's shape and surface area to volume ratio affect its ability to regulate temperature. You learnt about surface area to volume ratio in Chapter 2. Organisms that are able to generate internal heat do so throughout their entire volume. However, organisms can only lose heat through their surface.

As an organism grows, its surface area and volume do not increase proportionally.

- When an animal's body structure is small, its surface area is larger than its volume. It has a high surface area to volume ratio. The higher the surface area, the more heat is lost (Figure 9.2.7)
- When an animal's body structure is large, its surface area is smaller than its volume. It has a low surface area to volume ratio. The lower the surface area, the less heat is lost (Figure 9.2.8).

Body structures with a high surface area to volume ratio allow animals to rapidly lose heat, while body structures with a low surface area to volume ratio allow animals to conserve heat.

An animal that lives in a hot environment needs to be able to lose heat rapidly to prevent overheating. Structures with a high surface area to volume ratio allow heat to be quickly lost to the external environment. Examples of such structures are the long thin body of a black-handed spider monkey (Figure 9.2.7a), the long thin legs of a camel (Figure 9.2.7b) and the long ears of a fennec fox (Figure 9.2.7c).

An animal living in a cold environment needs to conserve body heat. Structures with a low surface area to volume ratio reduce the amount of heat lost to the external environment. The red panda (Figure 9.2.8c), which lives in cold high-altitude areas of Central Asia, has small ears to reduce heat loss.

FIGURE 9.2.7 Animals can increase their surface area to volume ratio to lose heat by having (a) long thin bodies (black-handed spider monkey), (b) long thin legs (camel) or (c) long ears (fennec fox).

FIGURE 9.2.8 Animals can decrease their surface area to volume ratio to conserve heat by having (a) a short large body (snow leopards), (b) short thick legs (emperor penguins) and (c) short ears (red panda).

Body coverings for temperature regulation

The emperor penguin (*Aptenodytes forsteri*) has many structural adaptations to cope with life in the harsh Antarctic climate. Penguins have four layers of thick, scale-like feathers, creating a windproof coat (Figure 9.2.8b). They also have thick blubber to keep them warm while swimming in the icy ocean. Juvenile penguins have soft down for insulation, which is a more effective insulator than adult feathers on land but is of little use in the sea. They must moult before they can swim. Penguins and other animals in cold climates tend to have bodies with a small surface area to volume ratio to assist them in conserving body heat.

Body coverings for camouflage

Body coverings are not only important to keep organisms warm, they also function to provide protection from predators. In particular, the colour of an organism's body covering, fur or feathers, may allow it to take cover and not be seen. To protect itself from predators, the Mary River turtle has algae growing on its back to help it blend in with the vegetation growing from the river bed (Figure 9.2.9).

FIGURE 9.2.9 The Mary River turtle has green algae growing along its back to camouflage it from predators.

PHYSIOLOGICAL ADAPTATIONS

Physiological adaptations relate to the functioning of an organism at the different levels of organisation, from biochemical to cell, tissue, organ, system and organism level.

Physiological adaptations of plants

Plants inhabit a wide range of environments, from the hottest deserts to high mountain peaks, fast-flowing rivers and even the coastal intertidal zone. So they need an equally impressive range of adaptations to cope in what are often stressful conditions. Physiological adaptations play an important role in enabling plants to meet environmental challenges.

Crassulacean acid metabolism (CAM)

Crassulacean acid metabolism, also known as **CAM photosynthesis**, is an example of a physiological adaptation that enables greater efficiency in water storage and use in plants. It is most commonly found in plants living in dry environments, such as succulent plants in deserts. Some xerophytes and some plants adapted to saline conditions can minimise water loss during the heat of the day by using the CAM pathway.

In CAM plants the stomata open only at night to collect carbon dioxide. Rather than using the carbon dioxide immediately, as non-CAM photosynthesising plants do, it is stored as malic acid in cell vacuoles. During the day the malic acid is transported to the chloroplasts, where it is used to produce the carbon dioxide needed for photosynthesis (Figure 9.2.10). By storing the carbon dioxide required for photosynthesis, the plant is able to close its stomata during the heat of the day to reduce water loss. This physiological adaptation allows plants to survive in environments of extreme heat and aridity.

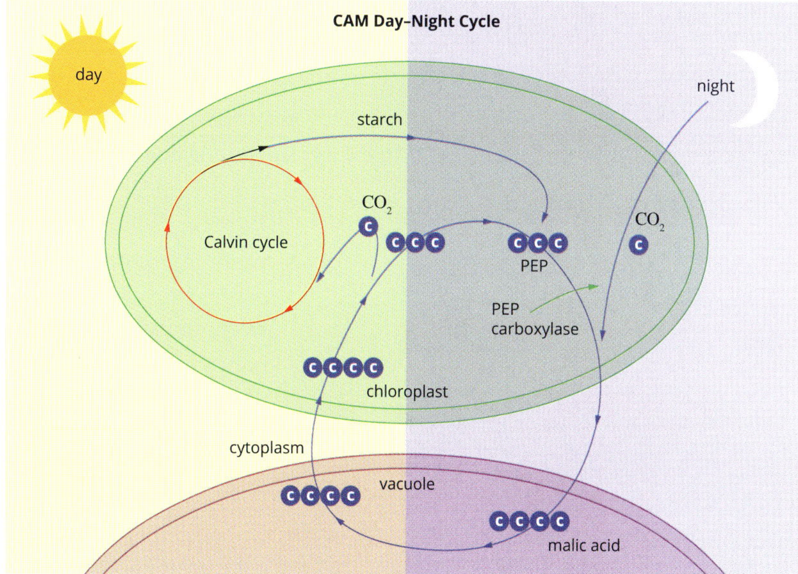

FIGURE 9.2.10 A summary of the complex CAM pathway of some plants in hot and dry or saline environments. This metabolic pathway enables the plants to absorb and store carbon dioxide at night to avoid losing precious water during the day.

Frost tolerance

FIGURE 9.2.11 Physiological adaptations, such as increasing solute concentrations, producing frost-inhibiting proteins and changing plasma membrane composition, allow some plants to survive in extremely cold climates.

Extreme cold can be very damaging, and even lethal, to plants that are not adapted to cope with such conditions. Ice crystal formation inside cells causes plant plasma membranes to burst, killing the cells. Cold temperatures can also decrease enzyme activity and change the fluidity of plasma membranes, both of which effect a wide range of physiological processes in the plant. To overcome these problems, plants living in cold climates have evolved strategies that enable them to tolerate freezing temperatures (Figure 9.2.11).

A high concentration of solutes such as sugars and salts lowers the freezing point of water. Plants that can accumulate high concentrations of these solutes in their leaves are therefore less likely to be damaged by freezing temperatures.

Some plants produce proteins that reduce the risk of cell damage from freezing. Antifreeze proteins inhibit the growth and recrystallisation of ice crystals by binding to them. Dehydrin proteins bind to water molecules inside the cell, changing the structure of the water and stabilising the plasma membrane.

In addition to adjusting solute concentrations and producing proteins to tolerate freezing, plants can also change the lipid composition of their plasma membranes to optimise functionality in cold temperatures.

Enzymes catalyse metabolic reactions only within certain temperature ranges. They denature (break down) at high temperatures and are inactive at low temperatures.

Regulation of salinity

Salinity is a major problem for many agricultural crops. In many areas, over-irrigation of agricultural land has resulted in highly saline soils, which most food crops cannot tolerate. Saline soils disrupt water and nutrient uptake by the roots, suppressing plant growth. When salt enters the plant's cells, it causes ion imbalance, inhibiting metabolic processes, and eventually leads to cell death. Plants living in saline environments such as coastal dunes, salt marshes or salt lakes have evolved physiological mechanisms to cope with the salinity (Figure 9.2.12). Plant species that can survive high salinity are known as **halophytes** (from the Greek *halos*, meaning salt and *phyton*, meaning plant). These plants use a variety of mechanisms to exclude or regulate the concentration of salt in their tissues.

FIGURE 9.2.12 *Arthrocnemum indicum* is a coastal halophyte adapted to living in a highly salty environment. This species is able to control salt levels by increasing water uptake.

Some physiological adaptations that plants have evolved to cope with salinity are:

- Compartmentalisation of ions within the cells and tissues of the plant by transporting excess salt to vacuoles or old tissue. This avoids the toxic accumulation of salt in the cytoplasm.
- Excluding salt at the roots and leaves by:
  - shedding leaves that are overloaded with salt
  - excreting salt from salt glands
  - pumping salt out of the roots
  - controlling transpiration to avoid excess salt being delivered to the shoots from the soil
  - balancing the rate of growth with the uptake of soluble ions to maintain a constant salt concentration in tissues
  - increasing water uptake to dilute salt concentrations in tissues (Figure 9.2.12).

Physiological adaptations of animals

Animals display an astounding range of physiological adaptations. With these adaptations some species are able to overcome extreme conditions and exploit seemingly uninhabitable environments.

Examples of physiological adaptations in animals include:

- the ability to produce concentrated urine to conserve water in desert animals such as the spinifex hopping-mouse
- venom production for prey capture or defence, as seen in most snakes, wasps, spiders, many marine animals, and even some mammals, such as the platypus
- colour changes in response to sunlight for thermoregulation, as in Namaqua chameleons
- shivering to maintain body temperature when cold, in endothermic animals (including humans).

Further examples of physiological adaptations in animals are discussed below.

Camouflage

Camouflage enables many organisms to blend in with their environment. This adaptation has many advantages, but it is particularly useful for avoiding predators or for capturing prey. One of the most amazing examples of camouflage is seen in the common octopus, *Octopus vulgaris* (Figure 9.2.13). This octopus can change colour and texture to match its underwater environment, blending in with corals, sand or kelp to hide itself from predators and prey.

The common octopus has specialised colour-changing cells called chromatophores that enable it to change colour to match its surroundings. Physiological mechanisms move pigment to and from the cells and change their reflection to produce the effect. In addition, tissues under the octopus's skin can create textures to match its environment.

FIGURE 9.2.13 The common octopus (*Octopus vulgaris*) uses specialised cells called chromatophores to camouflage itself.

Heat exchange for cooling

Heat exchangers work in different ways in different animals. In desert ungulates such as the gemsbok oryx (*Oryx gazella*) (Figure 9.2.14a), a heat exchanger is used to keep the brain cool. If the oryx is dehydrated and can no longer afford to lose water, it stops sweating. This causes its body temperature to rise, sometimes as high as 43°C. If blood at such high temperatures entered the brain, the animal would die. To avoid this, the arterial blood travels through a network of smaller arteries that intertwine with a network of veins before it enters the brain. This network of veins and smaller arteries is called the carotid rete system (Figure 9.2.14b).

The venous blood in the carotid rete system has travelled through the nasal sinuses and has been cooled using evaporative cooling in the nostrils. As the cooler blood from the nostrils passes in the opposite direction to the warmer blood from the body, the heat flows from the hotter to the cooler blood. This process is known as **countercurrent heat exchange**. This cools the blood entering the brain by several degrees, enabling the animal to survive in extreme heat and drought.

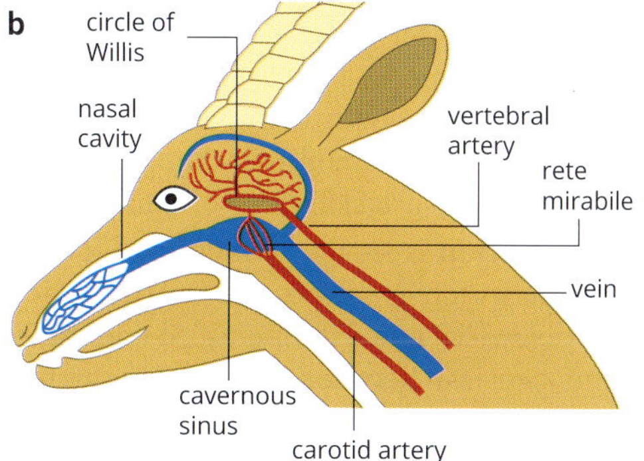

FIGURE 9.2.14 (a) A gemsbok oryx (*Oryx gazella*) in the Kalahari desert, South Africa. (b) The oryx's carotid rete system cools the hot arterial blood from the body before it enters the brain.

Heat exchange for heating

Countercurrent heat exchange also occurs in animals living in extremely cold climates, to reduce heat loss and maintain body temperature.

Penguins have heat exchangers in their flippers, feet and tails. These extremities have a relatively large surface area and are exposed to the cold, so they lose heat quickly. Blood from the feet flows back to the heart through veins close to the arteries. The warm blood in the arteries transfers heat to the veins so that blood moving back towards the heart is warmed, maintaining the penguin's body temperature (Figure 9.2.15). The blood travelling to the feet is cooled, so heat loss is minimised.

The diameter of the arteries flowing through the feet is also reduced to decrease the flow of blood to the extremities and further reduce heat loss. In this way the cells in the feet receive oxygen and nutrients and remain warm enough to function, but less heat is lost to the environment.

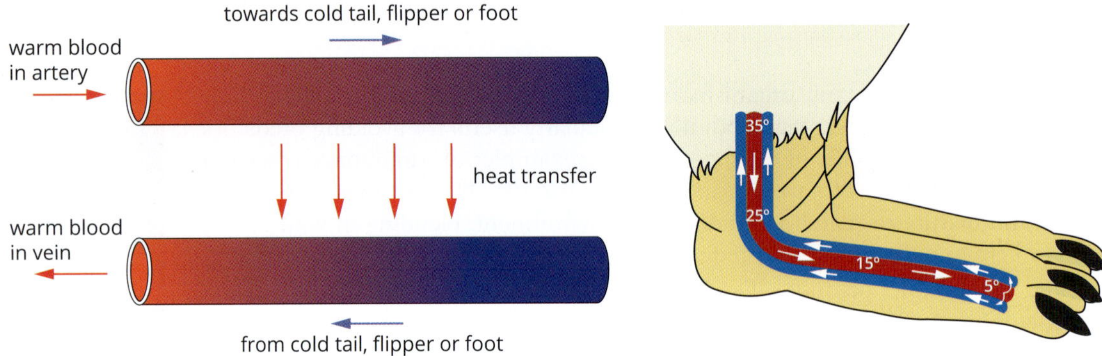

FIGURE 9.2.15 The heat exchange mechanism in the circulatory system of penguins ensures that heat loss at the extremities is minimised, while the core body temperature is maintained.

FIGURE 9.2.16 Antarctic cod (*Notothenia sp.*) use a variety of physiological mechanisms to survive extreme cold. One of these mechanisms is the production of antifreeze proteins.

Antifreeze proteins

Some fish that inhabit very cold water, such as the Antarctic cod (*Notothenia coriiceps*), manufacture a type of protein that prevents tissue from freezing. These antifreeze proteins, similar to those produced by plants, circulate in the blood of the fish and prevent the growth of ice crystals, keeping their blood liquid (Figure 9.2.16).

Torpor

Torpor is a physiological state in which the metabolic rate is lowered to save energy. This enables an organism to cope with environmental stresses such as extreme cold or heat or decreased food or nutrient availability, and can occur over short or long periods.

A long period of torpor is often called dormancy. Hibernation, brumation and aestivation are different forms of prolonged torpor.

- **Hibernation** is prolonged torpor that occurs in winter. Over summer and autumn the animal builds up a thick layer of body fat that will provide them with energy during the hibernation period in winter. During hibernation the animal can decrease its body temperature and heart rate to conserve energy. Hibernation occurs mostly in mammals, but some species of birds also hibernate. Bears, bats and squirrels are examples of animals that hibernate (Figure 9.2.17).
- **Brumation** is similar to hibernation but involves different metabolic processes. Reptiles such as snakes and lizards undergo brumation. Brumation is triggered by decreases in air temperature and daylight hours. It begins just before winter and can last between one and eight months. How long a reptile remains in brumation depends on the air temperature and the size and age of the animal. Once brumation begins, the reptile eats less or not at all, but wakes regularly to drink.

FIGURE 9.2.17 Pair of greater mouse-eared bats (*Myotis myotis*) hibernating in a cave.

- **Aestivation** is prolonged torpor in hot and dry conditions. Examples of aestivating animals are snails, frogs, crocodiles, tortoises, lungfish and some birds.

Green-striped burrowing frogs (*Cyclorana alboguttata*; Figure 9.2.18) inhabit semi-arid to arid regions of eastern Queensland and northern New South Wales. These frogs spend up to nine months of the year in aestivation. During this time they live underground in small burrows and do not eat. Research has shown that during this time they can reduce their metabolic rate by up to 80%. This allows them to survive these underground periods just on their store of body fat.

Torpor is an example of both a behavioural adaptation (retiring to a cave or seeking shelter and going to sleep) and a physiological adaptation (the slowing of the heart, breathing and metabolic rates associated with periods of torpor).

FIGURE 9.2.18 The green-striped burrowing frog (*Cyclorana alboguttata*) aestivates for up to nine months of the year underground.

Bioluminescence

Bioluminescence is a physiological adaptation in which light is produced by an organism to attract attention, frighten enemies or lure prey. Bioluminescence is a form of chemiluminescence, which involves the release of light energy following a chemical reaction. Fireflies (Figure 9.2.19), deep sea fish and sea jellies are some of the organisms that are bioluminescent, producing the chemicals luciferin (a pigment) and luciferase (an enzyme). The luciferin reacts with oxygen to create light. The energy system for bioluminescence is highly efficient, with no excess heat being produced.

FIGURE 9.2.19 Fireflies, also known as lightning bugs, regulate the production of flashes of light from their abdomen to signal their presence to one another, which is particularly important during the mating season.

BIOFILE

Diving in the deep

Some diving mammals, such as the crab-eater seal (*Lobodon carcinophagus*) (see figure), can stay submerged at depths of 430 metres for more than 10 minutes. Diving mammals are able to store oxygen much more efficiently than other mammals. Some seals can store 70% of their oxygen in their blood; humans can store only 51%. These larger oxygen stores are possible because of increased levels of haemoglobin in the blood and myoglobin in muscles (haemoglobin and myoglobin are proteins that bind to oxygen), along with larger blood volumes in these animals.

Crab-eater seals (*Lobodon carcinophagus*) are able to dive to depths of 430 metres, remaining submerged for more than 10 minutes. They have a range of physiological adaptations that make this possible.

Diving mammals have a high tolerance for lactic acid build-up, so their muscles can still function efficiently when oxygen stores have been depleted. These animals also have excellent control over their organs, reducing blood flow to those that are not needed for immediate survival, such as the digestive organs, while conserving precious oxygen for vital organs such as the heart and brain. This also reduces the work of the heart, slowing the heart rate dramatically and further conserving oxygen.

MOVEMENT AND BEHAVIOURAL ADAPTATIONS

Movement and **behavioural adaptations** are actions that an organism takes to improve survival or reproduction. Plants have movement adaptations that allow them to move toward favourable conditions and away from unfavourable conditions. In animals, behaviours may be learnt, such as the use of tools in chimpanzees and crows, or instinctive, such as a spider spinning a web. The behaviour of animals can be incredibly complex, but even the simplest behaviours can be critical for survival and reproduction.

Adaptations for movement in plants

Although plants do not have muscles or a nervous system like animals, they can still move in response to their environment. In most cases the mechanisms for plant movement are controlled by chemicals such as hormones, or by turgor pressure, both of which are physiological adaptations. Plants can undergo two types of movement in response to environmental stimuli. One is called tropism and the other is called nastic movement.

Tropism

FIGURE 9.2.20 The growth of these seedlings towards the light is an example of tropism.

Tropism is plant growth in response to an environmental factor such as gravity, light or water (Figure 9.2.20). The response depends on the direction of the **stimulus**—the plant will either grow towards (positive tropism) or away from (negative tropism) the stimulus. Tropisms are controlled by hormones such as auxin, gibberellin, ethylene and cytokinin.

Types of tropisms include:

- phototropism—growth in response to light
- geotropism or gravitropism—growth in response to gravity
- chemotropism—growth in response to chemicals
- thigmotropism—growth in response to touch
- hydrotropism—growth in response to water concentration.

Nastic movement

Nastic movement is the movement of plant tissue in response to an environmental stimulus (but not necessarily in the direction of the stimulus). This allows a plant to adapt to changes in its environment by changing its orientation. Some nastic movements in plants are:

- thigmonasty—movement in response to touch
- photonasty—movement in response to a change in light intensity
- thermonasty—movement in response to a change in temperature.

FIGURE 9.2.21 The Venus flytrap (*Dionaea muscipula*) uses mechanosensors (hairs) on the leaf surface to trigger cell pores to open in the lower side of the leaf. Water rushes into these cells, causing them to expand and forcing the trap to close.

Thigmonastic movements include the rapid opening and closing of plant parts in response to touch, such as those observed in the Venus fly trap, *Dionaea muscipula* (Figure 9.2.21). The Venus fly trap is a carnivorous plant that is adapted to low levels of nitrogen in the soil. It obtains nitrogen by trapping prey such as flies, which it attracts by secreting a sweet sap. When a fly touches the tiny hairs (mechanosensors) on the leaves, an electrical signal is sent to the centre of the trap. This signal opens pores in the trap's lower layer of cells, allowing water to rush in from the cells in the upper layer of the trap. The rapid change in pressure (turgor) causes the cells on the lower side of the trap to expand, forcing the trap to snap shut, trapping the fly inside. Enzymes released by the plant then digest the insect. About one third of the energy-carrying molecules, adenosine triphosphate (ATP), in the cells are used up in each movement. This is why after repeated touches a leaf will not respond until its energy reserves have been replenished.

The flowers and leaves of many plants respond to changes in light intensity, opening during the day and closing at night or on cloudy days (Figure 9.2.22). This is an example of photonasty.

An example of thermonastic movement is the opening and closing of tulips in response to air temperature. The petals open as the air temperature rises and close when the temperature falls. This behaviour allows the pollen to be exposed only in warmer weather, when pollinators are more likely to visit the flower, and protects it during cooler weather. As in the thigmonastic movement of the Venus fly trap, this movement is a result of turgor pressure.

Plants that are capable of rapid movement rely on internal changes in turgor. Changes in turgor are usually initiated by contact with objects outside the plant. The cells involved are in the parenchyma tissue of the cortex or specialised swellings (pulvini) at the base of leaves or leaflets. Some movements may be very fast, occurring in less than a second.

FIGURE 9.2.22 A bloodroot plant (*Sanguinaria canadensis*) displaying photonastic movements on a cloudy morning. In low light this plant closes its leaves and flowers.

BIOFILE

Trigger plant

Trigger plants (*Stylidium* species) have one of the fastest movements in the plant world. The flowers have a structure called a column (the trigger), where the male anthers as well as the female stigma are located. When triggered by the touch of an insect, the column flicks back against the insect's body, depositing or picking up pollen from its back (Figures 9.2.23a and b). This mechanism ensures cross-pollination between plants. Some other plants have similar mechanisms in which the stamens are pulled in towards the centre of the flower, usually hitting the pollinator (Figure 9.2.23c).

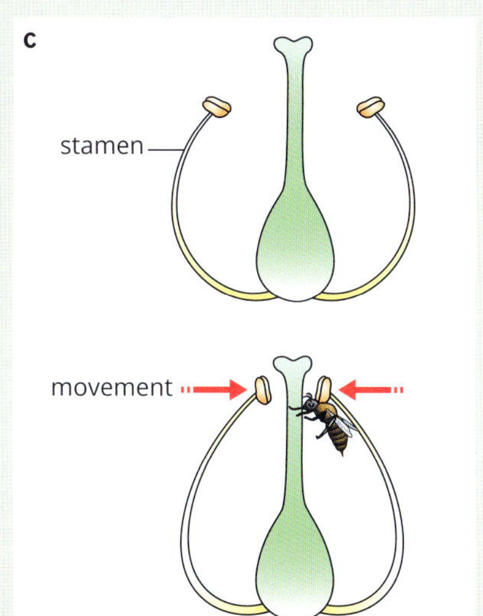

FIGURE 9.2.23 (a) The column of a trigger plant flower before it is triggered. (b) The column after it has been triggered. (c) In some plants, the touch of a pollinating insect can cause the stamens to be pulled rapidly inwards, depositing pollen on the insect.

Behavioural adaptations of animals

Behavioural adaptations in animals that help them to survive in extreme environmental conditions include:

- seeking or leaving shade or shelter
- evaporative cooling to lower body temperature
- huddling to maintain body temperature
- migration.

Seeking or leaving shade or shelter

Many desert animals have behavioural adaptations that are very important in regulating the rate of heat exchange with their environment. An example is the central netted dragon (*Ctenophorus nuchalis*). To raise its body temperature, this lizard emerges from under a rock and basks in the sunshine, spreading itself out at right angles to the Sun's rays. To lower its body temperature or reduce the rate of increase in body temperature, the lizard orientates its body parallel to the Sun's rays, minimising the exposed surface area, or simply retreats beneath a rock or into a burrow (Figure 9.2.24).

Some animals such as desert snakes and tortoises adopt nocturnal behaviour during summer to prevent overheating. They move only in the cooler evening, avoiding the extreme heat of the day. Animals may also seek shelter to increase their body temperature when it is cold or windy.

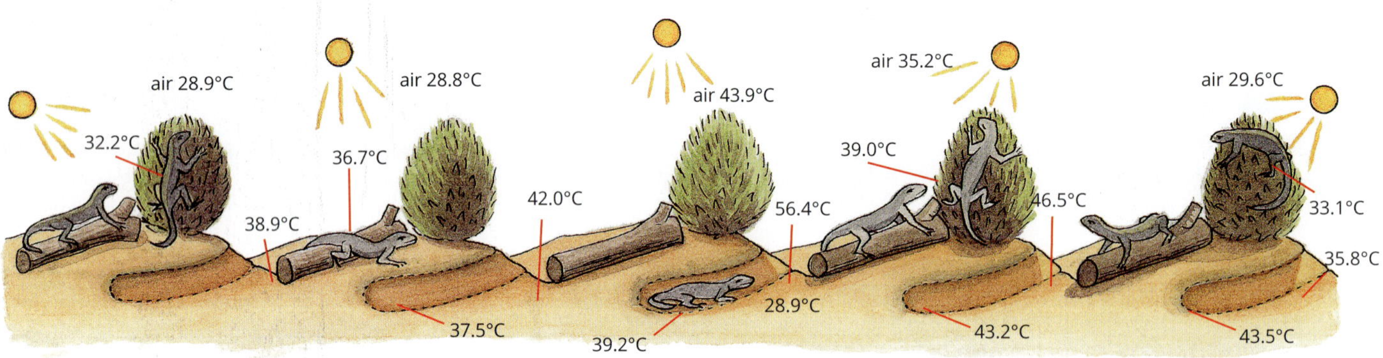

FIGURE 9.2.24 The central netted dragon (*Ctenophorus nuchalis*) has adapted its behaviour to desert conditions, regulating its temperature throughout the day by seeking shade or basking in sunlight.

Evaporative cooling

Evaporative cooling is used by many land animals to lower their body temperature by releasing heat into the environment. Although evaporative cooling is a physiological adaptation, it is achieved by behavioural adaptations, such as:

- panting or licking limbs
- spraying water on the body
- wallowing in mud or water
- mouth gaping
- gular fluttering
- urohydrosis.

Panting or licking limbs enable animals to release heat effectively using evaporative cooling. For example, kangaroos lick their paws, and animals such as dogs, gazelles and foxes pant. The fennec fox (*Fennecus zerda*) has been observed panting at a rate of 690 times per minute after chasing prey. The rate of panting is proportional to the amount of air flowing over the tongue. If animals can flatten their tongue, increasing its surface area, while increasing their panting rate, then the cooling effect is greater. Sometimes even penguins have to pant. In warmer weather, they also hold their flippers out of the water so that both surfaces are exposed and can release heat via evaporative cooling.

FIGURE 9.2.25 A female African elephant (*Loxodonta africana*) splashes water over her body, making use of evaporative cooling to control her body temperature. Mud remaining on the elephant's skin provides protection against solar radiation.

Spraying water on the body is commonly seen in elephants (Figure 9.2.25) but is also a behaviour used by many other animals.

Wallowing in mud or water is a very common behaviour. Animals such as pigs, elephants, rhinoceroses and deer wallow in mud; the wet mud acts like sweat to cool the skin. Animals such as hippopotamuses, tapirs, bison, horses and cattle wallow in water to cool down.

Mouth gaping is seen in many animals, such as crocodiles and alligators (Figure 9.2.26). When air moves across the moist surface of an open mouth, evaporative cooling from the membranes inside the mouth reduces the temperature of blood being supplied to the brain.

Gular fluttering is a cooling behaviour in which birds flap membranes in their throat to increase evaporation from the moist buccal (mouth) region; as the air temperature increases, birds increase the amount of gular fluttering.

Urohydrosis is a cooling behaviour exhibited by some birds, including vultures and storks. They urinate on their legs, creating an evaporative cooling effect.

FIGURE 9.2.26 A saltwater crocodile (*Crocodylus porosus*) opens its mouth to allow water to evaporate from its moist tongue.

FIGURE 9.2.27 Emperor penguin chicks huddling together for warmth

Huddling

Huddling is used by many animals, such as penguins, to cope with cold temperatures (Figure 9.2.27). Thousands of emperor penguin chicks may huddle together for warmth in the spring, when they begin to develop their adult plumage. By huddling, penguins decrease the surface area of the group exposed to the harsh environment. They continually rotate which animals are on the outside, each taking a turn in the freezing cold winds.

Migration

Some animals move extremely long distances each year to inhabit a different area. This type of seasonal pattern of relocation is known as migration. The purpose of migration is usually to seek better food availability, or to move to a better site for breeding along with suitable climatic conditions. Birds navigate their migratory paths using the position of the Sun and Moon, as well as topographical details and cues from the Earth's magnetic field. Migration is an innate behaviour prompted by cues from the environment, such as the length of daylight. These cues are closely coordinated with an animal's biological clock and trigger biological responses, such as increased feeding before migration.

BIOFILE

Wallowing in mud is cool

Wallowing in mud has many advantages for animals, including skin maintenance, camouflage, parasite control, protection from solar radiation, and social play. One of the more common reasons is thermoregulation. Many animals, such as hippopotamuses, elephants and pigs wallow in mud to lower their body temperature. Like sweating, the evaporation of the water in the mud cools the animal's skin by carrying heat away from the body. It can cool the animal's body by up to 2°C, making it more efficient than sweating. Wallowing in mud has an advantage over water, too; the water in the mud evaporates more slowly than water alone, keeping the animals cooler for longer.

A pig wallows in mud to cool down.

CASE STUDY

Adaptations of mangroves

Mangroves grow in the intertidal zone on shallow, muddy shores (Figure 9.2.28). This environment presents them with constantly changing and challenging conditions that they need to adapt to, including:

- fluctuating salinity levels with the movement of the tide or from freshwater entering a tidal river
- lack of oxygen for their roots because they are growing in waterlogged soil
- boggy, unstable soil that makes anchorage difficult
- seed dispersal in an aquatic environment.

FIGURE 9.2.28 Mangroves grow in the constantly changing and challenging environment of the intertidal zone.

Getting rid of salt

Mangroves have three methods of ridding themselves of salt: exclusion, excretion and accumulation. Some mangroves exclude salt by actively pumping it out across membranes at their root surface. Other mangroves, including *Ceriops* (in Queensland) and *Avicennia* (which grows as far south as Victoria), also have salt-excreting glands on their leaves (Figure 9.2.29).

FIGURE 9.2.29 Salt crystals on a mangrove leaf. The salt was excreted in solution through specialised salt glands, and evaporation formed the crystals. This physiological adaptation allows the plant to regulate internal salt concentration.

Specialised roots

Oxygen normally enters roots through lenticels, which are rough spots consisting of loose, corky tissue through which gas exchange can occur. Mangroves have evolved a range of aerial roots, all of which have lenticels. These aerial roots include peg roots, pneumatophores and stilt roots. Pneumatophores increase the surface area exposed to the air at low tide for maximum oxygen uptake. These types of aerial roots, together with cable roots that spread laterally, also help stabilise the plant in the soft mud (Figure 9.2.30). The cable roots have a mat of fine, hair-like roots that absorb nutrients and water.

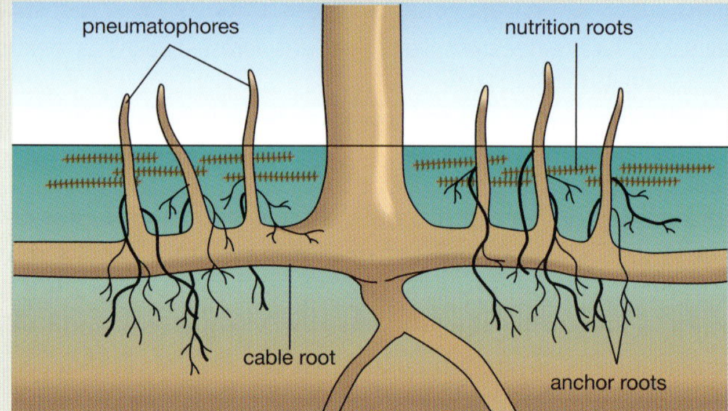

FIGURE 9.2.30 The structure of the root system of a typical mangrove plant, showing cable roots and pneumatophores.

Seed dispersal

The seeds of mangroves are buoyant and are adapted for dispersal by water. Some mangroves are viviparous. Viviparity in botany (plant science) means that the seed germinates and the young plant starts to develop while still attached to the parent plant. When the germinating seed falls into the water it already has a developing root system (Figure 9.2.31). This enables it to quickly anchor itself before it can be washed away by wave action.

FIGURE 9.2.31 Mangrove seeds have adaptations that enable them to disperse in their aquatic environment. In some species the seeds already have a developing root system before they leave the parent plant, and the seeds of most species can float in water.

CASE STUDY **ANALYSIS**

Camels—adaptation marvels

Dromedaries (*Camelus dromedarius*; Figure 9.2.32) combine numerous adaptations to survive in the desert. Desert conditions push homeostatic mechanisms to the limit, creating huge physiological variations that would kill most other animals. A fully hydrated camel has a body temperature range of 36–38°C. When dehydrated and exposed to high environmental heat, a camel's body temperature may fluctuate up to 41°C. Other animals also allow their body temperature to increase, but not to the same extent.

Most of the adaptations seen in camels are for coping with high temperatures, but they also have adaptations that allow them to walk on sand and withstand sandstorms.

FIGURE 9.2.32 Dromedary (single-humped) camel (*Camelus dromedarius*).

Structural adaptations

The following structural adaptations help camels to cope in their environment:

- wide toes that spread their body weight, allowing them to easily walk on sand
- long, thin legs with a high surface area to volume ratio that help them to lose heat more efficiently
- long eyelashes, and nostrils that can easily close, to protect them from wind and sand particles.

Physiological adaptations

The following physiological adaptations help camels to cope in their environment:

- very efficient kidneys that produce extremely concentrated urine
- a fat hump, which acts as insulation, and a total lack of fat on the rest of the body to permit cooling
- countercurrent blood flow in their heads, which protects the brain from overheating.

Behavioural adaptations

The following behavioural adaptations help camels to cope in their environment:

- When camels eat a plant, they take small bites, not consuming the entire plant. That way the plant is not killed so that it will grow back and produce more plants for the camel to eat.
- Camels allow their urine to run down their legs so that it cools them due to evaporative cooling.
- As it is cooler at night, camels spend that time searching for food to reduce energy and water loss.
- During the day, camels orientate themselves towards the Sun, presenting the least possible body area for the absorption of radiant heat.

Analysis

1 Describe two ways in which camels reduce water loss during the day.

2 Use the data provided about camels to answer the following questions.

   a State the optimum range for a camel's body temperature.

   b State the tolerance range for a camel's body temperature.

   c Use your answers to parts **a** and **b** to draw a labelled tolerance range graph of the camel's body temperature. Label the body temperature values, and mark the tolerance range, optimum range, zone of physiological stress and zone of intolerance.

3 Environmental temperatures in the desert may fluctuate and range from 40°C during the day to 20°C at night on average. However, a camel's body temperature when well hydrated usually lies between 36°C and 38°C. Using this information, determine if a camel is an endotherm or an ectotherm. Explain your reasoning.

9.2 Review

OA ✓✓

SUMMARY

- An adaptation is an inherited characteristic that increases the likelihood of survival and reproduction of an organism in a particular environment.
- The optimum range is the range of abiotic factors within which an organism thrives and functions at its best.
- The tolerance range is the range of abiotic factors within which an organism can survive but not thrive. The tolerance range includes the optimum range.
- There are three main types of adaptation: structural, physiological and behavioural.
- Structural adaptations are anatomical or morphological features that help an organism to survive in its environment.
- Examples of structural adaptations of plants are:
  - fewer, sunken or protected stomata
  - stomatal hairs to create a humid microclimate
  - thick, waxy cuticle
  - leaf shape: rolled leaves; reduced surface area.
- Examples of structural adaptations of animals are:
  - thick fur and blubber
  - large or small ears
  - spines for protection against predators
  - body coverings for camouflage.
- Physiological adaptations relate to the functioning of the animal at the biochemical, cellular, tissue, organ, system and whole organism levels.
- Examples of physiological adaptations of plants are:
  - CAM photosynthesis
  - frost tolerance
  - salinity tolerance.
- Examples of physiological adaptations of animals are:
  - countercurrent heat exchange mechanisms
  - dormancy, hibernation, torpor and aestivation
  - production of venom or poisons
  - shivering
  - production of antifreeze proteins.
- A movement or behavioural adaptation is how an organism acts or moves in response to its environment.
- Examples of adaptations for movements in plants are:
  - phototropism
  - geotropism
  - thigmotropism
  - nastic movements.
- Examples of behavioural adaptations of animals are:
  - huddling
  - panting, licking skin, wallowing and gular fluttering
  - seeking shade or sunlight
  - nocturnal activity.

KEY QUESTIONS

Knowledge and understanding

1. What are the three main types of adaptations?
2. List three types of abiotic factors and give an example of how each can affect an organism.
3. How does the ratio of surface area to body volume affect the ability of an animal to regulate its body temperature? Provide examples for a hot and cool environment.
4. What is countercurrent heat exchange?
5. Why is excessive soil salinity a problem for plants? List two physiological adaptations that allow plants to survive in highly saline environments.

Analysis

6. Salinity is an example of an abiotic factor for which organisms have a specific tolerance range. Consider a freshwater fish and a saltwater fish, and compare their tolerance ranges for salinity.
7. Draw a diagram showing three different types of leaves that are adapted to different environments. Label the important structures. In what sort of environments would you find these plants?
8. **a** What is the crassulacean acid metabolism (CAM) pathway? How does it differ from the normal photosynthetic process?
   b What are some of the disadvantages of the CAM pathway? Why would this be a poor strategy in a damp environment?

9.3 Interdependencies between species

Every day, organisms interact with one another and their physical environment in a variety of ways (Figure 9.3.1). No organism lives in complete isolation, meaning they have adapted to intermingle in many ways: they eat one another, and they also compete with one another for food, space, mates and nest sites. This section explores the importance of interdependencies between species within an ecosystem, including the impact of changes to keystone species and predators and the ecological roles that each have in structuring and maintaining the distribution, density and size of a population. You will also learn how Aboriginal and Torres Strait Islander peoples' knowledge and perspectives have contributed to our understanding of species' adaptations and interdependencies in Australian ecosystems.

FIGURE 9.3.1 Many flowering plants depend on their ability to attract insects and birds to pollinate flowers and disperse seeds. Bees feed on the nectar in blossom, and at the same time transfer pollen from one flower to another.

INTERACTIONS IN THE ENVIRONMENT

Within an environment, there are many different levels of organisation, including individual organisms, populations, species, communities and ecosystems.

- A **population** is a group of organisms of the same species, living together in a defined geographic area.
- A **species** is a group of organisms that can interbreed to produce viable, fertile offspring.
- A **community** is different species living together in a particular place at a particular time, interacting with one another.
- An **ecosystem** is a self-sustaining complex that includes organisms, the physical environment and the interactions between them in a particular area. The complexity of ecosystems varies depending on the number of populations, species and communities within them and the interactions between them. A tropical rainforest contains many different species and is an example of a complex ecosystem, whereas a desert ecosystem is simpler because it contains fewer species. The role that a species occupies in an ecosystem is called its **ecological niche**.

INTERACTIONS BETWEEN SPECIES

Species in an ecosystem are interdependent; that is, they rely on each other. If one species is removed from an ecosystem, any species that interacted with it are affected. Interactions between species are usually classified according to how the interaction affects the survival and reproduction of the species involved. Interactions between species can be classified as beneficial, benign or harmful. Interactions can also be classified as feeding and non-feeding interactions. For example, predation is a feeding interaction.

Interactions between different species are known as **interspecific** interactions. Interactions between individuals of the same species are known as **intraspecific** interactions.

Symbiosis

Sometimes two quite different organisms live and function together in a close association, to the benefit of at least one of them. Different species living together in a close partnership is called **symbiosis**. Each species is called a symbiont. Mutualism, commensalism and parasitism are all examples of symbiotic relationships.

BIOFILE

Symbiosis in the sea

Cleaner fish and cleaner shrimps have a symbiotic relationship with their hosts. They eat dead skin and parasites on the surface of larger marine animals. This photograph shows a suckerfish attached to the shell of a sea turtle. The suckerfish eats food scraps from the feeding activity of the turtle, as well as parasites on the turtle's shell.

A suckerfish attached to the shell of a sea turtle in the Red Sea.

Beneficial interactions

Beneficial interactions are those in which both species involved benefit. Mutualism is a type of beneficial interaction between species.

Mutualism

Mutualism is a partnership between two different kinds of organism where both of them benefit. There are two types of mutualism: obligate mutualism and facultative mutualism. **Obligate mutualism** is when both species are completely dependent on each other for survival and reproduction. One cannot survive without the other. Many pollination relationships are examples of obligate mutualism. Yucca plants (Figure 9.3.2) rely exclusively on yucca moths to move pollen from one plant to another, and in turn the moths depend on the flowers for a safe place to hatch their eggs.

FIGURE 9.3.2 A Yucca plant in Texas, USA. Yucca plants have an obligate mutualistic relationship with yucca moths.

Facultative mutualism is when both species benefit from interacting but do not rely on each other for their survival. This is sometimes called 'proto-cooperation'. Some species can rid themselves of harmful parasites through facultative mutualism. For example, tick-birds and oxpeckers (Figure 9.3.3) stand on cattle or other large animals and feed on the parasites in the hair and on the skin of their host. The birds benefit from easy access to their food source.

FIGURE 9.3.3 A yellow-billed oxpecker (*Buphagus africanus*) cleans ticks from a giraffe (*Giraffa camelopardalis*).

Benign interactions

Benign interactions are interactions in which no species is harmed. Commensalism is a type of benign interaction.

Commensalism

Commensalism is an interaction between species in which only one species benefits but the other species is not harmed. Animals such as birds or possums nesting in a tree hollow is an example of commensalism. In this case the bird or possum benefits and the tree is not harmed (Figure 9.3.4).

Trees are also often host to epiphytes: smaller plants such as orchids, ferns, mosses, liverworts and lichens that live on the trunk or in the crown of the tree (Figure 9.3.5). The epiphyte receives sunlight and rainwater. This relationship is usually benign for the tree because it is neither helped nor harmed (unless it becomes overloaded with the weight of the epiphytes on its branches).

FIGURE 9.3.4 Many owl species nest in tree hollows where it is relatively safe during the day. This is an example of commensalism, because the owl benefits and the tree is not harmed.

FIGURE 9.3.5 Epiphytes form a commensal relationship with the tree they grow on.

CASE STUDY

Bacteria and coral bleaching

Corals live in a symbiotic relationship with unicellular algae called zooxanthellae. The photosynthetic algae provide food for the coral polyps and the polyps produce a hard skeleton of calcium carbonate, which is a home for both organisms (Figure 9.3.6a). Corals are bleached (become pale) and eventually die when they lose their zooxanthellae (Figure 9.3.6b). This can occur when sea temperatures rise, causing the coral polyps to become stressed and to expel their algal partner. Increased coral bleaching has been connected with rising sea temperatures associated with climate change.

It has been suggested that the increased bleaching of corals may be caused by a bacterial infection that occurs when the seawater temperature rises. In 2004 Eugene Rosenberg found a new species of bacterium, *Vibrio shiloi*, that is always associated with bleached and dying coral (*Oculina patagonica*) in the Mediterranean Sea. This bacterium is closely related to *Vibrio cholerae*, the bacterium that cause cholera.

Experiments found that coral did not bleach in water at 29°C, but did if *Vibrio shiloi* was present. No bleaching occurred even when the bacterium was present if the water temperature was only 16°C. Rosenberg's research team found a temperature-dependent protein produced by the bacterium that enables it to stick to the coral. Once the bacterium has stuck to the coral, it multiplies and penetrates the polyp. Inside the polyp it synthesises a toxin that inhibits photosynthesis, killing the zooxanthellae. The temperature-dependent protein explains why the bacterium can only enter and grow in the coral polyp in the warm months of the year, causing coral bleaching at that time.

This is a remarkable example of the interactions between species (coral polyps, zooxanthellae and bacteria) and their dependence on a particular characteristic of the physical environment (water temperature).

FIGURE 9.3.6 (a) Coral polyps have a mutualistic relationship with unicellular algae (zooxanthellae). (b) Coral bleaching.

BIOFILE

Lichens

Lichens look like they are one organism, but they are in fact two organisms in an obligate mutualistic relationship. They consist of an alga and a fungus that live and work together. The alga photosynthesises to produce food, which the fungus can share. The fungus shelters the alga and absorbs mineral nutrients and water. Lichens can break down and extract minerals from solid rock.

This close symbiotic relationship enables lichens to colonise areas that are hostile to other organisms. They can grow on tree trunks, on bare rocks, in the cold of Antarctica and in the heat of the desert—places where the algae and fungi would not survive on their own.

A lichen is the result of an obligate mutualistic relationship between an alga and a fungus. The orange pigment that colours this lichen is produced by the fungus.

BIOFILE

Viruses are obligate parasites

All forms of life—even bacteria—are susceptible to viruses. A virus is characterised by its ability to infect a living organism; it invades a living cell and uses the cell's own structures and metabolism to replicate. Viruses cannot grow or reproduce outside of a living host cell. This makes them obligate parasites, because they rely completely on their host for their survival.

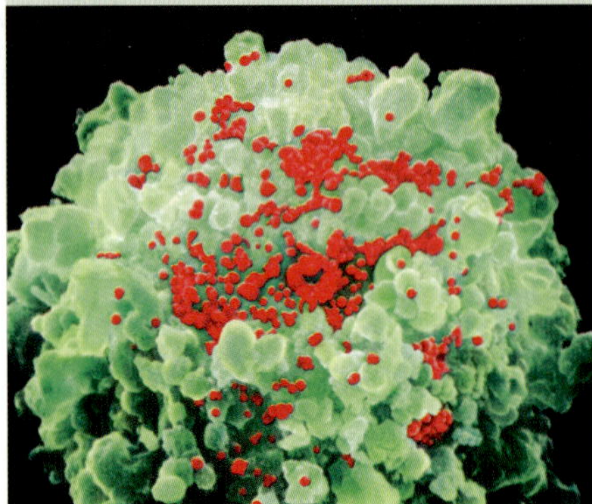

This living T lymphocyte blood cell (green) is infected with human immunodeficiency virus (HIV) (red). The virus is a parasite, using the blood cell to reproduce.

Harmful interactions

Harmful interactions are those in which only one species benefits, and the other species is harmed as a result of the interaction. Parasitism and amensalism are types of harmful interactions.

Parasitism

In **parasitism**, one species (the parasite) benefits and the other species (the host) is harmed. Ectoparasites such as ticks and mistletoes live on or outside the host (Figure 9.3.7). Endoparasites such as parasitic fungi and wood-borers live inside the host (Figure 9.3.8). A **parasite** obtains its food from the host but does not necessarily kill it. The type of harm the parasite causes the host varies, but may include shortened lifespan, impaired functions such as digestion, photosynthesis or reproduction, reduced ability to withstand stresses such as drought or cold, and greater vulnerability to predators.

FIGURE 9.3.7 Mistletoes are ectoparasites; they grow on other plant species and take nutrients and water from them.

FIGURE 9.3.8 This stick insect has been invaded, and is being killed, by a parasitic fungus. The fruiting bodies of the fungus are erupting from the insect's body and will release spores that will help the parasite to spread.

Amensalism

Amensalism refers to an association between species in which one is inhibited or killed and the other species is unaffected.

A simple model of amensalism is the way in which animals can inadvertently damage vegetation around them but are unaffected by the relationship. For example, animals such as sheep and cattle often trample grass. The grass may be damaged or killed, but the animals receive no benefit from having done so (Figure 9.3.9a). Similarly, some waterbirds such as cormorants kill vegetation in places where they roost or nest (Figure 9.3.9b). This is because their droppings have a high nitrogen, phosphate and potassium content, which plants cannot tolerate.

FIGURE 9.3.9 (a) Larger animals such as sheep and cattle trample and damage grass, but are unaffected by the relationship. (b) Cormorant droppings can kill vegetation, but this does not benefit the birds.

Competition

All organisms have a set of biological requirements for their survival and reproduction. All organisms need resources such as nutrients and water to sustain themselves, shelter for protection, and mates for reproduction. **Competition** occurs when two organisms require the same resource and there is limited access to this resource. Competition can be intraspecific (between individuals of the same species) or interspecific (between different species).

For example, populations of two species of protists, *Paramecium aurelia* and *Paramecium caudatum*, will grow rapidly in separate but identical cultures, and then both populations will level off (Figure 9.3.10.) However, if these species are grown together, the population of *P. caudatum* grows initially, but then its population decreases to extinction. In other words, *P. aurelia* outcompetes *P. caudatum*. *P. aurelia* continues to reproduce to reach a similar population density as in the first experiment, in which the two species were separated.

Population growth of two species in competition

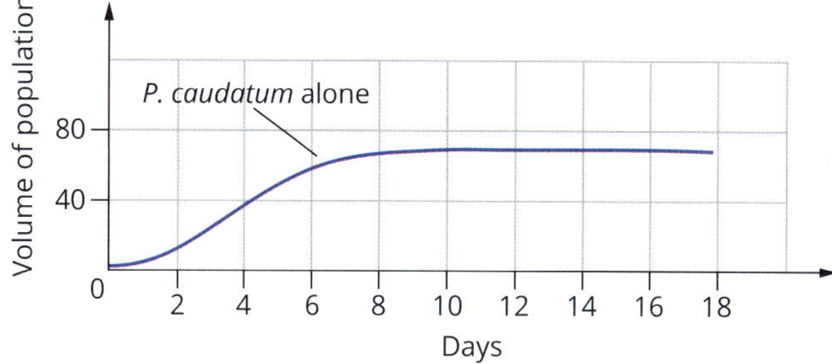

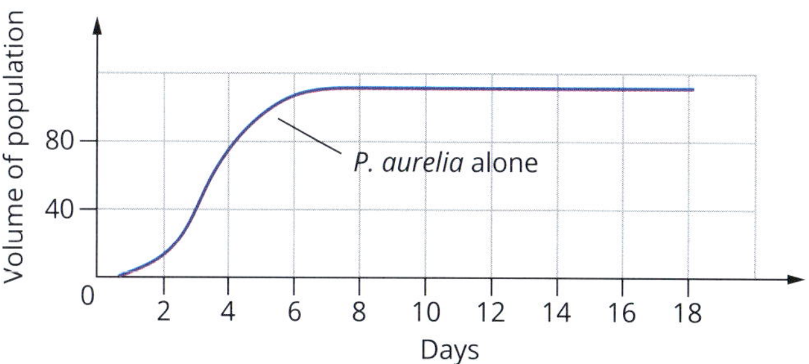

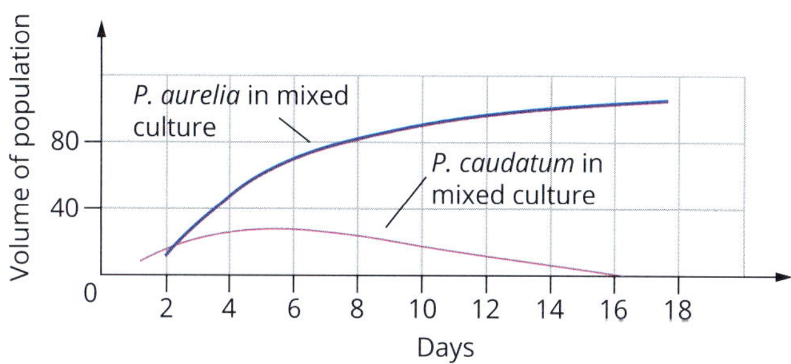

FIGURE 9.3.10 When two species of *Paramecium* are cultured together, the population growth of both species is slowed until eventually *P. aurelia* outcompetes *P. caudatum*.

Feeding interdependencies

All organisms in an ecosystem require energy to survive. Organisms can be divided into two groups depending on the strategies they use to obtain energy. **Autotrophs** (self-feeders) make their own energy from sunlight using photosynthesis. Plants, algae and cyanobacteria are types of autotrophs. **Heterotrophs** (other-feeders) obtain energy by eating other organisms. Heterotrophs include all animals and fungi. Some heterotrophs feed on plants, some feed on insects or other animals, and others feed on dead and decaying material. Almost all organisms are consumed by at least one other organism. Even organisms that do not have any predators are consumed by other organisms when they die.

Species are interconnected within food chains and food webs in ecosystems. **Food chains** and **food webs** link organisms according to their feeding relationships. The position of an organism in a food chain or food web is known as its **trophic level**. Unsurprisingly, when a change occurs to one species, other species and sometimes even the entire food web are affected. The complexity of a food web gives an ecosystem its stability. In a simple food web, the loss of one species would have a disastrous effect on the other organisms. In a complex food web, the loss of one species has less effect, since alternative food sources are often available. Examining the feeding relationships between organisms in an ecosystem can help in understanding why some species are affected by such changes. For example, sea otters along the coasts of the northern and eastern North Pacific Ocean feed on sea urchins and keep the sea urchin numbers in balance. If sea otters were suddenly removed from the ecosystem, sea urchin numbers would increase, and the kelp that the urchins feed on would be overgrazed (Figure 9.3.11).

FIGURE 9.3.11 Sea otters (*Enhydra lutris*) have a feeding relationship with sea urchins that is important for ecosystem balance.

Predation

FIGURE 9.3.12 The (a) Canadian lynx (*Lynx canadensis*) and (b) snowshoe hare (*Lepus americanus*) form a predator–prey relationship.

Predation occurs when one animal (the **predator**) kills and feeds on another animal (the **prey**). If the density of the prey species increases, predators will have more access to this food source and their population will increase. The increased predation will then reduce the population of the prey species. As the number of prey falls, intraspecific competition in the predator population will reduce its population size.

Predator and prey relationships constantly fluctuate in this way. For example, the Canadian lynx (*Lynx canadensis*) preys almost exclusively on the snowshoe hare (*Lepus americanus*) (Figure 9.3.12). The population of hares varies according to factors such as climate, disease, and availability of food. An increase in the hare population leads to an increase in the lynx population. When there are more lynxes, the hare population may decline and this in turn may cause the lynx population to decline again. The graph in Figure 9.3.13 shows this repeating cycle.

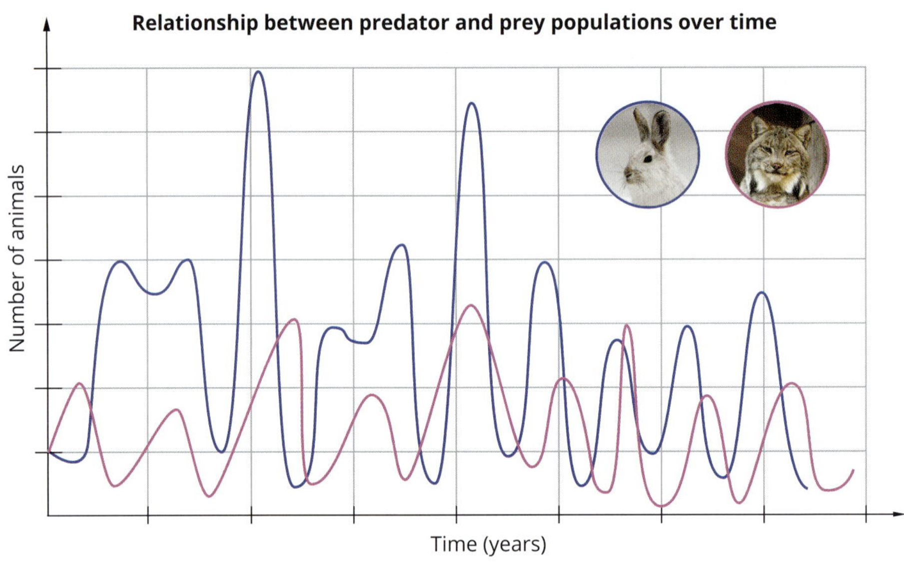

FIGURE 9.3.13 As the number of prey (snowshoe hares, indicated by the blue line) increases, so does the number of predators (Canadian lynxes, indicated by the pink line). Over time the prey population decreases as a result of predation, which leads to a decrease in the predator population.

KEYSTONE SPECIES

Some species in an ecosystem can be identified as keystone species. A **keystone species** is a species that plays a critical role in maintaining the structure and functioning of their ecosystem. This species is key to the existence of many other species and is so named after the stone at the top of an arch (a keystone) that holds the arch together and gives it strength. The keystone species does not need to be abundant or a top predator to have a large effect, but without the keystone species, the arch (or the ecosystem) may collapse (Figure 9.3.14). When a keystone species is removed from an ecosystem, the ecosystem becomes much less stable and its structure changes. Keystone species have a much greater impact on their ecosystems than might be expected from their numbers.

keystone

FIGURE 9.3.14 A keystone is the central stone that supports the arch structure in a stone archway. A keystone species is so named because it maintains the structure of its ecosystem in a similar way.

The first keystone species

The term keystone species was first applied during a study of food webs in rock pools. The carnivorous sea star *Pisaster ochraceus* was identified as the top predator in the rock pools, feeding on the mussel *Mytilus californianus* (Figure 9.3.15). As an experiment, all the *P. ochraceus* were removed from one rock pool (Figure 9.3.16a). A nearby rock pool was left undisturbed as a control (Figure 9.3.16b). In the test rock pool, the remaining species competed with each other to occupy the extra space and use the additional resources made available. The mussel species began to dominate and quickly overran the site, crowding out the other species. The mussels consumed so much of the limpets' algal food source that the limpet population declined. Within a year, the diversity of species decreased from 17 to just 7. In the control rock pool, there was no change in species number or distribution. Consequently, the interaction between the sea star and mussel was shown to support the structure and species diversity of these communities. As these significant changes resulted from the removal of the sea star, *P. ochraceus* was named a keystone species.

FIGURE 9.3.15 Sea star *Pisaster ochraceus* feeds on a mussel, *Mytilus californianus*, in a rockpool.

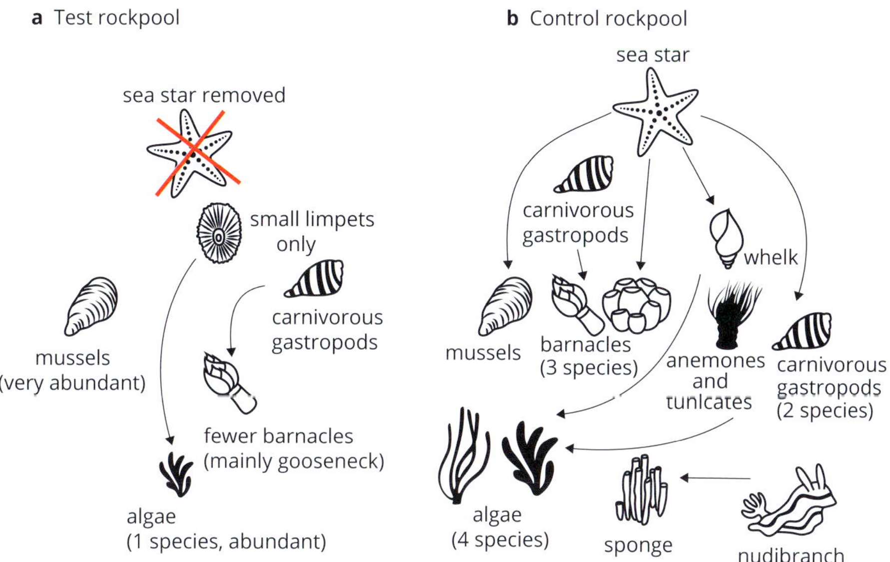

FIGURE 9.3.16 Schematic representation of the keystone role of predatory sea star *Pisaster ochraceus* in an intertidal ecosystem. (a) Removal of *P. ochraceus* allows mussels (*Mytilus californianus*) to dominate, and reduces species diversity. (b) Predation by the sea star *P. ochraceus* maintains a diverse community.

This experiment demonstrated the impact of removing a keystone species from an ecosystem. As keystone species usually have a greater impact on an ecosystem than their numbers would indicate, identification of keystone species is very difficult. Often it is only after a keystone species has been removed from an ecosystem or their numbers have been greatly reduced that their importance to an ecosystem is noticed.

FIGURE 9.3.17 The great white shark (*Carcharodon carcharias*)

Example of a keystone species: Great white shark

The number of great white sharks (*Carcharodon carcharias*) has been declining rapidly over the past 50 years, mostly because they are caught in fishing nets or are hunted for their fins (Figure 9.3.17). The great white shark is a keystone species that helps maintain the stability of marine food webs, and the decline of this species has had far-reaching effects on marine ecosystems. The great white shark is a predator at the top of the food chain (apex predator), keeping populations of fish, seals and sea lions in check. This in turn affects the populations of organisms that the fish, seals and sea lions consume.

The importance of the great white shark to marine ecosystem stability had not been recognised until shark numbers had significantly declined. The loss of apex predators such as the great white shark often results in a trophic cascade. A **trophic cascade** is a top-down effect where the loss of an apex predator results in a large increase in the number of mid-level predators, also called mesopredators. In 2002, scientists proposed that a trophic cascade led to a severe reduction in the abundance of bivalve molluscs in the waters off the coast of North Carolina, USA. In this area, the apex predators (several species of large sharks) preyed upon the mesopredators (smaller species of sharks and rays). The removal of the apex predators was hypothesised to have caused a trophic cascade resulting in a complete collapse in the numbers of bivalve molluscs, which were a food source of the now abundant mesopredators.

This resulted in the loss of the scallop (a bivalve) fishing industry, which had operated successfully for more than 100 years. Figure 9.3.18 shows the changes in abundance measured over a period of time of the large sharks (apex predators, top row of graphs), the smaller sharks and rays (mesopredators, middle row of graphs) and the bay scallops (bottom row of graphs) that were the prey of the mesopredators.

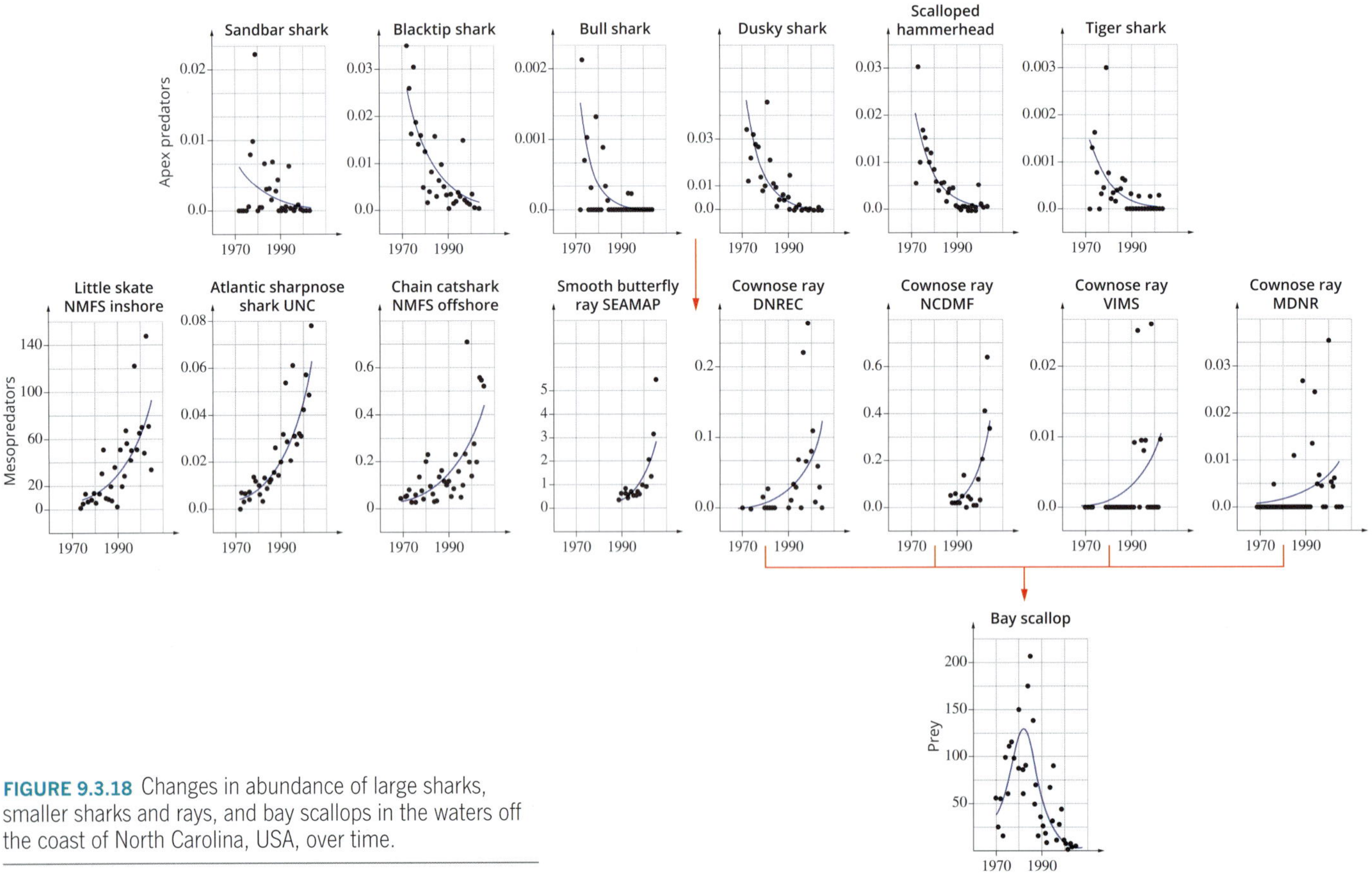

FIGURE 9.3.18 Changes in abundance of large sharks, smaller sharks and rays, and bay scallops in the waters off the coast of North Carolina, USA, over time.

Keystone species and habitat

Some species are keystone species because they affect the habitats of an ecosystem. For example, elephants preserve the grasslands of African savannas by eating any young trees that grow (Figure 9.3.19). Without the elephants, the savannas would be dominated by trees and shrubs and eventually become forests or shrublands, and the many smaller grazing herbivores such as wildebeests and zebras would starve.

Elephants also play an important role in forest ecosystems and are often referred to as the 'gardeners of the forest'. They disperse more seeds of a diverse range of plant species over greater distances than any other animal. While foraging on tree sprouts and shrubs, they carve pathways through forests, opening up gaps and allowing sunlight to reach the forest floor. This elephant behaviour helps develop a more productive ground layer for a variety of plant species to grow and also provides food for animals that live under the forest canopy.

FIGURE 9.3.19 Elephants are a keystone species in African savannas because they maintain the grassland ecosystem.

Keystone species and human impacts

Human activities can have a detrimental effect on ecosystems, particularly where these activities affect a keystone species. One example is the culling of grey wolves from Yellowstone National Park, USA (Figure 9.3.20).

The wolves were originally seen as a pest, but their eradication led to a rapid increase in the elk population, which resulted in overgrazing and the loss of aspen and willow plants. This led to a loss of habitat and food for many smaller species such as beavers and songbirds, as well as stream bank erosion and water sedimentation. In 1995, grey wolves were reintroduced to Yellowstone National Park and the ecosystem is slowly recovering.

FIGURE 9.3.20 Grey wolves are a keystone species in Yellowstone National Park because they keep the elk population in check.

BIOFILE

Keystone quoll

One well-known keystone species is the northern quoll (*Dasyurus hallucatus*), also known as the native cat. This species has become endangered for many reasons, including bushfires and feeding on poisonous cane toads. The quoll feeds on a large variety of foods, including fruit, insects, birds, mammals and reptiles. Through feeding, the quoll helps control the numbers of its prey species, and with the quoll's decline, the delicate balance of those populations is being disrupted.

The northern quoll (*Dasyurus hallucatus*) is an endangered species of carnivorous marsupial found in Queensland, the Northern Territory and Western Australia.

FIGURE 9.3.21 (a) The pink rock orchid (*Dendrobium kingianum*) and (b) its geographic distribution.

Biomass is an amount of organic matter. It is usually expressed as mass per unit area, such as kg/m^2.

FIGURE 9.3.23 The density of plant lice can be measured by counting the number of lice per leaf.

POPULATION DYNAMICS

In theory, populations should continually increase in size as more individuals are produced. However, this is rarely the case in an ecosystem. Instead, population distribution, density and size are determined by a variety of factors that influence rates of births, migration and deaths.

Population distribution

Distribution or range is all the places where a particular species can be found. For example, emus are found only in Australia, and kiwis are found only in New Zealand. The pink rock orchid (*Dendrobium kingianum*) is restricted to parts of eastern Australia (Figure 9.3.21). **Endemic** species are those that are found only in one defined geographic region (e.g. Australia).

Distributions may change over time. For example, humans have reduced the distribution of many species by clearing habitat and disrupting ecological processes. Conversely, humans have increased the distribution of some species by accidentally or deliberately transporting them to new ecosystems.

There are three basic patterns of distribution (Figure 9.3.22): random, uniform and clumped (or clustered).

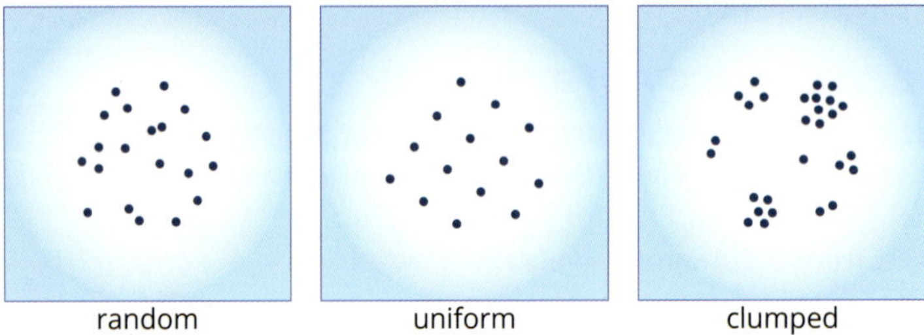

FIGURE 9.3.22 The three patterns of population distribution: random, uniform and clumped.

Population density

The **density** of a population is the number of individuals per unit of area or volume. For example, this might be the number of rabbits in a given area or the number of fish in a particular volume of water.

If it is difficult to count individuals, the density of a population can be measured in **biomass**. For example, counting blades of grass is a tedious process, so a grass population is often measured in kilograms per unit area. A small area of grass can be cut and weighed, and then this value can be used to calculate the total biomass of the population.

The area used to measure density might sometimes be represented by a less conventional unit. For example, the population density of plant lice can be expressed as the number of individuals on one leaf (Figure 9.3.23).

Populations can also be discussed in terms of abundance. **Abundance** is the total number of individuals in a population.

Population size

The size of a population is affected by four processes:

- **natality** (births or germination)
- **mortality** (deaths)
- **immigration** (organisms moving into a population)
- **emigration** (organisms moving out of a population).

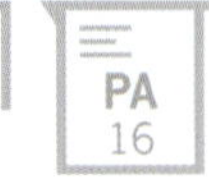

Birth and immigration bring new individuals into a population and increase the population size. Death and emigration remove individuals from a population and decrease the population size. Immigration and emigration are collectively known as **migration**. These four processes determine the rate of change in a population over time (Figure 9.3.24).

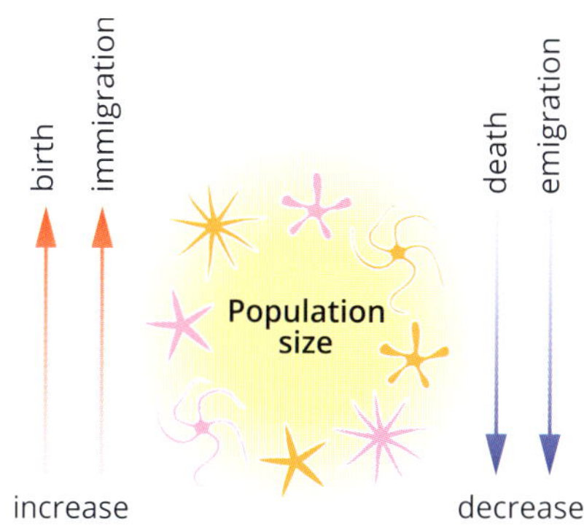

FIGURE 9.3.24 The size of a population depends on birth rate, death rate and migration.

Exponential population growth

Ecologists use mathematical formulae to model the theoretical growth of a population over time. The graph in Figure 9.3.25 shows a theoretical growth curve for a population in an ideal environment.

The graph assumes that the number of immigrants equals the number emigrants over time. Change in this population is therefore a function of births and deaths only. This type of growth is known as **exponential growth**. The J-shaped curve of the graph is characteristic of exponential population growth.

As long as the birth rate is higher than the death rate, a population will grow. If the birth rate remains consistently higher, then the population may grow exponentially.

Year	0	1	2	10
Size of population	100	200	400	102 400

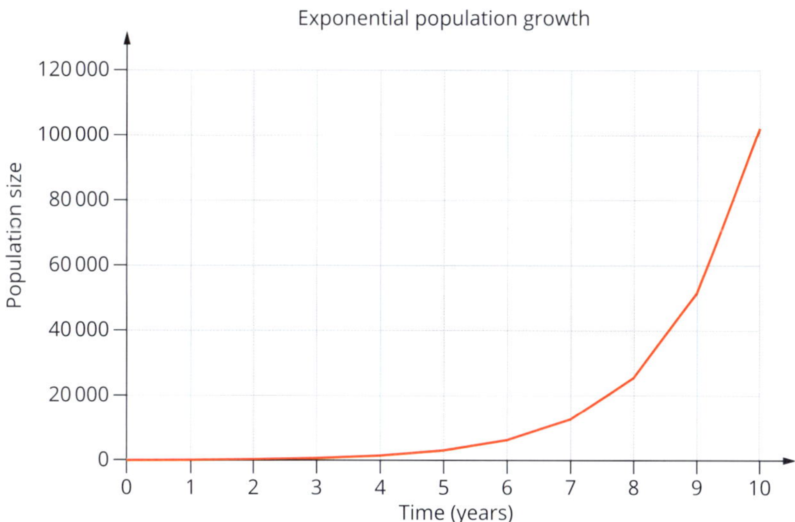

FIGURE 9.3.25 The size of a population over time (years), showing exponential population growth.

Species that tend to experience exponential population growth are those that have a short generation time and give rise to large numbers of offspring. Examples of these are bacteria, many weed species and some types of insects. In most instances exponential population growth occurs only for relatively short periods of time.

Exponential growth is normal for some plants and animals when environmental conditions are favourable and resources are abundant (plentiful). Because these conditions generally last only a short time, exponential growth is usually short-lived. But if favourable conditions continue then a 'population explosion' may occur.

BIOFILE

Invasive population growth

Salvinia fern is a free-floating aquatic weed that often has population explosions. It can survive for up to 20 months in dry conditions, but under favourable environmental conditions, such as high nutrient levels, it can double its population every two to five days. If these conditions continue, salvinia will form a dense mat on top of a waterway, preventing other aquatic plant life from receiving sunlight. Due to its growth rate and damage to aquatic habitats, this introduced species is a declared prohibited weed throughout Australia.

Salvinia fern (*Salvinia molesta*) is one of the world's most invasive aquatic weeds.

CASE STUDY

Crown-of-thorns sea star population explosion

The crown-of-thorns sea star (*Acanthaster planci*) is a large echinoderm that lives on coral reefs of the Indian and Pacific oceans, close to the coasts of eastern Africa, Japan, Hawaii and tropical Australia (Figure 9.3.26). Adults have up to 23 arms, all of which are covered on the top with poisonous spines. Adult crown-of-thorns sea stars eat anemones and the soft-bodied polyps of corals, leaving behind only the skeleton of the coral.

It is likely that the crown-of-thorns sea star has been on the Great Barrier Reef for as long as the reef has been in existence, keeping fast-growing corals under control. However, population explosions occasionally occur, and they are caused by several factors.

- The abiotic environment*:* The population explosions may be triggered by heavy rains and cyclones, which result in increased nutrient levels being washed from the land into the sea. This in turn leads to an increase in phytoplankton, which is the food of crown-of-thorns larvae. With an abundant food source, more larvae survive into adulthood.
- Reproduction: The crown-of-thorns sea star can reproduce at a great rate when conditions are favourable. A female produces over one million eggs in a spawning season, so even a small increase in the survival of offspring will result in a much larger adult population.
- Predator control: The crown-of-thorns sea star has some natural predators, such as large molluscs (tritons), but it is protected by its large size, poisonous spines and nocturnal behaviour. Fishing may also have reduced the number of predators.

FIGURE 9.3.26 The crown-of-thorns sea star (*Acanthaster planci*)

- Dispersal: The crown-of-thorns sea star can swim and spread for the first 3–4 weeks of its life before settling at the bottom of the sea.

Population explosions of the crown-of-thorns sea star reduce populations of coral (Figure 9.3.27). Previously coral has recovered over time, but there is concern that coral will be destroyed faster than it can regrow. Research by the Australian Institute of Marine Science shows that the coral population has declined by approximately 50% in the past 30 years. The crown-of-thorns sea star is one of the major causes of this decline, along with cyclones and ocean warming.

FIGURE 9.3.27 When numbers of crown-of-thorns sea stars increase, coral is killed, leaving only bleached coral skeletons.

BIOFILE

An explosion of algae

An algal bloom is a population explosion of aquatic phytoplankton (algae or cyanobacteria), causing water to change colour and become toxic. A common culprit in lakes and ponds is the cyanobacterium *Anabaena*.

The algal bloom outbreak in this pond is very evident in the dense green colour of the water.

FIGURE 9.3.28 During a drought, water becomes the limiting factor for population growth.

Limiting factor

The **limiting factor** of a population's size, density or growth is the scarcest of the resources needed by a population. For example, food, water, shelter, nutrients and light are essential for a population's growth. If all of these resources except water are available in large quantities, water is the limiting factor and organisms will have to compete for it (Figure 9.3.28).

Factors that affect population size and density

All populations have the potential for exponential growth, but a number of factors can affect population size and density and prevent initial or continued exponential population growth. The size of a population is affected by many factors interacting in complex ways. However, the effects of one factor can often amplify (increase) the effects of other factors.

Factors that influence population size are either density-independent factors or density-dependent factors.

Density-independent factors

Density-independent factors affect a population's size regardless of the size or density of the population. They include:

- the conditions in which the species can survive; that is, its daily and seasonal tolerance range for various abiotic factors
- major changes or disturbances to the environment, such as bushfires, droughts, floods or human-made changes.

Natural disasters and human-made changes can have wide-ranging effects on plant and animal species. The destruction of habitat can displace many organisms, and some major changes to the environment can kill organisms (Figure 9.3.29).

Density-dependent factors

The effects of **density-dependent factors** on population size increase as the population density increases. Density-dependent factors include:

- competition for resources, such as food, water, shelter and mates
- predation
- crowding
- parasitism
- infectious disease.

a

b

c

FIGURE 9.3.29 (a) Bushfires affect many animal and plant populations. (b) Clearing rainforests for palm oil plantations in Malaysia destroys habitat for many species, including elephants, tigers and orangutans. (c) Pollution of waterways reduces oxygen levels and causes the death of fish and other aquatic organisms.

Carrying capacity

In the absence of a limiting factor, the population growth of a species will be continuously exponential. However, in the real world, population growth is affected by density-dependent factors such as competition for resources.

When a species' population reaches **equilibrium** and becomes relatively constant, with the number of births and deaths in the population cancelling each other out, the species has reached the maximum population size that the ecosystem can sustain indefinitely. This is called the ecosystem's **carrying capacity** for that species.

The S-shaped graph in Figure 9.3.30 shows the initial exponential growth of a population, which then flattens out as it begins to be affected by density-dependent factors. The population growth rate may decline until births and deaths balance each other and the population is limited to the carrying capacity of the environment. This pattern of growth is known as **logistic growth**.

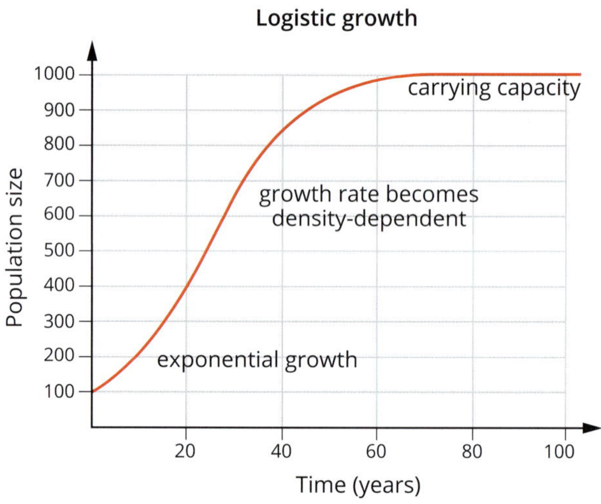

FIGURE 9.3.30 As a population increases, density-dependent factors begin to limit population growth and the population reaches its carrying capacity.

The carrying capacity of an environment is dynamic; that is, it varies over time. Factors that can affect carrying capacity include:

- weather and climate changes
- major changes to an environment
- fluctuation in populations of food species or competitors.

The factors that can affect carrying capacity can be abiotic or biotic (Table 9.3.1). Water availability is an example of an abiotic factor that can affect carrying capacity. For example, during a drought water availability might become the limiting factor for a population of kangaroos, and the number of individuals that the environment can support will be reduced.

TABLE 9.3.1 Abiotic and biotic factors affecting carrying capacity

Abiotic	• soil • water • space • shelter
Biotic	• fluctuation of prey species • fluctuation of predator species • fluctuation of species that compete for resources

INDIGENOUS AUSTRALIAN CONNECTION TO THE LAND

Indigenous Australians, as the traditional custodians of the land, have a rich and sophisticated relationship with the landscape, established over more than 65 000 years. A deep sense of belonging to Country connects the land to all aspects of identity, purpose and belonging. Rather than owning land, Indigenous people are custodians of land which holds the story of their culture, spirituality, language, lore and creation. They are entrusted with the knowledge and responsibility to care for that land. Indigenous peoples' understandings of the concept of Country is informed by the idea that all things are interconnected and interdependent, and they believe that if you look after Country, then Country will look after you.

This deep, nurturing relationship between people and the land is often described as 'connection to Country'. The relationship between many Indigenous people and the land is one of reciprocity and respect—the land sustains and provides for people, and people manage and sustain the land through culture and ceremony. Because of this close connection, when the land is disrespected, damaged or destroyed, this can impact the wellbeing of Indigenous people.

This intimate relationship with the land has allowed Indigenous people to sustain their ecosystems and environments for thousands of years, and to this day, Indigenous people continue to draw on their Indigenous knowledge for sustainable practices.

FIGURE 9.3.31 The Chinnup season in the Gariwerd (Grampians) seasonal cycle is known as the season of cockatoos and occurs between late May and late July.

As custodians of the land, it is important to understand the ways in which different species interact. While some species live together in harmony, others will compete or prey on one other. These interactions can be indicators of weather conditions and the health of the ecosystem. Observing the behaviours of species provides insight into the changes occurring in an ecosystem, such as the signs of a season beginning and ending. For example, the Jardwadjali and Djab Wurrung peoples of the Grampians in Victoria have six distinct seasons characterised by the behaviour observed in animals, the flowering of plants and the weather patterns (Figure 9.3.31). Knowing the breeding cycles and relationships between species is also important for sustainable resource management, such as when to hunt particular species to ensure food is available year-round.

CASE STUDY

The innovation of Indigenous ecosystem management

The Booderee National Park is located in the Jervis Bay Territory, a three hour drive south of Sydney. 'Booderee' is the Dhurga word for Bay of Plenty and the park is owned by the Wreck Bay Aboriginal community on the traditional lands of the Yuin nation. It is jointly managed by Parks Australia, the Wreck Bay Aboriginal community and other key stakeholders. Each year the Joint Board of Management implements a range of sustainability programs, including fire management. The use of fire to manage local ecosystems is a cultural practice that has been used by Indigenous peoples for over 50000 years. In fact, recent research indicates that fire management is more than just land management—it is an active intervention that increases the productivity of the land for agricultural practices. As such, land management with 'firestick farming' is a highly complex and selective operation. For example, fire was used on land with more fertile soils to maximise food production, and poorer soil was left for forest. Indigenous fire-management practitioners employ careful and considered works, taking cues from observations of the seasons, environment and weather conditions, based on five key principles:

1. The majority of agricultural lands are burnt on a rotating mosaic. This involves burning small patches of vegetation with a low-intensity fire to create a mosaic of burnt and unburnt areas, allowing plants and animals to survive in refuges.
2. The type of Country to be burnt and the condition of the bush informs the time of year for the fires to be lit.
3. The timing of the burn is informed by the prevailing weather.
4. The growing season of particular plants are avoided at all costs.
5. Clans work together applying mutually agreed rules to coordinate the burning or non-burning. Neighbouring clans are advised of all fire activity.

The planned application of fire management eliminated the risk of uncontrolled fire. This strategy broke the land into mosaics of cleared grassland and areas of forest, increasing both the diversity and productivity of the land. The application of Indigenous peoples' knowledge to manage the land using fire systematically over the landscape purposefully shaped conditions that favour certain crops, increase yields and suppress weeds. The relationship between Indigenous Australians' active role in caring for Country influenced the adaptations and interdependencies between species and has enabled them to act like keystone species of the ecosystems for tens of thousands of years.

One example is the domestication of the yam daisy, or murong (*Microseris lanceolata*), a tuber that provided a staple source of fibre and protein in the diet of Indigenous people (Figure 9.3.32). The careful maintenance of the yam fields enabled bountiful harvests, and cultivation of the soils enabled establishment of rich pastures or croplands.

FIGURE 9.3.32 The yam daisy or Murong (*Microseris lanceolata*) is a highly valued food crop cultivated by Indigenous Australians over many centuries.

In many instances areas are burnt after plant species produce seeds. Many native plants require intense heat to assist in seed production or germination and the ash in the soil provides nutrients for plant growth (Figure 9.3.33). Extensive knowledge of the interdependencies between species in an ecosystem is vital to ensure that the timing, intensity and size of the fire applied allows the ecosystem to recover.

FIGURE 9.3.33 Firestick farming is used by Indigenous Australians for a variety of reasons, including to assist in seed production or germination.

Indigenous fire management practices are now being integrated with other methods to more sustainably and proactively support Australian ecosystems. For example, Booderee National Park's fire management program includes the adoption of fire regimes, fire intensities and fire frequencies to maintain the park's ecosystem and manage bushfire risk.

9.3 Review

SUMMARY

- An ecosystem is a self-sustaining complex that includes organisms, the physical environment and the interactions between them in a particular area.
- Interactions between different species are known as interspecific interactions. Interactions between individuals of the same species are known as intraspecific interactions.
- Species in an ecosystem are interdependent; that is, they rely on each other. If one species is removed from an ecosystem, any species that interacted with it are affected.
- Interactions between species can be classified as beneficial (mutualism), benign (commensalism) or harmful (parasitism or amensalism). Interactions can also be classified as feeding and non-feeding interactions.
- Keystone species are those that are critical to the stability and structure of an entire ecosystem. If they are removed, the entire ecosystem is affected.
- Keystone species often have a much greater impact on an ecosystem than their numbers would suggest.
- Distribution or range is all the places where a particular species can be found.
- The density of a population is the number of individuals per unit of area or volume.
- Population size is affected by four processes:
  - births (natality)
  - deaths (mortality)
  - immigration
  - emigration.
- The limiting factor of a population's size, density or growth is defined as the scarcest of the necessary resources needed by a population to grow.
- Factors that influence population size are either density-independent factors or density-dependent factors.
  - Density-independent factors affect population size regardless of population density and include the conditions in which the species can survive (i.e. its tolerance range) and major changes or disturbances to the environment.
  - Density-dependent factors limit population size more as the population density increases and include competition, predation, crowding, parasitism and infectious disease.
- Carrying capacity is the maximum population size of a species that its environment can sustain indefinitely. Carrying capacity is not fixed.
- Indigenous Australians have a close relationship with the land and place a strong emphasis on taking care of the land. This is known as caring for Country.
- Indigenous Australians' extensive knowledge of interdependencies between species and their adaptations has allowed them to successfully manage and live in close connection with Australian ecosystems for thousands of years.

KEY QUESTIONS

Knowledge and understanding

1 What is an ecosystem? Give two examples of an ecosystem.

2 State whether each of the following interactions is parasitism, mutualism or predation.
   a lice living in human hair and feeding on human blood
   b orangutan feeding on durian fruits
   c a sea eagle catching and eating a fish
   d a lizard eating an earthworm
   e clownfish feeding among sea anemones
   f vampire bats feeding on the blood of live birds or mammals
   g bees feeding on a flower and taking pollen to other flowers
   h a crown-of-thorns sea star feeding on coral polyps
   i *Plasmodium* protozoans living in humans and causing malaria

3 Describe interspecific and intraspecific competition and give an example of each.

4 Define the term 'keystone species'.

continued over page

9.3 Review *continued*

5 Label the blanks in this diagram with the four processes that affect population size for every species.

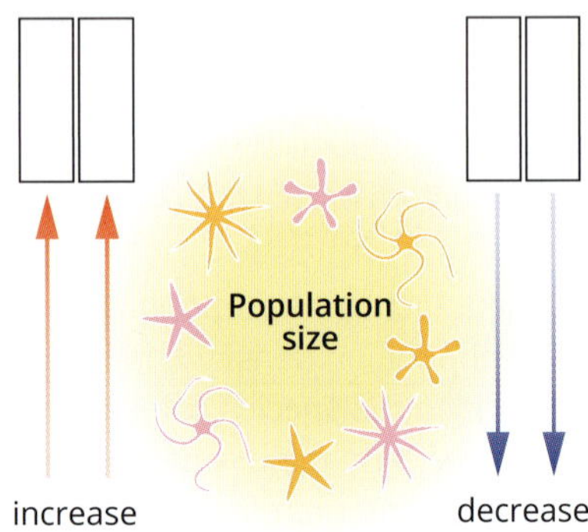

6 Use a specific example to explain how one abiotic factor can affect population growth.

Analysis

7 How would you measure the density of species in each of these situations? Explain your answer in each case.

a sheep in a paddock

b grass in a field

c leaves on a plant

8 The graph below gives information about three lizards and the termites that are their prey. Using the information in the graphs, explain how the three lizard species are able to exist together in an ecosystem.

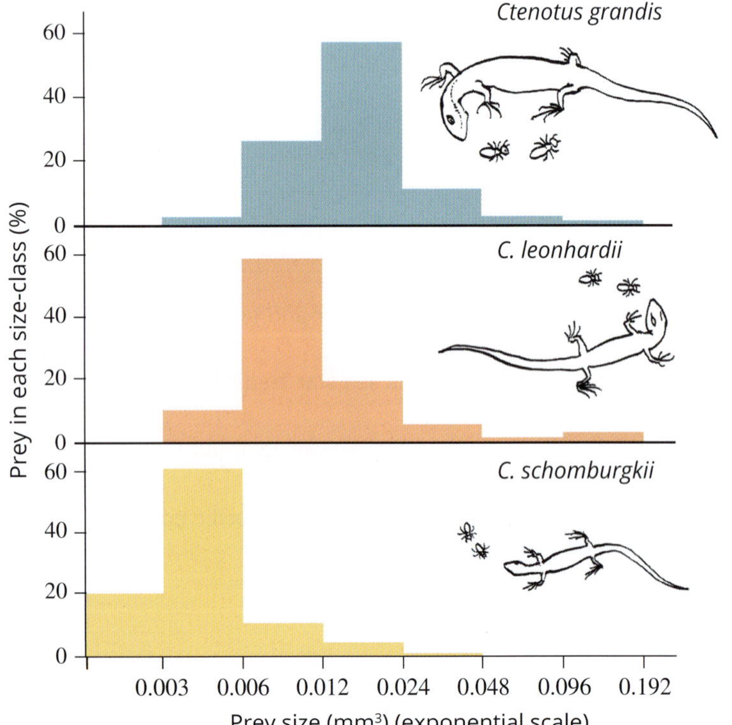

9 The food web below shows a simple marine ecosystem.

a Identify a potential keystone species in this food web. Explain your choice.

b Predict three possible consequences of the removal from this ecosystem of the keystone species that you identified.

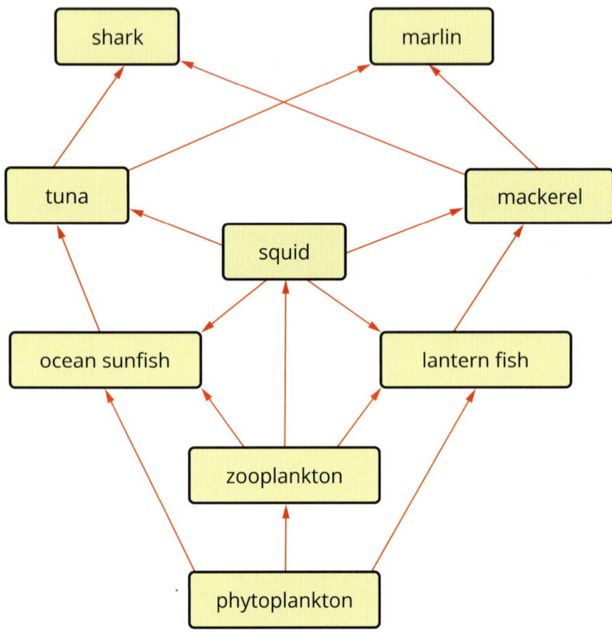

10 How do Indigenous Australians' understandings of Country differ to non-Indigenous understandings of Country?

Chapter review

OA ✓✓

09

KEY TERMS

abiotic factor
abundance
adaptation
aestivation
allele
amensalism
autotroph
behavioural adaptation
biological fitness
biomass
biotic factor
bottleneck effect
brumation
CAM photosynthesis
carrying capacity
commensalism
community
competition
countercurrent heat exchange
density
density-dependent factor
density-independent factor
distribution
ecological niche
ecosystem
ectotherm
emigration
endemic
endotherm
equilibrium
evaporative cooling
evolutionary potential
exponential growth
facultative mutualism
food chain
food web
founder effect
gene flow
gene pool
genetic diversity
genetic drift
halophyte
heterotroph
hibernation
hydrophyte
immigration
inbreeding
inbreeding depression
interspecific
intraspecific
keystone species
limiting factor
logistic growth
mesophyte
migration
mortality
mutualism
nastic movement
natality
natural selection
obligate mutualism
optimum range
parasite
parasitism
physiological adaptation
physiological stress
population
predation
predator
prey
selection pressure
species
stimulus
structural adaptation
symbiosis
tolerance range
torpor
trophic cascade
trophic level
tropism
xerophyte

REVIEW QUESTIONS

Knowledge and understanding

1 Which of the following is a source of new alleles in a species?
   A gene flow
   B crossing over
   C genetic drift
   D mutation

2 Define genetic diversity.

3 Explain how zoos increase or maintain genetic diversity in captive populations.

4 Does genetic drift have a greater impact on the genetic diversity of small or large populations? Explain your answer.

5 Define the term 'adaptation'.

6 The ability of the octopus to camouflage by changing colour is an example of a:
   A structural adaptation
   B behavioural adaptation
   C physiological adaptation
   D limiting factor

7 Describe how adaptations benefit individuals, populations and species.

8 Copy and complete the following table by listing adaptations that animals have to survive in the environments listed.

Environment	Adaptation		
	Structural	**Physiological**	**Behavioural**
desert			
snow			
deep underwater			
long, cold winter			

9 Describe how CAM (crassulacean acid metabolism) photosynthesis is a beneficial adaptation for plants living in the desert.

10 Describe how an emperor penguin is adapted to life in the harsh Antarctic climate, mentioning at least one structural, one physiological and one behavioural adaptation.

11 The birds in a backyard aviary include galahs, corellas and sulphur-crested cockatoos. Is the aviary an ecosystem? Explain your answer.

12 A keystone species is critical to the stability of a whole ecosystem. Give one example of such a species and explain what makes it a keystone species.

13 A species of fish on a coral reef undergoes a rapid decline in population growth after the arrival of another species of fish that uses the same sources for food. What kind of competition is this? Why would the population growth of the first fish decline?

14 Name three limiting factors that affect population growth.

15 Explain how low genetic diversity in Tasmanian devils relates to an increased risk of extinction due to factors such as devil facial tumour disease.

Application and analysis

16 Complete the following table, where S is a structural adaptation, P is a physiological adaptation and B is a behavioural adaptation. More than one adaptation may be correct for each organism.

Organism	Feature	Type of adaptation (S, P or B)	Benefits to organism
mangrove	pneumatophore		
honey possum	long, brush-like tongue		
kangaroo	sleeps in shade during the day		
echidna	goes into torpor		
saltbush	salt-secreting glands in leaves		

17 a Describe three adaptive advantages of bioluminescence, providing one example of each.

b Bioluminescence is particularly common in the ocean, suggesting that it is an advantageous trait, but it is comparatively rare on land. Why would bioluminescence be more advantageous in the ocean than on the land?

18 The spotted jelly, *Mastigias papua*, needs to incorporate free-living algae called zooxanthellae, because it obtains its energy mainly from the carbon fixed by the algae. The spotted jelly is also able to obtain energy by feeding on phytoplankton, tiny invertebrates and microbes. What type of interspecific relationship do the spotted jelly and algae have?

A predatory
B commensalism
C obligate mutualism
D facultative mutualism

19 Which of the following situations describes a population that is increasing in size?
(B = birth, D = death, I = immigration, E = emigration)

A (B + D) > (I + E)
B (B + E) > (D + I)
C (D + E) < (B + I)
D (B + E) < (D + I)

20 What is the relationship between evolution and biodiversity? Define the term 'evolutionary success' and propose why this term does or does not reflect the true relationship between evolution and biodiversity.

21 Endothermic and ectothermic animals regulate their body temperatures in different ways. Consider the following graph, which shows the body temperatures of a bobcat (pink line) and a snake (blue line) for different ambient (environmental) temperatures.

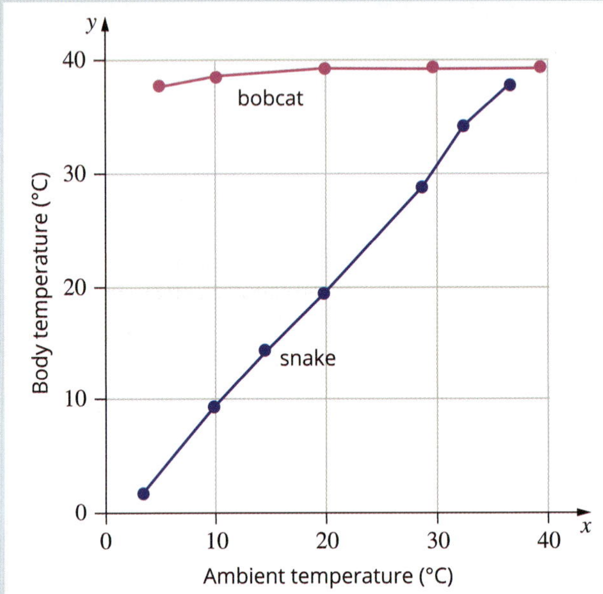

a When the ambient temperature is 30°C, what is the body temperature of the snake?

b Use this graph to identify which animal is endothermic and which animal is ectothermic. Explain how you reached this conclusion.

22 Mammals that hibernate are generally quite small (bats, rodents and pygmy possums) and yet bears (brown, black, grizzly or polar) are also said to be hibernators. A bear's body temperature may only drop a few degrees, compared to other hibernators whose temperatures can go down to as little as 5°C. Based on what you know of torpor and hibernation, judge whether bears are true hibernators.

23 During hibernation, the oxygen–haemoglobin dissociation curve of squirrels shifts to the left, as shown on the graph below. This means that oxygen binds more readily to haemoglobin in the blood. Determine how this might be an advantage to a hibernating animal.

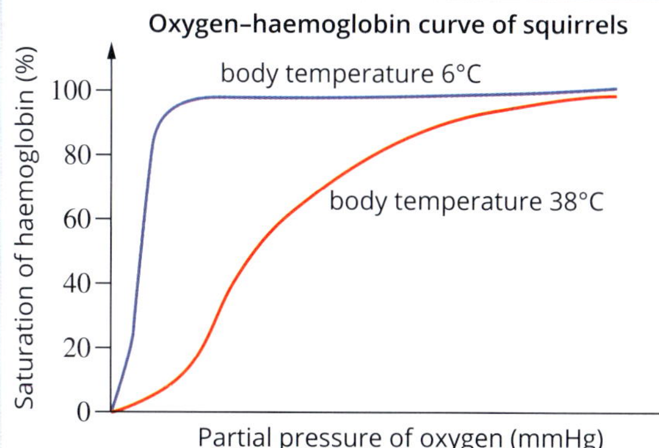

24 Consider the following graph, which shows changes in the rabbit and fox populations on an isolated area of open farmland in western Victoria.

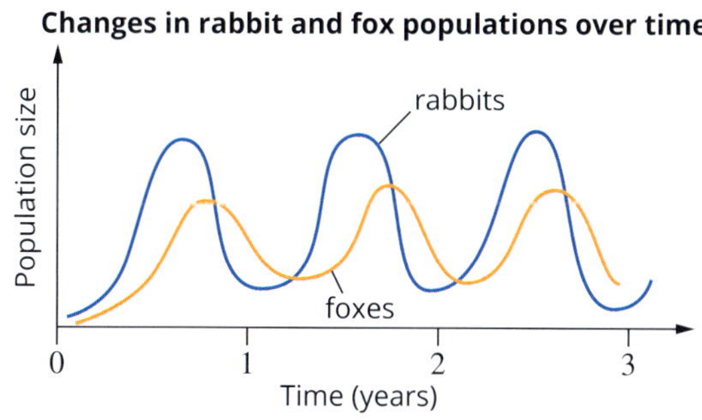

- **a** What factors might have caused the initial growth in the rabbit population?
- **b** Why does the growth in fox numbers follow that of rabbits?
- **c** What factors could cause the decline in the rabbit population?
- **d** How are the rabbit numbers able to build up again in the following year?
- **e** What would happen to the fox population if the rabbit population suddenly crashed, for instance from the effects of calicivirus? Show this by extending the graph.

25 On their website, along with scientific data and analysis of weather patterns, the Australian Bureau of Meteorology has a section dedicated to Indigenous weather knowledge. They acknowledge that through storytelling and ceremony, many Aboriginal and Torres Strait Islander communities have continued to propagate millennia of knowledge of how various plants and animals react to the weather and environment around them.

Many Aboriginal and Torres Strait Islander nations have seasonal calendars that reference the environmental events of their Country, which has helped them to survive on the land for thousands of years—this is called 'reading the Country'.

- **a** How did the Indigenous nations maintain historical knowledge of their environment without access to modern methods of record keeping?
- **b** The people of the Gariwerd (Grampians) area in Victoria recognise six distinct weather periods that reference climate changes as well as patterns of plant flowering and fruiting and animal behaviour. Some of this knowledge is outlined below. Analyse it to answer the questions that follow.

Chinnup is the season of cockatoos, around late May to the end of July. During Chinnup, possums have pouch young; fungi and certain orchids appear on the ground; mistletoe, early wattles and eucalypts start to flower (attracting insects and nectar-feeding birds); and yellow-tailed black cockatoos seek new feeding grounds.

Kooyang is the eel season. It falls around January to March at the hottest and driest time, when bushfire risk is highest. During Kooyang, many insects are in larval form or hatching, snakes are seen basking in the morning sun, eels and native trout are on the move in the streams, and herbal medicines like tannins from gums and wattles are most available.

  - **i** Propose why the natural events of Chinnup were the signal for people to prepare possum-skin cloaks and move to areas with rock shelters.
  - **ii** Predict some of the foods that would have been eaten in each of these two seasons.
  - **iii** Identify three examples of species interdependencies.
  - **iv** Use a table to classify all the biotic and abiotic factors in the seasonal descriptions above.
- **c** Assess whether Indigenous knowledge is more relevant to local seasonal cycles than the four-season Western model that is usually applied across the whole of Australia.

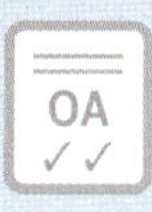

REVIEW QUESTIONS

How do inherited adaptations impact on diversity?

Multiple-choice questions

1 Select the option that is not an example of asexual reproduction.
- A reproduction via outgrowths of the cell in baker's yeast
- B fertilisation of orchids resulting in formation of a fruit
- C formation of plantlets on specialised leaves of kalanchoe
- D formation of spores during sporogenesis without meiosis in red algae

2 Which of the following does not contribute to variation in future offspring?
- A recombination
- B germline cell mutations
- C somatic cell mutations
- D independent assortment

3 Select the most accurate definition of a clone.
- A a genetically identical organism
- B an exact copy of a different species
- C an organism produced by genetic engineering
- D a genetically modified organism

4 Artificial embryo splitting is a technique used for:
- A whole organism cloning by in vitro separation of an early stage embryo into two
- B human IVF programs to provide twins for the parents
- C tissue culture of endangered plant species
- D somatic cell nuclear transfer research to create identical clones of valuable livestock

5 Four adaptations of the Australian red kangaroo are:
1 a dense network of blood vessels close to the skin in the forelimbs
2 licking the forelimbs in hot weather
3 a powerful tail that acts as a counterbalance when hopping
4 widening (dilation) of the blood vessels in the forelimbs in hot weather.

Which one of the following choices correctly classifies these four adaptations?
- A 1 = physiological, 2 = behavioural, 3 = structural, 4 = physiological
- B 1 = structural, 2 = behavioural, 3 = structural, 4 = physiological
- C 1 = structural, 2 = behavioural, 3 = structural, 4 = behavioural
- D 1 = physiological, 2 = behavioural, 3 = physiological, 4 = structural

6 Organisms display a variety of survival adaptations. Which row in the following table has them identified in the correct categories?

	Structural	Physiological	Behavioural/ movement
A	blubber layer	CAM photosynthesis in plants	tropisms in plants
B	frost tolerance	small ears	sunken stomata
C	rolled leaves	hibernation	salinity regulation in plants
D	webbed feet	feathers	camouflage markings

7 Which of the following statements regarding parasitism is false?
- A Parasitism benefits one species and the other species is not harmed.
- B Many mould species are parasites.
- C Parasitism is not a limiting factor for the host organism.
- D Parasites usually live on or in their host.

8 A tree branch encrusted with lichen is seen here with a mountain pygmy possum.

The relationship between lichen and tree is an example of:
- A parasite–host
- B symbiosis
- C mutualism
- D commensalism

9 Identify the density-dependent factor.
- A rainfall
- B fire
- C temperature
- D food supply

10 What type of environment are sclerophyll plants adapted to?

A wetlands and estuaries

B harsh, dry climates and nutrient-deficient soils

C tropical climates with nutrient-rich soils

D hot, arid deserts with sandy or rocky surfaces

11 The following two graphs show the population growth of species A and species B. Which of the following statements is correct?

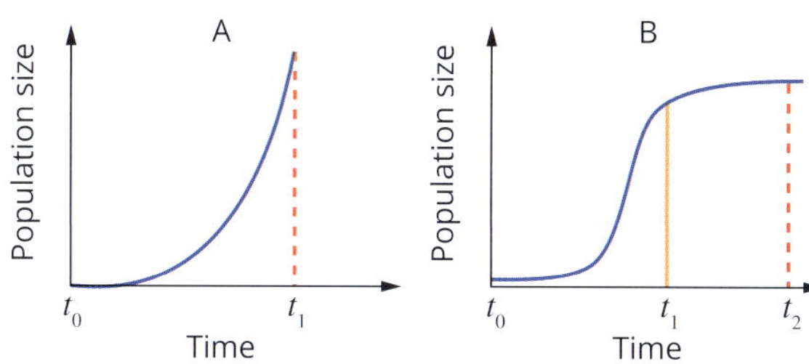

A Species A shows exponential population growth between times $t = t_0$ and $t = t_1$, but species B does not.

B Species B has reached the carrying capacity of the ecosystem by time $t = t_2$.

C No deaths occurred in species A or B between times $t = t_0$ and $t = t_2$.

D The graph for species B is typical of an insect population such as a locust plague.

Short-answer questions

12 Some species of aphids are capable of reproducing either asexually (via parthenogenesis) or sexually. Some other organisms have the capacity to reproduce asexually by fragmentation.

a Compare parthenogenesis and fragmentation.

b Explain the advantage of the following:

i asexual reproduction when the environment is favourable

ii switching to sexual reproduction after a period of asexual reproduction or when the environment becomes unfavourable.

13 Cultivated bananas are sterile and the fruit develops without seeds. However, cultivated bananas can be propagated by tissue culture. Each resulting new plant is a clone of the parent plant. The overall banana production around the world has decreased recently because of banana freckle disease.

a How is the process of propagating plants by tissue culture different from cutting and grafting?

b What is the advantage of using tissue culture to grow plants?

c Discuss the advantages and disadvantages of cloning crops such as bananas.

14 Today, scientists are using genetic technologies that artificially manipulate reproduction for many organisms.

a Reproductive technologies include cloning techniques. Outline the methods used for two different cloning techniques.

b Predict how the use of cloning techniques would alter the genetic diversity of a population.

c Analyse the benefits of using reproductive cloning technologies. Support your answer with examples.

15 Biodiversity provides resilience to change. Interpret this statement, using examples to support your answer.

16 Water is essential for photosynthesis and has a structural role in plant cells. Therefore, a large number of structural adaptations found in plants enable them to conserve water. This is particularly true for xerophytes found in arid environments.

The distribution of stomata across the upper and lower epidermis of four different plants was tabulated and the results are shown in the table below.

Plant	Number of stomata per cm² on the upper epidermis	Number of stomata per cm² on the lower epidermis
J	0	22 500
K	46 000	0
L	1190	28 000
M	0	0

The plants investigated grow:

- in a forest (*Eucalyptus pilularis*)
- floating on the surface of lakes (*Nymphaea alba*)
- in deserts (*Pistacia mexicana*)
- totally submerged in fresh water (*Elodea canadensis*).

a Match each of the plants, labelled J–M in the table, with the environment in which it grows, and in each case justify your choice.

b Identify three other adaptations that can be found in xerophyte plants such as *Pistacia mexicana* that assist in maintaining water balance.

c Water conservation is also vital for the survival of animals living in hot, dry environments. Name one such animal, and summarise the range of relevant adaptations available to them, including a structural, a physiological and a behavioural adaptation.

17 Scientists investigated the effect of temperature on the rate of photosynthesis in two different plants, plant A and plant B. The graph shows the results of the experiment.

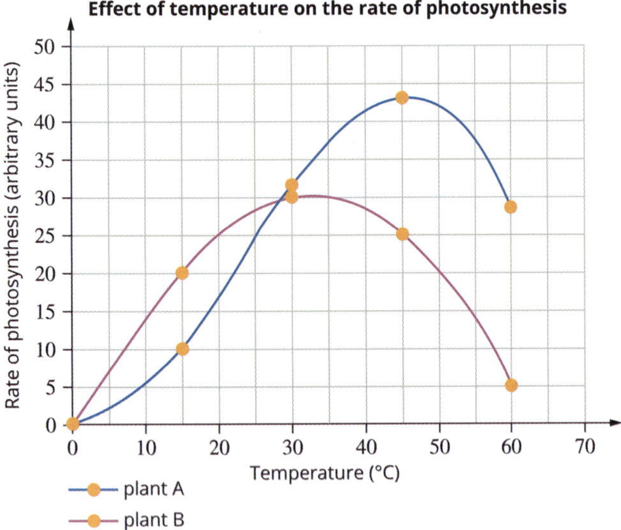

a Compare the effect of temperature on the rate of photosynthesis in plant A and plant B.

b Suggest why the rate of photosynthesis falls when temperature is above 50°C.

c Deduce which plant is more suited for a desert environment and support your answer with reference to the graph.

18 Auxins are a group of plant hormones that are responsible for phototropism.

a i Define the term 'positive phototropism'.

ii Identify the stimulus and response for phototropism in a plant.

b i Nastic movement is the term used for another category of plant responses to environmental stimuli. Recall two examples of nastic movement.

ii Distinguish between tropism and nastic movement with reference to examples.

19 The following diagram shows a kelp forest food web of the far northern Pacific Ocean.

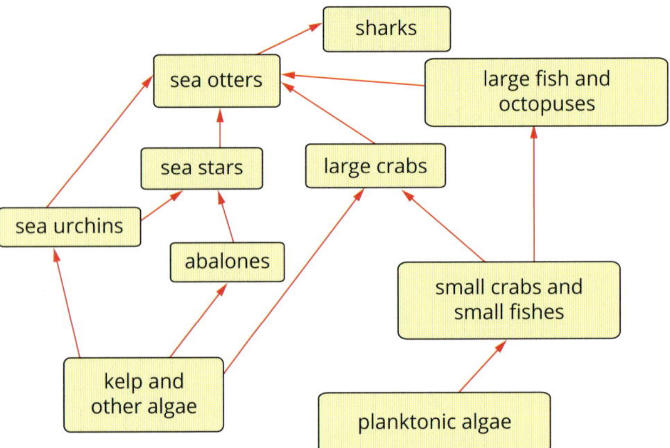

The sea otter is a keystone species in this kelp forest of the Northern Hemisphere.

a Define the term 'keystone species'.

b Explain what would happen if sea otters were removed from the kelp forest.

20 Limpets in the genus *Lottia* are aquatic molluscs that feed on the green algae which grow on rocks on seashores. Black oystercatchers (*Haematopus bachmani*) are birds that feed on these limpets.

A study was conducted in the Monterey Bay area in the US state of California to determine the density of two species of limpets (*Lottia digitalis* and *Lottia scabra*) on a vertical surface of sandstone. The results of the study are shown in the following graph.

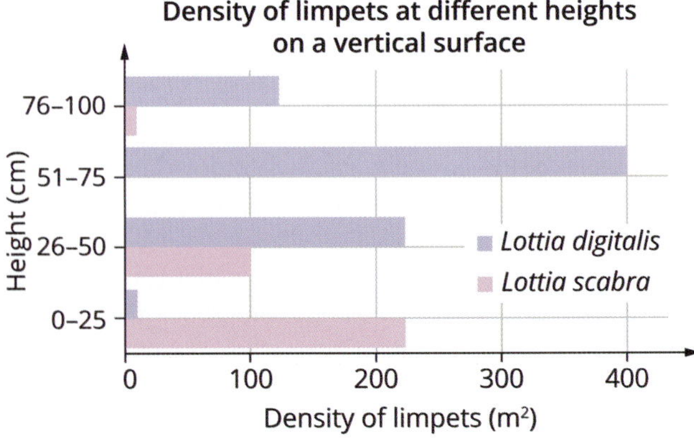

a Which species tend to dominate the upper areas of the vertical surface?

Black oystercatchers live at sea level and are unable to climb vertical surfaces, so when they feed they tend to stay at the bottom of the vertical surface. The typical height of a black oystercatcher is approximately 35 cm.

b Suggest which species of limpet is the preferred prey of black oystercatchers. Explain your answer.

The following diagram shows a cross-sectional view of how each species of limpet adheres to sandstone. *Lottia scabra* secretes a substance that dissolves sandstone. By scraping the softened rock with its radula (a rough tongue-like structure), the limpet creates a 'home scar' in the rock surface.

Lottia scabra *Lottia digitalis*

sandstone surface

home scar

c i Identify the types of adaptation displayed by *Lottia scabra*.

ii Analyse how the presence of a home scar would be an advantage for this limpet.

21 Boodies (*Bettongia lesueur)* are burrowing marsupials, also known as bettongs or rat kangaroos. They were reintroduced into a feral-free animal sanctuary called Faure Island, part of the Shark Bay World Heritage Site in Western Australia. The boodie population was monitored using a trap-and-release method. The graph below shows the changes in the population from July 2002 to January 2013.

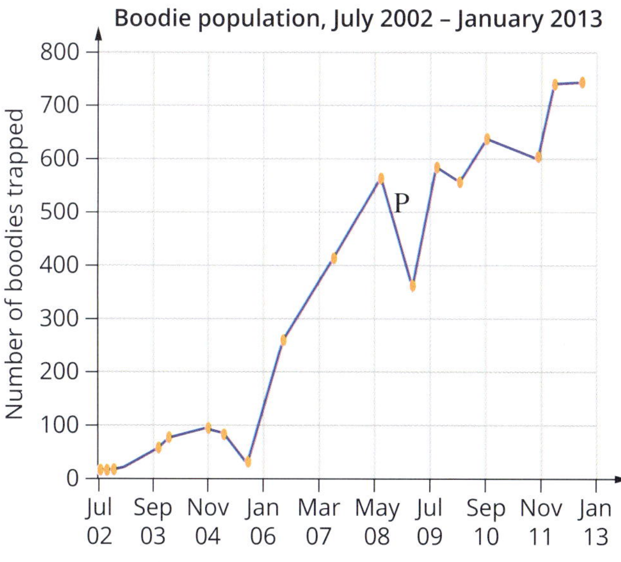

a Suggest why there is a slow increase in population on the island from July 2002 to November 2004.

b Propose a reasonable explanation for the fall in the boodie population at point P.

c Do you think that the population has reached carrying capacity?

i Explain your answer.

ii Predict what the graph line would look like after 2013 if the environment remains favourable.

d Fossil evidence suggests that at least three mammal species were present on Faure Island before European arrival: the western barred bandicoot (*Perameles bougainville*), the woylie or bush-tailed bettong (*Bettongia penicillata*) and the Shark Bay mouse (*Pseudomys fieldi*). It is thought that competition with introduced herbivores may have caused the extinction of the native mammals. A pastoral lease was granted in 1873 that meant the land carried sheep and goats for over one hundred years until the Australian Wildlife Conservancy (AWC) bought the lease in 1999. In preparation for the reintroduction of the threatened native mammals, the AWC conducted ecological surveys and eliminated feral animals such as goats, cats and horses. Continuing field programs conducted by the AWC include integrated weed control, fire management and monitoring for feral animals.

i Briefly outline and assess the importance of each of the actions taken by the AWC.

ii Recommend whether or not the AWC should work in partnership with local Aboriginal people.

e Compare the benefits of the reintroduction of boodies to Faure Island to the captive breeding programs of the boodie in sanctuaries.

22 a Nudibranchs such as the one pictured below are marine organisms. Some species of nudibranchs consume algae, belonging to the group dinoflagellates, and maintain them within their tissues where they use sugars produced by the algae.

i Identify the three possible relationships between the dinoflagellates and the nudibranchs.

ii Define the conditions for each relationship.

b Assess whether knowledge of the types of species interactions that occur is important for understanding how to maintain biodiversity.

23 Mistletoe is a semi-parasitic plant that grows on eucalypts and some other native plants. The diagram below shows how a mistletoe bird is able to transfer mistletoe seeds to the surface of the branch of a new host tree.

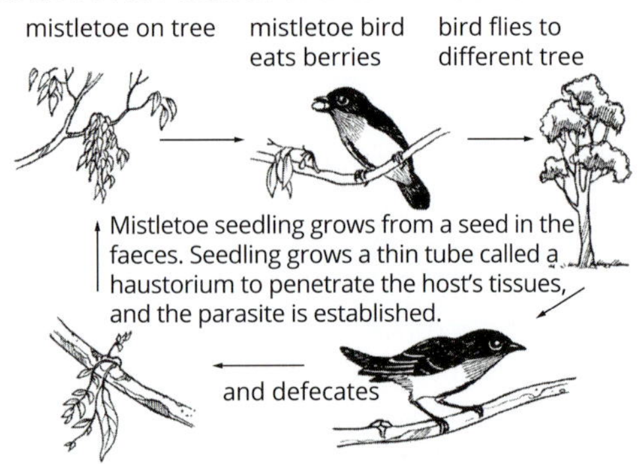

a Name the relationship between the mistletoe bird and the mistletoe. Explain your answer.

b Use a table to classify the following species interactions:

- **i** senior male gorillas stop junior males from mating
- **ii** a flea feeds on a dog
- **iii** a cassowary eats fruits from various rainforest trees and then disperses the seeds in its droppings
- **iv** two magpies fight over a worm
- **v** a lion steals the kill from a leopard
- **vi** a hermit crab uses the discarded shell of a mollusc.

24 Saltmarsh in Victoria was once regarded as useless land, with serious attempts made to eliminate it and allow access for ports, jetties and marine vessels. In more recent times, the importance of saltmarsh for protecting shorelines, as feeding grounds for local and migratory birds, as nurseries for a wide range of marine animals and as ecosystems of significance in their own right, has become better understood. The substantial areas of saltmarsh and mangrove in Westernport, of about 1000 hectares, are now among the biggest, least disturbed and most diverse versions of this ecosystem in southern Australia. Westernport is the coastline and large bay, including French and Phillip Islands, to the east of Port Phillip Bay and the Mornington Peninsula.

This diagram shows a transect of a typical Westernport saltmarsh. Plant species found in each zone marked in the transect are listed below the drawing.

a Propose a plausible reason for the relatively sharp boundaries between the different species of plants found in the zones shown in the transect.

b The long-nosed potoroo (*Potorous tridactylus*) is a small native marsupial with a conservation status of Vulnerable. It is found adjacent to saltmarsh in coastal heath–woodland that has a dense understorey of sedges, ferns and low shrubs, where its main diet is the fungi growing on tree roots. These fungi are important for the health of forest trees, and the potoroos are part of an interdependent relationship as they disperse the fungal spores through the forest. Historical records indicate the long-nosed potoroo occurred throughout Westernport, and today the species can still be found on islands in Port Phillip Bay. Discuss if this potoroo is, or was, a keystone species of the inland Westernport ecosystem.

c The Boonwurrung people are traditional owners of the Westernport land. Their Country, which also includes part of the land area now called Melbourne, extended right out to the ocean when the adjacent Port Phillip Bay was a large flat plain where the Boonwurrung hunted kangaroos and cultivated yam daisy. Through thousands of years of observation, passed on as oral cultural knowledge, the Boonwurrung were able to predict the availability of their seasonal resources by certain changes in plant growth and animal behaviour. The foreshores provided seafood and saltmarsh plants. Creeks provided fresh drinking water and brought animals to the area, and nourished plants and trees. Plants provided ingredients for medicine and painting, materials for clothing and baby-carriers, and tools for hunting and food-gathering. All of these resources were found locally or traded with neighbouring groups.

- **i** Suggest three major points of difference between the traditional Boonwurrung lifestyle and that led by the larger, denser population living in the Westernport area today.
- **ii** Deduce the extent of traditional knowledge that would have been held about the saltmarsh ecosystem and contrast it with the approach taken at first by Europeans to reclaim the saltmarsh for other land use.

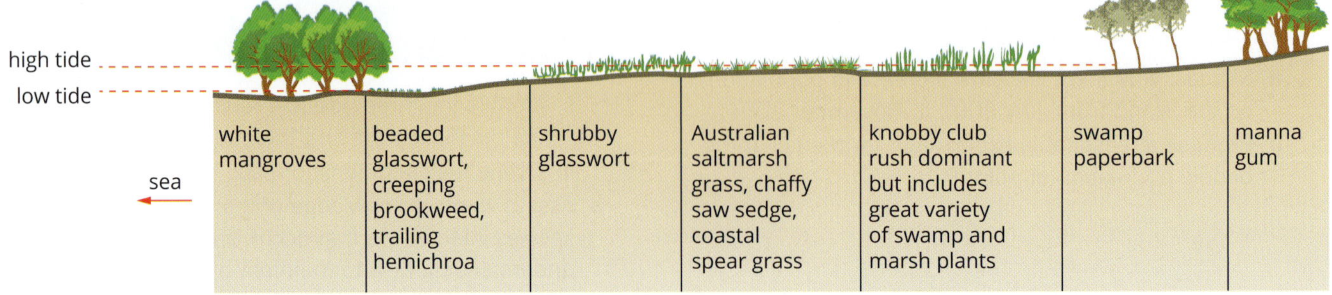

25 The Australian Wollemi pine (*Wollemia nobilis*), shown in the figure, was once thought to be extinct, and was known only from fossil records going back 200 million years. In 1994 some bushwalkers discovered a living grove of these unusual trees in a remote and deep canyon of the Blue Mountains in New South Wales. The 100 mature trees are up to 40 m high and the oldest has since been dated at about 1000 years. To protect the ancient grove, the location remains secret. *W. nobilis* was cloned for commercial sale using tissue culture and micropropagation. The funds raised are used for research and protection of the unique pine in its natural habitat. These asexually propagated pines have proved to be popular with gardeners and are remarkably easy to grow.

a Outline the steps most likely used in the process of propagating the Wollemi pine for commercial sale.

b i Discuss any biological advantages or disadvantages of having a protected natural population that is genetically isolated.

ii There are now tens of thousands of the propagated Wollemi pines growing in botanic and private gardens around the world. Examine the implications of this in respect to their genetic diversity and the long-term survival of the species.

The severe bushfires of December 2019 moved ever closer to the canyon in Wollemi National Park where the 100 mature 'living fossils' grow, raising critical concerns. A secret emergency plan swung into action to try to save the ancient pines with a temporary watering system. The pines are susceptible to introduced diseases so even at this critical time all the equipment had to be sterilised on site. The volunteer crew and equipment were helicoptered into the dangerous fire zone and had to work under extreme pressure. Aerial fire retardant and water-bombing was conducted when planes and helicopters could be diverted from other firefighting emergencies. Through their heroic efforts, the dedicated team ultimately saved the main grove of mature Wollemi pines, although all new seedlings at ground level were burnt to a cinder.

c i Suggest at least two reasons why the fire rescue effort was kept secret until after the pines were saved.

ii Investigate the bioethical issue of using scarce resources and putting volunteer crew at risk in order to save the wild grove of Wollemi pine.

d The Sydney Royal Botanic Garden launched a citizen science project online to gather information about the many cultivated specimens of *Wollemia nobilis.* Assess the scientific value of survey information gathered in this way.

Glossary

RED ENTRIES VCE Biology Study Design extracts © VCAA (2021); reproduced by permission

A

abiotic factors The non-biological parts of an environment that influence ecosystems and the organisms that live in them.

abundance The number of individuals in a population.

accuracy The accuracy of a measurement relates to how close it is to the 'true' value of the quantity being measured.

active transport The movement of substances across membranes that involves the use of energy.

adaptation (1) An inherited characteristic that increases the likelihood of survival and reproduction of an organism or species. (2) The process by which a species becomes well-suited to its lifestyle and environment.

adenine (A) A nitrogen-containing base (a purine) that occurs in nucleotides of DNA and RNA.

ADH See *antidiuretic hormone.*

adult stem cell A stem cell that is present in some adult tissues. Adult stem cells can give rise only to a limited range of cells.

aestivation A long period of torpor in hot and dry conditions. See also *hibernation.*

aim A statement describing in detail what will be investigated.

aldosterone A steroid hormone produced by the adrenal cortex. Its main function is to regulate blood pressure and sodium and potassium levels.

allele One of the alternative forms of a gene. Most genes have two alleles, but more than two alleles are possible.

alternation of generations The alternation between haploid (n) and diploid ($2n$) life cycles in sexually reproducing organisms. In most animals, the diploid stage is the body of the animal and the haploid stage is the internally produced gametes (sperm and eggs). Some plants alternate between a haploid gametophyte stage and a diploid sporophyte stage. See also *gametophyte* and *sporophyte.*

amensalism A relationship between organisms of different species in which one of the organisms benefits and the other is harmed or killed. An example is a paralysis tick and its host. The tick benefits by feeding on the blood of its host, and the host suffers by becoming ill or possibly dying from the effect of neurotoxins injected by the tick.

amino acid An organic compound containing an amino group ($-NH_2$) and a carboxyl group ($-COOH$) at opposite ends of the molecule. Linked amino acids form the peptide chains in protein molecules.

amylase An enzyme that breaks down starch molecules.

anaphase The stage of mitosis after metaphase and before telophase. During anaphase, choromosome pairs separate at the centromeres and move to the opposite poles of the cells.

antidiuretic hormone (ADH) A hormone that increases the permeability of the collecting duct of the kidney to water. This increases the amount of water reabsorbed, resulting in a smaller volume (and therefore more concentrated) urine. ADH is secreted by the pituitary gland. Also called vasopressin.

apoptosis A process of cell death that involves a characteristic series of steps. Also called programmed cell death.

apoptotic body A structure that is formed during apoptosis when cellular components are broken down and the cell's plasma membrane bulges to form a bud.

asexual reproduction Reproduction in which one parent gives rise to a new individual from its body cells. The resulting offspring are genetically identical to their parent.

autoimmune disease A disorder or disease resulting from the persistent presence of antibodies directed against particular parts of the body. It occurs as a result of an impaired ability of the immune system to recognise self.

autosomal dominant The form of genetic inheritance where a dominant trait is passed on to offspring via an autosomal gene. Only one copy of the allele from one parent is needed to express a dominant phenotype.

autosomal recessive The form of genetic inheritance where a trait can be inherited by offspring when neither parent displays the trait. Two copies of the allele (one from each parent) are needed to express a recessive phenotype.

autosome Chromosome that is not a sex chromosome.

autotroph An organism that is able to produce its own food from inorganic materials, using light or chemical energy. Plants that photosynthesise are the most common autotrophs. All autotrophs are producers.

B

bar graph A graph in which categorical data are represented by horizontal bars. Each bar represents one category of independent variable (such as a range of values, or a particular type of thing), and the length of the bar represents the value of the dependent value for that range or thing.

baroreceptor A receptor that detects blood pressure in vertebrates, sending the information to the brain to regulate blood pressure.

base Any of the four compounds adenine (A), thymine (T), guanine (G) and cytosine (C) present in the nucleotides of the nucleic acids DNA and RNA, and uracil (U) in RNA, forming the linking points between strands.

behavioural adaptation Any action that an organism takes to improve its ability to cope with abiotic and biotic factors in its environment, increasing its chances of survival and reproduction.

BGL See *blood glucose level.*

bile A secretion produced by the liver and stored in the gall bladder, from where it is released into the small intestine. Bile acts as an emulsifying agent, physically breaking up large fat droplets into smaller droplets to increase the surface area of food being digested.

binary fission A form of asexual reproduction in unicellular organisms, in which the parent cell divides into two approximately equal parts.

biogenesis The principle that cells are formed only from pre-existing cells.

biological fitness An organism's ability to survive and reproduce in its natural environment.

biomass The mass of living matter per unit area (e.g. kg/m^2), or the equivalent amount of chemical energy bound in the mass of tissue (e.g. kJ/m^2). Biomass measurements may be for total biomass, or for the biomass of a particular group of organisms such as plants.

biotic factor Any factor relating to the biological parts of the environment, as opposed to the abiotic (physical and chemical) parts.

bladder A muscular organ that receives urine from the kidneys and holds it before it is excreted through the urethra.

bleb A bulge in the plasma membrane caused by degradation of the internal structure of the cell.

blood glucose level (BGL) The amount of glucose (sugar) in the blood. Glucose is used for cellular respiration and its uptake by cells is regulated by the hormone insulin.

bottleneck effect The resulting impact when a large portion of a population is removed from the habitat by chance, typically as a result of a natural disaster. The effect of genetic drift is more significant on the smaller population, as the remaining gene pool has reduced diversity.

Bowman's capsule The region of a nephron into which filtered plasma flows from the glomerulus.

brumation A type of torpor undergone by many reptiles. It is similar to hibernation but differs in the metabolic processes involved. Brumation begins just before winter and lasts between 1 and 8 months.

budding A form of asexual reproduction in which a new individual arises as an outgrowth or bud from the parent.

bulb A large, often spherical underground bud of a plant, surrounded by fleshy leaves. The bulb stores food for the growing shoot of the plant.

C

caecum An intestinal pouch at the junction of the small and large intestine. In some herbivores, such as koalas, it is very enlarged and acts as a fermentation chamber for the digestion of cellulose.

callus A mass of undifferentiated cells that form when plant cells are grown in a medium as part of the cloning process.

CAM photosynthesis A form of photosynthesis that occurs in many plants growing in hot, dry environments. Stomata open at night to take in carbon dioxide, which is incorporated into malate. During the day the stomata are closed to reduce transpiration, and malate is metabolised to release carbon dioxide, which is then used by cells. CAM stands for crassulacean acid metabolism.

cancer A group of diseases characterised by uncontrolled cell division.

carbohydrate An organic compound consisting only of carbon, hydrogen and oxygen atoms, with the hydrogen and oxygen atoms in the same proportion as in water (2 to 1). Carbohydrates include sugars and starches.

carrier An individual that has an allele for a condition but does not express the condition because it is masked by a dominant phenotype. The carrier can pass the allele to its offspring, who will express the condition if they receive the same allele from the other parent.

carrier protein A transport protein that changes shape when molecules bind to it, so that the molecules can pass through the plasma membrane. Carrier proteins take part in facilitated diffusion and active transport. See also *channel protein*.

carrying capacity The maximum population of a species that can be supported indefinitely by an ecosystem.

Casparian strip A water-resistant strip in the endodermis of roots that regulates the entry of water and solutes.

caspase One of a group of enzymes involved in protein and DNA cleavage. Caspases are involved in apoptosis.

cell The smallest structural and functional unit in a living thing. All cells have a plasma membrane and contain cytoplasm, organelles and genetic material (DNA). In plants and fungi, cells also have a cell wall.

cell compartmentalisation The formation in the cytosol of specialised structures enclosed by membranes, including the nucleus, mitochondria, endoplasmic reticulum, Golgi apparatus, endosomes, lysosomes and chloroplasts.

cell cycle The events in the life of a cell, from its formation by cell division through its growth and function until it divides again. It begins with the G_1 stage and proceeds through S, G_2, mitosis and finally cytokinesis. A G_0 resting stage may also be entered during G_1.

cell cycle control system A system within the eukaryotic cell that operates cyclically to trigger and coordinate the events of the cell cycle. The system is controlled by a set of molecules.

cell differentiation The process by which a cell changes from one type to another. This is usually an unspecialised cell becoming a specialised cell.

cellular respiration The energy-releasing processes that occur in cells. In particular, the aerobic stage in the complete breakdown of glucose to produce ATP, which occurs in mitochondria and produces 36 or 38 molecules of ATP per molecule of glucose.

cellular response The activation of a cellular activity or process.

cellulose A complex carbohydrate molecule consisting of a chain of many glucose molecules. It is the main component of plant cell walls. Its formula is $(C_6H_{10}O_5)_n$.

centriole A small cylindrical organelle consisting of a group of microtubules, and occurring as a pair in the centrosome in the cells of animals and some other organisms. Centrioles are replicated in the S phase, and the two pairs formed separate during mitosis and move towards the opposite ends (poles) of the cell.

centromere A part of the chromosome that attaches to spindle fibres during mitosis, and where the two sister chromatids of a double-stranded chromosome are joined.

channel protein A transport protein that molecules do not usually bind to. Channel proteins allow specific molecules to pass through the plasma membrane, and are used in facilitated diffusion. See also *carrier protein*.

chemical digestion The action of enzymes in breaking down complex compounds into simple compounds that can be used for metabolism.

chemoreceptor A sensory receptor that detects and responds to specific chemical substances.

chiasma (pl. chiasmata) A point of crossing of strands of non-sister chromatids observed during the first division of meiosis.

chloroplast A green organelle in plant cells, in which photosynthesis takes place. A chloroplast consists of many folded layers of membrane and contains chlorophyll.

cholesterol A steroid lipid found in most body tissues. Cholesterol is an important component of plasma membranes in animals and is used to form other steroid compounds.

chromatid One of two copies of a chromosome formed during the S stage of interphase. The two copies, called sister chromatids, are joined at a centromere. See *sister chromatids*.

chromosome In eukaryotes, a complex structure consisting of DNA strands coiled around histone proteins, carrying the hereditary information of the cell in the form of genes. All body cells in a particular species have the same number of chromosomes.

cisternae The flattened, sac-like membranes found in the Golgi apparatus and endoplasmic reticulum.

cleavage furrow The early division of a zygote into smaller cells by mitosis.

clone A biological entity (such as a gene, cell, tissue or organism) that is a genetically identical copy of another entity.

cloning (1) The process of replication that creates a new biological entity, such as a gene, cell, tissue or organism. (2) In animals, the creation of a new individual by transferring the nucleus of a somatic cell into an enucleated egg, which is then implanted for development. The resulting individual will be genetically identical to the parent that provided the nucleus.

codominance The occurrence of a phenotype in a heterozygote that results from the expression of both alleles. An example is the AB blood group in humans.

cohesive bond A bond between molecules of a substance resulting from the shape and structure of the molecules.

column graph A graph in which categorical data are represented by vertical columns. Each column represents one category of independent variable (such as a range of values, or a particular type of thing), and the length of the column represents the value of the dependent value for that range or thing.

commensalism A relationship between two organisms in which only one benefits, but the other organism is not harmed.

community A group of species that occur in the same area and interact, or could interact, with each other.

companion cell A specialised cell within the phloem tissue of vascular plants. Companion cells are a type of parenchyma cell that gives metabolic support to the sieve tube cells.

competition An interaction between organisms that are seeking to use the same resource, such as food, water, shelter, sunlight or mates.

competitive inhibition The inhibition of an enzyme due to a molecule that binds to the active site of the enzyme, preventing the substrate from binding.

complementary base pair Either of only two possible pairs of bases in a double-stranded DNA molecule. Adenosine (A) paired with thymine (T), or guanine (G) paired with cytosine (C).

complete dominance In sexual reproduction, the expression of only one phenotype in all heterozygous individuals.

concentration gradient A difference in the concentration of a solute between one region and another; for example, across a membrane.

conclusion A summary that outlines how the results of an investigation support or do not support the original hypothesis.

conduction In relation to thermal conduction in biology, the transfer of heat via physical contact.

conidia (sing. conidium) The fungal spores that are produced asexually by budding.

continuous variable A variable that can have any number value within a given range.

control centre The part of a feedback mechanism that determines the response required and sends an appropriate signal to the effector.

control group A group of subjects in an experiment that is identical to the experimental group and is treated in an identical way, except that the variable of interest (the independent variable) is kept constant.

controlled variable A variable that is kept constant during the investigation.

convection The transfer of heat via currents in a liquid or gas.

corm A solid, bulb-like underground stem that stores food for a plant and also sends down a root at the start of each growing season.

countercurrent heat exchange A physiological adaptation of animals in which blood flowing in adjacent blood vessels exchanges heat. This adaptation allows animals living in hot environments to cool blood before it enters the brain, and it allows animals living in cold environments to heat blood flowing from the extremities to the heart, maintaining the core body temperature.

crossing over The exchange of chromosomal material between non-sister chromatids of a homologous chromosome pair during prophase I of meiosis.

cultivar A variety of plant that has been selectively bred, usually for its fruiting, flowering or aesthetic properties.

cuticle A protective waxy coating on the surface of plant organs (e.g. leaves).

cutting A form of asexual reproduction in plants, where a section of the plant is removed from the parent and placed in soil or water to grow.

cytokinesis The division of a cell following mitosis or meiosis, when the cytoplasm divides and the cell splits into two daughter cells.

cytoplasm The contents of a cell, enclosed by the plasma membrane, including the fluid (cytosol) and all organelles except the nucleus.

cytoplasmic pathway One of two possible pathways for movement of water and mineral ions absorbed from the soil through the roots of vascular plants. The cytoplasmic pathway is where most mineral ions and some water passes through the cytoplasm of living root cells. See also *extracellular pathway*.

cytosine (C) A nitrogen-containing base (a pyrimidine) that occurs in nucleotides of DNA and RNA.

cytosol The fluid inside a cell in which the cell's organelles, proteins and other structures are suspended.

D

daughter cell A new cell formed by cell replication.

death receptor pathway One of two pathways by which apoptosis can be initiated. Death receptor molecules on the surface of cells that are under stress initiate a series of reactions that lead to programmed cell death (apoptosis). See also *mitochondrial pathway*.

density In population studies, the number of individuals per unit of area or volume.

density-dependent factor A limiting factor whose effect depends on the size of the population.

density-independent factor A limiting factor whose effect does not depend on the size of the population.

deoxyribonucleic acid (DNA) A nucleic acid made up of a sequence of nucleotides, each with a deoxyribose sugar, phosphate and base (adenine, cytosine, guanine or thymine), linked by phosphodiester bonds. DNA is the carrier of genetic information in all living things and most viruses. It occurs in chromosomes in the nucleus or nucleolus, and also in mitochondria and plastids.

dependent variable The variable that is measured to study the effect of changes in the independent variable.

diffusion The passive movement of a solute from a region of higher concentration to a region of lower concentration.

digestion The breakdown of food into a form that can be used by an organism for metabolism. Digestion involves mechanical digestion and chemical digestion.

digestive enzyme An enzyme that assists in the digestion of otherwise indigestible matter.

digestive system The system of organs that breaks down and absorbs nutrients from food and removes waste products.

dihybrid cross A cross between pure lineages that exhibit two different phenotypes. An example is a cross between a pea with dominant phenotypes of yellow seeds and red flowers, and a pea with recessive phenotypes of green seeds and white flowers.

diploid Having two sets of chromosomes ($2n$). All somatic cells are diploid.

discrete variable A variable that can have only certain values. For example, the number of individuals in a population can only be whole numbers.

distribution The geographic extent of a group of organisms. It is commonly applied to the extent of a population or species.

DNA See *deoxyribonucleic acid*.

DNA replication The process in which a DNA molecule is copied to produce two identical DNA molecules.

dominance (adj. dominant) The expression of one allele of a gene in the phenotype of an individual rather than another allele of the same gene. See also *complete dominance*, *codominance*.

dominant phenotype The phenotype expressed in a heterozygous individual; that is, an individual carrying different alleles of the same gene.

double helix The structural shape of a DNA molecule, consisting of two linear lengths of nucleotides twisted spirally about each other, and connected by phosphodiester bonds.

E

ecological niche The role of an organism or species in an ecosystem, including its position in the food web, how it obtains its food and how it reproduces.

ecosystem A system formed by organisms interacting with one another and their physical environment.

ectotherm An animal that depends on external heat sources to regulate their body temperature.

effector A cell or tissue that responds to a stimulus.

egestion Elimination of food that has not been absorbed by the gut.

embryo The stage in the development of a vertebrate between the fertilisation of the ovum and the development of the characteristics of the adult organism (the foetus).

embryonic stem cell A stem cell that can be obtained from blastocysts. Embryonic stem cells are pluripotent and can differentiate into any of the three germ layers (ectoderm, endoderm and mesoderm).

emigration The movement of individuals out of a population.

endemic Occurring only in a particular area. For example, the Tasmanian devil is endemic to Tasmania.

endocrine system The animal body system that is responsible for the production of hormones.

endocytosis The movement of material into a cell by enclosing it in plasma membrane, which then pinches off to form a vesicle within the cell. Endocytosis includes phagocytosis (the entry of solids) and pinocytosis (the entry of liquids).

endotherm An animal that maintains a more or less constant body temperature, which is usually higher than the temperature of the surrounding environment.

enterocyte A specialised cell that lines the small intestine to increase absorption of nutrients.

enzyme A protein molecule that catalyses (speeds up) biochemical reactions.

epigenetics The study of molecular events such as methylation that occur on DNA but do not alter the DNA sequence, and result in different phenotypes.

epiglottis A thin flap of cartilage that covers the entrance to the larynx, preventing food from entering the trachea during eating.

epithelium (adj. epithelial) A thin layer of tissue covering the external surfaces of a multicellular organism, and also lining the inner surfaces of internal structures such as intestines and lungs.

equilibrium A state of stability in which opposing factors are balanced.

eukaryotic cell A cell that contains a membrane-bound nucleus and other membrane-bound organelles. Protists, fungi, plants and animals are eukaryotes.

evaporation The process of liquid turing into vapour.

evaporative cooling The release of heat from the body via evaporation. Sweating and panting are forms of evaporative cooling.

evolutionary potential The ability of a population to adapt or evolve in response to environmental change, thus increasing their chances of survival and reproduction.

excretion The removal of waste substances from the body of an organism.

excretory system The system of organs that removes waste products from the body.

exocytosis A type of active transport in cells in which molecules such as proteins are expelled from a cell. The molecules are enclosed by a vesicle, which then fuses with the plasma membrane and expels the contents into the extracellular fluid.

experimental group The group of subjects in an experiment in which one variable (the independent variable) is altered in order to measure its effect on another variable (the dependent variable). See also *control group*.

exponential growth The growth of a population in which rate of growth is proportional to population size. A graph of exponential growth shows an increasing gradient over time.

exponential relationship A mathematical relationship in which the rate of change of one variable is proportional to the value of the other variable.

external environment The environment immediately surrounding an organism.

exteroreceptor A sensory receptor that detects external stimuli.

extracellular digestion Chemical digestion in which the enzymes are secreted into a cavity, where digestion takes place.

extracellular fluid The fluid outside the cells in a multicellular organism.

extracellular pathway One of two possible pathways for movement of water and mineral ions absorbed from the soil through the roots of vascular plants. The extracellular pathway is where most water and some mineral ions pass through or between the cell walls. See also *cytoplasmic pathway*.

extremophile An organism that lives in an extreme environment, such as somewhere with a very high pressure, temperature or salinity.

F

F1 generation The offspring of a cross between members of the parental generation.

F2 generation The offspring of a cross between members of the F1 generation.

facilitated diffusion The diffusion of ions and molecules through a plasma membrane via ion channels and channel proteins. Facilitated diffusion does not require chemical energy from the conversion of ATP to ADP.

facultative mutualism A form of mutualism in which the individuals do not depend on each other for survival, but both benefit from the relationship.

fermentation The stage in the breakdown of glucose that follows glycolysis when there is no oxygen present. Fermentation produces either lactic acid (in most animals) or alcohol (in most plants and micro-organisms).

fertilisation Penetration of an egg by sperm and fusion of the egg and sperm nuclei.

fieldwork A biological investigation that takes place outside the classroom or laboratory and involves investigating organisms in the natural environment.

filtration In the kidney, the process by which the primary kidney filtrate is formed, from fluid passing from Bowman's capsule into the nephron.

fission A form of asexual reproduction in unicellular organisms, in which the parent cell divides into two parts. See also *binary fission.*

fluid mosaic model A model that describes the structure of the plasma membrane, in which phospholipids and unanchored proteins are free moving, giving the membrane fluidity. Anchored proteins are scattered throughout, giving the membrane a matrix pattern.

food chain A sequence of feeding relationships, beginning with a producer and ending with a higher order consumer. The producer is eaten by a first-order consumer, the first-order consumer is eaten by a second-order consumer, and so on.

food web A network of interlinked food chains that describes the feeding relationships between all organisms in an ecosystem.

founder effect Occurs when a small portion of a population disperses to a new location and becomes genetically isolated from the main population. The allele frequencies of the founding population are completely dependent on those of the specific individuals that were relocated, and therefore may be significantly different from those of the original population.

fragmentation A form of asexual reproduction in which an organism breaks into two or more parts, each of which regenerates the missing pieces to form a complete new organism.

G

G_0 phase One of the phases of interphase in the cell cycle. Cells may enter the G_0 phase, also known as the resting phase, at the start of the G_1 phase. The cell carries out its normal functions and does not change its internal structure or size. See also *G_1 phase*, *G_2 phase* and *S phase.*

G_1 checkpoint A checkpoint that occurs towards the end of G_1 in interphase. It checks that there are adequate resources for the cell to divide, the cell is large enough to divide, and the DNA has not been damaged. Also known as the restriction point. See also *G_2 checkpoint* and *M checkpoint.*

G_1 phase The first phase of interphase after mitosis. The new daughter cell gains energy, undertakes metabolic processes and almost doubles in size. See also *G_0 phase*, *G_2 phase* and *S phase.*

G_2 checkpoint A checkpoint that occurs towards the end of G_2 in interphase. It checks that there are adequate resources for the cell to divide, the cell is large enough to divide, and the DNA has not been damaged. If any of these checks are failed the cell does not enter the mitotic phase. See also *G_1 checkpoint* and *M checkpoint.*

G_2 phase The last phase of interphase before mitosis. The cell undergoes a secondary stage of growth, metabolism and energy acquisition to prepare for mitosis. See also *G_0 phase*, *G_1 phase* and *S phase.*

gall bladder An organ that stores and concentrates bile before releasing it to the small intestine.

gamete A haploid cell capable of fusion with another haploid cell to form a zygote. In vertebrates the gametes are sperm and egg cells. Also known as sex cells.

gametophyte The gamete-forming haploid stage in the life cycle of some plants.

gastric chief cell A specialised cell that secretes proteases into the stomach.

gene A section of DNA that contains instructions for making a protein or RNA molecule. Particular genes have specific locations on chromosomes. Genes are copied and passed from one generation to the next during reproduction.

gene expression The process by which genetic information (DNA or RNA) is used to synthesise a functional gene product (usually a protein or RNA molecule).

gene flow The movement of alleles between different populations; includes the dispersal of pollen and seeds in plants.

gene mapping The determination of the location of genes, and the distance between them, on a chromosome.

gene pool All the alleles possessed by members of a population, which may potentially be passed to the next generation.

genetic cross The intentional breeding (or crossing) of two genetically different organisms to determine the inheritance pattern of particular traits.

genetic diversity The variation in genes or alleles within a population or species. Also called genetic variation.

genetic drift The random changes to allele frequencies in a gene pool as the result of a chance event. This has a more significant impact on smaller populations, as the chance death of one individual could eliminate an allele from the gene pool.

genome The DNA in one full set of chromosomes present in the nuclei of normal cells of a species, plus the DNA in mitochondria and (in plants) chloroplasts.

genotype (1) The total set of genes of an organism. (2) The combination of alleles for a trait carried by an individual.

genotypic ratio The frequency of genotypes expected in the offspring of a genetic cross. A genotypic ratio is written as homozygous dominant : heterozygous : homozygous recessive.

germ cell Any cell in an organism that gives rise to gametes.

germ layer The primary layer of cells that is formed during embryogenesis. Animals with bilateral symmetry have three layers: endoderm, mesoderm and ectoderm. Animals with radial symmetry have two layers: endoderm and ectoderm.

gland A specialised endocrine tissue that synthesises, stores and secretes hormones.

glomerulus A cluster of capillaries in the renal corpuscle of the kidney nephron. Filtration occurs through the walls of the capillaries that form the glomerulus, into the Bowman's capsule.

glucagon A hormone produced in the pancreas that causes glycogen to be broken down in the liver, releasing glucose into the blood, thus opposing the effect of insulin.

glucose A simple sugar (formula $C_6H_{12}O_6$) that is a product of photosynthesis. It is the main source of energy for cells in living things, and is essential for cellular respiration.

glycogen A complex carbohydrate molecule consisting of glucose subunits. Glycogen is the main carbohydrate storage molecule in animals.

glycolipid A lipid with a carbohydrate group attached. It is a component of the plasma membrane and is a marker for cell recognition.

glycoprotein A protein that has a carbohydrate group attached to the polypeptide chain. Glycoproteins are components of the plasma membrane and are receptors for molecules such as hormones.

goblet cell A specialised cell that produces and secretes mucin to protect epithelial tissue.

Golgi apparatus (also known as Golgi body, Golgi complex) An organelle composed of a stack of cisternae in which proteins are assembled and then packaged in vesicles for exocytosis.

graft A form of asexual reproduction in plants, where a section of the plant is removed from the parent and transferred onto the cut stem of another plant (the rootstock).

guanine (G) A nitrogen-containing base (a purine) that occurs in nucleotides of DNA and RNA.

guard cells The specialised epidermal leaf cells bordering the stomata. Stomata open when the guard cells are turgid (swollen).

H

haploid Containing one set of chromosomes (half the normal number of chromosomes of a diploid cell).

hemizygote (adj. hemizygous) A diploid cell or organism with only one copy of a particular chromosome. Human males are hemizygotes because they have one X chromosome rather than two.

hereditary Able to be passed from parent to offspring, or from one generation to the next.

heterogametic Having different sex chromosomes; for example, human males with XY.

heterotroph An organism that must obtain nutrients from other organisms.

heterozygote (adj. heterozygous) A diploid individual with different alleles for a particular gene.

hibernation A long period of torpor during the colder months of the year. See also *aestivation.*

homeostasis The maintenance of a more or less stable internal environment, even when external conditions change.

homogametic Having two similar sex chromosomes; for example, human females with XX.

homologous chromosomes Matching pairs of chromosomes in a diploid organism. Homologous chromosomes carry the same genes in the same loci.

homozygote (adj. homozygous) A diploid individual with two identical alleles at a particular genetic locus.

hormone A molecule that regulates the growth or activity of those cells capable of responding to it (target cells). Hormones are produced by specialised groups of cells within an organism.

hybrid An individual produced by a cross between parents with different genotypes.

hydrophilic Water attracting. Polar ions and molecules that dissolve easily in water.

hydrophobic Water repelling. Non-polar molecules that are relatively insoluble in water.

hydrophyte A plant that lives in water.

hyperglycaemia A state of having a higher than normal blood glucose level.

hypersecretion An excess in secretion of a substance, such as insulin.

hyperthyroidism A condition in which the thyroid produces more T3 and T4 than the body needs.

hypoglycaemia A state of having a lower than normal blood glucose level.

hyposecretion A deficiency in secretion of a substance, such as insulin.

hypothalamus In vertebrates, the base and part of the sides of the brain immediately below the thalamus. In mammals the hypothalamus directly or indirectly controls aspects of the internal environment, particularly through the secretion of various hormones, such as anterior pituitary hormones.

hypothesis A suggested explanation for observed facts. An experimental hypothesis is used to make predictions that can be tested experimentally.

hypothyroidism A condition in which the thyroid produces less T3 and T4 than the body needs. This condition results in a range of serious health effects, including low heart rate and depressed ovarian function.

I

immigration The movement of individuals into a population.

immuno-suppressant drug A drug that inhibits the immune response against foreign particles or tissues. Immuno-suppressant drugs are used to prevent the rejection of transplanted organs or tissues.

inbreeding The loss of genetic variation within a population or species due to matings between closely related or genetically similar individuals.

inbreeding depression The reduction of biological fitness and chances of survival and reproduction in a population due to inbreeding. See *inbreeding*.

incomplete dominance A form of inheritance in which neither the dominant nor the recessive phenotype is expressed completely. In heterozygotes, both alleles are partially expressed, producing an intermediate phenotype.

independent assortment The independent arrangement of chromosomes during metaphase I in meiosis. The independent assortment of chromosomes during meiosis results in genetic variation in the gametes.

independent variable The variable that is altered during an experiment to test its effect on another variable (the dependent variable). Also called experimental variable.

inorganic compound Any compound that does not include carbon. In addition, oxides, carbonates, bicarbonates, carbides and cyanides are usually also considered to be inorganic compounds.

insulin A hormone secreted by β cells in the pancreas, controlling the concentration of glucose in the blood. Insulin is secreted in response to high glucose levels, and acts by suppressing the breakdown of glycogen to glucose in the liver, stimulating the storage of glucose as glycogen in the liver and muscles, and stimulating the formation of fat using glucose.

integral protein A protein that is a permanent part of the plasma membrane.

internal environment The watery extracellular fluid that surrounds the cells of a multicellular organism.

interoreceptor A sensory receptor that detects internal stimuli.

interphase The phase in the cell cycle when the cell is not undergoing mitosis.

interspecific Occurring between or involving two or more species.

intracellular fluid The fluid inside a cell.

intraspecific Occurring between or involving two or more individuals of a species.

inverse relationship A mathematical relationship in which one variable increases when the other decreases.

in vitro fertilisation (IVF) The fertilisation of an ovum outside the body, under laboratory conditions. IVF is used particularly if normal fertilisation cannot occur.

IVF See *in vitro fertilisation*.

K

karyotype A visual depiction of the number, size and shape of chromosomes in an individual.

keystone species A species on which the entire structure and functioning of an ecosystem depends. Without the keystone species, the ecosystem structure would change significantly, and the ecosystem would function in a very different way.

kidney An organ involved in filtering the blood and producing urine in complex animals.

L

large intestine The last portion of the digestive system in complex animals. The large intestine absorbs water and stores faeces before egestion.

law of dominance The principle, first stated by Gregor Mendel, that explains the relationship between two alleles. The law of dominance states that in an individual with two different alleles for a trait (heterozygote), the allele for the dominant trait will mask the allele for the recessive trait and the dominant trait will be expressed in the phenotype. Also called Mendel's third law of inheritance.

law of independent assortment The principle, first stated by Gregor Mendel, that individual inherited traits assort independently, so that the occurrence of a trait (such as brown eyes) in an offspring is independent of the occurrence of any other trait (such as attached ear lobes). Because of linkage, the law applies only to alleles on different chromosomes (or far enough apart on the same chromosome). Also called Mendel's second law of inheritance.

law of segregation The principle, first stated by Gregor Mendel, that explains the inheritance of alleles. The law of segregation states that each gene has two alleles that separate (or segregate) during meiosis (haploid gamete formation) and randomly unite during fertilisation, resulting in offspring inheriting one allele from each parent. Also called Mendel's first law of inheritance.

lenticel A porous group of cells that allows gas exchange across the otherwise airtight and waterproof cork layer covering the stems and roots of woody plants.

lignin A complex organic compound deposited in the cell walls in the xylem vessels, tracheids and supporting tissue of vascular plants. Lignin gives strength to the stem and other plant parts. It is not present in non-vascular plants such as mosses.

limiting factor Any factor that prevents a population from growing larger. Common limiting factors are the availability of water, food, shelter, nesting sites and mates.

line graph A graph in which the relationship between the variables is represented by a straight line, curved line, or series of line segments.

linear relationship A mathematical relationship between variables in which a change in one variable produces a proportional change in the other variable. The graph of a linear relationship is a straight line.

linkage The tendency for two or more genes on the same chromosome to be inherited together because they are close together on a chromosome. Linked genes may be separated if crossing over occurs between them.

linkage group A group of genes located close together on the same chromosome that is inherited as one unit. Linkage groups do not assort independently but genes within the group may be separated by crossing over during meiosis.

lipase An enzyme that digests lipids.

lipid An organic compound that is insoluble in water but soluble in alcohol, ether or chloroform. Lipids include fats, oils, sterols, some hormones, fat-soluble vitamins, glycerides and phospholipids.

liver A large organ in vertebrates that is involved in many important metabolic processes, including protein manufacturing, fat storage and processing, bile secretion, and metabolism of toxins.

locus (pl. loci) The site on a chromosome where a particular gene is located.

logistic growth Population growth in which the growth rate decreases as the population approaches the carrying capacity. A graph of logistic growth is an S-shaped curve.

loop of Henle A U-shaped loop in a mammalian kidney between the proximal and distal convoluted ducts, dipping into the medulla. Its main function is to recover water and sodium chloride from urine, thus making the urine more concentrated and reducing the amount of water that needs to be taken in.

lower epidermis A layer of cells on the lower side of the leaf that protects the inner cells and allows the stomata to open and close depending on the needs of the plant.

lysosome An organelle vesicle containing digestive enzymes used in the digestion of waste and foreign material.

M

M checkpoint Checkpoint that occurs towards the end of metaphase. It checks that all the spindle fibres have correctly attached to the sister chromatids and the chromosomes are correctly aligned at the cell equator. Also known as the spindle checkpoint. See also *G_1 checkpoint* and *G_2 checkpoint*.

mark–recapture A type of study in which animals are captured, marked and then released.

mean The average value of a set of values, calculated by dividing the sum of the values by the number of values.

mechanoreceptor A type of receptor that detects hearing, balance, pressure and touch.

median The value in the middle of an ordered list of values.

meiosis A division of a nucleus that results in one copy of each homologous chromosome and one sex chromosome in each daughter cell. Meiosis produces four genetically unique daughter cells, each with half the number of chromosomes of the parent cell.

meniscus The curved upper surface of liquid in a tube or container, caused by surface tension. A meniscus can be concave (as in water in a glass tube) or convex (as in mercury in a thermometer).

meristem A type of tissue in plants that contains undifferentiated cells and is the site of cellular differentiation and specialisation. Meristem tissue usually occurs at the tips of roots and shoots of plants, where most tissue growth occurs.

mesophyll cell Cell that makes up the thin-walled, loosely packed photosynthetic plant tissue that forms most of the interior of leaves.

mesophyte A plant adapted to a medium water availability.

messenger RNA (mRNA) An RNA molecule that is transcribed from DNA in the nucleus, then passes into the cytoplasm and binds to a ribosome, where it is used to build an amino acid sequence (polypeptide).

metabolism The total of the physical and chemical processes by which energy and matter are made available by an organism for its own use. Metabolism is controlled by enzymes.

metaphase The stage of mitosis between prophase and anaphase, in which chromosomes align at the equatorial plane of the cell and spindle fibres attach to the centromeres of chromosomes.

method The specific steps taken to collect data during a scientific investigation.

methodology A brief description of the general approach taken to investigate a research question or hypothesis and the reasons why this approach is taken.

microvillus (pl. microvilli) A microscopic fold of the inner surface of intestinal epithelial cells. Microvilli increase the surface area for the absorption and secretion of substances.

migration The geographic movement of organisms.

mineral Any naturally occurring inorganic substance. In nutrition, important minerals include elements such as magnesium, potassium, calcium, iron and sodium. Minerals in foods are essential for maintaining biological functions.

mitochondrial pathway One of two pathways by which apoptosis can be initiated. The mitochondrial pathway is triggered when cell components are damaged, initiating a series of reactions that lead to programmed cell death (apoptosis). See also *death receptor pathway*.

mitochondrion (pl. mitochondria) The organelle in which cellular respiration occurs. Each mitochondrion is composed of many layers of folded membrane.

mitosis A division of a nucleus that results in two cells that are genetically identical to the parent cell. Asexual reproduction and cell replication for growth occur by mitosis.

mitospore A spore produced asexually by mitosis.

mitotic spindle A network of fibres formed by the centrioles that attach to the centromeres of choromosomes to separate the strands during mitosis.

mode The value that occurs most often in a data set.

monohybrid cross A cross between individuals that have different pairs of alleles of a particular gene. For example, one individual might have *T* and *t* alleles, and the other might have *t* and *t* alleles. Monohybrid crosses are used to study the inheritance of one characteristic.

mortality The death rate in a population, usually expressed as number of deaths per unit of population in a given time period. For example, the death rate in Australia in 2019 was 5.3 per 1000 population.

mouth The first part of the digestive system of animals, where food is mechanically digested by chewing and saliva is produced for chemical digestion.

mRNA See *messenger RNA*.

multicellular organism An organism consisting of more than one cell.

multipotent Describes a cell that can develop only into cells of a similar type. For example, stem cells in bone marrow are multipotent because they can develop into different blood cells but not into other types of cells.

murein A giant molecule that forms a mesh-like layer on the outside of the plasma membrane of most bacteria. Each molecule consists of glycans (large-molecule sugars) linked by chains of amino acids. Also called peptidoglycan.

mutagen A physical, chemical or biological agent that can cause mutations in DNA.

mutant (1) A cell or organism carrying an altered (mutated) gene. (2) An individual with a phenotype that is different from the wild type.

mutation A permanent change in the base sequence of DNA. Mutations may occur spontaneously or in response to radiation or harmful substances.

mutualism A symbiotic relationship between two organisms in which both organisms benefit. An example is pollination of flowers by insects, in which the insect receives nutrition and the plant is able to reproduce.

N

nastic movement A movement of plant tissues in response to an environmental stimulus, such as a change in humidity or temperature. Nastic movements are independent of the direction of the stimulus.

natality The birth rate in a population, usually expressed as number of births per unit of population in a given time period. For example, the birth rate in Australia in 2018 was 12.6 per 1000 population.

natural selection The mechanism by which evolution is believed to occur. Some individuals in a population have inherited characteristics that make them more likely to survive and reproduce than others in the population. These individuals pass these characteristics on to their offspring. Over time this removes less suitable variations, so that evolutionary change gradually occurs.

negative feedback loop A control system in which the response produced by a stimulus reduces the size of the original disturbance. This eventually leads to homeostasis.

neoplasm Abnormal tissue growth, caused by unusually rapid cell replication. Neoplasms may be malignant (cancerous) or benign (not cancerous).

nephron The functional unit of the kidney; consisting of a Bowman's capsule surrounding a glomerulus and a tubular region leading into a collecting duct. About one million nephrons are found in each human kidney.

nervous system The network of nerve cells that transmits signals throughout the body in response to internal and external stimuli.

neurotransmitter A group of signalling molecules produced by neurons and used to carry a signal across synapses between cells.

nitrogenous waste The waste products from the breakdown of proteins, including ammonia, urea and uric acid.

nominal variable A categorical variable in which there is no inherent order. Nominal variables can be counted but not ordered.

non-shivering thermogenesis The production of body heat by an increase in metabolic rate in brown fat. Brown fat is rich in mitochondria and is capable of high rates of aerobic metabolism.

nucleolus (pl. nucleoli) A dark-staining body in the nucleus, where ribosomal RNA is synthesised.

nucleotide A molecule consisting of a 5-carbon sugar (ribose in RNA, or deoxyribose in DNA), a nitrogenous base (purine or pyrimidine) and a phosphate group. Nucleotides are the building blocks of nucleic acids such as DNA and RNA.

nucleus An organelle that contains genetic information (used for the synthesis of proteins) and directs the activities of the cell.

O

obligate mutualism A form of mutualism in which one of the organisms cannot survive without the other.

observation A value or other information obtained during an experiment.

oesophagus The muscular tube that transports food from the mouth to the stomach in complex animals.

oncogene A gene that induces uncontrolled cell divisions, leading to the development of a neoplasm.

optimum range The range of an abiotic condition that best suits survival and reproduction of a species.

organ A structure, consisting of different tissues, that carries out one or more specific functions.

organelle Any specialised structure in the cytoplasm of a cell, including Golgi apparatus, mitochondrion, endoplasmic reticulum, vacuole, ribosome, chloroplast and nucleus.

organic compound Any chemical substance containing carbon, once thought to come from living organisms. Common organic compounds are proteins, carbohydrates and lipids. However, oxides, carbonates, bicarbonates, carbides and cyanides are usually not considered to be organic compounds.

origin In prokaryotes, the point at which the chromosome is attached to the plasma membrane.

osmoconformer A marine organism that maintains a concentration of solutes (dissolved substances) in their body that is equal (isotonic) to their surroundings.

osmolality The osmotic pressure of a liquid, measured in osmoles of solute per kilogram of water.

osmoreceptor A sensory receptor that detects changes in osmotic pressure in the internal environment. Most osmoreceptors are located in the hypothalamus.

osmoregulation The maintenance of water balance in an organism.

osmoregulator An animal that maintains the internal osmotic concentration of its body fluids regardless of changes in the external concentration.

osmosis The passive diffusion of free water molecules across a semi-permeable membrane from a more dilute solution to a more concentrated solution.

osmotic gradient A difference in the concentration of a solute (dissolved substance) on each side of a semi-permeable membrane.

osmotic pressure The pressure that causes free water molecules to move along a concentration gradient (osmotic gradient) across a semi-permeable membrane. It is caused by a difference in concentration of the solutions on each side of the membrane.

outlier A reading that lies a long way from other results. Repeating readings may be useful in further examining an outlier.

P

pancreas An organ of the digestive system of complex animals that produces digestive enzymes.

pancreatic acinar cell A specialised cell found in the pancreas that produces and secretes digestive enzymes that are transported to the small intestine.

parasite An organism that lives in or on another organism and benefits by feeding on nutrients.

parasitism An interaction between species where one organism, the parasite, feeds from another organism, the host. The host is harmed and the parasite benefits.

parenchyma cell A specialised cell within the phloem tissue of vascular plants. Parenchyma cells make up the soft tissue of a plant and contain chloroplasts.

parental type A gamete that has the same alleles that are present in the parent. Also called parental gamete.

parthenogenesis The development of an egg in the absence of fertilisation by sperm. It is a normal part of the life cycle in some insects and crustaceans.

pedigree analysis The determination of the pattern of inheritance of a characteristic or condition by reference to a pedigree chart (also known as a family tree) in which the presence or absence of the characteristic is recorded over generations.

pedigree chart A diagram in which the presence or absence of a heritable characteristic or condition is recorded over generations. Also known as a family tree.

peer-reviewed Other scientists have checked the information and have agreed that it is appropriate for publication.

peripheral protein A protein that is a temporary part of the plasma membrane. Peripheral proteins bind to integral proteins or penetrate the periphery (outside layer) of the plasma membrane.

peristalsis Coordinated muscular contractions and relaxations of the wall of the digestive tract that move a bolus of food from the oesophagus to the intestines.

personal errors Include mistakes or miscalculations.

phagocytic cell A cell capable of engulfing pathogens or foreign particles to destroy them.

phagocytosis The process by which a solid particle in the extracellular fluid is taken into a cell. The particle is enclosed by a section of plasma membrane, which then pinches off to form a vesicle within the cell's cytoplasm. Phagocytosis is a type of endocytosis.

phenotype (1) An observable character or trait of an organism. (2) The overall appearance of an organism.

phenotypic ratio The frequency of phenotypes expected in the offspring of a genetic cross. A phenotypic ratio is written as dominant phenotype : recessive phenotype.

phloem The plant tissue through which sugars and other organic compounds are distributed to different parts of a plant. In flowering plants, phloem consists of sieve tubes, companion cells and fibres.

phospholipid A fat-like substance, usually based on glycerol. Phospholipids are essential components of plasma membranes. They are involved in the uptake of fats and fatty acids from the products of digestion.

phospholipid bilayer A membrane made of two layers of phospholipid molecules, with a hydrophilic exterior and hydrophobic interior. The plasma membrane that surrounds cells consists of a phospholipid bilayer.

photoreceptor A form of chemical receptor that is sensitive to light.

photosynthesis (adj. photosynthetic) The process by which plants and other photosynthetic organisms convert energy from sunlight into chemical energy for biological functions. It occurs in plastids.

physiological adaptation Any functional or biochemical reactions that take place in organelles, cells, tissues, organs, systems or the whole organism to improve an organism's ability to cope with abiotic and biotic factors in their environment, increasing their chances of survival and reproduction.

physiological stress A state brought on by internal or external factors that disrupt the homeostatis of an organism.

pie chart A circular diagram divided into sectors, with each sector representing the value of one set of data as a proportion of the total data set.

piloerection The erection of hair cells in response to cold, fright or shock. The sympathetic nervous system triggers this reflex.

pinocytosis The process by which a mass of fluid is taken into a cell. The fluid is first surrounded by a section of plasma membrane, which then pinches off to form a vesicle within the cell. Pinocytosis is a type of endocytosis.

pituitary gland An endocrine gland found in the brain that secretes hormones.

placebo A treatment that has no effect but is presented as genuine treatment. Placebos are used in control groups of experiments.

plasma membrane A bilayer (double layer) of phospholipids that encloses the contents of a cell and controls the movement of substances into and out of the cell. Also called cell membrane or plasmolemma.

plasmid A fragment of DNA that is outside the chromosomes, in the cytoplasm. Plasmids usually include genes and can replicate independently. In genetic engineering, bacterial plasmids can be used to produce recombinant DNA.

plasmodesmata (sing. plasmodesma) The microscopic channels that connect the cytoplasm of adjacent cells in plants and some algae.

ploidy The number of full sets of chromosomes in an organism's karyoptye. Haploid (one set, or n) and diploid (two sets, or $2n$) are the most common ploidy states, but other states are possible.

pluripotent Describes a cell that can develop into several different cell types. An example is a human embryonic stem cell, which can form all adult cell types.

podocyte cell A specialised cell that is found in the nephron that assists in filtration.

point sampling A method of sampling that involves counting organisms only at selected points.

population A group of organisms of the same species that interact with each other.

positive feedback loop A control system in which the response produced by a stimulus increases the size of the original disturbance.

potency In regards to stem cells, their potential to differentiate into different cell types.

precision Refers to how closely a set of measurement values agree with each other. Precision gives no indication of how close the measurements are to the true value and is therefore a separate consideration to accuracy.

predation The killing and consumption of an animal by another animal.

predator An animal that kills and consumes other animals.

prey An animal that is a food source for another animal.

primary data Data you have collected yourself.

primary source A source that includes first-hand information, such as the results of an original experiment. See also *secondary source*.

principle A scientific theory that is so strongly supported by evidence that it is considered unlikely to be shown to be untrue in the future.

processed data Data that has been mathematically manipulated.

prokaryotic cell A cell that does not have a membrane-bound nucleus and that lacks most organelles. All prokaryotes are bacteria.

prophase The first phase in mitosis in which chromosomes condense and become visible, centrioles move to opposite sides of the nucleus and form poles, the nuclear membrane breaks down and centrioles form spindle fibres between the two poles.

protease An enzyme that digests proteins.

protein A nitrogenous organic compound consisting of one or more long chains of amino acids.

proto-oncogene A normal cellular gene which could become a gene that triggers molecular events that lead to cancer.

provisional data Preliminary data that is subject to revision.

Punnett square A visual method of showing all possible combinations of alleles that could be seen in offspring from two parents.

purine A group of chemical bases that include adenine (A) and guanine (G), which are present in the nucleotides of DNA and RNA.

pyrimidine A group of chemical bases that include thymine (T), cytosine (C) and uracil (U), which are present in the nucleotides of DNA (C and T) and RNA (C and U).

Q

quadrat An area (usually a square) within which a biological survey (such as counting plants or identifying species) is carried out.

qualitative data Data that consists of categorical variables.

quantitative data Data that consists of numerical variables.

R

radiation The transfer of heat via electromagnetic waves, specifically infrared.

random errors Affect the precision of a measurement and are present in all measurements except for measurements involving counting. Random errors are unpredictable variations in the measurement process and result in a spread of readings. The effect of random errors can be reduced by making more or repeated measurements and calculating a new mean and/or by refining the measurement method or technique.

random selection A selection that is not affected by bias.

range The difference between the highest and lowest values.

raw data The data recorded during an experiment.

reabsorption In the kidney, the process by which the primary kidney filtrate is taken back into the tissues via nephrons.

reception Detection of a hormone by a cell's receptor. See *receptor*.

receptor A specialised structure that can detect a specific stimulus and initiate a response.

recessive Relating to a trait or phenotype (encoded by an allele or gene) whose appearance is subordinate to a dominant trait.

recessive phenotype A phenotype that is observed only in homozygous individuals.

reciprocal cross A cross in which a male of strain A is crossed with a female of strain B, and a female of strain A is crossed with a male of strain B.

recombinant chromosome A chromosome that has exchanged genetic material by crossing over during meiosis.

recombinant gamete A gamete (sex cell) carrying a combination of alleles not observed in the parent, as a result of crossing over during meiosis.

reduction division A nuclear division that halves the number of chromosomes in the daughter cells. Reduction division occurs in meiosis.

renin An enzyme that is produced and stored in the kidneys. It plays a role in regulating blood pressure by catalysing the conversion of angiotensinogen to angiotensin I. This is then converted to angiotensin II, an effective vasoconstrictor.

repeatability The closeness of the agreement between the results of successive measurements of the same quantity being measured, carried out under the same conditions of measurement. These conditions include the same measurement procedure, the same observer, the same measuring instrument used under the same conditions, the same location, and repetition over a short period of time.

repeat trial An experiment that is conducted again, in exactly the same manner as a previous experiment.

replication (1) Experimentation carried out on more than one set of subjects at the same time. (2) The production of new cells by cell division.

reproducibility The closeness of the agreement between the results of measurements of the same quantity being measured, carried out under changed conditions of measurement. These different conditions include a different method of measurement, different observer, different measuring instrument, different location, different conditions of use, and different time.

reprogramming The conversion of one cell type into another cell type, such as a somatic cell into a pluripotent cell.

research question A statement that defines what is being investigated.

response A physiological or behavioural change in an organism as a result of receiving a stimulus.

rhizome A creeping stem from which vertical stems arise from buds. Rhizomes enable plants to regenerate when the above-ground parts have died.

ribonucleic acid (RNA) A nucleic acid that is a single strand made up of a sequence of ribose sugars and bases (adenine, cytosine, guanine and uracil) linked by phosphodiester bonds. There are three forms: messenger RNA (mRNA), ribosomal RNA (rRNA) and transfer RNA (tRNA).

ribosomal RNA (rRNA) The RNA part of a ribosome. It is synthesised in the nucleolus and is essential for protein synthesis.

ribosome A small organelle composed of protein and RNA. Ribosomes are often attached to rough endoplasmic reticulum and are the site of protein synthesis.

risk assessment A systematic way of identifying the potential risks associated with an activity.

root hair A very thin extension of an epidermal cell of a root. Root hairs increase the root's surface area, making the absorption of water and minerals from soil more efficient.

root pressure Osmotic uptake of water that accompanies the active uptake of mineral salts and contributes to the movement of water up xylem in some plants.

rootstock A plant that has well developed roots and has its stem or a branch cut and a graft of another plant added to it. See *graft*.

rough endoplasmic reticulum Layers of intracellular membranes associated with ribosomes. Rough endoplasmic reticulum is involved in protein synthesis.

RNA See *ribonucleic acid*.

rRNA See *ribosomal RNA*.

S

S phase The synthesis phase of interphase where chromosomes are replicated in the nucleus. See also *G_0 phase*, *G_1 phase* and *G_2 phase*.

safety data sheet (SDS) A document that contains important information about the possible hazards in using a substance and how the substance should be handled and stored.

scatterplot A graph in which two variables are plotted as points. The x coordinate of a particular point is one measured value of the independent variable and the y coordinate is the corresponding measured value of the dependent variable.

scientific method The systematic, objective collection of data by experiment in order to determine whether predictions based on a particular hypothesis are correct.

sclerenchyma cell A specialised cell within the phloem tissue of vascular plants. Sclerenchyma cells have very thick cell walls to provide structural support for the plant. These are most commonly found in the more mature stems of a vascular plant.

secondary data Data you have not collected yourself.

secondary source A source of information that does not include first-hand information. See also *primary source*.

secretion (1) The release of specific substances from a cell or group of cells. (2) In kidneys, the active excretion of particular substances by the cells of the duct walls.

selection pressure An environmental factor that affects the survival and reproductive success of an individual based on their particular phenotype.

semi-permeable membrane A membrane that allows only some molecules to pass across it by osmosis or diffusion. Also called partially permeable membrane, selectively permeable membrane.

separation division The division in meiosis II where the sister chromatids are separated and four genetically different haploid daughter cells are produced.

sex chromosome A chromosome that is involved in sex determination. In humans the sex chromosomes are the X and Y chromosomes. Also called an allosome.

sex-limited inheritance The inheritance of a trait that is expressed only in one sex, even though both sexes carry the gene. An example is haemophilia A, an X-linked trait that occurs almost exclusively in males (cases in females are extremely rare).

sex-linked inheritance The form of inheritance related to genes that occur on the sex-chromosomes (X and Y in humans). An example is red-green colour blindness, which is caused by a mutation in a gene on the X chromosome. See also *X-linked*, *Y-linked*.

sexual reproduction Reproduction involving the fusion of two gametes (egg and sperm), which are the haploid products of meiosis.

shivering thermogenesis The production of heat energy through the involuntary movement of muscles (shivering).

sieve tube A specialised cell within the phloem tissue of vascular plants. Sieve tube cells lack a nucleus and lignin in their cell walls and form linear rows of elongated cells to transport substances directly from one cell to another through small pores. They are usually found with adjoining companion cells. See *companion cell*.

signal transduction The process of a cell detecting and responding to a signalling molecule (hormone).

signalling molecule A chemical message (e.g hormone) that is used to communicate information between cells.

simple diffusion The movement of molecules across a membrane without the use of channel proteins or carrier proteins.

sink A site where something is stored or consumed.

sister chromatids Two joined chromatids that form the familiar 'X' shape of most chromosomes. See *chromatid* and *chromosome*.

small intestine Part of the digestive system in complex animals. The small intestine is the site of nutrient absorption.

smooth endoplasmic reticulum A continuous membrane system that forms a series of flattened sacs within the cytoplasm of eukaryotic cells. It is not associated with ribosomes and is involved in the synthesis of lipids.

solute A substance dissolved in a fluid (the solvent).

solvent A fluid in which a substance (the solute) is dissolved.

somatic cell Any cell of an organism except a cell that gives rise to gametes (eggs and sperm). Somatic cells in mammals are diploid ($2n$); that is, they contain a full set of paired chromosomes.

somatic cell nuclear transfer (SCNT) A cloning technique that involves transferring the nucleus from a somatic cell into an egg cell with the nucleus removed. The egg cell is then induced to divide as though under the natural fertilisation process. The resulting embryo is genetically identical to the donor somatic cell.

source (1) A site where something is produced. (2) A document or person from which information has been obtained.

specialised cell A cell with features that allow it to perform a specific function. Examples of specialised cells are red blood cells, nerve cells and muscle cells in animals, and root hair cells and guard cells in plants.

species The basic category or group in the binomial system. Organisms that are grouped into species usually closely resemble each other and can interbreed. Different species usually do not interbreed with one another in the wild, and if they do interbreed then their offspring are not capable of reproduction.

sporangium (pl. sporangia) The structure in a spore-bearing plant in which spores develop. Also called spore case, spore capsule.

spore A haploid cell that can develop into a new organism without sexual reproduction. In plants, algae and fungi, spores are the products of meiosis, but fungi and some algae can also produce spores by mitosis.

spore formation A form of asexual or sexual reproduction that involves the production of spores. See *spore*.

sporophyte The asexual, spore-forming diploid stage in the life cycle of a plant. The spores are haploid, are produced by meiosis and develop into haploid gametophytes. In seed plants (gymnosperms and angiosperms), the sporophyte stage is the prominent stage (i.e. the plant structure), while the gametophyte stage is the less conspicuous pollen and embryo sac.

stem cell A cell that can differentiate into a specialised cell.

stimulus An environmental factor that an organism can detect and respond to.

stimulus–response mechanism A response which reduces the effect of the original stimulus by reversing its direction. Stimulus–response mechanisms maintain homeostasis by adjusting conditions to the optimal state.

stimulus–response model A series of events involved in the reception, transduction and cellular response to a signalling molecule (hormone).

stolon A thin, horizontal branch that serves to propagate a plant. Stolons usually grow along the ground, giving rise to roots and aerial branches at node points. An example of a plant with stolons is the strawberry. Also called a runner.

stoma (pl. stomata) A pore in the leaf epidermis, bounded by specialised guard cells that open and close the pore. Stomata are the main routes through which gas exchange occurs in plants, and through which water loss is regulated.

stomach A muscular organ of the digestive system in complex animals.

stomata *See stoma*.

structural adaptation Any anatomical or morphological feature that improves an organism's ability to cope with abiotic and biotic factors in their environment, increasing their chances of survival and reproduction.

surface area to volume ratio The relationship between the surface area and volume of a structure. As the size of a structure increases, its surface area to volume ratio decreases.

symbiosis A close association between two different organisms, from which at least one of the organisms benefits. Symbiosis includes mutualism, commensalism and parasitism.

synapsis The process of pairing of two homologous chromosomes during meiosis.

system A group of organs in animals that work together for a particular purpose. The major systems in humans are the integumentary system, skeletal system, circulatory system, muscular system, digestive system, nervous system, endocrine system, respiratory system and excretory system.

systematic errors Affect the accuracy of a measurement. Systematic errors cause readings to differ from the true value by a consistent amount each time a measurement is made, so that all the readings are shifted in one direction from the true value. The accuracy of measurements subject to systematic errors cannot be improved by repeating those measurements.

T

target cell The cell a hormone specifically affects.

telomere An area of repetitive DNA at either end of the DNA molecule of a eukaryotic chromosome. Telomeres protect the end of the chromosome from deteriorating and from clumping together with other chromosomes.

telophase Final stage of mitosis, between anaphase and cytokinesis. During telophase a nuclear membrane re-forms around the chromosomes at each pole, the spindle is dismantled and disappears, and the chromosomes become longer and thinner.

test cross A type of backcross in which an individual with the dominant phenotype is crossed with an individual of the recessive phenotype for the character being studied. A testcross is used to identify whether the individual with the dominant trait is homozygous or heterozygous.

theory A hypothesis that is supported by a great deal of evidence from a wide variety of sources.

thermoreceptor A sensory receptor that detects and responds to temperature.

thermoregulation A process used by some animals to maintain their internal (core) temperature.

thymine (T) A nitrogen-containing base (a pyrimidine) that occurs in nucleotides of DNA, but not RNA.

thyrotropin releasing hormone (TRH) A hormone secreted by the hypothalamus, which acts on the anterior pituitary to secrete TSH (thyroid stimulating hormone).

tissue A group of similar cells functioning together.

tissue culture A method of growing cells or tissues in an artificial medium containing essential nutrients, salts and growth factors.

tolerance range The range of environmental conditions (such as temperature or salinity) that an organism can tolerate without injury.

torpor A state of inactivity or dormancy in animals, in which the body temperature is lower and the metabolism is slower than normal.

totipotent Describes a cell that can give rise to any cell type and potentially a complete new organism.

tracheid A long, hollow cell with a thickened wall and tapering ends, found in the xylem of vascular plants. Tracheids transport water and nutrients to the living cells of the plant.

trait A particular characteristic or feature of an organism.

transduction The series of events that takes place within a cell as a response to the binding of a signalling molecule (hormone) with a receptor of a target cell.

transect A line along which a biological survey is conducted. Transects are used mainly in botanical surveys.

translocation The transport of organic substances in the phloem of a vascular plant.

transmembrane protein An integral protein that spans both phospholipid layers of the plasma membrane.

transpiration The loss of water from the leaves of plants through stomata. Transpiration causes suction, which draws water up xylem vessels from the roots.

transpiration stream The flow of water within a plant, from the uptake by the roots to the loss of water to the environment via the leaves.

TRH See *thyrotropin releasing hormone*.

trisomy An abnormality in which a cell has three copies of a particular chromosome. An example is trisomy of chromosome 21, which causes Down syndrome in humans.

trophic cascade The series of ecological changes that occur in the food web when a top level predator is added or removed from an ecosystem.

trophic level The position of an organism in a food chain. For example, an organism that feeds on producers is a first-order consumer, and an organism that feeds on that first-order consumer is a second-order consumer.

tropism Plant growth in response to an external stimulus such as gravity, light or water. The plant might grow towards the stimulus (positive tropism) or away from it (negative tropism).

true-breeding Producing only progeny with a particular characteristic or trait seen in the parent. Also called pure-breeding.

true value The value, or range of values, that would be found if the quantity could be measured perfectly.

tuber A starchy, underground stem that stores food for the plant. Tubers often grow just under the surface of the soil. Examples of plants with tubers are the potato and cassava.

tumour-suppressor gene A gene whose protein product inhibits cell division, thus preventing uncontrolled cell growth that would result in a cancer.

turgor (adj. turgid) The rigid state or fullness of a plant cell caused by internal fluid pressure (turgor pressure) acting on the cell wall. Turgor is maintained by the osmotic intake of water into the cell.

type 1 diabetes An autoimmune disease in which the pancreas does not make insulin. Type 1 diabetes has a genetic basis, but appears to also require an environmental factor such as a viral infection in order to be expressed.

U

uncertainty The range of possible values within which the true value lies. Uncertainty results from errors during measurement. See *true value*.

unicellular organism An organism consisting of a single cell.

unipotent Describes a stem cell that can differentiate only into one cell type.

upper epidermis A layer of cells on the upper side of the leaf that protects the inner cells, prevents water loss and allows sunlight to penetrate

urea A water-soluble molecule (CH_4N_2O) that is a major product of protein breakdown. It is excreted by many vertebrates, including mammals.

ureter The tube that carries urine from a kidney to the bladder for storage, before release via the urethra.

urethra The tube that carries urine from the bladder to the exterior for excretion.

urine The liquid by-product of metabolism in complex animals.

V

validity A measurement is said to be valid if it measures what it is supposed to be measuring. An experiment is said to be valid if it investigates what it sets out and/or claims to investigate.

variable A factor or condition that can change during an experiment.

vascular bundle A grouping of vascular tissues in vascular plants, containing both xylem and phloem. Vascular bundles are continuous, from the roots into the stem, branches and leaves.

vascular plant A plant that has vascular tissues (xylem and phloem) in which the cell walls contain lignin. All living plants, except bryophytes, are vascular plants.

vascular tissue The tissue that conducts water and nutrients from the roots to the leaves in vascular plants. It consists of two types of tissue: xylem and phloem. Vascular tissue also provides structural support to a plant.

vasoconstriction The narrowing of the internal diameter of a blood vessel.

vasodilation The widening of the internal diameter of a blood vessel.

vegetative propagation A form of asexual reproduction found in plants, in which a piece of a plant (usually a growing tip of a stem) is separated from the plant and grows into a new plant. It is a method used widely in horticulture to produce new plants.

vesicle A small organelle consisting of a membrane filled with fluid. Vesicles are often involved in transport within the cell, but may have other functions.

villus (pl. villi) A tiny fold in the lining of the intestine. Villi increase the surface area available for the absorption of food.

vitamin Any organic compound required in small amounts for cell processes. In humans there are 13 such compounds, called vitamins A, B group (eight vitamins), C, D, E and K.

WXYZ

wild type The phenotype most commonly observed in a natural population.

X-linked Resulting from the inheritance of a gene on the X chromosome. An X-linked trait may be inherited from either parent, because both have an X chromosome.

xerophyte A plant adapted to dry conditions.

xylem The tissue in vascular plants that transports water and nutrients upwards from the roots. It consists of hollow chains of dead cells.

xylem vessel A long tube consisting of cells joined end to end, through which water and nutrients are transported from the roots to the leaves in a vascular plant.

Y-linked Resulting from the inheritance of a gene on the Y chromosome. A Y-linked trait is passed from father to son. It is never observed in females because they do not have a Y chromosome.

zygote The diploid cell resulting from the fusion of an egg and sperm. It is the first stage of the development of a unique new organism.

Index

Bold page numbers refer to where glossary terms are bolded in the text.

D

E

I

J

K

L

M

N

O

P

Q

R

S

T

U

V

W

X

Y

Z

Attributions

We thank the following for their contributions to our textbook:
The following abbreviations are used in this list: t = top, b = bottom, l = left, r = right, c = centre.

COVER: **Science Photo Library:** Steve Gschmeissner

123RF: Akhararat Wathanasing, p. 22c; Aleksandr Markin, p. 225br; alila, pp. 121tr, 146tr, 265bl; aquafun, p. 373tr; Brian Lasenby, p. 97tr; Cathy Yeulet, p.256tl; Daniel Bueno, p. 272cr; Dario Lo Presti, p. 380bl; Dean Fikar, p. 372tl; Denis and Yulia Pogostins, p. 324bl; designua, pp. 131tc, 224br, 356tr; Deyan Georgiev, p. 23tl; Dmitry Lobanov, p. 226cl; Duncan Noakes, p. 366cl; Ekaterina Smirnova, p. 282tl; gator, p. 360tl; Geoffrey Whiteway, p. 24bc; Irina Tischenko, p. 382tr; Ivan Mikhaylov, p. 37br; ivolodina, p. 215cr; Jacoba Susanna Maria Swanepoel, p. 361bl; Johan Swanepoel, p. 379tr; kasto, p. 23br; Kirill Kurashov, p. 80tr; Kriangkrai Kantepa, p. 326bl; Marcin Ciesielski, p. 372bc; Marilyn Barbone, p. 340bl; Martin Maritz, p. 372cl; my123rf88, p. 356tr; nialat, p. 376tl; Ramzi Hachicho, p. 333c; Rido, p. 206tl; Roberto Biasini, pp. 97bl, 98tc; Robyn Mackenzie, p. 247bl; Sakarin Sawasdinaka, p. 272tr; Sarah–Jane Allen, p. 247br; shihina, p. 357c; Stuart Jenner, p. 215cl; Szilagyi Palko Pal, p. 373br; Tatiana Makotra, p. 206bl; Tracy Fox, p. 338cl; Valentina Vasilieva, p. 383br; Vira Dobosh, p. 326bc; Waroot Tangtumsatid, p. 333cl; Wong Sze Yuen, p. 295tr; Zlikovec, p. 326cr.

AAP: Elaine Thompson, p. 328br.

Age Fotostock: P & R Fotos, p. 121bl.

Alamy Stock Photo: Bill Bachman, p. 27tl; Chris Howes/Wild Places Photography, p. 25br; Chronicle, p. 350br; Duncan Usher, p. 346tl; Encyclopaedia Britannica Inc., p. 348br; EPA, p. 213cr; Jeremy Sutton–Hibbert, p. 335cr; Mark Kerrison/Alamy Live News, p. 351br; MShieldsPhotos, p. 24cr; Nigel Cattlin, p. 157cl; Pat Canova, p. 268br; Sabena Jane Blackbird, p. 380tl; Science History Images, p. 150cl; Science Photo Library, pp. 9bl, 11tr, 42br, 115bc, 133cr, 265tr, 277cl, 315bl, 322l, 324c, 325br, 325cr, 327tr; Stephen Barnes/Business, p. 15br; Thomas Chamberlin, p. 377tr; Universal Images Group North America LLC, p. 129tr; Volodymyr Burdiak, p. 352bl.

Andrijich, Frances: p. 5c.

Asian Pacific Journal of Reproduction: Adapted from 'Assessment of telomere length during post–natal period in offspring produced by a bull and its fibroblast derived clone' by Donatella Balduzzi, Anna Pozzi, Roberto Puglisi and Roberta Vanni, *Asia Pacific Journal of Reproduction*, 3(1): pp. 1–7, 2014, p. 342cr.

Auscape International Photo Library: Densey Clyne, p. 368bl; Jean–Paul Ferrero, p. 392cr.

Australian National Botanic Gardens: pp. 365bl, 365cl.

Batterham, Philip & McKenzie, John: p. 270cl.

Borangi, Sahar: p. 327c.

Biodiversity Heritage Library: p. 269bl.

Clarke, Philip: p. 386tr.

Department of Health: Adapted from 'Vaccine preventable diseases in Australia, 2005 to 2007' by Chiu C1, Dey A, Wang H, Menzies R, Deeks S, Mahajan D, Communicable Disease Intelligence, 34 Supp, pp. S1–S167, 2010, pp. 49tl, 52bl.

Diabetes Education Online: These illustrations are made available by written permission of The Regents of the University of California. All rights reserved., p. 227tr.

DK Images: Kevin Jones/Dorling Kindersley, p. 325cl.

Donegan, Ailish: p. 371tr.

Expandable Mind Software: Image courtesy of Expandable Mind Software – www.emindweb.com, p. 183tr.

Fotolia: 2436digitalavenue, p. 378tl; adpePhoto, p. 376bl; barbara_foto, p. 367br; borisoff, p. 154cr; Denys Rudyi, p. 262tl; Edward Westmacott, p. 326bl; KAR, p. 38cl; Kokhanchikov, pp. 323cr, 340br; Kurhan, p. 134cr; Leah–Anne Thompson, p. 374bc; Photozi, p. 383cr; powell83, p. 134tcr; Richard Carey, p. 383cr; Romolo Tavani, p. 134bcr; Sondem, p. 134tr; szymanskim, p. 240tl; wongtanarak, p. 331cr.

Fuhrer, Bruce: p. 28r.

Getty Images: BSIP, p. 326tl; Ed Reschke, pp. 124l, 157c, 162c, 170cr, 194bcr; Encyclopaedia Britannica, pp. 94bl, 94br; Kristian Septimius Krogh, p. 331br; Leonello Calvetti, p. 202bc; Mint Images/Art Wolfe, p. 316tc; Victor Shupletsov, p. 203bl.

ImageFolk: Biodisc, p. 168cl; Charles O'Rear, pp. 79br, 93br; Frans Lanting, p. 367cr; Michael Peuckert, p. 160bl; Michael Rosenfeld/Science Faction, p. 4br; Nature Connect, p. 194tl; Zephyr, p. 299br.

iStockphoto: lukaves, p. 171tr.

Katsis, Andrew: p. 64l.

Mulder, Raoul: pp. 64r, 89cr.

Museums Victoria: Based on data from 'The Human Story' by C. Lockwood, 2008., pp. 49bl, 52cr.

Library of Congress: Library of Congress Prints and Photographs Division Washington, p. 106cr.

National Geographic Society: Michael Nichols, p. 167br; Jason Edwards, p. 172br.

Nature Picture Library: Jordi Chias, p. 362cl.

Nobel Committee for Physiology or Medicine: © The Nobel Committee for Physiology or Medicine. Illustratraed by Annika Röhl, p. 5br.

Pearson Education Ltd: pp. 36cl, 122tr.

PLOS: Based on data from 'Sexual verus asexual reproduction: Distinct outcomes in relative abdundance of parthenogenetic Mealybugs following recent colonisation' by Ryoko Ichiki, Daisuke Kageyama, Jun Tabata and Hirotaka Tanaka, PLOS ONE, 11(6): e0156587, 2016, p. 341cr.

Oxford Designers & Illustrators Ltd: pp. 96c, 194bcl, 241tr, 277br.

Pearson Education Inc: *Biological Science* 3rd edn by Lizabeth Allison, Scott Freeman and Kim Quillin, fig. 37.20, 2008, pp. 204c, 238tl.

Royal Botanic Gardens Victoria: p. 170cl.

Safe Work South Australia: Commonwealth of Australia © 2020, p. 19.

Science Photo Library (SPL): p. Alfred Pasieka, pp. 9tr, 118bc, 145tr; AMI Images, p. 76cl; Bsip, Gillies, p. 116cr; BSIP, LAURENT, p. 116tl; Chris Hellier, p. 356tr; Chris Knapton, p. 244–245; Clinical Photography, p. 231tr; CNRI, p. 146cl; Cordelia Molloy, pp. 272br, 364cl; Dan Dunkley, p. 55tr; David Parker, pp. 242–243, 314, 392; Dept. of Clinical Cytogenetics, Addenbrookes Hospital, p. 258br; Dennis Kunkel Microscopy, pp. 70–71, 83cr; Dirk Wiersma, p. 360cl; Dr. David Furness, p. 92tl; Dr. Gopal Murti, pp. 112–113, 130tl; Dr. Jack Bostrack, p. 76tl; Dr. Jeremy Burgess, pp.73cl, 111cl, 125tc, 148bl, 172cr, 327cr; Dr. Kari Lounatmaa, p. 91tr; Dr. Keith Wheeler, p. 92cl; Dr. Lothar Schermelleh, p. 73br; Dr. P. Marazzi, pp. 132cl, 226tr; Dr. Torsten Wittmann, p. 92b; Dr. Yorgos Nikas, p. 114tl; Eddy Gray, p. 93cl; Eye of Science, pp.76c, 77c, 77cl, 93bc, 137cr, 356tr; Francis Leroy, Biocosmos, p. 149cr; Gail Junkus, p. 365tr; Garo / Phanie, p. 226cl; Georgette Douwma, p. 116cl; Gunilla Elam, pp. 74bl, 129cr; Herve Conge, ISM, pp. 120tl, 173bl; Ian Hooton, p. 216tl; Jacopin, p. 214cl; James Stevenson, p. 130cl; Javier Trueba/MSF, p. 18cl; John Durham, p. 147tr; John Sibbick, p. 155cr; Jose Calvo, p. 152–153; Joseph F. Gennaro Jr, pp. 68–69, 144, 236; Louise Hughes, p. 93tr; Lucas Bustamante/Nature Picture Library, pp. 344–345; Manfred Kage, pp. 115cr, 115tc; Martin Dohrn, p. 214tl; Matthew Oldfield, p. 138bl; Maurizio De Angelis, p. 131tr; Medical RF.Com, p. 293c; NIBSC, p. 374cl; Overseas/Collection CNRI, p. 180br; Pascal Goetgheluck, p. 304l; Philippe Psaila, p. 334tl; Power and Syred, pp. 122br, 326c, 356tr; Prof. P. Motta & T. Naguro, pp. 87cr, 88cl, 89tl; 90tl; Prof. Philippe Vago, ISM, pp. 261cr, 277bl, 277cr, 277tr; Prof. J.C. Chermann, p. 134br; Prof. P.M. Motta, S. Makabe & T. Naguro, p. 88tl; RIA Novosti, p. 213tr; Rosenfeld Images Ltd., p. 332cl; J.C. Revy, p. 145cr; Sci–Comm Studios, pp. 30tl, 139tr; Science Photo Library, pp. 24tl, 33tr, 111tl, 196–197; Sinclair Stammers, p. 155br; Stephen Ausmus/US Department of Agriculture, p. 25cr; Steve Gschmeisner, pp. 82cl, 135tl, 137tr, 218br, 224tl, 238cr, 280–281, 320–321; TEK Image, pp. 1–3; Thomas Deerinck, p. 131br; Trevor Clifford Photography, p. 23bl; Wim Van Egmond, p. 181cr; Wim Van Egmond, Visuals Unlimited, p. 154tl; Zephyr, p. 132bl.

Science Source – Photo Researchers, Inc: Biophoto Associates, p. 246tl; Charles V. Angelo, p. 159cl; Jacob W. Frank, p. 26bc; Michael McCoy, p. 363tr; Michael Tyler, p. 335tr.

Shutterstock: 467173, p. 396tl; Adventuring Dave, p. 165br; Aleksandr Denisyuk, p. 358bl; Alila Medical Media, pp. 117cl, 117tc, 191t, 259bl, 259br; All–stock–photos, p. 362bl; Anders Peter Photography, p. 269tl; Andreea Dragomir, p. 374br; applo2499, p. 376tl; ARENA Creative, p. 264bl; arka38, p. 357tr; attem, p. 22tc; Bill Frische, p. 356bl; Birdiegal, p. 221cr; Blamb, p. 238tl; Bork, p. 23tr; Casa nayafana, p. 231tr; Cathy Keifer, p. 354tl; chiqui, p. 122tr; Christian Musat, p. 358cr; covenant, p. 352cr; Creativa Images, p. 383tr; Designua, pp. 183br, 185bl, 216cr, 262c; Dimarion, p. 239cl; Dr. Morley Read, p. 160tl; Eric Isselee, p. 390cr; Erni, p. 269cl; Ethan Daniels, p. 373cr; Fer Gregory, p. 363cr; Gentoo Multimedia Limited, p. 358bc; Gerasymovych Oleksandr, p. 347tl; Greg Brave, p. 104cl; Jan Bruder, p. 139tr; Jerry Lin, p. 368br; Jim Cumming, p. 316tr; Jody, p. 358cl; Joe Ravi, p. 358br; John Carnemolla, p. 379bc; Jubal Harshaw, p. 236bl; Juliet Photography, p. 267tl; Jun Zhang, p. 391c; KentaStudio, p. 22bc; Kirsten Wahlquist, p. 375br; Kozorez Vladislav, p. 156cl; KuLouKu, p. 118cl; Laborant, p. 358c; Lebendkulturen.de, p. 21cr; LightField Studios, p. 34tl; ligio, p. 79tr; LSkywalker, pp. 145l, 146tl; Lucasdm, p. 349cr; magnusdeepbelow, p. 361cr; mandritoiu, p. 394cr; Marc Witte, p. 367cl; Marco Uliana, p. 364bl; Marioner, p. 149tl; Mariusz Potocki, p. 363c; Martin Fowler, pp. 333cr, 397tr; Michael Felix Photography, p. 76cr; Millenius, p. 347cr; Monkey Business Images, p. 18tl; motorolka, p. 227br; irin-k, p. 167tr; PeJo, p. 239br; Peter Cripps, p. 313tr; Photoanto, p. 28l; photong, p. 23tc; Pimonpim w, p. 212bl; Provasilich, p. 336cl; Quarz, p. 156bl; Rainer Lesniewski, p. 25cl; Richard Whitcombe, p. 382bl; robuart, p. 212bc; Roman Vintonyak, p. 371br; Ron Rowan, p. 264tl; Ronsmith, p. 34cl; RTimages, p. 21cl; Ryan M. Bolton, pp. 374cr, 379cr; Sauletas, p. 383bc; Shelli Jensen, pp. 374c, 391cr; SJ Allen, p. 381br; Skumer, p. 90br; Susan Flashman, p. 189br; Swellphotography, p. 394c; Tanor, p. 372br; Tefi, p. 238tl; The Drone Zone, p. 350tl; Timonina, p. 180tr; Trevor Scouten, p. 385tr; VEK Australia, p. 386cr; Vilainecrevette, p. 368tl; wang song, p. 186tr; Warren Metcalf, p. 23br; Wolfgang Zwanzger, p. 369tr; Zharate, p. 28c.

Sinauer Associates Inc: *The Cell: A molecular approach* 2nd edn by Geoffrey M. Cooper, 2000, pp. 51br, 51tr.

The United Nations Economic Commission for Europe (UNECE): p. 19.

Victorian Curriculum and Assessment Authority (VCAA): Selected VCE extracts from the VCE Biology Study Design (2022-2026) are copyright Victorian Curriculum and Assessment Authority (VCAA), reproduced by permission. VCE® is a registered trademark of the VCAA. The VCAA does not endorse this product and makes no warranties regarding the correctness or accuracy of its content. To the extent permitted by law, the VCAA excludes all liability for any loss or damage suffered or incurred as a result of accessing, using or relying on the content. Past VCE exams, current VCE Study Designs and related content can be accessed directly at www.vcaa.vic.edu.au

Walter & Eliza Hall Institute of Medical Research: p. 133tr.

Western Australian Museum: p. 4tr.

Wikimedia Commons: Alexbateman, p. 30bc; Dan Gustafson, p. 328bl.

Wyss Institute for Biologically Inspired Engineering at Harvard University: Images by Huh, Dongeun and Donald E. Ingber, p. 164tr.

Zoological Society of London: © ZSL/ChrisVanWyk, p. 359cr.